Solvents and Self-Organization of Polymers

NATO ASI Series

Advanced Science Institutes Series

A Series presenting the results of activities sponsored by the NATO Science Committee, which aims at the dissemination of advanced scientific and technological knowledge, with a view to strengthening links between scientific communities.

The Series is published by an international board of publishers in conjunction with the NATO Scientific Affairs Division

A **Life Sciences** B **Physics**	Plenum Publishing Corporation London and New York
C **Mathematical and Physical Sciences** D **Behavioural and Social Sciences** E **Applied Sciences**	Kluwer Academic Publishers Dordrecht, Boston and London
F **Computer and Systems Sciences** G **Ecological Sciences** H **Cell Biology** I **Global Environmental Change**	Springer-Verlag Berlin, Heidelberg, New York, London, Paris and Tokyo

PARTNERSHIP SUB-SERIES

1. **Disarmament Technologies**	Kluwer Academic Publishers
2. **Environment**	Springer-Verlag / Kluwer Academic Publishers
3. **High Technology**	Kluwer Academic Publishers
4. **Science and Technology Policy**	Kluwer Academic Publishers
5. **Computer Networking**	Kluwer Academic Publishers

The Partnership Sub-Series incorporates activities undertaken in collaboration with NATO's Cooperation Partners, the countries of the CIS and Central and Eastern Europe, in Priority Areas of concern to those countries.

NATO-PCO-DATA BASE

The electronic index to the NATO ASI Series provides full bibliographical references (with keywords and/or abstracts) to more than 50000 contributions from international scientists published in all sections of the NATO ASI Series.
Access to the NATO-PCO-DATA BASE is possible in two ways:

– via online FILE 128 (NATO-PCO-DATA BASE) hosted by ESRIN, Via Galileo Galilei, I-00044 Frascati, Italy.

– via CD-ROM "NATO-PCO-DATA BASE" with user-friendly retrieval software in English, French and German (© WTV GmbH and DATAWARE Technologies Inc. 1989).

The CD-ROM can be ordered through any member of the Board of Publishers or through NATO-PCO, Overijse, Belgium.

Series E: Applied Sciences - Vol. 327

Solvents and Self-Organization of Polymers

edited by

Stephen E. Webber

Petr Munk

Department of Chemistry and Biochemistry,
The University of Texas at Austin,
Austin, Texas, U.S.A.

and

Zdenek Tuzar

Institute of Macromolecular Chemistry,
Academy of Sciences of the Czech Republic,
Prague, Czech Republic

Kluwer Academic Publishers

Dordrecht / Boston / London

Published in cooperation with NATO Scientific Affairs Division

Proceedings of the NATO Advanced Study Institute on
Solvents and Self-Organization of Polymers
Belek, Antalya, Turkey
July 31–August 11, 1995

A C.I.P. Catalogue record for this book is available from the Library of Congress

ISBN 0-7923-4222-4

Published by Kluwer Academic Publishers,
P.O. Box 17, 3300 AA Dordrecht, The Netherlands.

Kluwer Academic Publishers incorporates the publishing programmes of
D. Reidel, Martinus Nijhoff, Dr W. Junk and MTP Press.

Sold and distributed in the U.S.A. and Canada
by Kluwer Academic Publishers,
101 Philip Drive, Norwell, MA 02061, U.S.A.

In all other countries, sold and distributed
by Kluwer Academic Publishers Group,
P.O. Box 322, 3300 AH Dordrecht, The Netherlands.

Printed on acid-free paper

Printed in the Netherlands

CONTENTS

PREFACE

The idea of an international conference on self-organization of polymers resulted from a six-years' cooperation between the groups of the University of Texas at Austin (P. Munk and S. E. Webber) and the Czech Academy of Sciences and Charles University (Z. Tuzar and K. Procházka). Our projects dealing with block copolymer micelles and their properties have been supported by a number of agencies including the Czech-U.S. Joint Science and Technology Fund and by a NATO Collaborative Grant (ZT and PM), the U.S. Army Office of Research (PM and SW), and the National Science Foundation Polymers Program (SW). The impetus for planning a meeting at a Turkish locale grew from discussions during the stay of Prof. E. Arca (Hacettepe University, Ankara) in the laboratory of P. Munk. We feel very fortunate to have secured NATO funding for the Advanced Studies Institute that forms the basis for this book, with some important supplementary funding from the European Office of Research of the U.S. Army.

Naturally the scope has been extended from our own primary interest in the area of polymer micelles to the more general area of the role of solvents in promoting polymer self-organization. It was our aim to bring together scientists who are experts in macromolecular chemical synthesis, in the physical chemistry, and in the physics of polymer self-organization. The lectures also included experimental results and methods, analytical theory, and computer simulations. We feel that we were able to bring together scientists who would not often attend the same conferences and the cross-fertilization of ideas was most gratifying.

While the meeting concentrated on problems of basic science, we have not excluded practical applications such as pharmacology. Besides the main lectures, many interesting new results were presented by the participants in the form of posters and in lively discussions following lectures, during the formal poster sessions, and at many opportunities for informal discussion. Finally, topical subjects were discussed in five parallel round-table sessions at the end of the meeting. This discussion provided the participants a sense of where this field is at the present moment and some educated guesses about where it might go in the future.

Finally, special mention must be made of the hospitality of the Turkish people who we encountered outside the meeting and the efforts of the Turkish scientists who attended the ASI to smooth the way for the visitors and explain the fascinating mosaic that is modern Turkey. Certainly for the organizers, and we hope the participants, this was a once in a lifetime experience.

Petr Munk, Stephen E. Webber(Austin, Texas)
Zdenek Tuzar (Prague)

June, 1996

OVERVIEW OF POLYMER MICELLES

Z. TUZAR

Institute of Macromolecular Chemistry
Academy of Sciences of the Czech Republic
162 06 Prague 6, Czech Republic

1. Introduction

Micelles of block and graft copolymers belong to a broad family of polymer colloids. This family covers self-associating, self-assembling or self-organizing polymeric systems that possess surface-active or colloidal properties. The term "polymer colloids" may sound controversial. At the beginning of the polymer science as a new field of chemistry, most chemists claimed that large organic molecules, i.e., macromolecules or polymers, did not exist. Seventy years ago, even the most enlightened organic chemists believed that cellulose, proteins and polymers synthesized in laboratory were only colloids, i.e., aggregates of small organic molecules. Later, when the existence of macromolecules was undisputably recognized, polymer science was treated, in many textbooks of physical chemistry in chapters about colloids. That is why some polymer scientists dislike the word "colloid" and are reluctant to accept the name of a 30 years old branch of polymer science, polymer colloidics.

Polymer colloidics is based on block and graft copolymers. In selective solvents (good solvents for one block and poor for the other), block and graft copolymers self-assemble into various organized structures, the most prominent ones being multimolecular spherical micelles, cores of which are formed by insoluble blocks and shells by soluble blocks (Fig. 1). These micelles can solubilize otherwise insoluble substances, both low-molecular and polymeric. Micelles with extremely swollen cores form microemulsions, microlattices, or stable polymer dispersions.

S.E. Webber et al. (eds.), Solvents and Self-Organization of Polymers, 1–17.

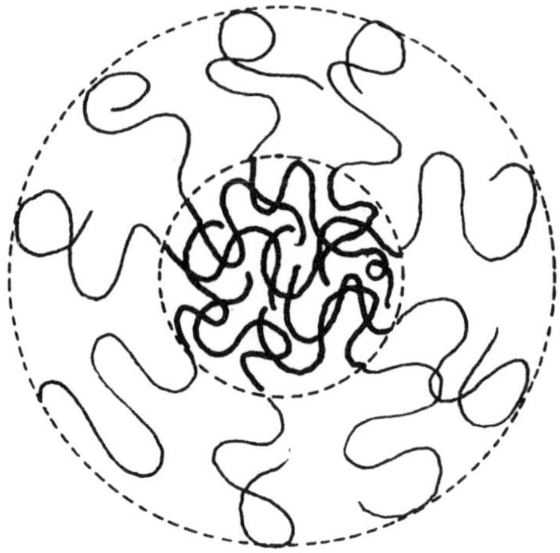

Figure 1. Sketch of a block copolymer micelle.

1.1. SHORT HISTORICAL SURVEY

Block copolymer micelles have been studied for more than 30 years. First observations, however, have been made even earlier: Many authors (e.g. [1-3]) observed a very stable turbidity when trying to carry out fractional precipitation of graft and block copolymers. Later, this turbidity had been ascribed to multimolecular micelles. Their shape, size, molar mass, and other properties have been studied by numerous experimental methods (see Table 1).

Both experimentalists and theorists show increasing interest in this field, and the number of papers dealing with block and graft copolymer micelles grows dramatically in last several years. Various aspects of polymer colloidics have been surveyed from both experimental [4,5] and theoretical [6] point of view.

References in [4] and [5] show that in sixties and seventies experimental data had been accumulated primarily concerning micellar molar mass, size, shape, unimer/micelle mass ratio, and hydrodynamic properties of micelles. Later, various authors focused on dynamics of micellar equilibria, kinetics of micelle formation and dissociation, as well as on structure and dynamics on segmental level. Most of these studies were performed with block copolymer micelles in organic selective solvents. Typical micellar systems are listed in Table 2.

TABLE 1. Experimental methods for micelle characterization

	Method	Information obtained
(a)	Static light scattering	Weight-average molar mass Radius of gyration
(b)	Small-angle X-ray and neutron scattering	Weight-average molar mass Radius of gyration Core radius Macrolattice structure
(c)	Dynamic light scattering	Diffusion coefficient Hydrodynamic radius
(d)	Electron microscopy	Size and shape
(e)	Sedimentation velocity	Micelles/unimer mass ratio
(f)	Size exclusion chromatography	Dynamics of micellar equilibrium
(g)	Stopped flow	Kinetics of micelle formation and dissociation
(h)	Steady-state and time-resolved fluorescence	Dynamics of micellar equilibrium Dynamics on segmental level
(i)	Nuclear magnetic resonance	Dynamics on segmental level

TABLE 2. Some typical micellar systems

Copolymer	Solvent (selectively good for)
PS-*b*-PMMA	p-xylene (PS) acetone (PMMA)
PS-*b*-PI	dimethylformamide (PS) hexane (PI)
PS-g-PI	decane (PI) tetrahydrofuran/heptane (PI)
PS-*b*-PB-*b*-PS	methyl ethyl ketone (PS)
PS-*b*-PDMS	decane (PDMS)
PS-*b*-hPI	heptane (PhI) 1,4-dioxane/heptane (PS)
PS-*b*-hPB-*b*-PS	1,4-dioxane (PS)

PS..polystyrene, PMMA..poly(methyl methacrylate), PI..polyisoprene, PB..polybutadiene, PDMS..poly(dimethylsiloxane), hPI..hydrogenated polyisoprene, hPB..hydrogenated polybutadiene, *b* block, *g* graft

Only in last ten years the effort shifted to micelles formed by hydrophobic/hydrophilic block copolymers in aqueous media. Due to its importance, this topic will be a subject of a separate Chapter (Z. Tuzar: Copolymer micelles in aqueous media).

1.2. BASIC DIFFERENCES BETWEEN LOW-MOLAR-MASS AND POLYMER MICELLES

Micelles of soaps and surfactants in water have been known and studied in detail for many decades. Low-molar-mass (LMM) and polymer micelles (PM) exhibit both similarities and differences:

(a) Molar mass of PM is at least two orders of magnitude larger then that of LMM.

(b) While PM with only a few exceptions are of spherical shape, LMM assume, depending on concentration, various shapes (spheres, rods, sheets, lamelae).

(c) Critical micelle concentration in PM systems is much smaller than in LMM systems.

(d) Dynamics of micellar equilibria and kinetics of micelle formation and dissociation are substantially slower processes in PM systems than in LMM systems. In PM systems a micellar equilibrium may be even kinetically frozen.

(e) Micelle formation in LMM and Pluronics aqeous systems is an entropy driven process, micelle formation in PM in organic solvents is an enthalpy driven process.

2. Molar Mass and Size of Micelles

Although this topic is trated in more detail in the Chapter of P. Munk, three items are briefly mentioned here: Association mechanism, determination of micellar molar mass and size, and unimolecular micelles.

2.1. OPEN AND CLOSED ASSOCIATION

Block copolymer micellization can proceed in principle by two mechanisms: by open or closed association. Open association is characterized by a series of equilibria between unimers, dimers, trimers, up to n-mers. Closed association is characterized by a dynamic equilibrium between molecularly dissolved copolymer (unimer) and an n-mer. It has been proved that block copolymer micellization, like micellization of soaps and surfactants, is a closed association. Since micellar equilibrium is shifted in most cases strongly towards micelles, standard static light scattering (SLS) measurements provide true weight-average micellar molar mass. Due to generally different contrasts of micellar core and shell, SLS, small-angle X-ray and neutron scattering (SAXS and SANS) give only an apparent radius of gyration of micelles. As we have demonstrated, a true value of radius of gyration, as well as geometrical radius of the core and the thickness of the shell can be obtained from SLS and SAXS (or SANS) [7]. Calculated value of the geometrical radius of the whole micelle (based on a simple core/shell concentric-sphere model) agreed quite well with the experimental value of the hydrodynamic radius from dynamic light scattering (DLS) [7].

2.2. UNIMOLECULAR MICELLES

The concept of a unimolecular micelle, in which an insoluble block is collapsed and soluble block forms a protective outer layer (Fig. 2), was

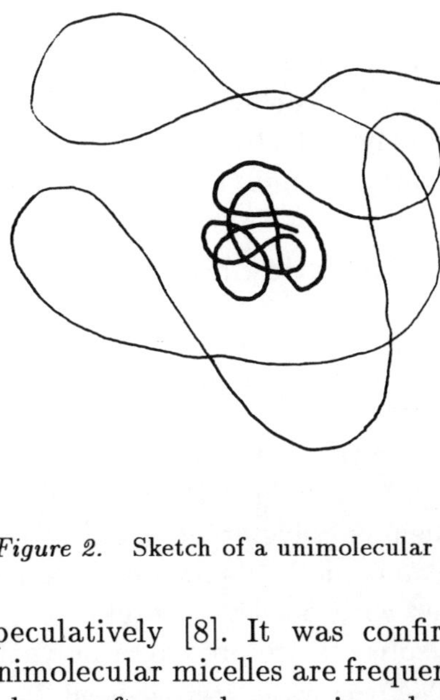

Figure 2. Sketch of a unimolecular micelle.

first formulated speculatively [8]. It was confirmed later by numerous experiments that unimolecular micelles are frequently formed by multiblock copolymers [9] or by graft copolymers in selective solvents for grafts, e.g., in [10,11]. Unimers of a diblock copolymer assume a conformation of unimolecular micelles if they are in an equilibrium with polymolecular micelles, as it was proved by NMR [12] and SAXS [13].

3. Dynamics of Micellar Equilibria

Unimer $\Longleftrightarrow$ micelle equilibrium represents a dynamic process, in which copolymer molecules migrate at a given rate between micellar and unimer states, i.e., also between micelles themselves. Two experimental methods employed to study the micellar equilibrium are steady-state fluorescence and sedimentation velocity.

In fluorescence studies [14,15] block copolymer micelles with core-forming blocks labeled by a fluorescence donor were mixed with identical micelles labeled by a fluorescence acceptor. Exchange of copolymer molecules between micelles, i.e. formation of mixed micelles with both fluorescence moieties in their cores, was monitored as the non-radiative energy transfer. In both papers, [14,15], equilibrium process proved to be very complex and was roughly characterized by two relaxation times: a faster (on time scale of 10^2 s) and a slower (10^3 s), the actual values of which depended on the degree of swelling of micellar cores. Activation energies led the authors [15] to a hypothesis, that the equilibration process might include, besides a transfer of free chains between independent micelles, also another mechanism, such as micelle-micelle collisions.

Sedimentation velocity approach will be mentiond only briefly, here, since it will be dealt with in detail in the Chapter of P. Munk. The method consists in mixing two kinds of micelles with different molar masses (and thus with different sedimentation coefficients) and monitoring the formation of mixed micelles by an analytical ultracentrifuge. Results of two studies, [16] and [17], lead to conclusions similar to those mentioned above.

4. Kinetics of Micelle Formation and Dissociation

There has been so far only one study on the kinetics of the formation and dissociation of block copolymer micelles [18]. The method employed was stopped-flow with SLS detection. Micellar systems were a diblock copolymer polystyrene-*b*-hydrogenated polyisoprene ($M_w = 105 \times 10^3$ g mol^{-1}, $w^{PS} = 0.4$) and a triblock copolymer polystyrene-*b*-hydrogenated polybutadiene-*b*-polystyrene ($M_w = 74 \times 10^3$ g mol^{-1}, $w^{PS} = 0.28$), both in dioxane and in a mixture dioxane/20 vol.% heptane. Characteristics of micelles with aliphatic cores and polystyrene shells of both copolymers in both solvents are presented in Table 3.

The process of micelle formation was realized by a fast mixing of a molecular copolymer solution ($c = 1 \times 10^{-2}$ g ml^{-1}) in a thermodynamically good solvent mixture, dioxane/40 vol.% heptane, with the same amount of dioxane. The resulting solution contained micelles in the mixture dioxane/20 vol.% heptane. Dissociation of micelles was realized by mixing a micellar solution ($c = 1 \times 10^{-2}$ g ml^{-1}) in dioxane with the same amount of heptane. Resulting solution in dioxane/50 vol.% heptane contained only molecularly dissolved copolymer.

The dead time (i.e., the time between the mixing of a solvent and a solution, and the start of SLS measurement) of the instrument used [18]

TABLE 3. Characteristics of two micellar systems used in the kinetic investigation

Copolymer	dioxane $M_w^{(m)} \times 10^{-6}$ g mol^{-1}	dioxane n	dioxane/20 vol.% heptane $M_w^{(m)} \times 10^{-6}$ g mol^{-1}	dioxane/20 vol.% heptane n
PS-*b*-hPI	20.7	197	18.0	171
PS-*b*-hPB-*b*-PS	5.1	69	4.8	65

$M_w^{(m)}$ is micellar weight-average molar mass, n is association number, PS is polystyrene block, hPB is hydrogenated polybutediene block. After [18].

was 1 ms. The micelle formation process was monitored as an increase in the intensity of scattered light, the micelle dissociation as a decrease in the intensity of scattered light. It has been found that the decay curves (Fig. 3) could best be fitted by a double-exponential function

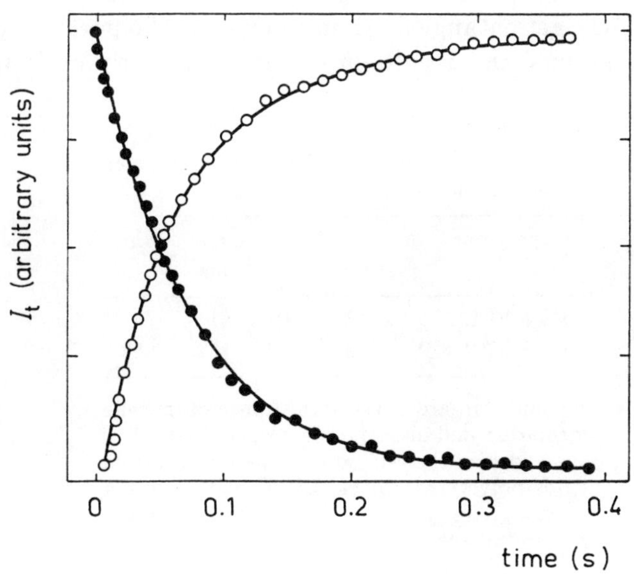

Figure 3. Time dependence of the intensity of scattered light for micelle formation (open circles) and micelle dissociation (full circles).

$$S(t) = a_1 e^{-t/\tau_1} + a_2 e^{-t/\tau_2}, \tag{1}$$

where a_1 and a_2 are normalized amplitudes, τ_1 and τ_2 are relaxation times, and S(t) is $(I_\infty - I_t)/I_\infty$ and I_t/I_0 for micelle formation and dissociation, respectively. Without giving τ_1 and τ_2 any specific physical interpretation, the double-exponential representation was used only as an acceptable approximation of apparently continuous spectrum of relaxation times, and the mean value $\tau = a_1\tau_1 + a_2\tau_2$ was assumed to characterize the rates of micelle formation, τ_F, and micelle dissociation, τ_D (Table 4). The quantity W, defined as

$$W = (a_1\tau_1^2 + a_2\tau_2^2)^{1/2}/\tau \tag{2}$$

was used as a measure of the width of the relaxation times distribution.

Table 4 illustrates the differences between the kinetics of the formation and dissociation of diblock and triblock copolymer systems. Similar values of τ_F and very different values of W in both systems have been explained as follows: Micelle formation proceeds in two stages. In the first stage, a good solvent is changed by mixing into a selective one and copolymer molecules with one block collapsed aggregate and form a system of polydisperse particles. In the second stage collapsed unsolvated blocks relax into entropically more favourable conformations. In this process the core-forming blocks may get entangled. Simultaneously, the particles equilibrate by exchanging unimers; this results in a system of nearly uniform micelles.

TABLE 4. Kinetics of block copolymer micellization

Copolymer	τ_F ms	W_F	τ_D ms	W_D
PS-*b*-hPI	43	0.02	1	–
PS-*b*-hPB-*b*-PS	69	0.57	137	0.83

τ_F and τ_D are relaxation times of micelle formation and dissociation, respectively, W is the width of the relaxation time distribution. PS is polystyrene block, hPI is hydrogenated polyisoprene block, hPB is hydrogenated polybutadiene block. After [18].

While τ_F reflects only the first stage, W monitors the equilibration process, which is obviously more complex in case of a triblock copolymer system.

Analogous reasoning holds also for the data on micelle dissociation in Table 4. Here, the first step is a fast influx of solvent molecules into the micellar core and the second step a solvent-assisted separation of copolymer molecules from a micelle. The small value of τ_D (< 1 ms) for the diblock copolymer can be explained by an easy escape of unimer molecules from a micelle, since each core-forming block crosses the core/shell interface only once. In case of a triblock copolymer, any unimer molecule crosses the core/shell interface twice. This process involves disentanglements that are slowed down by unfavourable interactions between non-compatible blocks. As a result, the relaxation time of triblock copolymer dissociation is at least two orders of magnitude longer than that of diblock copolymer micelles.

5. Thermodynamics of Micelle Formation

5.1. THEORIES OF MICELLE FORMATION

Only in last twelve years, theories of micelle formation have been developed, after structure and various properties of polymer micelles were established experimentally. Some authors (e.g., in [19–22]) determined the Gibbs energy of micelle formation assuming uniform segment concentrations in the core and in the shell, other authors (e.g., [23–25]) developed scaling theories of polymer micelles with compact cores and extended shells. Various aspects of these theories are amply discussed in other Chapters.

5.2. GIBBS ENERGY OF MICELLE FORMATION

An important question arising in the thermodynamics of micelle formation is: Which term, entropic or enthalpic, contributes more to the decrease of Gibbs energy. It has been shown [26] that, for the closed association and large association number, n, the standard Gibbs energy and the standard enthalpy change per mol of the solute in a micelle, $\Delta G°$ and $\Delta H°$, respectively, can be expressed by approximate relations

$$\Delta G° \cong RT \ln cmc \tag{3}$$

$$\Delta H° \cong RT^2 (\mathrm{dln}\ cmc/\mathrm{d}T) \tag{4}$$

Under assumption that n and $\Delta H°$ are independent of temperature (unfortunately, this is not always true), integration of equation (2) gives

$$\ln cmc = \Delta H°/RT + \mathrm{Const} \tag{5}$$

Price *et al.* [27-31] obtained, using equation (4) heats of micellization for several copolymer/organic selective solvent systems. The data confirmed that micellization is an enthalpy-driven process. The same result was obtained by direct calorimetric measurements [32,33].

Nagarajan's theoretical model-based results [22] show that the main contribution to the exothermic process is the formation of the micellar core.

It will be shown in another chapter (Z. Tuzar: Copolymer micelles in aqueous media) that micellization of both low-molar-mass and block copolymer micelles in water is an entropy driven process.

6. Concentrated Micellar Systems

Two systems will be discussed here: regular micelles (AB or ABA types in selective solvents of A blocks) and the so called tentacled micelles (ABA in a selective solvent of B block). In our detailed study [34,35] we approached the problem of concentrated micellar solutions of both types with a combination of various experimental techniques: SLS, DLS, SAXS, sedimentation velocity and rheology. Our findings are summarized below.

6.1. REGULAR MICELLES

Unlike soap and surfactants which can form at higher concentrations micelles of various kind (rods, sheets, lamelae, etc.), concentrated equilibrated systems of AB or ABA block copolymers in selective solvents of A blocks form regular spherical micelles with cores of B blocks and shells of A blocks. SAXS and rheology studies (e.g. [36]) showed that at higher concentrations micelles can form so-called macrolattices.

In our study [35,36] we have used a triblock copolymer with polystyrene outer blocks and hydrogenated polybutadiene middle block ($M_w = 7 \times 10^4$ g mol^{-1}, mass fraction of polystyrene 0.28) In a solvent mixture dioxane/25 vol.% heptane, micelles with aliphatic cores and polystyrene shells were formed by a direct dissolution at concentrations up to 15 wt.%. At low concentrations (up to several per cents), SLS provided micellar molar mass (4.35×10^6 g mol^{-1}) and DLS hydrodynamic radius (24.8 nm). SAXS data showed that the core radius was constant (20 nm), independent of concentration. Zero-sheer viscosity of micellar solutions, $\eta°$, (Fig. 4) showed a steep increase at concentrations above 10 wt.%. In this concentrated region, SAXS gave evidence of the macrolattice formation. It means, that micelles keep the same association number, independent of concentration, and at higher concentrations they get space-organized.

6.2. TENTACLED MICELLES

Triblock copolymers ABA in selective solvents for the middle block B, i.e. in selective precipitants of the outer blocks A, may behave differently (in both dilute and concentrated solutions) from those in selective solvents for

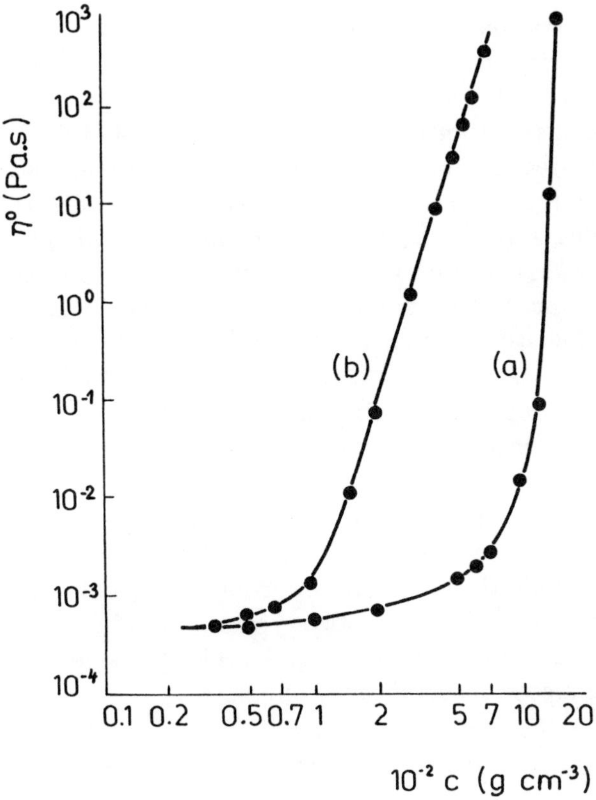

Figure 4. Concentration dependence of zero-shear viscosity for regular micelles (a) and tentacled micelles (b).

the outer blocks (Section 6.1.). Depending on the lengts of the blocks and the selectivity of the solvent, block copolymers may or may not associate, and those associates may assume various forms, as described in [37–42]. In concentrated solutions, a tendency to the formation of physical gels has been reported, e.g., in [43]. It has been claimed that the knots of such a gel are formed by insoluble outer blocks and the bridges by soluble middle blocks.

In our study [34,35], the same copolymer as in the Section 6.1. has been used. The selective solvent for the middle aliphatic block and a precipitant for the outer polystyrene blocks was heptane. Concentration dependence of η° (Fig. 4) shows two regions, like in case of regular micelles, but a steep increase in η° starts at about ten times lower concentration.

In the low concentration region (up to ca. 1×10^{-2} g ml^{-1}), SLS and sedimentation velocity showed a normal closed association pattern with unimer and polymolecular micelles (n = 32) independent of concentration. Negative slope of the concentration dependence of the diffusion coefficient (from DLS) indicated strong intermicellar attractive forces. SAXS data ruled out the model of a simple physical network described in [43]: It showed unambiguously that polystyrene formed homogeneous, highly packed spheres with a radius 6.4 nm (corresponding to the core molar mass evaluated from SLS), independent of concentration in the whole concentration region studied, i.e., up to 7 wt.%, only at concentration above ca 2 wt.% micelles were space-correlated, i.e., were forming a macrolattice.

From all data obtained (including rather complex spectra of correlation times from DLS in concentrated solutions, not discussed here) we offered the following model:

(a) Dilute solutions (up to 1 wt.%). Micelles with the association number 32 consist of cores accomodating most of 64 polystyrene end blocks and shells formed by aliphatic loops. A few copolymer molecules acquire an extended conformation with one polystyrene block in the core and the second stretched out as a tentacle (Fig. 5). The tentacles are responsible for the attractive forces observed by DLS.

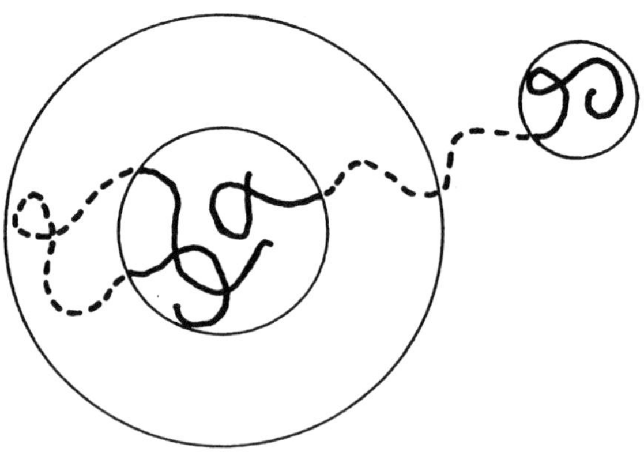

Figure 5. Model of a micelle formed by a triblock copolymer in a selective solvent for the middle block.

(b) Concentrated region (1–7 wt.%). Micelles survive in this region, but due to tentacles and unimers and dimers (existence of which was indicated by sedimentation velocity at low concentrations) micelles get space-organized in a rather complex network (Fig. 6).

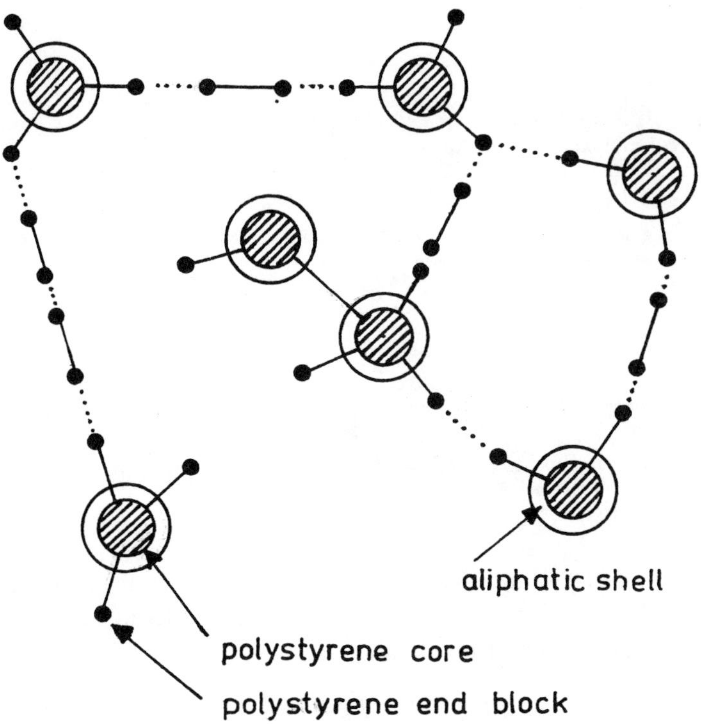

Figure 6. Model of a network formed by tentacled micelles.

The model of tentacled micelles was corroborated by both thermodynamic reasoning [41] and by computer modelling [44]. Theory of the formation of physical gels for ABA block copolymers in selective solvents for middle block [45], as well as the computer modelling [46] that appeared recently, more or less corroborate our speculative model [35].

7. Hydrodynamic Behavior of Micelles

One of our study mentioned earlier [7] leads to a conclusion that micelles of a certain type behave like hydrodynamic hard spheres. A question emerged, how a shell swelling influences the hydrodynamic character of micelles. The experiment was performed with micelles having aliphatic cores and polystyrene shells in mixed solvent dioxane/heptane [47]. Dioxane is a selective solvent for polystyrene shells, while heptane is a selective solvent for aliphatic cores. An increase in heptane content in a solvent mixture leads to the swelling of cores and deswelling of shells. Experimental SLS and DLS data in solvent mixtures containing 10, 20 and 30 vol.% heptane are presented in Table 5.

TABLE 5. SLS and DLS data for micelles of polystyrene-*b*- hydrogenated polybutadiene-*b*-polystyrene in dioxane

Heptane vol.%	$M_w^{(m)} \times 10^{-6}$ g mol^{-1}	$R_h^{(m)}$ Å	$A_2 \times 10^6$ cm^3 g^{-2} mol	$A_2^{HS} \times 10^6$ cm^3 g^{-2} mol	k'_D cm^3 g^{-1}	k_D
0	4.80	192	2.1	3.07	4.02	1.09
10	4.44	223	4.2	5.66	6.63	1.05
20	4.20	232	7.5	7.19	7.54	1.01
30	3.77	249	12.7	10.97	11.95	1.15

$M_w^{(m)}$ is the micellar molar mass, A_2 is the second virial coefficient, $R_h^{(m)}$ is the hydrodynamic radius of micelles; A_2^{HS}, k'_D and k_D are calculated values (for definitions see text). After [47].

SLS provided $M_w^{(m)}$ and A$_2$, DLS provided $R_h^{(m)}$ and the difusion virial coefficint, k'_D. The last value, k_D, has been obtained from the concentration dependence of the diffusion coefficient, D_c:

$$D_c/D_0 = 1 + k_D\Phi = 1 + k'_D c \tag{6}$$

where D_0 is the diffusion coefficient at infinite dilution, Φ is the volume fraction of micelles and k_D is a value which is not directly accessible from experiment but can be calculated (see below).

Only small changes in values of $M^{(m)}$ and $R_h^{(m)}$ with the increased amount of heptane in solvent mixtures in Table 5, indicate that the effects of swelling of the cores and deswelling of the shells compensate each other. However, an increase in A_2 and k'_D means that the effect of the core swelling predominates.

The evidence of the hard sphere behavior of micelles in all solvent mixtures in Table 5 can be deduced from a good agreement between experimental values of A_2 and values for the hard sphere model, A_2^{HS}. The latter value has been calculated from the relation

$$A_2^{HS} = 4N_A V/(M^{(m)})^2 \tag{7}$$

where V is the excluded volume of a micelle, $V = 4\pi(R_h^{(m)})^3/3$.

Even a more convincing proof of the hard sphere behavior can be seen in the almost perfectly constant value of k_D, which was calculated from the relation

$$k_D = k'_D/\bar{v} \tag{8}$$

where $\bar{v}$ is the partial specific volume of swollen micelles, approximated by their volume *per* unit of $M^{(m)}$, defined as $N_A V/M^{(m)}$. Using again, like in equation (5), R_h for the calculation of V, equation (6) can be written as

$$k_D = 3k'_D M^{(m)}/[4\pi N_A (R_h^{(m)})^3] \tag{9}$$

The k_D value from Table 5 (1.08 ± 0.07), independent of the solvent composition, is close to the value 1.3, obtained experimentally for a dispersion of silica particles [48], which may be considered as close to hard spheres.

In spite of the evidence of hydrodynamic hard sphere behavior of block copolymer micelles given in [47] and discussed above, it can be expected that micelles with extremely loose shells, like those with relatively small cores and long ('hairy') shell-forming blocks or micelles with polyelectrolyte shells, may show different hydrodynamic behavior.

Acknowledgements. This research was supported by the Collaborative Research Grant 920166 from the Scientific and Environmental Division of the NATO and by the U.S.-Czech Science and Technology Joint Fund No. 95010.

References

1. Merrett, F.M. (1957) Graft copolymers with preset molecular architecture, *J. Polym. Sci.* **24**, 467–477.
2. Hartley, F.D. (1959) Graft copolymer formation during the polymerization of vinyl acetate in the presence of polyvinylalcohol, *J. Polym. Sci.* **34**, 397–417.
3. Schlick, S. and, Levy, M. (1960) Block-polymers of styrene and isoprene with variable distribution of monomers along the polymeric chain, *J. Phys. Chem.* **64**, 883–886.
4. Riess, G., Hourtrez, G., and Bahadur, P. (1985) Block copolymers, in H.F. Mark and N.M. Bikales (eds.), *Encyclopedia of Polymer Science and Engineering*, 2nd Edition, Wiley, New York, pp. 324–434.
5. Tuzar, Z. and Kratochvíl, P. (1993) Micelles of block and graft copolymers in solutions, *Surface and Colloid Science* **15**, 1–83.
6. Halperin, A., Tirrell, M., and Lodge, T.P. (1992) Tethered chains in polymer microstructures, *Adv. Polym. Sci.* **100,** 31–71.
7. Tuzar, Z., Pleštil, J., Koňák, Č., Hlavatá, D., and Sikora, A. (1983) Structure and hydrodynamic properties of poly[styrene-*b*-(ethene-*co*-butene)-*b*-styrene] micelles in 1,4-dioxane, *Makromol. Chem.* **184**, 2111–2121.
8. Gallot, B. and Sadron, Ch. (1971) A new kind of polymeric material, *Macromolecules* **4**, 514–515.

9. Bresler, S.E., Pyrkov, L.M., Frenkel, S.Ya., Laius, L.A., and Klenin, S.I. (1962) Molecular conformation of 4/5 block copolymer of styrene and isoprene, *Vysokomol. Soyed.* **4**, 250–255.
10. Selb, J. and Gallot Y. (1981) Comportement de copolymeres greffes en milieu solvant selective des greffons, *Makromol. Chem.* **182**, 1775–1786.
11. Price, C. and Woods, D. (1973) Synthesis and solution behavior of polystyrene-g-polyisoprene copolymers, *Polymer* **14**, 82–86.
12. Spěváček, J. (1982) ^{1}H NMR study of styrene-butadiene block copolymer micelles in selective solvents, *Makromol. Chem., Rapid Commun.* **4**, 697–703.
13. Pleštil, J. and Baldrian, J. (1975) Determination of the structure parameters of styrene/butadiene block copolymer in heptane by means of small-angle X-ray scattering, *Makromol. Chem.* **176**, 1009–1028.
14. Procházka, K., Bednář, B., Mukhtar, E., Svoboda, P., Trněná, J., and Almgren, M. (1991) Energy transfer in block copolymer micelles, *J. Phys. Chem.* **95**, 4563–4568.
15. Wang, Y., Kausch, C.M., Chun, M., Quirk, R.P., and Mattice, W.L. (1995) Exchange of Chains between micelles of labeled polystyrene-*block*-poly(oxyethylene) as monitored by nonradiative singlet energy transfer, *Macromolecules* **28**, 904–911.
16. Tian, M., Qin, A., Ramireddy, C., Webber, S.E., Munk, P., Tuzar, Z. and Procházka, K. (1993) Hybridization of block copolymer micelles, *Langmuir* **9**, 1741–1748.
17. Pacovská, M., Procházka, K., Tuzar, Z. and Munk, P. (1993) Formation of block copolymer micelles: A sedimentation study, *Polymer* **34**, 4585–4588.
18. Bednář, B., Edwards, K., Almgren, M., Tormod, S., and Tuzar, Z. (1988) Rates of association and dissociation of block copolymer micelles: Light-stattering stopped-flow measurements, *Makromol. Chem., Rapid Commun.* **9**, 758–790.
19. Leibler, L., Orland, H., and Wheeler, J.C. (1983) Theory of critical micelle concentration for solutions of block copolymers, *J. Chem. Phys.* **79**, 3550–3557.
20. Noolandi, J. and Hong, K.M. (1983) Theory of block copolymer micelles in solution, *Macromolecules* **16**, 1443–1448.
21. Munch, M.R. and Gast, A.P. (1988) Block copolymers at interfaces 1. Micelle formation, *Macromolecules* **21**, 1360–1366.
22. Nagarajan, R. and Ganesh, K. (1989) Block copolymer self-assembly in selective solvents: Spherical micelles with segregated cores, *J. Chem. Phys,* **90**, 5843–5856.
23. Zhulina, Y.B. and Birshtein, T.M. (1985) Conformations of molecules of block copolymers in selective solvents (micellar structures), Vysokomol. Soyed. **27**, 511–517.
24. Halperin, A. (1987) Polymeric micelles: A star model, *Macromolecules* **20**, 2943–2946.
25. Marques, C.N. (1988) Adsorption of block copolymers in selective solvents, *Macromolecules* **21**, 1051–1059.
26. Corkill, J.M., Goodman, J.F., and Harrold, S.P. (1964) Thermodynamics of micellization of non-ionic detergents, *Trans. Faraday. Soc.* **64**, 202–207.
27. Price, C., Kendall, K.D., Stubbersfield, R.B., and Wright, B. (1983) Thermodynamics of micellization of polystyrene-b-poly (ethylene/propylene) block copolymer in n-decane, *Polym. Commun.* **24**, 326–328.
28. Price, C., Booth, C., Canham, P.A., Naylor, T.V., and Stubbersfield, R.B. (1984) The thermodynamics of micelle formation by a polystyrene-b-polyisoprene Block copolymer in N,N′-dimethylacetamide, *Br. Polym. J.* **21**, 311–313.
29. Price, C., Chan, E.K.M., Mobbs, R.H., and Stubbersfield, R.B. (1985) Thermodynamic investigation of micelle formation by a polystyrene-b-polyisoprene block copolymer in n-hexadecane, *Eur. Pclym. J.* **21**, 355–360.
30. Price, C., Chan, E.K.M., and Stubbersfield, R.B. (1987) The effect of block length on the thermodynamic stability of micelles formed by polystyrene-b-polyisoprene copolymers in n-hexadecane *Eur. Polym. J.* **23**, 649–651.
31. Price, C. (1983) Micelle formation by block copolymers in organic solvents, *Pure*

Appl. Chem. **55**, 1563–1572.
32. Price, C., Chan, E.K.M., Pilcher, G., and Stubbersfield, R.B. (1985) A calorimetric study of the enthalpy of micelle formation by a polystyrene-b-polyisoprene block copolymer in n-hexadecane, *Eur. Polym. J.* **21**, 627–628.
33. Procházka, K., Dalcros, H., and Delmas, G. (1988) A light scattering and calorimetric study of micelle formation by a polystyrene-b-hydrogenated polyisoprene block copolymer in a binary solvent (pentane/cyclopentane) *Can. J. Chem.* **66**, 915–918.
34. Pleštil, J., Hlavatá, D. Hrouz, J., and Tuzar, Z. (1990) Dilute and semidilute solutions of ABA block copolymer in solvents for A or B blocks: 1. Small-angle X-ray scattering study, *Polymer* **31**, 2112–2117.
35. Tuzar, Z., Koňák, Č., Štěpánek, P., Pleštil, J., Kratochvíl, P., and Procházka, K. (1990) Dilute and semidilute solutions of ABA block copolymer in solvents for A or B blocks: 2. Light scattering and sedimentation study *Polymer* **31**, 2118–2124.
36. Watanabe, H. and Kotaka, T. (1984) Rheology and structure of styrene-butadiene diblock copolymers dissolved in selective solvents, *Polym. Eng. Rev.* **4**, 73–122.
37. Kotaka, T., Tanaka, T., and Inagaki, H. (1972) Thermodynamic and conformational properties of styrene-methyl methacrylate block copolymers in dilute solution. IV. Behavior of diblock and triblock copolymers in selective solvents, *Polymer J.* **3**, 327–337.
38. Tanaka, T., Kotaka, T., and Inagaki, H. (1972) Thermodynamic and conformational properties of styrene-methyl methacrylate block copolymers in dilute solution. V. Light scattering analysis of conformational anomalies in p-xylene solution, *Polymer J.* **3**, 338–349.
39. Kotaka, T., Tanaka, T., Hattori, M., and Inagaki, H. (1978) Block copolymer micelles in dilute solution, *Macromolecules* **11**, 138–145.
40. ten Brinke, G. and Hadziioannou, G. (1987) Topological constraints and their influence on the properties of synthetic macromolecular systems. 2. Micelle formation of triblock copolymers, *Macromolecules* **20**, 486–489.
41. Balsara, N.P., Tirrell, M, and Lodge, T.P. (1991) Micelle formation of BAB triblock copolymers in sovents that preferentially dissolve the A block, *Macromolecules* **24**, 1975–1986.
42. Chu, B. (1995) Structure and dynamics of block copolymer colloids, *Langmuir* **11**, 414–421.
43. Ceresa, R.J. (1969) Effect of sequential arrangements on block copolymer properties, *J. Polym. Sci., Polym. Symp. (C)* **26**, 201–206.
44. Rodrigues, K. and Mattice, W.L. (1992) Intraparticle distribution functions for a micelle formed by a small symmetric triblock copolymer in a poor solvent for the terminal blocks, *Langmuir* **8**, 456–460.
45. Semenov, A.N., Joanny J.-F., and Khokhlov, A.R. (1995) Associating polymers: Equilibrium and linear viscoelasticity, *Macromolecules* **28**, 1066–1075.
46. Nguyen-Misra, M. and Mattice, W.L., Micellization and gelation of symmetric triblock copolymers with insoluble end blocks, *Macromolecules* **28**, 1444–1457.
47. Koňák, Č., Tuzar, Z., Štěpánek, P., Sedláček, B., and Kratochvíl, P. (1985) Interaction between block copolymer micelles in solutions, *Progress in Colloid and Polym. Sci.* **71**, 15–19.
48. Kops-Werkhoven, M.M. and Fijnaut, H.M. (1981) Dynamic light scattering and sedimentation experiments on silica dispersions at finite concentrations, *J. Chem. Phys.* **74** 1618–1625.

EQUILIBRIUM AND NONEQUILIBRIUM POLYMER MICELLES

PETR MUNK
Department of Chemistry and Biochemistry
and Center for Polymer Research,
The University of Texas at Austin, Austin, Texas 78712, U.S.A.

1. Introduction

Block copolymer micelles and detergent micelles have many common features. The detergent micelles are equilibrium systems. If their equilibrium is disturbed by some outside action, they reequilibrate within a very small fraction of a second. Consequently, in many studies of block copolymer micelles, it is tacitly assumed that they are also at equilibrium and that their properties faithfully reflect the underlying equilibrium thermodynamics. It is the purpose of this article to show that truly equilibrium block copolymer micelles are rare and that the equilibrium is difficult to recognize.

The true nature of the micelles depends strongly on the method of their preparation. Some micellar systems may approach the equlibrium reasonably fast, some very slowly, some may proceed toward a local minimum of the Gibbs energy that does not represent the true equilibrium, and finally some may stay in the nonequilibrium state indefinitely because the kinetics of any conceivable equilibrating process is so slow as to prevent any change within experimentally reasonable time-scales.

2. Micellar Origins

In the laboratory, block copolymer micelles are produced in one of two ways. In the first technique, the copolymer is dissolved molecularly in a solvent that is good for both blocks and then the conditions (temperature, composition of the solvent) are changed in a way that requires formation of micelles. In the other technique, a solid sample of the copolymer is directly dissolved in a selective solvent; sometimes the micellar solution is let to anneal by standing; sometimes

S.E. Webber et al. (eds.), Solvents and Self-Organization of Polymers, 19–32.

the annealing process is helped by some thermal treatment. We need to understand how these procedures operate on the molecular level.

2.1. AGGREGATION OF MOLECULARLY DISSOLVED COPOLYMERS

Honda *et al.* [1] explored the former approach. They employed poly(α-methylstyrene)-*block*-poly(*p*-vinylphenethyl alcohol) (PαMS-*b*-PVPA) in benzyl alcohol that is a good solvent for the PVPA blocks and a nonsolvent for the PαMS blocks. The PVPA blocks were much shorter than the PαMS blocks. This system has the critical micelle temperature (CMT) slightly above ambient temperature. The solutions were first kept 10-15 °C above CMT then quenched to a temperature below CMT and followed by static and dynamic light scattering. Under these conditions, the micellization is a relatively slow process and the micellar system develops on a time scale of hours to weeks. The authors found that the development of the system proceeded in two steps. In the first step, the rate of which increased with increasing concentration of the copolymer, both the apparent molecular weight and apparent size first increased and then tended to level off with the size levelling off sooner then the molecular weight. In the second step, that was much slower and the rate of which was independent of the copolymer concentration, both the molecular weight and the size increased again; this time in a more or less parallel way.

The authors interpret the first step as the birth of the micelles when clusters of unimers collapse to form the original micelles. Only clusters that are large enough form stable micelles. The apparent size of the particles is the z-average over all particles in the solution. As such, it is dominated by the micelles and represents their size; it does not change appreciably during the first step. The observed molecular weight is the weight average and consequently increases as the unimers population is converted to micelles. The first phase is completed when the supply of the unimers is exhausted. Thus, the first step of the micellization is essentially a typical nucleation process with a slow rate of nucleation. Unlike other types of nucleation, the early micelles cannot grow indefinitely, because micellization is a closed association process. Consequently, new nuclei are formed continuously and they grow only to a limited size.

The growth of the micelles slows down as they approach their equilibrium size. Before they can reach it, the supply of unimers is exhausted: their size is smaller than the equilibrium one. They reach the equilibrium size during the second step of the micellization process, during which the supply of unimers is replenished by the disintegration of the smaller, less stable micelles.

Thus, the second micellization step is similar to a conversion of smaller crystals into larger ones when left in contact with the mother liquid.

2.1.1 *Anomalous Micellization*

There are many reports that the formation of micelles from molecularly dissolved unimers is not a simple process. When the conditions (temperature, solvent composition) are changed from those favoring molecular dissolution to those that are just beyond the borderline of the homogeneous phase, very large particles are often formed. The solution may get a milky appearance (in contrast to the bluish tint of typical micellar solutions), the envelope of the scattered light becomes extremely asymmetric, and the viscosity may increase dramatically. Such a behavior is referred to as anomalous micellization.

Mandema *et al.* [2,3] postulated formation of large, cylindrical or lamellar, but otherwise undefined structures. Their argument was based on reversible transitions between the regular and anomalous solutions. Canham *et al.* [4] used light scattering and electron microscopy for studying the anomalous micellization of polystyrene-*block*-polybutadiene-*block*-polystyrene copolymers in ethyl acetate. At lower temperatures, regular spherical micelles are formed. At intermediate temperatures, the anomalous structures appear. Finally at higher temperatures, the copolymer dissolves molecularly. All transitions are fully reversible when temparature is changed between adjoining regions. Electron micrographs of the anomalous solutions revealed a presence of coiled fibers the diameter of which was of the same order of magnitude as the diameter of the spherical micelles (that were formed at lower temperature). Accordingly, the authors interpreted their results as a tendency of their copolymers to form cylindrical micelles as the most stable form within some range of experimental variables.

Nagarajan and Ruckenstein [5] compared the contributions to the Gibbs energy of a single molecule of a surfactant while embedded into a spherical micelle or an infinite cylindrical aggregate (both micelles having the diameter leading to the Gibbs energy minimalization). They found that the contributions are quite similar and that it depends on the finer detailes of the structure of the surfactant and on other thermodynamic factors involved in the micellization whether spherical or cylindrical aggregates are more stable.

Returning to block copolymer micelles, for which the basic thermodynamic considerations are the same as for the surfactants but the importances of various factors may be different, we find that all experimental studies report that the spherical micelles are formed predominantly and thus are presumably the more stable ones. All the exceptions occur when the systems observed are in the close vicinity of critical conditions that lead to the segregation of the insoluble parts of the molecules. What is important from the perspective of the present paper is the expectation that the micelles (of whatever shape) have a rather swollen insoluble core and, consequently, that the kinetics of their equilibration is quite fast.

Several more recent studies ascribe the anomalous micellization to the chemical heterogeneity of the copolymers. Zhou and Chu [6] studying ethylene oxide - propylene oxide block copolymers found that anomalous micellization does not occur again in solutions that were filtered when the strongly scattering particles occured first time. Tuzar and Kratochvíl [7] propose a model in which the small amount of a homopolymer or a copolymer with a high content of the insoluble block (they are always present in real block copolymer samples) forms an emulsion when the solvent is sufficiently poor for it but not poor enough for the bulk copolymer to form micelles. When the solvent gets poorer and the micelles are formed, the precipited homopolymer may be solubilized by them and the anomaly will disappear. Under some circumstances, the large particles may remain in solution even after that [8].

In our opinion, the term anomalous micellization is actually used for two different phenomena, both of which sometimes appear in the vicinity of the critical point for micellization.

2.2. MICELLE FORMATION FROM SOLID COPOLYMERS

Bulk block copolymers that have mutually incompatible blocks have a characteristic domain structure. When the volume fraction of one component is less than about 0.2 then, at equilibrium, it forms spherical domains embedded in the matrix of the major component. When the volume fraction is about 0.2 to 0.35 then the domains are cylindrical. Finally, when the volume fractions are about equal then a lamellar structure is formed. However, different morphologies may be displayed by samples that are not at full thermodynamic equilibrium due to the method of preparation of the bulk sample. Molau and Wittbrodt [9] in a classical study dissolved aliquots of a sample of styrene-butadiene block copolymer in two solvents; one selective for one block and the other for the second block. Both solutions were micellar. When they evaporated the solvents and made films of the copolymer, in one case they obtained spherical domains of polybutadiene in the polystyrene matrix, in the other case polystyrene spherical domains in the polybutadiene matrix. Thus, the morphology of the domains was preserved during the process of solvent evaporation and at least one of the resulting films was a nonequilibrium one.

We believe that a similar phenomenon may prevail also during the dissolution of a bulk block copolymer in a selective solvent. Then the resulting micelles will depend on the morphology of the bulk sample as well as on the interactive properties of the selective solvent with respect to the two blocks and on the glassy or elastic nature of the blocks. Let us consider several possible scenarios.

1. The solvent is only a moderate nonsolvent for one of the blocks. Then the soluble blocks will swell at the beginning extensively, the insoluble blocks will also swell enough for allowing considerable segmental mobility within their domains. These domains (of any shape) may then break up into smaller entities that together with adjacent soluble chains will form individual micelles. It is quite conceivable that the aggregation number of these micelles will be related to the thickness of the original lamellae, diameter of the cylinders, or that it will remain the same as the number of molecules in the original spherical domain (which was possibly formed by evaporating a different selective solvent that was related to micelles with a different equilibrium aggregation number). Thus, the micelles formed by this dissolution process may not be the equilibrium micelles corresponding to the solvent used for the dissolution. In the subsequent step, the micelles may either reequilibrate, or this reequilibration may be quite slow, or it may be kinetically totally frozen.

2. The solvent swells the insoluble domains only marginally. Initially the dissolution may proceed as in the case sub 1. However, the insoluble blocks may have difficulties while extricating themselves from their original domains, the breakup will be less complete resulting in much bigger micelles. These micelles will probably be kinetically frozen.

3. The solvent does not swell the insoluble domains at all. If the domains are glassy, the sample will not dissolve at all.

4. The bulk sample was prepared by evaporation or freeze-drying of a solvent from a micellar solution in which the micellar cores were not swollen by the solvent and, consequently, are not entangled in the bulk sample. In this case, the micellar cores will preserve their identity and the sample will dissolve in the same or different solvent to a micellar solution with the original micelles still present.

5. A micellar solution in a solvent that swells the cores appreciably was evaporated. As the solution becomes more and more concentrated the micellar

cores are getting closer to each other and may start to interpenetrate and form micellar clusters. When the evaporation is sufficiently slow (or when the solvent is good for both blocks) the bulk polymer may end up in its equilibrium (spherical, cylindrical, or lamellar) morphology. Otherwise the micellar clusters may be preserved in the bulk and then reappear when the sample is redissolved in a solvent that does not swell the cores appreciably.

6. A micellar solution in a core-swelling solvent is quickly frozen and then freeze-dried. During the freezing (i. e., crystallization of the solvent) the micelles are concentrated in the intercrystalline space and may partially interpenetrate as in the case sub 5. Again micellar clusters may appear after redissolution. The size of the clusters will depend on the quality of the solvent before freeze-drying as well as of the one after freeze- drying: different solvents may break up the clusters to a different degree.

3. Transfer of Micelles to a Different Medium

Some bulk block copolymers would not dissolve in some selective solvents. Typical cases are copolymers insoluble blocks of which are glassy under the experimental conditions and do not swell in the selective solvent. Polystyrene and poly(methyl methacrylate) as insoluble blocks and water or a buffer as the selective solvent are characteristic examples. However, micelles with such cores in aqueous media constitute an interesting and possibly useful class of systems. Several strategies have been developed for preparation of such micellar solutions. In all of them the micelles are formed in a selective solvent that allows their formation, this solvent is then replaced by the solvent of choice. Distillation, dialysis, and freeze-drying were used for this purpose.

Riess and Rogez [10] dissolved polystyrene-poly(ethylene oxide) block copolymers in tetrahydrofuran-water mixture. Then THF was removed from the resulting micellar solution by azeotropic distillation under reduced pressure. This procedure is applicable only when the solvent component to be removed is sufficiently volatile.

The dialysis procedure was pioneered by Tuzar *et al.* [11]. It consists of the dialysis of the solution of the copolymer in the micelle-forming solvent to the desired one. Sometimes the dialysis may be performed in a single step but usually it is recommended that it is performed in several steps. We have been using this procedure for block copolymers of styrene and methacrylic acid. These copolymers form micelles in water-dioxane mixtures containing 60 to 80 volume percent of dioxane. When aiming at solutions in water we dialyze these solutions against a succession of dioxane-water mixtures that differ by about 20 percent of dioxane in their composition. Some of the micellar systems are not soluble in water but are soluble in buffers. In such a case it is necessary to use the buffer instead of water already in the last one or two solution mixtures against which the micelles are being dialyzed.

The molecular processes during the dialysis are sometimes complex. During the early stages the micellar solution may still be in a dynamic equilibrium and the aggregation number of the micelles may change with the changing quality of the solvent mixture. However, once the solvent becomes sufficiently poor with respect to the core, the reequilibration is effectively frozen and the molecular weight of the micelles does not change any more. Qin and Munk [12] have found that the observed changes in the micellar weight were within experimental errors when the styrene-methacrylic acid micelles were transferred by the stepwise dialysis from the 80 % dioxane-water mixture to aqueous buffers.

Freeze-drying of micellar solutions is a very complicated process. Because the melting point of dioxane is about 12 °C and the eutectic point of dioxane-water mixtures is about -13 °C it is possible to freeze-dry solutions in this solvent mixture having any composition. Hoping that the integrity of individual micelles would not be compromised by freeze-drying we tried to develop a procedure for preparing the copolymer in a dry form that would be soluble in aqeous solvents. We prepared micellar solutions in our standard 80 % dioxane-water mixture and freeze-dried them. Indeed, the resulting powder was soluble in aqeous buffers while the original sample (prepared by freeze-drying from a solution in wet dioxane) was not. However, the micellar weights (measured by static light scattering) were typically two orders of magnitude larger the the weights of the parental micelles, were not reproducible, and could be slightly but not appreciably reduced by sonication. When the powder was redissolved in the original 80/20 solvent mixture, the micellar weight returned to its original value. However, solutions in mixtures with lower dioxane content exhibited also high values of micellar weight.

Apparently, the mechanism we described at the end of the preceding section was operative in this case. We visualize that only a few polystyrene chains connect the neighboring micellar cores while the micellar solution is being concentrated during the freezing. They then become trapped when the solvent sublimes away. When dissolved in a nonswelling selective solvent these chains act as tie-lines between the micelles. The number of tie-lines surviving the dissolution process depends on the thermodynamic quality of this solvent with respect to the core as well as on other details of the whole procedure. In reasonably good solvents the tie-lines are released and the clusters may revert to the original micelles.

These results led us to a decision to study the dialysis and freeze-drying transfers in more detail. [13] We have selected a copolymer with a molecular weight of 39 000 with 0.52 weight fraction of polystyrene. This copolymer formed relatively small, well-behaving micelles. Our starting material was freeze-dried from wet dioxane in which it was molecularly dispersed. We dissolved it in three dioxane/water mixtures that were designated as 80/20, 70/30, and 60/40. The numerator in these labels represents the volume percent of dioxane in the mixture. The solutions were micellar and stable. We have measured them by static light scattering. Then their aliquots were dialyzed against the other solvents and measured again. The results are collected in Table 1.

Table 1. Molecular Weights of Micellar Solutions Prepared by Dialysis from a Micellar Solution in a Different Solvent Mixture.

Dialysed from	Dialysed to	$M_w \times 10^{-6}$ Before Dialysis	$M_w \times 10^{-6}$ After Dialysis
80/20	70/30	3.5	3.1
80/20	60/40	3.5	3.6
70/30	80/20	4.8	3.9
70/30	60/40	4.8	7.3
60/40	80/20	13.0	4.4
60/40	70/30	13.0	10.3

The Zimm light scattering diagrams were normal for the micelles in all solvent mixtures. While the molecular weights of the original micelles in the three solvents were quite different, their weight did not change much when dialyzed to a worse solvent. However they decreased when dialyzed toward the 80/20 solvent, in which the micelles seem to be fully equilibrated.

We used the same copolymer sample for the study of the freeze-drying procedure. The solutions in the three solvent mixtures (80/20, 70/30, 60/40) were freeze-dried, then each was redissolved in all three mixtures again and measured. The results are presented in Table 2. The samples shown as freeze-dried from 98/2 are the original samples freeze-dried from wet dioxane.

It is apparent that the breakup of the original structure freeze-dried from the molecularly dissolved solution leads to quite different sizes of the micelles depending on the composition of the mixed solvent. Further, the nature of the solid copolymer after freeze-drying from a mixed solvent depends on the composition of the solvent from which it was freeze-dried: when a series of samples prepared by this procedure is dissolved in the same solvent (70/30) the resulting micellar weight increases with the worsening solvent power of the original mixture. In other words, the original breakup of the reference structure in the first dissolution is still preserved after the freeze-drying and second dissolution. However, this preservation is not complete: the dissolution in 60/40 tends to increase the micellar weight and dissolution in 70/30 tends to reduce it. Dissolution in 80/20 leads again to the equilibrium state.

Table 2. Molecular Weights of Micellar Solutions Prepared from a Copolymer Freeze-dried from a Previously Prepared Solution in Mixtures of Dioxane and Water with Different Composition.

Freeze-dried from	Dissolved in80/20 $M_w \times 10^{-6}$	Dissolved in 70/30 $M_w \times 10^{-6}$	Dissolved in 60/40 $M_w \times 10^{-6}$
98/2	3.5	4.8	13.0
80/20	3.5	7.1	10.0
70/30	3.5	9.0	13.2
60/40	3.6	9.9	14.0

4. Hybridization of Block Copolymer Micelles

When micellar solutions are prepared that are not at thermodynamic equilibrium they may gradually approach that equilibrium and during this process to provide us with a valuable information about the dynamics, kinetics and thermodynamics of micellar systems. Hybridization of micelles proved itself to be an excellent method for obtaining this information.

For this technique, two different copolymers of the same family are micellized separately in a given selective solvent. The two solutions are mixed and the changes of their properties are followed as a function of time by some convenient technique. Cantú *et al.* [14] used light scattering, Procházka *et al.* [15] followed the fluorescence of labeled copolymers. Sedimentation velocity was the method of choice in our experiments. [16,17]

In our first study [16], block copolymers of styrene and methacrylic acid were used. Our two micellar solutions exhibited different sedimentation coefficients. After mixing, we repeatedly performed the sedimentation velocity experiment. Depending on the characteristics of the two types of the micelles and of the solvent mixture, we obtained various results. Typically, in the first run the two types of the micelles preserved their identity and two sedimentation peaks were observed having the original sedimentation coefficients. In the subsequent runs the two peaks approached each other and eventually they merged. During this process the slower peak (corresponding to smaller micelles and smaller unimers) usually grew at the expense of the faster peak.

What was remarkable was the strong dependence of the kinetics of this process on the details of the experiments. In the first series of experiments, the micelles were dissolved in dioxane containing 20 volume percent of water. In this mixture, the polystyrene cores are extensively swollen containing about 50 % of the solvent. When both micelles were relatively small, the whole hybridization process was completed before the first sedimentation run could be made. When one or both of the micellar species were relatively large, no hybridization was observed within twenty days. The kinetics could be followed for the intermediate cases. The rate of the hybridization decreased significantly with increasing size of the insoluble block as well as with the size of the soluble block. When the solvent mixture contained 30 % of water or more no change in the sedimentation pattern was observed.

The driving force for the hybridization is the increase of entropy when the two types of the unimers are mixed within the micelles. The mechanism of the hybridization consists primarily from the transfer of unimers among the micelles of both types. The rate of the transfer depends mainly on the ease with which the unimers escape from the micelles. During this process the insoluble blocks must first extricate themselves from the central part of the core and locate themselves at the core boundary. This extrication is the slower the longer the insoluble block and the tighter (less swollen) is the core. Once the unimer is at the core boundary, it has to separate itself from the core; this incurs contact enthalpy penalty that also increases with the length of the insoluble block. The separated unimer then has to negotiate its way trough the shell the thickness of which increases with the length of the soluble block. The longer lasts this travel the larger the chance that the unimer will be recaptured by the core. The combination of these effects explains the steep dependence of the hybridization rate on the sizes of both blocks and the composition of the solvent.

The fact that the slower sedimentation peaks corresponding to smaller unimers grows at the expense of the larger is somewhat surprising considering that the smaller unimers are more mobile than the larger ones. Seemingly, the reentry of the small unimers into the larger micelles incurs an entropy penalty because the short entering insoluble block has difficulties reching the central part of the larger core thus forcing the resident blocks to stretch beyond their preferred conformation. No such penalty is incured when a longer insoluble block enters the smaller micelles. Consequently, the more mobile small unimers are primarily recaptured by the small miceles and the transfer of the larger unimers becomes the dominant process.

We observed a different phenomenon in several experiments in which we mixed small triblock micelles with substantially larger diblock micelles. A third peak appeared between the original peaks and gradually grew at their expense. This was apparently a nucleation phenomenon. In our opinion, the equilibrium hybrid micelles that contain quite different numbers of both unimers than their homomicelles are strongly preferred thermodynamically. However, the unimer transfer process described above is blocked by the entropy penalties incurred when unimers enter micelles of appreciably different size from their parental micelles. Consequently, the solution outside of the micelles that is saturated by unimers of both types with respect to their homomicelles is oversaturated with respect to the equilibrium hybrid micelles and the nucleation ensues. It is interesting to note that this process allows for the change of the total number of micelles in the system and thus may lead to a fuller equilibrium than the unimer transfer procedure that leads to a system where the original number of micelles is conserved. This number may be thermodynamically inappropriate but no mechanism may exist for its readjustment. This is the same phenomenon that we described in the previous section, namely the unwillingness of the micellar solution to convert itself to a solution containing micelles of a different aggregation number even if this number is thermodynamically more appropriate and the mobility of the unimers is sufficient to produce a narrow distribution of aggregation numbers.

To confirm our conclusion that the hybridization in some cases is blocked by kinetic factors we tried to prepare the hybrid micelles of such copolymer pairs using a different procedure. We dissolved both copolymers in wet dioxane, in which the copolymers are dispersed molecularly, freeze-dried this solution, and redissolved it in the selective solvent. In most cases the resulting micellar solution produced a single sedimenting boundary. However, for two mixtures of a small triblock copolymer with a large diblock copolymer (one of the pairs in the original experiments did not hybridize at all, while the other exhibited the three-peak behavior) the resulting micellar solution produced two boundaries, both of which had sedimentation coefficients in-between the two original homomicellar values.

We consider this phenomenon to be an equivalent of the phase behavior of partially miscible liquids. When in such mixtures the thermodynamic interaction between the two components becomes unfavorable, they separate into two phases, both of which are mixtures. By analogy, when the dissimilarity between two copolymers (i. e., the differences between the block sizes) increases, the molecules of the two types find it increasingly difficult to be accomodated within the same micelles and they will form two types of hybridized micelles. The difference in their compositions, their sizes, and their sedimentation coefficients will increase with the dissimilarity of the copolymers. If this dissimilarity is small enough, uniform micelles of a single type will result.

In our next study [17] we followed the hybridization of a micellar solution containing micelles formed by two different but similar copolymers. Both of them had polystyrene as the core forming block. The shell block was polyisoprene in one of these copolymers and hydrogenated polyisoprene in the other. The sizes of the blocks in the two copolymers were well matched. The solvent was decane. Both copolymers formed uniform micelles, but these had different sedimentation coefficients: those with the polyisoprene shell sedimenting faster. This was probably a result of different thermodynamic interactions of the solvent with the aliphatic resp. olefinic shells.

These micelles indeed hybridized. The rate of hybridization was a very steep function of temperature. At 20 ^{o}C no change in the sedimentation pattern was observed during 40 days. At 29 ^{o}C the micelles after about one day induction period started hybridizing and formed a single narrow peak after about four days. At 31 ^{o}C no induction period was observed and hybridization was complete after about one day. Finally, at 35 ^{o}C a single peak appeared already during the first sedimentation run, i. e., after about 20 minutes.

In our opinion, this steep dependence on temperature is caused by a change of the mobility of the polystyrene chains. There is in the temperature range studied either a significant increase of the swelling of the core by the solvent or the core that is partially swollen undergoes a glass transition.

5. Conclusions

Well-behaved equilibrium block copolymer micelles are formed only in a narrow window of experimental conditions when the selective solvent is just poor enough with respect to the core forming block so that the anomalous micellization is avoided but good enough so that the core is sufficiently swollen and the core-forming blocks have adequate mobility to migrate among different micelles and to establish an appropriate number of micellar cores to produce a system with the lowest possible Gibbs energy. Under other conditions, the micellar properties are extensively influenced by the history of the system.

In nonequilibrium micellar systems two processes serve for lowering the Gibbs energy: migration of unimers and nucleation/disintegration of micelles. Both these processes become kinetically frozen when the mobility of the core segment decreases, but the unimer migration survives longer. Thus in many situations the migration produces micelles with a narrow distribution of aggregation numbers (a local minimum of the Gibbs energy) while the average micellar size is a nonequilibrium one as a result of the frozen nucleation process. The average size then depends on the history of the sample.

Two types of nonequilibrium micellar systems must be distinguished. In one of them individual micelles exist but have inappropriate size. Such systems originate either by a gentle (dialysis) transfer to a different medium or by a breakup of larger structures under conditions of intermediate mobility of the core-forming segments.

In the other type, the micelles form clusters in which individual micelles are probably interconnected by a few tie-lines formed by the insoluble blocks residing in two micellar cores. Such structures may be formed when a micellar solution is concentrated (by evaporation or crystallization of the solvent) so that some micellar cores are in a close contact while the core segments are still sufficiently mobile. Redispersion of such structures in a solvent not allowing segmental mobility preserves the tie-lines and produces clusters.

The technique of micellar hybridization enhances the rate of equilibration by adding a new component to the driving force toward unimer migration, namely the entropy of mixing. This may lead to the ressurection of the nucleation mechanism under circumstances when no nucleation would occur if the copolymer molecules were uniform. This technique also led to the discovery of a new phenomenon: the coexistence of two types of hybrid micelles when the properties of the two micelle-forming copolymers are sufficiently different. This coexistence is a micellar equivalent of the phase behavior of two partially miscible liquids.

Acknowledgment. This research was supported by the U.S. Army Research Office (Grants DAAL03-90-G-0147, DAAH04-93-G-0405, and DAAH04-95-10127), by the Collaborative Research Grant 920166 from the Scientific and Environmental Division of the North Atlantic Treaty Organization, and by the U.S.-Czech Science and Technology Joint Fund No 95010.

6. References

1. Honda, C., Hasegawa, Y., Hirunuma, R., and Nose, T. (1994) Micellization Kinetics of Block Copolymers in Selective Solvents, *Macromolecules* **27**, 7660-7668.
2. Mandema, W., Zeldenrust, H., and Emeis, C.A. (1979) Association of Block Copolymers in Selective Solvents, 1. Measurements on Hydrogenated Poly(styrene-isoprene) in Decane and in *trans*-Decalin, *Makromol. Chem.* **180**, 1521-1538.
3. Mandema, W., Emeis, C.A., and Zeldenrust, H. (1979) Association of Block Copolymers in Selective Solvents, 2. Viscosity, Diffusion and Light-Scattering Measurements on Hydrogenated Poly(styrene-isoprene) in *trans*-Decalin/Decane Mixtures, *Makromol. Chem.* **180**, 2163-2174.
4. Canham, P. A., Lally, T. P., Price, C., and Stubbersfield, R. (1980) Formation of Worm-like Micelles from a Polystyrene-Polybutadiene-Polystyrene Block Copolymer in Ethyl Acetate, *J. C. S. Faraday I* **76**, 1857-1867.
5. Nagarajan, R. and Ruckenstein, E. (1991) Theory of Surfactant Self-Assembly: A Predictive Molecular Thermodynamic Approach, *Langmuir*, **7**, 2934-2969.
6. Zhou, Z. and Chu, B. (1988) Anomalous Micellization Behavior and Composition Heterogeneity of a Triblock ABA Copolymer of (A) Ethylene Oxide and (B) Propylene Oxide in Aqueous Solution, *Macromolecules* **21**, 2548-2554.
7. Tuzar, Z. and Kratochvíl, P. (1993) Micelles of Block and Graft Copolymers in Solutions, *Surf. Colloid Sci.* **15**, 1-83.
8. Tuzar, Z., Kratochvíl, P., Procházka, K., and Munk, P. (1993) Block Copolymer Micelles in Aqueous Media, *Collect. Czech. Chem. Commun.* **58**, 2362-2369.
9. Molau, G.E. and Wittbrodt, W.M. (1968) Colloidal Properties of Styrene-Butadiene Block Copolymers, *Macromolecules* **1**, 260-264.
10. Riess, G. and Rogez, D. (1982) Micellization of Poly(styrene-b-ethylene oxide) Block Copolymers, *Polymer Preprints, Div. Polymer Chem., Am. Chem. Soc.,* **23**(1), 19-20.
11. Tuzar, Z., Webber, S.E., Ramireddy, C., and Munk, P. (1991) Association of Polystyrene-Poly(methacrylic Acid) Block Copolymers in Aqueous Media, *Polymer Preprints, Div. Polymer Chem., Am. Chem. Soc.,* **32**(1), 525-526.
12. Qin, A. and Munk, P., unpublished results.
13. Icduygu, M. G. and Munk, P., manuscript in preparation.
14. Cantú, I., Corti, M., and Salina, P. (1991) Direct Measurement of the Formation Time of Mixed Micelles, *J. Phys. Chem.* **95**, 5981-5983.
15. Procházka, K., Bednar, B., Svoboda, P., Trnena, J., Mukhtar, E., and Almgren, M. (1991) Nonradiative Energy Transfer in Block Copolymer Micelles, *J. Phys. Chem.* **95**, 4563-4568.
16. Tian, M., Qin, A., Ramireddy, C., Webber, S. E., Munk, P., Tuzar, Z., and Procházka, K. (1993) Hybridization of Block Copolymer Micelles, *Langmuir* **9**, 1741-1748.
17. Pacovská, M., Procházka, K., Tuzar, Z., and Munk, P. (1993) Formation of Block Copolymer Micelles: A Sedimentation Study, *Polymer* **34**, 4585-4588.

BLOCK COPOLYMERS : SYNTHESES, COLLOIDAL PROPERTIES AND APPLICATION POSSIBILITIES OF MICELLAR SYSTEMS

Gérard RIESS and Guy HURTREZ
Ecole Nationale Supérieure de Chimie de Mulhouse
3, rue Alfred Werner 68093 Mulhouse Cedex
Centre de Recherche sur la Physico-Chimie des Surfaces Solides
24, avenue du Président Kennedy 68200 Mulhouse (FRANCE)

1 Introduction

Block copolymers are defined as having a linear arrangement of blocks of varying monomer composition [1]. That is, a block copolymer is a combination of two or more polymers joined end-to-end.

The increasing importance and interest in block copolymers arises mainly from their unique properties in solution and in the solid state which are a consequence of their molecular structure. In particular, sequences of different chemical composition are usually incompatible and therefore have a tendency to segregate in space. Amphiphilic properties in solution and microdomain formation in the solid state are directly related to the specific molecular architecture, which can be designed by using existing monomers or polymers.

In this contribution, we intend to outline the synthesis of these copolymers, as well as their colloidal and interfacial properties, especially with respect to their behavior as polymeric surfactants in solution and in the solid state.

These aspects will be appoached by considering:

- some specific synthesis techniques to produce block copolymers of well-defined structure, composition and molecular weight
- different examples of micelle formation
- the surface activity and emulsifying properties of block copolymers in connection with some of their application possibilities.

Copolymers with branched structures, such as star-block or radial block, as well as graft copolymers, although their properties might somehow be similar to those of block copolymers, will not be considered in this survey.

2 Synthesis

The polymerization methods leading to diblock, triblock or segmented block copolymers are based on two general reaction schemes.

S.E. Webber et al. (eds.), Solvents and Self-Organization of Polymers, 33–51.

In the first one, α or α,β active sites are generated on a polymer chain poly A which then initiate the polymerization of a second monomer B. Such a polymerization can be free radical, anionic or cationic and may be shown as follows:

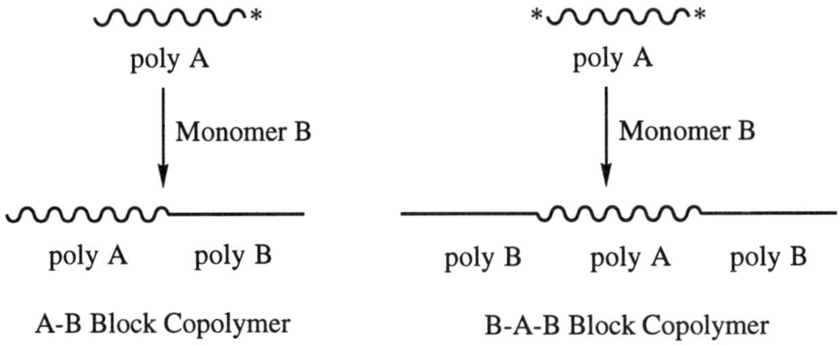

The second method, which is usually called condensation, is a reaction between chemical functional groups present at the end of different polymers. Such a situation for the synthesis of a diblock and triblock copolymer can be shown as:

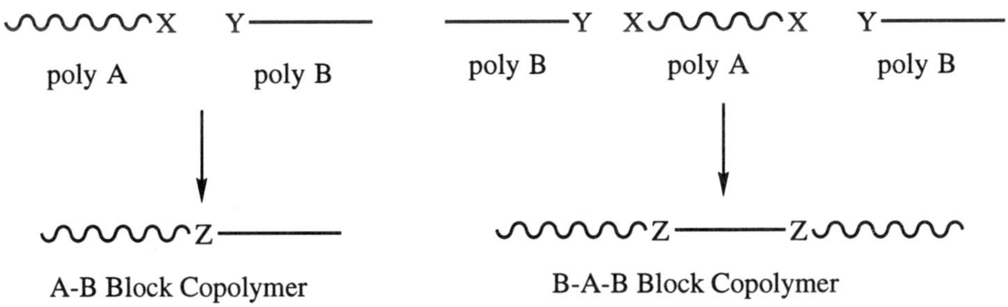

where X is a functional group like OH, NH_2...
Y is a functional group like COCl, NCO...
Z is the reaction product of X plus Y.

The selection of a given synthesis technique will depend on the following criteria:

- the polymerization mechanism, e.g. free radical, anionic or cationic polymerization for the monomers A and/or B; the most suitable case will be that where both monomers A and B are polymerizable by the same mechanism
- the structure of the copolymer which one wants to obtain, e.g. diblock, triblock or multiblock
- the desired molecular weight range, knowing that condensation reactions are usually prefered for the preparation of block copolymers of lower molecular weight, e.g. from 1 000 to 50 000
- the required purity of the end product (absence of homopolymers in a diblock or absence of diblock contaminant in a triblock copolymer)
- the required monodispersity of each block.

Anionic polymerization and to some extend cationic polymerization, are in this respect most suitable for the preparation of block copolymers, as these techniques lead to products:

- with well-defined structure, composition and molecular weight
- of narrow molecular weight distributions for the different sequences
- with low amounts of impurities (absence of homopolymers)
- which can be easily functionalized either in end position and/or at the junction of the blocks.

Some of these possibilities for anionic polymerization will be outlined in the following.

2.1 PREPARATION OF BLOCK COPOLYMERS BY ANIONIC POLYMERIZATION

Anionic polymerization, since its discovery by SZWARC [2], has proved to be one of the most useful techniques for the preparation of monodisperse polymers and well-defined block copolymers. It differs from other methods, and especially from the free radical polymerization, in that under necessarily stringent conditions required for polymerization, a termination or transfer step is absent; for this reason it is called "living" anionic polymerization. The possibility of preparing almost pure block copolymer of a specific molecular weight, composition and structure, becomes realized by anionic polymerization. Monomers of a diene, nonpolar vinylic, acrylic, methacrylic and cyclic ether nature may be polymerized by this technique.

A typical example is the preparation of polystyrene-block-poly(ethylene oxide) or polystyrene-block-poly(2-vinylpyridine)-block-poly(ethylene oxide) by using phenylisopropyl potassium as an initiator [3].

In addition to A-B or A-B-C type block copolymers, it is also possible by anionic polymerization to prepare B-A-B type triblock copolymers either by a coupling reaction or by using difunctional organometallic initiators.

In more recent years, anionic polymerization techniques have been adapted for the polymerization of acrylic and methacrylic monomers [4]. All acrylic di- and triblock copolymers, such as
poly(methyl methacrylate)-block-poly(t butyl acrylate) (PMMA -b-PtBA)
poly(t butyl acrylate)-block-poly(methyl methacrylate)-block-poly(nonyl methacrylate)
have been prepared by using specific complexing agents [5].

The interesting fact is that the PtBA sequence can be readily transformed in a water soluble poly(acrylic acid) (PAA) sequence, as shown in the following scheme:

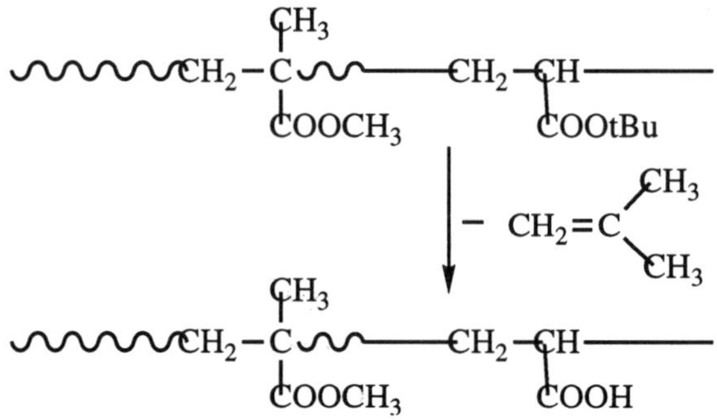

Block copolymers having two water soluble blocks could also be prepared according to the following reaction scheme [6]:

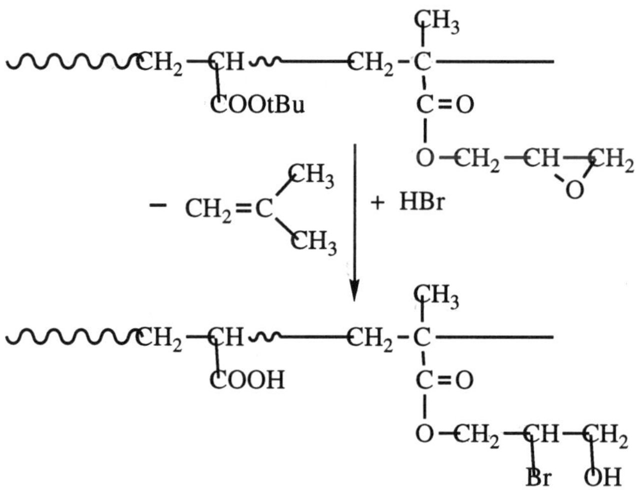

2.2 FUNCTIONALIZED BLOCK COPOLYMERS

An additional advantage of anionic polymerization is that functionalized block copolymers with specific end groups may be obtained by reacting a "living" polymer with suitable reagents.

Thus functionalized polystyrene-block-polybutadiene (PS-PBut) have been obtained as indicated by the following reactions:

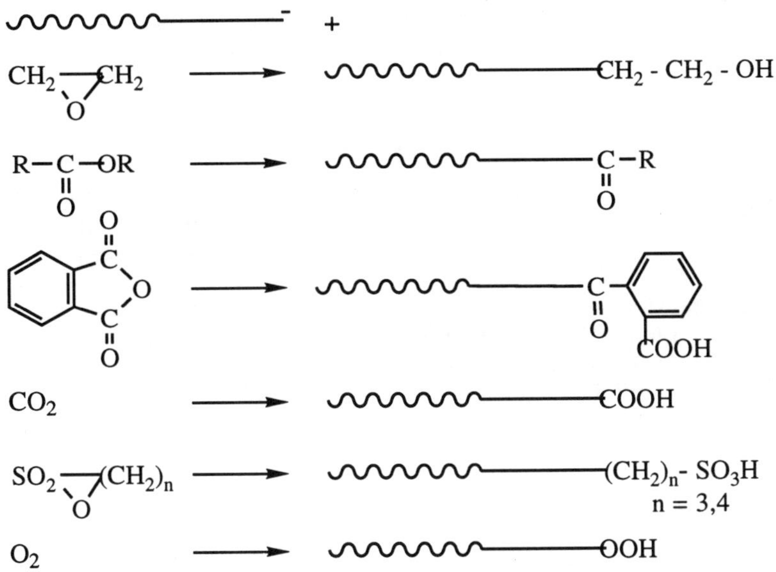

Block copolymers bearing fluorescent labels such as phenanthrene or anthracene groups at the block junction could be prepared by sequential anionic polymerization [7]. These products, specifically with one fluorescent group per block copolymer chain, are particularly useful for photochemical studies and for the direct determination of the number average molecular weight (M_n) of the copolymer [8,9].

2.3 ACTIVE CENTER TRANSFORMATION: ANIONIC - FREE RADICAL POLYMERIZATION

Due to specific limitations of anionic, cationic or free radical polymerization, an alternative approach to the block copolymer synthesis is to devise processes whereby the polymerization mechanism can be changed at will to suit the monomers being polymerized sequentially.

Such transformation, e.g. from anionic to cationic polymerization or vice versa, have been reviewed by RICHARDS et al. [10].

All transformations involving anions, cations and radicals are possible; to show the potential area of interest, this principle will be illustrated by some examples of "anion to free radical" transformation developped in our laboratory. This type of transformation is based on the deactivation of a "living" anionic polymer with different reagents able to initiate the radical polymerization of a second monomer.

For example "living" polystyrene or polyisoprene have been deactivated with 4,4' bis (bromomethyl) benzoyl peroxide leading to a polymeric peroxide.

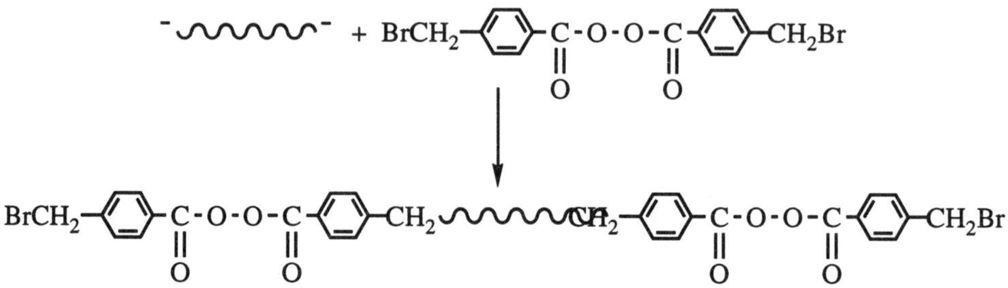

In an analogous way, peroxidized chain ends have been obtained by addition of a "living" polymer to a THF (tetrahydrofurane) solution saturated in oxygen [11] or by reaction with 2,2' azobisisobutyronitrile [12]. The reaction using azobisisobutyronitrile (AIBN) may be depicted thus:

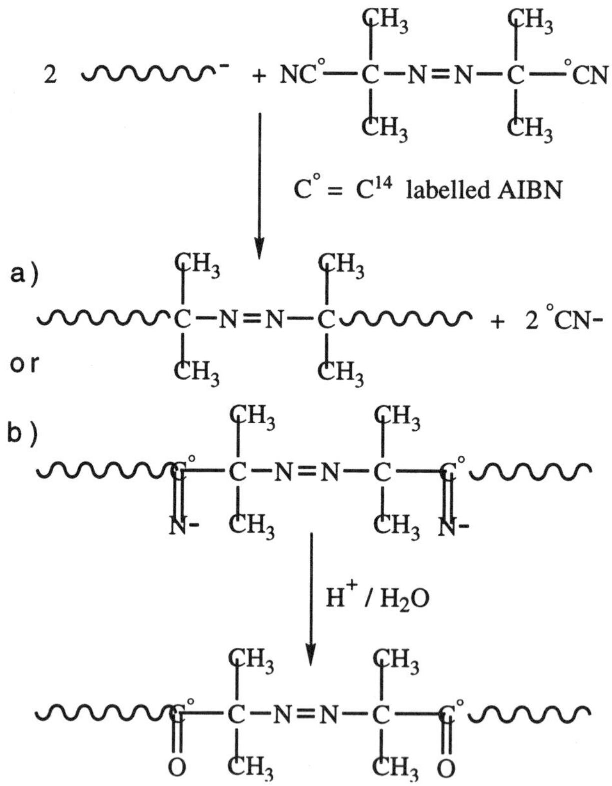

Starting with these polymeric peroxides or azo compounds, either by thermal or redox decomposition in the presence of a second monomer, diblock copolymers of the following type have been obtained:

- polyisoprene-poly(styrene-alt-maleic anhydride)
- polyisoprene-poly(styrene-co-acrylonitrile)

With respect to "living" anionic polymerization, these transformation reactions have a certain number of advantages as well as disadvantages. Thus the main advantage is that block copolymers based on polar vinyl monomers become accessible. However, as expected from the inherent characteristics of free radical polymerization such as:

- an initiation efficiency $f \leq 1$, with values of f decreasing with an increase in molecular weight of the polymeric initiator
- different types of chain termination (disproportionation or combination)
- various chain transfer reactions,

the resulting block copolymer is formed in a lower yield and its polydispersity (in structure, composition and molecular weight) is higher than those obtained by anionic "living" polymerization. Nevertheless, if this kind of copolymers might have certain limitations, it appeared that they are very valuable as additives due to their interfacial activity.

This synthesis route in a reverse order, e.g. free radical to anionic ring opening polymerization, has been recently developed on an industrial basis by GOLDSCHMIDT [13] for the preparation of polystyrene-poly(ethylene oxide) diblock copolymers (PS-PEO) according to the following reaction scheme:

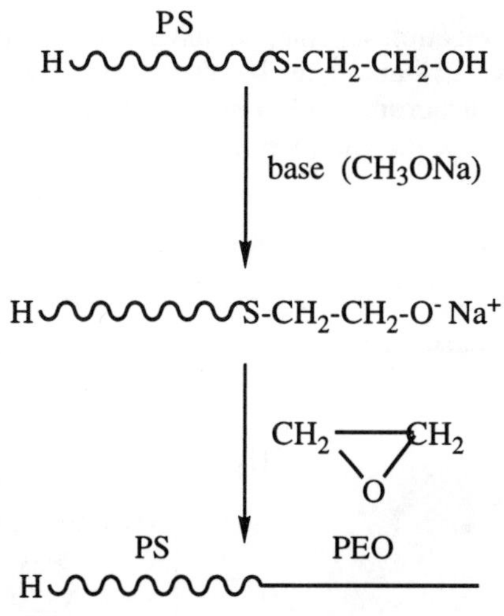

This reaction sequence involves a free radical polymerization of styrene in the presence of mercaptoethanol as chain transfer agent; the hydroxy functionalized PS is then the starting point for the ring opening polymerization of ethylene oxide.

2.4 CHEMICAL MODIFICATION OF BLOCK COPOLYMERS PRECURSORS

Modification of a block copolymer transforms at least one of the sequences and leads to a new copolymer.

A typical example of such a modification is the hydrogenation under catalysis of the polybutadiene block in a block copolymer, so that a polystyrene-block-poly(1-4 butadiene)-block-polystyrene could become polystyrene-block-polyethylene-block--polystyrene. Such examples have been reviewed [14]. The modification of the polybutadiene block in a block copolymer has also been achieved with thioglycolic acid in the presence of a free radical generator like AIBN [15].

Sequences other than polybutadiene sequences can be modified. The esterification of the maleic anhydride units of a polyisoprene-block-poly(styrene-alt-maleic anhydride) has been described [16].

A copolymer with a methacrylic acid sequence has been prepared from a trimethylsilyl methacrylate sequence by hydrolysis at room temperature with aqueous methanol [17].

The quaternization of polystyrene-block-poly(2-vinylpyridine) with ethyl bromide in preparing an amphiphilic copolymer has been reported [18]. In the same way, a quaternization with ethyl bromide followed by a sulfonation with H_2SO_4 transforms a styrene-(4-vinylbenzyl)dimethylamine-isoprene block copolymer into an amphoteric ion exchanger [19].

In the following reaction scheme, we have indicated the chemical modification of both blocks of a poly(1-2 butadiene)-block-poly(4-vinylpyridine) copolymer [20] leading by hydrogenation according to HAHN [21] to an hydrophobic ethylene-butene block and by quaternization of the vinylpyridine sequence to an hydrophilic sequence.

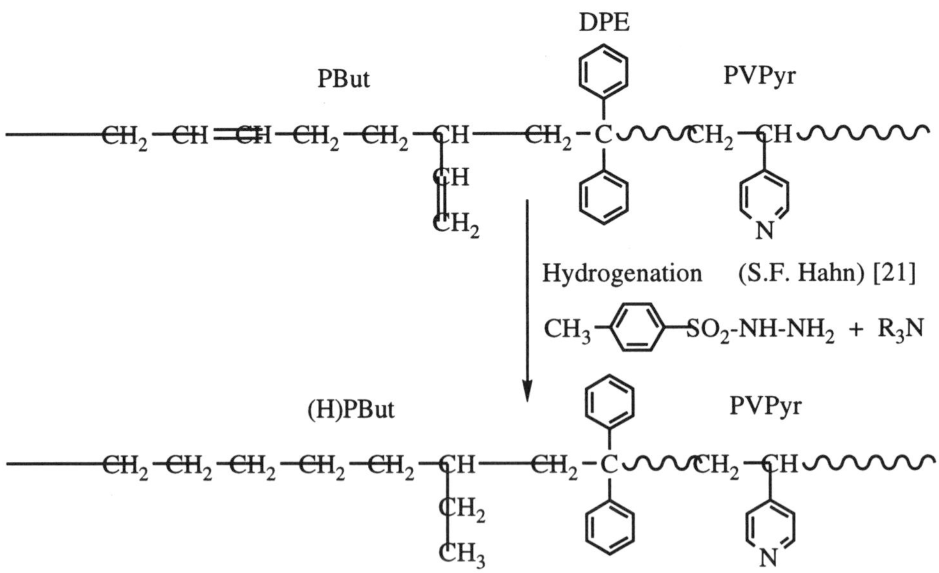

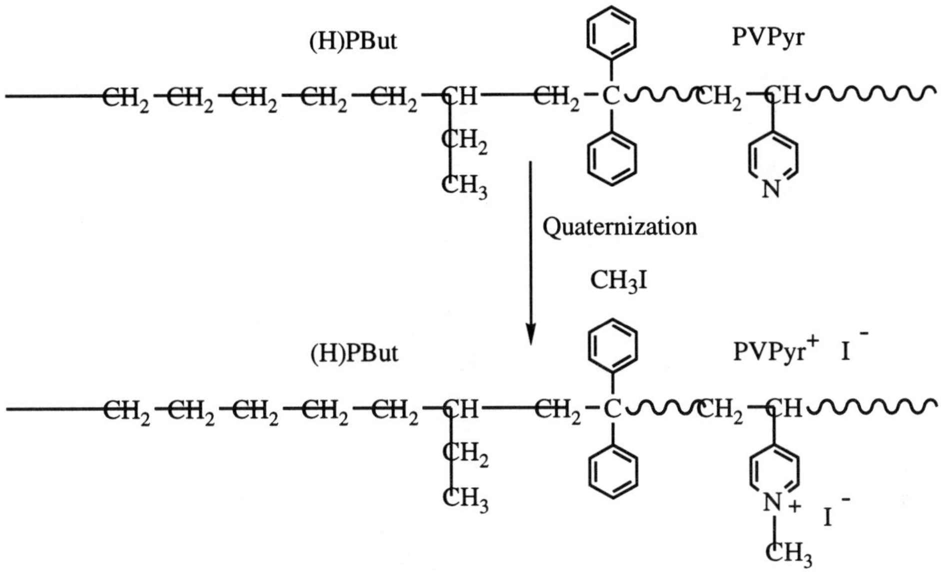

3 Colloidal Properties of Block Copolymers

The micellization of block copolymers having been extensively described by different authors of this symposia [22,26], this chapter attempts to outline some specific aspects of our recent research work in this area.

In connection with the problem of critical micelle concentration (CMC) of block copolymer micellar systems which is still controversed, our approach was to use fluorescent labelled copolymers [27] in a similar way as mentioned by WEBBER [26].

We have thus examined the micellization behaviour of:

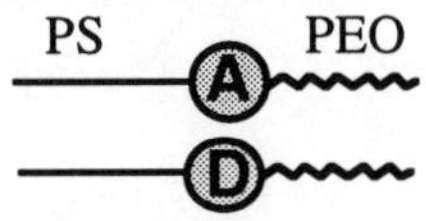

where: PS is a polystyrene sequence
PEO is a poly(ethylene oxide) sequence
(A) is an anthracenyl acceptor group
(D) is a phenanthrenyl donor group

These two groups are capable to produce a non radiative energy transfer if they are close together. In water, that means in a selective solvent for one block (PEO), and if the concentration is higher than CMC, these copolymers form mixed micelles with A and D groups located at the junction of the core and the shell of the micelle. In such micelles, the distance between A and D groups is low enough to induce a non radiative energy transfer according to the following scheme:

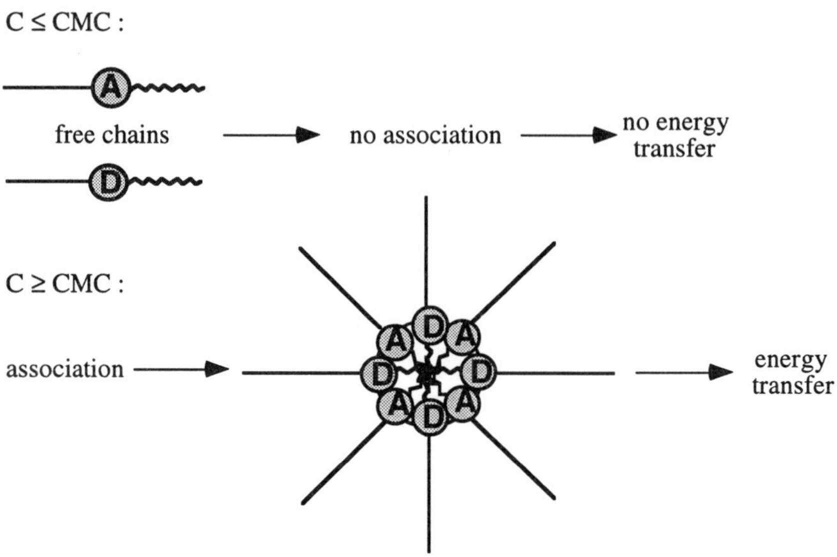

By measuring the fluorescence emission intensity If(A) and If(D) of each fluorescent probe at different concentrations, which is characteristic of the non radiative energy transfer, it is possible to estimate the critical micelle concentration of such PS-PEO copolymers in àqueous medium as shown in figure 1.

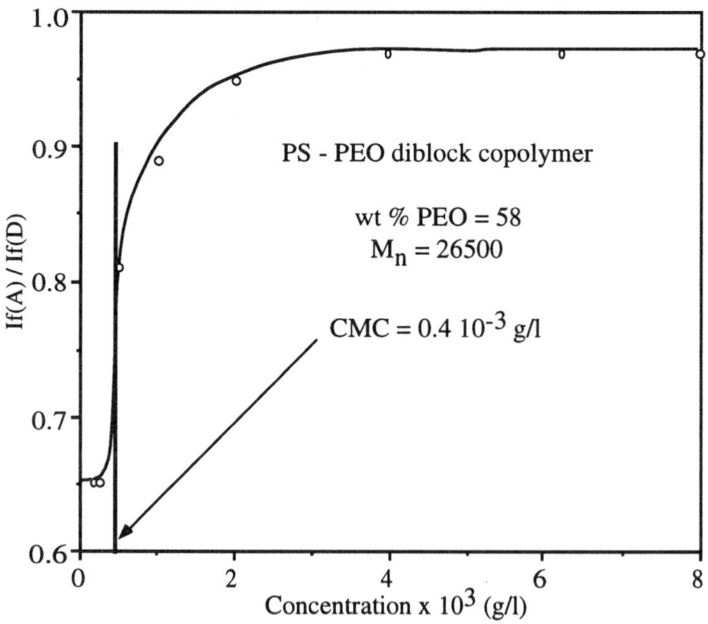

Figure 1: Energy transfer evolution versus concentration for an aqueous micellar system with PS-PEO labelled diblock copolymer. Determination of the CMC.

NMR as mentioned by TUZAR [22] is another valuable technique to distinguish the "mobile" part of the micellar fringe from the "frozen chains" in the core.

This technique has therefore been applied to study the micellization behaviour of poly(methyl methacrylate)-poly(acrylic acid) (PMMA-PAA) diblock copolymers of same composition (43 wt % AA) in CD_3OD as a function of the temperature and the total molecular weight.

For the copolymer of lower molecular weight (M_n = 28 000), the NMR spectra shows already at room temperature the characteristics peaks of PAA, as well as those of PMMA. In contrast, for the copolymers of higher molecular weight (M_n = 51 500 and M_n = 66 200), only the peaks of PAA show up at room temperature. The characteristics peaks of PMMA (CH_3 and OCH_3 peaks) only appear by increasing the temperature of the micellar systems, which is an indication that the PMMA chains of the micelle core are becoming "mobile".

By taking a characteristic peak of PMMA and one of PAA, it becomes possible to plot the ratio PMMA/PAA, normalized with respect to the corresponding ratioobtained for the copolymer in a common solvent ($CDCl_3$),as a function of temperature which is shown in the following figure 2.

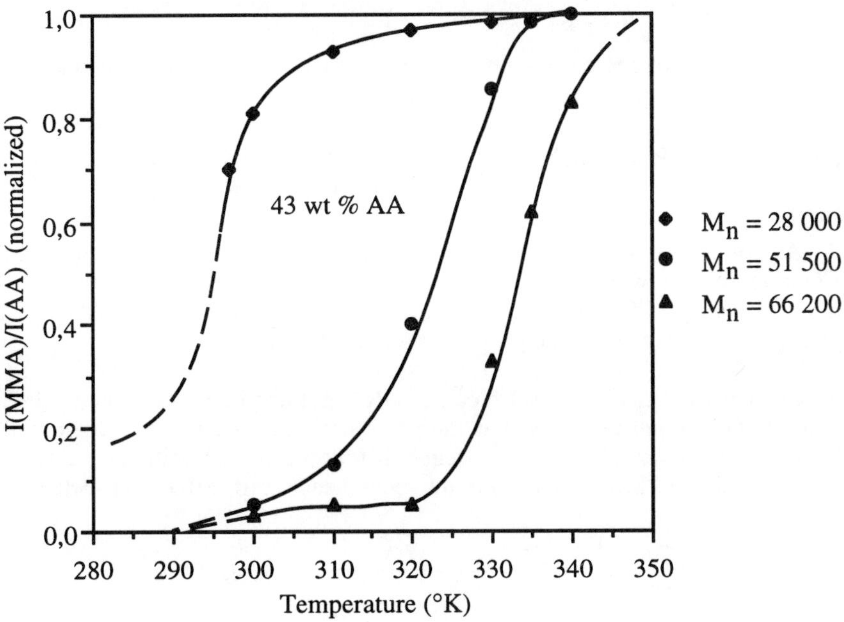

Figure 2: Ratio of NMR peak intensity versus temperature in CD_3OD for a micellar system with PMMA-PAA diblock copolymer.

As this phenomena is reversible with temperature, one can therefore estimate a critical micellization temperature for a given copolymer [6].

The micellization of PMMA-PAA block copolymer has also been studied in nbutylacetate which is a non-solvent for PAA and a Θ solvent for PMMA [28].

According to the model designed by HALPERIN [29] for this particular situation, it follows that the hydrodynamic radius R_H of the micelles scales such as :

$$R_H \sim K\, N_{PMMA}^{0.5}\, N_{PAA}^{0.09} \qquad (1)$$

with N being the number of the corresponding monomers units in a sequence.

The situation appears to be different for micelles formed by polystyrene-poly(ethylene oxide) di- or triblock copolymers (PS-PEO and PEO-PS-PEO) in water as studied by HURTREZ [30].

By applying the theory of HALPERIN [31] for a block copolymer series of diblock samples and triblock samples where $N_{PEO} > N_{PS}$, the following scaling relationships could be deduced:

for diblock PS-PEO copolymers: $$R_H = 1.77\, N_{PEO}^{0.31}\, N_{PS}^{0.09} \qquad (2)$$

for triblock PEO-PS-PEO copolymers: $$R_H = 0.78\, N_{PEO}^{0.42}\, N_{PS}^{0.14} \qquad (3)$$

A similar attempt has been made by applying NAGARAJAN's theory [32] where R_H scales as:

$$R_H = K_1\, N_{PEO}^{-0.17}\, N_{PS}^{\alpha} + K_2\, N_{PEO}^{0.74}\, N_{PS}^{\beta} \qquad (4)$$

With this approach, we could show that the characteristics constants are such as:

for diblock PS-PEO copolymers:	$K_1 = 1.01$	$K_2 = 0.78$
	$\alpha = 0.75$	$\beta = -0.53$
for triblock PEO-PS-PEO copolymers:	$K_1 = 0.69$	$K_2 = 0.22$
	$\alpha = 0.73$	$\beta = -0.17$

As pointed by MUNK [24] and by WEBBER [26], it seems that in the case of PS-PEO micelles in water, we are in a non-equilibrium situation. In fact, with chromophore labelled PS-PEO block copolymers, it was shown by HURTREZ [7], that there is almost no chain exchange for such micellar systems.

Poly(methyl methacrylate)-block-poly(t butyl acrylate) copolymers (PMMA-PtBA) have the unique property that their synthesis can be started either with the PMMA sequence or with the PtBA sequence. It follows that the initiator group will be in the first case at the PMMA end and in the second case at the PtBA end such as:

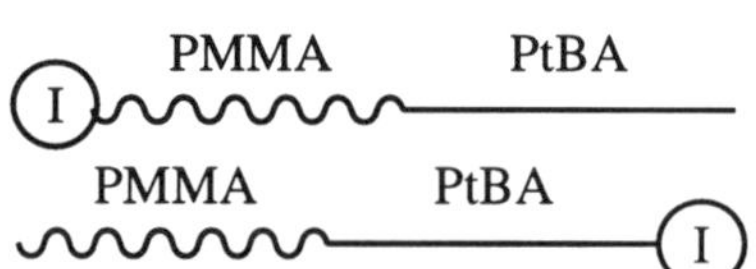

(type 1)

(type 2)

with 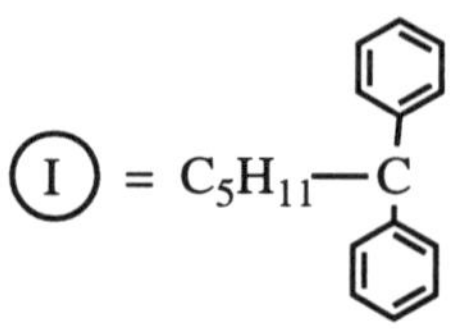

(addition of BuLi on diphenylethylene)

By micellization in methanol, which is a selective solvent for PtBA, the hydrophobic initiator end group (I) will therefore be either in the micellar core for copolymer type 1 or at the end of the PtBA fringe in the case of type 2. By following the diffusion coefficient D by dynamic light scattering versus the copolymer concentration, a dramatic difference appears between both cases as shown in figure 3 [6]:

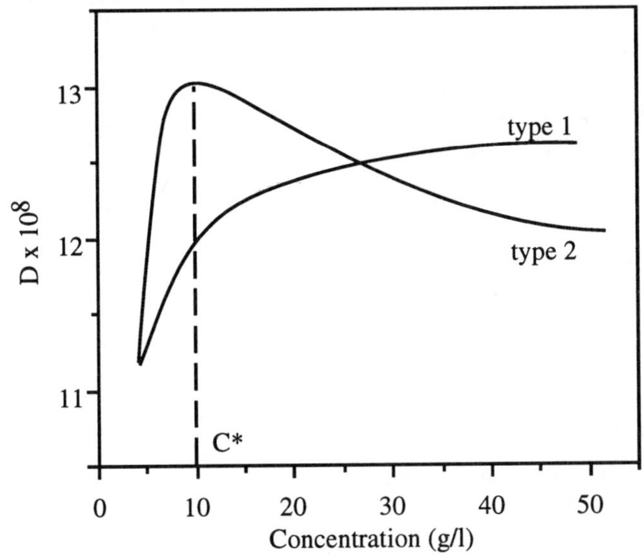

Figure 3: Evolution of the diffusion coefficient versus concentration for each type of micelle

For type 1 copolymer, D increases slightly with concentration at the beginning and then reaches a constant value when the micelles become predominant with respect to the unimers.

For type 2 copolymer, after a first increase of D, similar to that observed for type 1 copolymers, we have afterwards at a concentration C* a continuous decrease in D versus concentration. This is typically the situation where the more hydrophobic end groups (I) have a tendency to associate with each other.

This situation is depicted schematically as following:

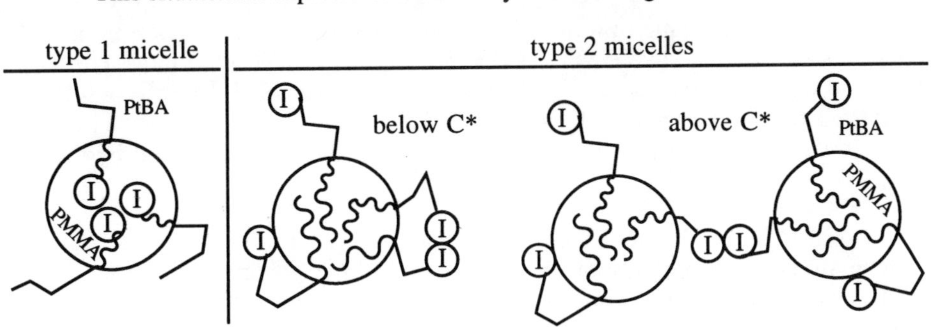

A typical example of the polyelectrolyte effect has been studied recently by CALDERARA [20] in the case of polystyrene-block-poly(4-vinylpyridine) copolymers (PS-P4VP), where the P4VP block can be quaternized with methyl iodide. The P4VP block is soluble in methanol, whereas the corresponding quaternized block $P4VP^+ I^-$ is soluble in water.

As an example, we have taken a PS-P4VP block copolymer with the following characteristics: $N_{PS} = 154$ $N_{P4VP} = 381$

The corresponding quaternized copolymer has therefore exactly the same degrees of polymerization: $N_{PS} = 154$ $N_{P4VP^+ I^-} = 381$

The hydrodynamic radius R_H, the aggregation number Z for the micelles and the density ρ of the corona are given in the following table.

TABLE 1: Characteristics of PS-P4VP and $PS\text{-}P4VP^+ I^-$ micelles in different media

	R_H (nm)	Z	ρ (g/cm^3)
PS-P4VP in methanol	22.7	75	0.14
$PS\text{-}P4VP^+ I^-$ in water	26.0	37	0.05
$PS\text{-}P4VP^+ I^-$ in water presence of KI (7.2 10^{-3} mole/l)	20.6	42	0.20

From this table, one can notice that:

- the aggregation number Z is highest for PS-P4VP in methanol, whereas Z is similar for the $P4VP^+ I^-$ micelles in water and after addition of an electrolyte such as KI

- the density of the micellar fringe is lowest for $P4VP^+ I^-$ micelles in water, which corresponds to an important stretching of the $P4VP^+ I^-$ chains; the addition of an electrolyte (KI) leads as expected to a denser packing of the chains in the micellar fringe and to a decrease of the overall dimension (R_H) of the micelle.

New micellar systems can be obtained with AB block copolymers for which the core is formed by the sequence poly A complexed either with metal ions or by another polymer due to the formation of an interpolymer complex.

A typical example of a micellar system having a metal complex as insoluble core has been described by OSSENBACH in the case of P2VP-PEO block copolymer [33]. By adding $CuCl_2$ solubilized in a small amount of THF to a solution of P2VP-PEO block copolymer (M_n = 184 500; PEO = 64.3 wt %) in benzene, micelles with a P2VP-copper complex core are formed. These micelles stabilized by the PEO fringe have an average diameter, determined by electron microscopy, of 34.0 nm.

More recently, we have shown that by free radical polymerization of 2-vinylpyridine in the presence of PMMA-PAA diblock copolymers in dioxane leads to micelles having a P2VP/PAA interpolymer complex core and a fringe of PMMA which remains soluble in dioxane [34].

4 Surface Activity of Block Copolmyers

The surface activity of block and graft copolymers, which is now a well established fact, make them useful as dispersants, emulsifiers, wetting agents, foams stabilizers, flocculants, demulsifiers, etc., in many industrial and pharmaceutical preparations [35]. Thus, an important application of block and graft copolymers is their use as stabilizers in the preparation of polymer or fillers dispersions, either in aqueous or nonaqueous media [36].

In emulsion or suspension polymerization, the most widely used block copolymer surfactants are of hydrophobic-hydrophilic type, based on poly(ethylene oxide) or poly(acrylic acid) such as:

poly(propylene oxide)-block-poly(ethylene oxide)	PPO-PEO
polystyrene-block-poly(ethylene oxide)	PS-PEO
poly(methyl methacrylate)-block-poly(acrylic acid)	PMMA-PAA

where the PEO or PAA sequence promotes the steric stabilization of the resulting latex.

The main interest of such "hairy latex" is that the thickness of the PEO or PAA fringe on the latex surface can be adjusted as a function of the molecular characteristics (structure, molecular weight, composition)of the block or graft copolymer. These PEO or PAA chains on the particle surface promote interesting application possibilities such as in paint systems and for the formation of core-shell polymer dispersions by controlled latex flocculation.

Furthermore, advantage of the emulsifying properties of PS-PEO block copolymers has been taken to prepare polymer particles with microvoids, which can be used as fillers and organic white pigments in a polymer matrix like in coatings [37,38].

This kind of polystyrene particles, with an average diameter of 1-10 μm and having microvoids in the range of 0.1-0.5 μm, are obtained according to the following steps:

- preparation of water in styrene emulsion (W/O: water in oil) by using a PS-PEO block copolymer of low PEO content as an emulsifier
- dispersion of this W/O emulsion in water by stabilizing the ternary emulsion (W/O/W: water in oil in water) with a second PS-PEO block copolymer having a high PEO content
- suspension polymerization of the W/O/W emulsion in a non agitated batch process
- filtration of the polystyrene particles and removal of the encapsulated water by drying under vacuum at room temperature.

By removal of the encapsulated water, microvoids in the PS particles having a PEO fringe are generated. Volume fraction of 50 % porosity could be obtained for such particles having an open or a closed cellular st`ructure.

With respect to classical surfactants, block copolymers have the unique property to act also under proper conditions as an emulsifier for oil-oil system. Thus block copolymers act as efficient emulsifiers for stabilizing emulsions composed of two immiscible organic liquids, generally when each of them is a selective solvent for one of the blocks of the copolymer.

On the basis of this concept, we have investigated the emulsifying effect of polystyrene-polyisoprene block copolymers for dimethylformamide-hexane emulsions and of polystyrene-poly(methyl methacrylate) copolymers for cyclohexane-acetonitrile emulsions [39,40]. From this study it was concluded that block copolymers, with a composition of about 50:50 and within the molecular weight range of 30 000 to 50 000, had the best emulsifying efficiency. This shows also that if oil-water emulsions can be obtained with relatively low molecular weight (1 000-10 000) block copolymers, the stabilization of oil-oil emulsions will require the use of block copolymers of higher molecular weight.

Polymeric oil in oil emulsions may be defined as two-phase systems which consist of two incompatible polymers, poly A and poly B, in the presence of a common solvent S. The presence of a corresponding block or graft copolymer A-B acts as an emulsifier for both types of emulsions which exist, namely polymeric water in water and polymeric oil in oil emulsions.

A typical example of such polymeric oil in oil emulsion that we have studied in detail, is the system consisting of polystyrene (PS) and polybutadiene (PBut) in a common solvent such as toluene and styrene [41,42]. By adjusting the concentration of both homopolymers, phase separation occurs due to the incompatibility of the polymeric species; one phase has PBut as main polymeric component, the other polystyrene. Emulsification of this system can be achieved by addition of polystyrene-polybutadiene block copolymers.

In a similar way, we have examined the polymeric oil in oil emulsion formed by poly(dimethylsiloxane) (PDMS), polystyrene and styrene in the presence of PDMS-PS block copolymers as stabilizers.

Thus by polymerization of styrene in the presence of silicon oil and PDMS-PS block copolymer, one obtains a composite material having a PS matrix in which small droplets (0.1-1 μm) of silicon oil are dispersed.

The kinetic friction coefficient of this material on steel is 0.1 as compared to 0.7 for pure PS.

The advantage of these liquid-filled polymers is to act as reservoir systems with the lubricant incorporated in the polymer matrix [43,44].

5 Concluding Remarks

This short review serves to illustrate some of the techniques by which it is possible to obtain "tailor made" block copolymers with controlled structure, molecular weight and composition. Ionic polymerization techniques (anionic, GTP, cationic...) are at present most suitable methods leading to well defined copolymers in a good yield. Numerous examples of A-B and A-B-A di- and triblock copolymers have been given in the literature. However, there is still a need for functionalized block copolymers, for A-B-C triblock copolymers with linear or star structure, for A-B-A' disymetric copolymers where A and A' are sequence of same chemical nature but of different molecular weights, etc...

Concerning the application possibilities of block copolymer micellar systems, one has to mention areas of interest such as:

- the encapsulation for controlled release systems
- viscosity improvers and sludge dispersants in base lubricating oil
- micellar catalysis
- oxygen transport systems, which has been described for PDMS-PEO micelles in an aqueous phase.

However, the main interest for block copolymers is their application as polymeric surfactants, as stabilizers for emulsion and suspension, where their surface and interfacial activity will depend directly on their micellar characteristics.

References

1 IUPAC Commission on Macromolecular Nomenclature (1974) Basic definitions of terms relating to polymers, *Pure Appl. Chem.* **40**, 477.

2 Szwarc, M., Levy, M., and Milkovich, R. (1956) Polymerization initiated by electron transfer to monomer. A new method of formation of block polymers, *J. Am. Chem. Soc.* **78**, 2656.

3 Park, Y.-D. (1995) Etude de copolymères triséquencés du type A-B-C, *PhD Thesis in preparation*, Université de Haute-Alsace, France.

4 Fayt, R., Forte, R., Jacobs, C., Jerome, R., Ouhadi, T., Teyssie, Ph., and Varshney, S.K. (1987) New initiator system for the "living" anionic polymerization of tert-alkyl acrylates, *Macromolecules*, **20**(6), 1442.

5 Vuillemin, B. (1994) Synthèse et caractérisation de copolymères blocs acryliques et méthacryliques. Application en tant que tensioactifs macromoléculaires, *PhD Thesis*, Université de Haute-Alsace, France.

6 Hosotte, Ph. (1995) Copolymères blocs acryliques. Synthèse et micellisation, *PhD Thesis in preparation*, Université de Haute-Alsace, France.

7 Hurtrez, G. (1992) Etude des copolymères séquencés poly(styrène-b-oxyde d'éthylène). Synthèses, propriétés colloïdales et tensio-actives, *PhD Thesis*, Université de Haute-Alsace, France.

8 Caldérara, F., Hruska, Z., Hurtrez, G., Nugay, T., and Riess, G. (1993) Synthesis of chromophore-labelled polystyrene / poly(ethylene oxide) diblock copolymers, *Makromol. Chem.* **194**, 1411-1420.

9 Hruska, Z., Vuillemin, B., and Riess, G. (1992) Synthesis of polystyrene-poly(tert-butyl acrylate) diblock copolymers bearing fluorescent groups at the junctions, *Makromol. Chem.* **193**, 1987.

10 Abadie, M.S.M., and Richards, R.H. (1980) Synthèses de copolymères nouveaux par t ransformation de centres actifs de polymérisation, *Inf. Chim.* **208**, 135.

11 Riess, G., and Palacin, F. (1973) Préparation de polymères peroxydés par voie anionique, *Inf. Chim.* **116**, 9.

12 Riess, G., and Reeb, R. (1981) Preparation of polymeric free-radical initiators by anionic synthesis: polymeric azo derivatives, *ACS Symp. Ser.* **166**, 477.

13 Esselborn, E., Fock, J., and Knebelkamp, A. (Goldschmidt AG Essen Germany) (Oct. 10-12, 1994) Surface active block copolymers, *EPF Symposium*, Basel, Switzerland.

14 Morton, M., Lee, N.C., and Terrill, E.R. (1982) Elastomeric polydiene ABA triblock copolymers with crystalline end blocks, *ACS Symp. Ser.* **193**, 101.

15 Llauro-Darricades, M.F., Banderet, A., and Riess, G. (1973) Séparation d'un copolymère greffé de l'homopolymère correspondant par la méthode des gels réversibles, 1 et 2, *Makromol. Chem.* **174**, 117.

16 Reeb, R. (1995) Synthèse de macroinitiateurs, *PhD Thesis in preparation*, Université de Haute-Alsace, France.

17 Morishima, Y., Hashimoto, T., Itoh, Y., Kamachi, M., and Nozakura, S. (1982) Synthesis of amphiphilic block copolymers. Block copolymers of methacrylic acid and p-N,N-dimethylaminostyrene, *J. Polym. Sci., Polym. Chem. Ed.*, **20**, 299.

18 Danicher, L., Lambla, M., and Leising, F. (1979) Quaternisation de copolymères séquencés polystyrène-poly(vinyl-2-pyridine) par le bromure d'éthyle, *Bull. Soc. Chim. Fr.*, **9-10**, 544

19 Fujimoto, T., Nagasawa, M., and Ohno, S. (to Toyo Soda Mfg. Co.) (May 29, 1981) Block copolymers containing segment with cation exchange groups, anion exchange groups and no ion exchange groups, *Fr. Demande*, **2,470,139**.

20 Caldérara, F., (1995) Synthèse et caractérisation de copolymères à blocs à base de vinylpyridine et de latex acryliques. Réalisation de revêtements protecteurs temporaires et lubrifiants sur substrats métalliques, *PhD Thesis*, Université de Haute-Alsace, France.

21 Hahn, S.F. (1992) An improved method for the diimide hydrogenation of butadiene and isoprene containing polymers, *J. Polym. Sci., Part A: Polym. Chem.* **30**, 397.

22 Tuzar, Z. (1995) Copolymer micelles in aqueous media, *NATO ASI: Solvents and selforganization of polymers*, Antalya, Turkey.

23 Tuzar, Z. (1995) Overview of polymer micelles, *NATO ASI: Solvents and selforganization of polymers*, Antalya, Turkey.

24 Munk, P. (1995) Equilibrium and nonequilibrium polymer micelles, *NATO ASI: Solvents and selforganization of polymers*, Antalya, Turkey.

25 Munk, P. (1995) Classical methods for the study of block copolymer micelles, *NATO ASI: Solvents and selforganization of polymers*, Antalya, Turkey.

26 Webber, S.E. (1995) Use of fluorescence methods to characterize the interior of polymer micelles, *NATO ASI: Solvents and selforganization of polymers*, Antalya, Turkey.

27 Caldérara, F., Hruska, Z., Hurtrez, G., Lerch, J.-P., Nugay, T., and Riess, G. (1994) Investigation of polystyrene-poly(ethylene oxide) block copolymer micelle formation in organic and aqueous solutions by nonradiative energy transfer experiments, *Macromolecules*, **27**(5), 1210.

28 Hosotte-Filbert, C. (1994) Etude de la micellisation et de l'adsorption de copolymères blocs acryliques. Application à la stabilisation de dispersions de TiO_2 en milieu organique, *PhD Thesis*, Université de Haute-Alsace, France.

29 Halperin, A. (1994) Private communication.

30 Hurtrez, G., and Riess, G. (1994) Polystyrene-poly(ethylene oxide) block copolymers micelles in water. Relation between micellar size and molecular characteristics of the copolymer, *Int. Polym. Colloids Group*, **25**(1), 60.

31 Halperin, A. (1987) Polymeric micelles: a star model, *Macromolecules*, **20**, 2943.

32 Nagarajan, R., and Ganesh, K. (1989) Block copolymer self-assembly in selective solvents: spherical micelles with segregated cores, *J. Chem. Phys.*, **90**, 5843.

33 Ossenbach- Sauter, M. (1981) Contribution à l'étude des propriétés émulsifiantes des copolymères séquencés. Emulsion polymère du type eau-eau, *PhD Thesis*, Université de Haute-Alsace, France.

34 Pousse, C. (1992) Micellisation d'un copolymère par complexation interpolymère, *Diploma Work*, Université de Haute-Alsace, France.

35 Riess, G., Bahadur, P., and Hurtrez, G. (1985) Block copolymers in *Encyclopedia of polymer science and engineering*, Second Edition, John Wiley & Sons, New York, **2**, 324.

36 Barett, K.E.J. (1975) *Dispersion polymerization in organic media*, John Wiley, New York.

37 Adam, H. (1989) Etude de la polymérisation en suspension d'émulsions triples stabilisées par des tensioactifs macromoléculaires. Préparation de pigments blancs organiques, *PhD Thesis*, Université de Haute-Alsace, France.

38 Adam, H., Riess, G., and Nicaud, C.G. (July 13, 1989) Particles of hydrophobic polymers containing voids, *Eur. Pat.*, **249554 B1**.

39 Periard, J., Banderet, A., and Riess, G. (1970) Emulsifying effect of block copolymers. Oil-in-oil emulsions, *Polym. Lett.*, **8**, 109.

40 Periard, J., Riess, G., and Neyer-Gomez, M.J. (1973) Etude de copolymères séquencés de faible masse moléculaire utilisés comme agents émulsifiants du type huile dans huile. Associations en milieu solvant sélectif, *Eur. Polym. J.* **9**, 687.

41 Molau, G.E. (1965) Heterogeneous polymer systems. I. Polymeric oil-in-oil emulsions, *J. Polym. Sci., Part A* **3**, 1267.

42 Gaillard, P., Ossenbach-Sauter, M., and Riess, G. (1982) *Polymer compatibility and incompatibility*, K. Solc. Ed., Vol. 2, Harwood Academic, New York.

43 Dieng, A.C., Lavielle, L., Brendlé, M., and Riess, G. (July 1992) *IUPAC Int. Symposium on Macromolecules*, Poster 5-P51, Prague.

44 Frischinger, I. (1989) Etude de matériaux composites polymères comportant une phase dispersée liquide. Extension aux matériaux alvéolaires polymères-charges minérales, *PhD Thesis*, Université de Haute-Alsace, France.

MICELLIZATION OF IONIC BLOCK COPOLYMERS IN THREE DIMENSIONS

MOFFITT, M., ZHANG, L., KHOUGAZ, K. and EISENBERG, A.
Department of Chemistry, McGill University
801 Sherbrooke Street West, Montreal, Quebec, Canada H3A 2K6

1. Introduction

The self-assembly of block copolymers in selective solvents has become a topic of considerable interest to a growing number of research groups. Analogous to the association of surfactants in solution, the formation of spherical and non-spherical aggregates of block copolymers is a phenomenon of both academic and industrial relevance, attracting the efforts of theorists and experamentalists alike. The diverse range of studies in this area of colloidal science is described in several review articles [1-3].

The present review will focus on the micellization of ion-containing block copolymers (or ionic block copolymers). These materials consist of hydrophobic blocks covalently linked to blocks containing a significant number of ionic moieties. A wide variety of ion-containing polymers are synthetically available, and several of these, such as poly(4-vinylpyridinium alkyl halides), poly(metal acrylates), poly(metal methacrylates), and sulfonated polystyrene, have been introduced into block copolymers. Much of the pioneering work on the colloidal behavior of ion-containing block copolymers was done by Selb and Gallot [4,5], and their review on the topic provides and excellent summary of early results. For the present discussion, we will be concerned primarily with the self-assembly of ion-containing diblocks, made up of a single hydrophobic strand linked to a single hydrophilic strand of ionic units.

The interest in ionic block copolymers arises mainly from the extreme incompatibility between the ionic and hydrophobic blocks, which results in a much stronger driving force for microphase separation than that of nonionic block copolymers in selective solvents. The strength of this driving force gives ionic block copolymers unusually low cmc's in both organic and aqueous solvents, compared with the cmc's of their nonionic counterparts. As well, aggregates of ionic block copolymers show impressive stability over long periods of time, even at extreme temperatures. This stability, though partially thermodynamic in nature, contains a predominant kinetic component which shall be discussed shortly.

Along with low cmc's and high aggregate stability, the large driving force for microphase separation in ionic block copolymers enables their micellization to be investigated over an impressively wide range of block copolymer compositions. In

S.E. Webber et al. (eds.), Solvents and Self-Organization of Polymers, 53–72.

organic solvents, the micellization of diblock ionomers has been observed for extremely short ionic block lengths, down to a single ionic unit [6-9]. In aqueous environments, decreasing the length of the ionic block will result in a transition from "star" micelles, with relatively small cores and large ionic coronae, to "crew-cut" micelles, with large cores and short ionic hair [10-12]. When the relative length of the ionic block (or ionizable PAA block) is decreased even further, a wide range of non-spherical morphologies, dependent on the block copolymer composition, become thermodynamically favourable [13]. Thus, an impressive array of aggregate sizes and shapes can be obtained from a single family of materials, simply by varying the relative lengths of the ionic and nonionic blocks.

In any discussion of ion-containing block copolymers, it is useful to distinguish between two classes of these materials: block ionomers and block polyelectrolytes. The difference between ionomers and polyelectrolytes has been the subject of considerable debate, and some workers have suggested a classification based on the ion content of the copolymer. Eisenberg and Rinaudo, however, have defined ionomers and polyelectrolytes according to the role of the ionic groups in determining the material properties, regardless of composition [14]. By their method of classification, block polyelectrolytes are materials in which the properties are governed by electrostatic interactions over the relatively large distances within the micelle coronae and between the aggregates, while block ionomers are materials in which the properties are governed by strong, short-range electrostatic interactions within ionic microdomains. These definitions account for the relative complexity of block polyelectrolyte systems, in which long-range electrostatic interactions are highly dependent on such solution properties as pH and ionic strength.

It follows from the above definitions that an ionic block copolymer may be either a block ionomer or a block polyelectrolyte, depending on the nature of the solvent in which self-assembly occurs. In organic solvents, ionic block copolymers can be classified as block ionomers, as microphase separation results in compact ionic cores, solubilized by hydrophobic coronae. In aqueous environments, it is the hydrophobic blocks which form the micelle core, and long-range electrostatic interactions among solvated ionic blocks allow us to classify water-soluble systems as block polyelectrolytes. An interesting off-shoot of the (solvent dependent) "dual nature" of ionic block copolymers is the possibility of morphology/phase inversion. For example, a diblock copolymer with a relatively short ionic segment will form a spherical, star-like micelle in organic solvents, while the same material in water may form “crew cut” spheres, rods, vesicles, or some other aggregate structure.

Before embarking on separate discussions of block ionomers and block polyelectrolytes, it is useful to compare a wide range of cmc data, in order to impress upon the reader the strong driving force for the micellization of ion-containing block copolymers in general. Figure 1 presents cmc values as a function of the insoluble block length for various organic and aqueous systems [15-20]. Data for a family of ionic block copolymers (polystyrene-*b*-poly(sodium acrylate), PS-*b*-PANa) in both aqueous [19] and organic [20] environments show cmc values which are low compared with those of surfactants and nonionic block copolymers in selective solvents. In particular, the cmc's of PS-*b*-PANa in organic solvent (~10^{-8} M) indicate an extremely strong driving force for the formation of block ionomer micelles.

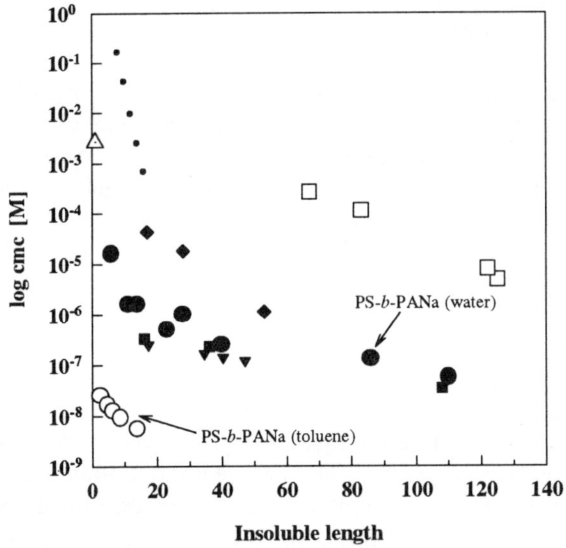

Figure 1. Typical cmc values for different micellar systems in aqueous media: sodium alkyl sufonate surfactants (●) [16], PEO-*b*-PBO-*b*-PEO (u)[18], PS-*b*-PANa (●) [19], PS-*b*-PEO (■) [17] and PEO-*b*-PS-*b*-PEO (▼) [17] and in nonaqueous media: AOT(Δ) [16], PS-*b*-PI (□) [15], PS-*b*-PANa (O) [20].

Since cmc values are directly related to the thermodynamics of micelle formation, it is tempting to translate this knowledge into a description of micelle-single chain equilibria at concentrations above the cmc. Such a description, however, though entirely sound from a thermodynamic perspective, does not properly consider the extremely slow kinetics of the system. In most aggregates of ionic block copolymers, the micelle cores are compact, glassy spheres, well below their glass transition temperature. On any reasonable time scale, therefore, the chains are effectively "locked" into the aggregate, precluding the existence of a true dynamic equilibrium between micelles and unimers. In fact, the term "micelle" must here be called into question, as it implies equilibrium associates such as those found in solutions of low-molecular weight surfactants. Although "micelle-like aggregate" (MLA) may be a more accurate term for the structures formed from ionic block copolymers, the term "micelle" is traditionally maintained, and will be used in the present review to describe aggregates of both block ionomers and block polyelectrolytes. It should also be mentioned that the micellization of ionic block copolymers may sometimes take place under equilibrium conditions, although subsequent treatment will result in kinetically-frozen aggregates. This is the case in the formation of aqueous "crew cut" micelles by the slow addition of water to solutions of block copolymers in DMF [11,12], a situation which will later be described in some detail.

The remainder of this review is divided into two main sections, concerning ion-containing block copolymers in organic (block ionomers) and aqueous (block polyelectrolytes) environments. The latter section will consider both block polyelectrolyte "star" micelles, and block polyelectrolyte "crew cut" micelles, along with a discussion of the morphological diversity which has been observed in water-soluble systems. While this chapter is concerned with the colloidal behavior of ion-containing block copolymers in three dimensions, a subsequent chapter by this group will present the phenomenon of two dimensional self-assembly on air-liquid interfaces [21]. Although we will do our best to mention all relevant studies in the field of ionic block copolymers in solution, the focus of our discussion will be on work done in this group.

2. Ionic Block Copolymers in Organic Solvents

As mentioned previously, ion-containing block copolymers in organic solvents are best described as block ionomers, with strong electrostatic interactions localized within compact, ionic cores. The aggregates are sometimes referred to as "reverse micelles", due to their structural similarity to reverse micelles of surfactants in oil, in which the ionic head groups form the core, surrounded by a corona of hydrocarbon chains.

Over the past five years, the aggregation behavior of block ionomers has been studied extensively by this group, in order to determine fundamental micellar parameters as a function of block copolymer composition. For the most part, these studies have focused on polystyrene (PS)-based diblock ionomers, with relatively long PS blocks covalently linked to short ionic segments. This composition regime is known to form star-like spherical micelles, with small ionic cores and large PS coronae. The work described in the present review has employed the following species of ionic blocks: poly(metal acrylates), poly(metal methacrylates), and quaternized poly(4-vinylpyridine) (P4VP). In other groups, systems such as quaternized P4VP-*b*-PS-*b*-P4VP triblock ionomers [22] and sodium-neutralized poly(tert-butylstyrene)-*b*-lightly sulfonated polystyrene [23] in nonpolar solvents have also been investigated.

The following summary of recent work on block ionomer reverse micelles will be divided into two parts. In the first part, the study of fundamental micellar parameters (aggregation numbers, core radius, etc.) by numerous techniques (size-exclusion chromatography, static and dynamic light scattering, small-angle x-ray scattering, etc.) will be described. A nuclear magnetic resonance (NMR) investigation of coronal chain mobility will also be discussed. We will then briefly mention the problem of determining the extremely low cmc's of ionic block copolymers in organic solvents. In the second part of this section, we will briefly describe the use of NMR in the study of water diffusion into the ionic core, along with the application of block ionomers as "microreactors" of controllable size in the synthesis of quantum-confined CdS-ionomer composites.

2.1 CHARACTERIZATION OF MICELLAR PARAMETERS IN BLOCK IONOMERS

In the first detailed investigation of ionic block copolymers in organic solvents, PS-*b*-PMANa and PS-*b*-PMACs of different compositions were studied in various solvents selectively good for polystyrene [24]. Size-exclusion chromatography (SEC) was used to demonstrate that the resulting micelles were extremely stable, and did not show an appreciable dynamic equilibrium over long periods of time.

For a wide range of PS-*b*-PMANa and PS-*b*-PMACs samples of various compositions, SEC coupled with viscometry was used to determine aggregation numbers and hydrodynamic radii. By this method, the micellar molecular weights (hence the aggregation numbers) were determined from a universal calibration curve of log([η]M) versus the SEC elution volume. Hydrodynamic radii were calculated from the following relationship:

$$[\eta]M = (10/3)N_a\pi r_h^3 \qquad (1)$$

Figure 2 shows plots of aggregation number and hydrodynamic radius versus the number of ionic repeat units for PS blocks of various lengths. Both of these parameters were found to increase with the length of the insoluble ionic block. Aggregation numbers were found to decrease as the length of the soluble block (PS) increased, while the hydrodynamic radius increased. This finding is not surprising, if we consider that the thickness of the corona increases with the length of the PS block; this results in an increase in the overall radius of the micelle, despite a lowering of the aggregation number. In a subsequent study, hydrodynamic radii from SEC and viscometry were confirmed by dynamic light scattering (DLS) [25].

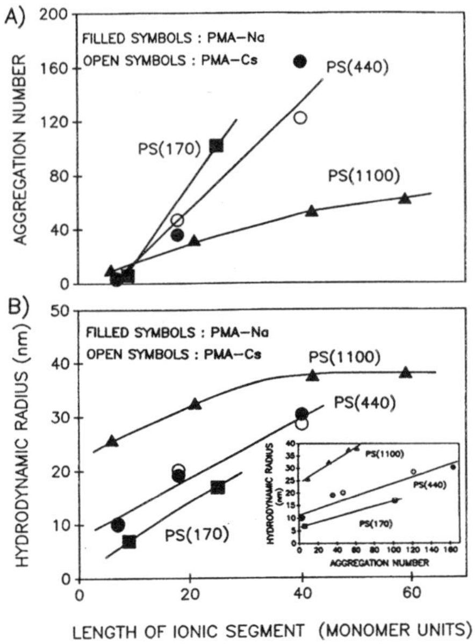

Figure 2. Aggregation number (A) and hydrodynamic radius (B) versus the length of the poly(metal methacrylate) block for three series of block ionomers.

Information on the internal structure of block ionomer micelles is not available using light scattering techniques (SLS and DLS), due to the limited length scales which can be probed with visible light. In order to probe the distances within reverse micelles (e.g., the radius of the ionic core), radiation with much shorter wavelengths must be applied. To this end, small-angle x-ray scattering (SAXS) has been used to determine micelle core sizes as a function of ionic block length for polystyrene-*b*-poly(cesium acrylate) (PS-*b*-PACs) and PS-*b*-PMACs block ionomers in toluene [26]. A typical SAXS profile for block ionomer micelles is a combination of a shape factor, which is attributable to scattering within the spherical ionic cores, and a structure factor, related to scattering between the cores. After the two components of the scattering profiles have been separated, form factors will yield information on the sizes of ionic cores.

The interpretation of form factors profiles for PS-*b*-PACs and PS-*b*-PMACs reveals that the ionic core radii scale roughly as $N_B^{3/5}$ (where N_B is the number of repeat units in the ionic block), independent of the PS block length. This is in agreement with the predictions of Halperin's model of "star" micelles formed from block copolymers [27]. The data were fitted to the scaling law and a proportionality constant of 6.5 Å was determined. In further work [28], core sizes were determined for quaternized PS-*b*-P4VP, and the same fit to the Halperin model was obtained. This seemed to indicate a universal relationship between the core radius and the length of the ionic block:

$$R_{core} = 6.5N_B^{3/5} \text{ Å} \tag{2}$$

Equation 2 is plotted in Figure 3, together with experimental data for different block ionomers in solution and in the solid state. SAXS data and computer modeling were used to determine radii polydispersity indexes (RPI), which were between 1.02 and 1.04, significantly less polydisperse than the polymer chains which make up the micelles.

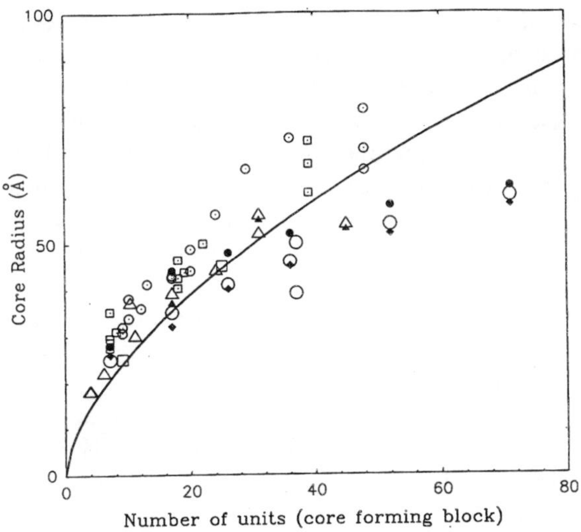

Figure 3. Plots of core radii of PS-based block ionomers vs. the number of units in the ionic block in toluene solution; P4VPMeI (○), PACs (Δ), PMACs (□), and in the bulk; PMACs solid (□), P4VPMeI solid (○), P4VPMeI cast from DMF (●), PACs cast from toluene (▲), P4VPMeI cast from toluene (◆). Solid samples are taken from ref. 27. The solid line is the best fit to the scaling relation $R_{core} \alpha aN_B^{3/5}$, where N_B is the length of the ionic block. The value of a is 6.5 Å.

A comparison of ionic core radii and contour (fully extended) lengths of the core-forming blocks revealed a high degree of chain extension within the cores of block ionomers. High extension of the core-forming blocks was also observed for nonionic cores within an ionic matrix. It was therefore concluded that extension within the cores of block ionomers is not due to electrostatic repulsion within the core, but to the minimization of interfacial energy between ionic and nonionic components.

Another micellar parameter which has been recently investigated in this group is chain mobility within the PS corona [29]. Using NMR, the coronal chain dynamics can be determined in some detail at the segmental level. This was demonstrated in systems of PS-*b*-PACs in CCl_4, by probing the mobility of ^{2}H-labeled PS "tags" at various distances from the ionic core. A comparative analysis of relaxation times revealed that segmental mobility is considerably restricted close to the ionic core, and approaches that of "free" homopolystyrene at ca. 40 from the ionic-nonionic junction. Segmental mobility within the corona was also found to become more restricted as the size of the ionic core increased.

A survey of literature on block copolymer micelles in selective solvents will reveal the wide range of techniques which have been used to measure cmc's in these systems [15-20]. All of these techniques (e.g. fluorescence, osmometery, viscometry) measure colloidal properties which undergo a discontinuous change when micellization occurs. Static light scattering (SLS) has also been employed with some success, though this method is sensitive to changes in the apparent molecular weight of the solution above the cmc [20]. To determine the cmc by SLS, the quantity Kc/R(0) (where K is the optical constant and R(0) is the Raleigh ratio extrapolated to 0 angle) is plotted as a function of concentration. Kc/R(0) is inversely proportional to the apparent molecular weight, and the plot is found to exhibit three distinct regions, attributable to unimers in solution, a transition region, and micelles in solution.

As we have seen, the cmc's of block ionomers in organic solvents are extremely low ($\sim 10^{-8}$ M), and the unimer region is often below the limit of detection by empirical methods. To determine the cmc's of such systems, therefore, empirical data were obtained for the micelle and transition regions, and these were fitted by computer software to a recent model of micellization. The results for PS-*b*-PANa in toluene are shown in Figure 1.

2.2 WATER SOLUBILIZATION AND NANOPARTICLE PRECIPITATION IN BLOCK IONOMERS

It is well established that reverse micelles of surfactants such as AOT are able to solubilize significant amounts of water within their ionic cores, resulting in water pools which can act as "microreactors" for hydrophilic reagents. The applicability of block ionomers for such localized, micro-scale (or nano-scale) chemistry is dependent on the ability of the ionic cores to solubilize polar species of low molecular weight.

The solubilization of water into the cores of block ionomers has been studied by proton NMR, in order to determine the distribution of water between the micelle and the nonpolar solvent [30,31]. In these experiments, a single proton chemical shift was observed-- a weighted average of the chemical shift of water in the ionic core and in the solvent. The single chemical shift indicated fast exchange of water between the micelle and the solvent, which suggested that the ionic cores are liquid-like in the presence of water.

From an analysis of the chemical shift of water protons in solutions of block ionomers, the equilibrium constants in favour of the reverse micelles were determined for various nonpolar solvents, and were found to occur in the following order: cyclohexane > benzene ~ toluene >> THF ~ DMF [30]. As a function of different ionic moieties, the equilibrium constants in favour of water solubilization were found to occur as follows: sodium sulfonates > sodium carboxylates >> methyl iodide pyridinium [31].

The desirability of block ionomers as "microreactors" for simple chemistry is heightened by the possibility of controlling the size of the core through variations in the length of the ionic block. In the formation of polymer-nanoparticle composites, for example, ultrasmall particles of controlled size can be precipitated within tailor-made ionic cores.

One such application has been demonstrated by the preparation of quantum-confined CdS nanoparticles within the ionic cores of PS-*b*-PACd [32]. The composites were prepared by first plasticizing the block ionomers with water, then treating with H_2S for a period of 8 hours. Particle sizes were determined from blue-shifts in the UV-vis spectra (related to CdS particle radii using an established correlation curve), and these were plotted against the radius of the original ionic core (calculated using equation 2). The results (Figure 4) show a linear relationship up to a particle radius of 50 Å, demonstrating that excellent size control has been achieved. Figure 4 also shows sizes of CdS nanoparticles prepared in the multiplets of random ionomers (dotted circles), which allow for size control down to 18 Å.

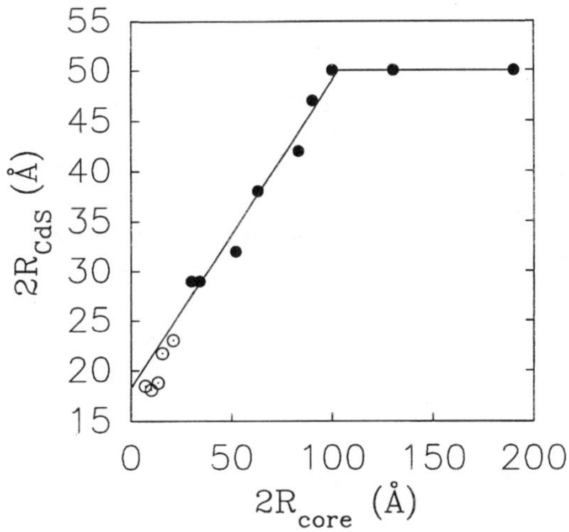

Figure 4. Plot of CdS particle diameter vs. diameter of the original core (calculated from $R_{core} = 6.5N_B^{3/5}$ Å). Dotted circles indicate CdS particles synthesized within the multiplets of random ionomers.

The solution stability of these CdS-block ionomer composites was also investigated, following re-neutralization of the poly(acrylic acid) (PAA) blocks with NaOH. The composites could be dissolved in organic solvents (e.g. toluene, THF) to form clear, yellow solutions of CdS-containing reverse micelles. The yellow powders were then repeatedly precipitated into MeOH and redissolved, without significant changes in the absorption spectra of the nanoparticles. The sizes of particles within the composites were also found to be stable to compression molding above the T_g of the PS matrix.

3. Ionic Block Copolymers in Aqueous Media

Ion-containing block copolymers dissolved in aqueous media form block polyelectrolyte micelles. Such micelles are the mirror image of the block ionomer micelles discussed in the previous section, as the nonionic blocks form the micelle core, surrounded by an ionic corona extending into solution. Two types of block polyelectrolyte micelles can be distinguished on the basis of relative block lengths: "star" and "crew-cut" micelles. Along with the studies discussed below considerable work has been done on block polyelectrolyte micelles [4, 33-37]. The present discussion of block polyelectrolytes will be divided into three parts. In the first part, we will discuss the micellization of water-soluble star-like micelles. The second part will be concerned with spherical "crew-cut" micelles. Finally, we will discuss novel morphological diversity of block polyelectrolytes with low ion content.

3.1 BLOCK POLYELECTROLYTE "STAR" MICELLES

Ionic block copolymers with long ionic blocks and relatively short PS blocks yield star-like micelles with small PS cores and thick ionic coronae. The detailed study of such systems has involved several series of PS-*b*-PANa; in each series, the PS block length is held constant while the PANa block length is varied. Such samples have allowed us to examine micellar characteristics (such as the cmc) as a function of soluble and insoluble block lengths and the addition of low molecular weight salts.

The cmc values of block polyelectrolyte micelles have been determined using fluorescence spectroscopy, by monitoring the fluorescence of a pyrene probe molecule [19]. When micellization occurs, the hydrophobic probe is preferentially solubilized in the micelle core, and changes in the fluorescence spectrum of pyrene allow the cmc to be determined. The effect of the insoluble PS block length on the cmc was investigated for a range of PS block lengths. The results are shown in Figure 1, in which the cmc values represent a constant interpolated PANa block length of 1000 units. From this data, we see that an increase of ca. 100 PS units resulted in a decrease in the cmc by a factor of 320. The effect of variations in the soluble block length was found to be much less pronounced. For example, when the PANa block length was increased from 300 to 1400, the cmc was lowered by less than a factor of 2.

The influence of the soluble PANa block length on the cmc was further explored in a subsequent publication [38]. At constant PS block length, it was found that the cmc passed through a maximum as the PANa block length was increased. This trend was explained on the basis of single chain solubility, and its dependence on the PANa block length. At relatively low ionic block lengths, it is obvious that the solubility of single chains will increase with the length of the soluble block. However, at higher ionic block lengths, an increase in the ion content of the block copolymer will effect an increase in the ionic strength of the solution, which will tend to decrease the solubility of single chains. The maximum which occurs in the cmc as a function of ionic block length can be regarded as a balance between these opposing trends.

For samples with relatively short PS blocks, it has been found that the cmc is more significantly influenced by the PANa block length than in samples with longer PS blocks [38]. This trend has also been seen in other micellar systems, and suggests that the micellization of block copolymers with short insoluble blocks is significantly affected by the soluble block, while longer insoluble blocks tend to dominate the micellization process.

Due to the polyelectrolyte nature of the corona, it was of interest to examine how the addition of low molecular weight salt affects the micellization of ionic block copolymers in water [38]. It was found that the cmc decreased linearly with the square root of NaCl concentration (C_s), which was varied from 0.10 to 2.5 M. This was explained on the basis of the solubility of single chains, which would tend to decrease with increasing salt content. d(log cmc)/d($C_s^{0.5}$) was plotted as a function of the PANa block length, and this yielded a sigmoidal curve, indicating that short PANa blocks are less affected by added salt than longer PANa blocks. The sigmoidal shape of the curve may be due changes in polyelectrolyte chain conformation as a function of ionic block length.

Other characteristics of PS-*b*-PANa "star" micelles were investigated by SLS measurements in aqueous solutions [39] and by SAXS measurements in the solid state [26]. In an investigation of aggregation numbers as a function of salt content, it was found that the aggregation number increased with added salt for low salt content, and then remained constant above ca. 0.10 M NaCl. This result, which has been predicted by theory, arises from the effect of added salt on the electrostatic repulsions of polyelectrolyte chains within the micelle corona [40]. At low salt content, electrostatic correlations dominate the chain configurations, significantly restricting micelle growth. As the salt content increases, the electrostatic repulsions are screened, and the micelle aggregation numbers increase. At moderate salt concentrations, the micellar properties are dominated by the core-forming block, and aggregation numbers do not vary significantly with C_s.

The effect of soluble and insoluble block lengths on aggregation numbers have been investigated in 2.5 M NaCl by SLS [39]. As predicted by theory, the insoluble block length was found to have a larger effect on aggregation numbers than the soluble block length. For instance, when the PS block length was varied from 11 to 50 units, the aggregation number increased from 54 to 270 (a factor of ca. 5). On the other hand, the effect of the PANa block length on aggregation number was dependent on the PS block length. For short PS block lengths, the aggregation number decreased with increasing PANa length, an effect which became much less significant for longer PS blocks. For example, for all samples in two series of block copolymers with PS block lengths of 11 and 23 units, the aggregation numbers were found to be 69 ± 9 and 148 ± 6, respectively, irrespective of varying PANa block lengths.

Since data was obtained for a wide range of block polyelectrolytes in 2.5 M NaCl, it was of interest to interpret the results according to various models which have been developed for star-like micelles of block copolymers [39]. Good agreement with experimental results was obtained with model predictions [27] of scaling relations for aggregation number and R_{core}. Experimental R_{core} values (calculated from aggregation numbers assuming unswollen cores) were also found to agree with those determined for similar samples in the solid state by an independent SAXS study [26]. The agreement between R_{core} values by these two methods suggests that the micelle cores of PS-*b*-PANa block polyelectrolytes are free of solvent. This result is entirely reasonable, considering the extremely unfavourable interactions between PS and the aqueous media.

3.2 BLOCK POLYELECTROLYTE "CREW CUT" MICELLES

"Crew-cut" micelles, formed from block copolymers such as PS-*b*-PAA with relatively long core-forming blocks (PS) and shorter hydrophilic corona-forming blocks (PAA), have recently received a great deal of attention in this laboratory. Such micelles are prepared in aqueous solution, by first dissolving the single chains in N,N-dimethylformamide (DMF), a good solvent for both blocks, then adding deionized water to induce micellization of the copolymers; finally, the DMF is removed by dialyzing the solution against water. At the point of micellization, DMF within the core will increase the mobility of unimers into and out of the micelle, and a dynamic equilibrium between micelles and unimers can be postulated [11]. As the micelles are dialyzed, however, the plasticizing effect is lost as DMF is forced out of the core, and the dynamic equilibrium eventually "freezes". Considering that the micelles are formed under equilibrium conditions, however, comparisons with thermodynamically-based models may be appropriate.

Static light scattering was used to study the phase behavior of PS-*b*-PAA block copolymers in DMF-water mixture [12]. Figure 5 shows some typical experimental curves of scattered light intensity from polymer/DMF/water solutions vs. water content. In these solutions, the scattered light intensity changed only slightly at relatively low water content, indicating that the polymer chain dimensions do not change appreciably. However, when the water content reached a critical value, the intensity increased rapidly as phase separation occurred. In the case of homopolystyrene, phase separation corresponded to the precipitation of the polymer, while the copolymer at the critical point underwent microphase separation, i.e. micellization.

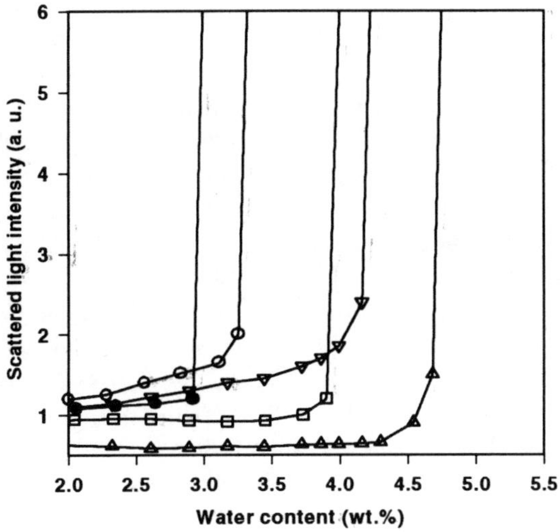

Figure 5. Plot of scattered light intensity vs. water content for solutions of different polymer and concentration of (●) PS(1140), 2.0E-3; (O) PS(1140)-*b*-PAA(165), 2.0E-3; (□) PS(500)-*b*-PAA(58), 2.0E-3; (∇) PS(386)-*b*-PAA(41), 2.0E-3; (Δ) PS(386)-*b*-PAA(41), 2.0E-4 g/g.

The critical water content (cwc), at which phase separation or micellization begins, can be obtained from the intercept of the horizontal and vertical straight line segments of each of the light scattering curves. For a typical polymer, it was found that the cwc increased as the polymer concentration decreased. A linear relationship was found between cwc values and the logarithm of the polymer concentration. For a particular polymer concentration, the cwc decreased with increasing polymer molecular weight; moreover, block copolymers showed a higher cwc value than that found for the homopolystyrene with the same PS chain length. These features of the cwc indicate that different block copolymers form micelles in solvents of different quality. Specifically, the value of χ parameter at the onset of micellization decreases with an increase in either the length of the core-forming PS block or the copolymer concentration.

The "crew-cut" micelle sizes were measured directly from the prints of negatives obtained by transmission electron microscopy (TEM) [11]. The polydispersity of such micelles is very low. The number-average core radius (R_{core}) was found to depend on the chain lengths of both PS (N_{PS}) and PAA blocks (N_{PAA}), and scales empirically according to the equation:

$$R_{core} \propto N_{PS}^{0.4} N_{PAA}^{-0.15} \qquad (3)$$

For a compact micelle core, some of the PS blocks must span the distance from the core/corona interface to the core center. Therefore, the core radius is directly related to PS chain stretching in the radial direction. According to equation 3, the radius of the spherical core increases with a decrease in the length of the PAA block at a constant PS block length, which implies a corresponding increase in the degree of PS stretching. This characteristic is considered very important, and will later be discussed in terms of morphological transitions in the aggregates of block copolymers with low PAA block content.

Dynamic light scattering (DLS) was used to determine the hydrodynamic radius (R_h) of "crew-cut" micelles [11]. From values of R_{core} and R_h, it was possible to estimate the thickness of the coronal shell. It was found that the PAA chains were highly extended in solutions at pH 7 with low ionic strength, and this effect was ascribed to the charged nature of the PAA block. However, in the presence of an inert electrolyte, the degree of the coronal chain extension was low. Also, the degree of coronal extension was found to be low when the degree of ionization of the PAA chains was low (i.e., at low pH).

Because of the chemical bond between the two blocks of the copolymer, the average surface area per coronal chain at the core/shell interface (A_c) can be calculated using the following equation:

$$A_c = 4\pi R_{core}^2/N_{agg} = 3V_S N_{PS}/R_{core} \tag{4}$$

where N_{agg} and V_S are the aggregation number and the volume per styrene repeat unit (0.167 nm^3), respectively. Since R_{core} is related to N_{PS} by an exponent which is less than one, equation 4 shows that A_c increases with the length of PS block. In star-like systems, the degree of chain stretching in the core increases, and the dependence of A_C on the PS block length is therefore weaker than that found in the present system. From equation 3, it can be shown that:

$$A_c \propto N_{PS}^{0.6} N_{PAA}^{0.15} \tag{5}$$

From this relationship, one can see that the surface area per coronal chain increases with the lengths of both PS and PAA blocks. Figure 6 shows a plot of A_c against copolymer compositions according to equation 5. Since the PS blocks are less stretched in the cores of "crew-cut" micelles than in those of "star" micelles, the surface density of coronal chains is quite low in the present system [13b].

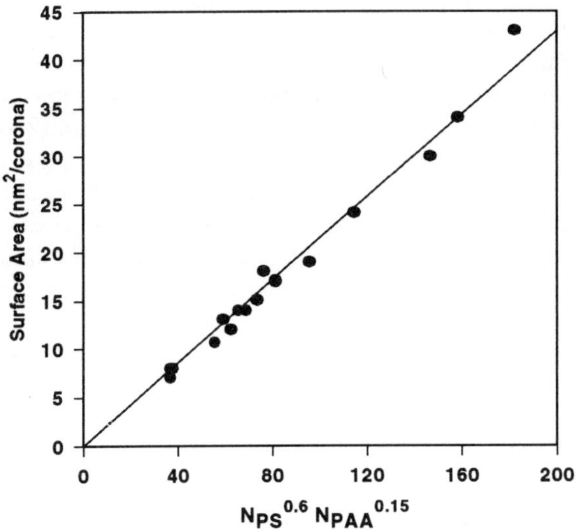

Figure 6. Plot of surface area per coronal chain vs. $N_{PS}^{0.6}N_{PAA}^{0.15}$, where N_{PS} and N_{PAA} are the degrees of polymerization of the PS and PAA blocks

3.3. MORPHOLOGICAL TRANSITIONS IN BLOCK POLYELECTROLYTES

In microphase separated systems, many examples of morphological transitions have been observed [13]. Small molecule surfactants, for example, have been found to form micelles of various morphologies. Several morphologies have been also observed in bulk block copolymers, and in two-dimensional micellar systems (see subsequent chapter for details). In block copolymer solutions, most of the systems which have been studied show spherical micelles; as well, some worm-like or rod micelles have been observed, and the existence of vesicles has also been postulated based on indirect evidence. In poly(styrene-*b*-butadiene)/homopolystyrene blends, it was reported that the morphology of the butadiene domains can change from spherical to cylindrical and eventually to vesicular, by increasing the butadiene block content of the copolymers, by increasing the homopolystyrene molecular weight, or by increasing the volume fraction of block copolymer in the blends. By extrapolating these reported trends, multiple morphologies of block copolymer micelles can be expected in low molecular weight solvents only if the block copolymer has a very short soluble block length.

Morphological transitions were observed in PS-*b*-PAA block polyelectrolytes as the PAA block length was decreased at constant PS block length (e.g. 200 styrene units) [13a]. While most of the copolymers with relatively short PAA blocks yielded spherical "crew-cut" micelles (e.g. PS(200)-*b*-PAA(21)), it was found that as the PAA block content decreased to a critical value (PS(200)-*b*-PAA(15)), the copolymers associated to form rod-like micelles. Vesicular aggregates were formed by copolymers with an even lower PAA content (PS(200)-*b*-PAA(8)). Also, lamellar micelles were occasionally observed in solutions of copolymers showing vesicles. Finally, when the PAA block content was extremely low (PS(200)-*b*-PAA(4)), the copolymers formed a new morphology, the so-called complex micelles. These structures are very highly polydisperse in size, with diameters exceeding one micron. A study of their internal structure showed that complex micelles consist of a large number of reverse micelles, in which polar PAA microdomains are dispersed in a PS matrix The surface of these complex micelles is hydrophilic, due to the presence of the short PAA chains. Such complex micelles therefore show considerable stability, and can be re-suspended in water after they have settled to the bottom of the container.

It is believed that the stretching of PS blocks in the micelle core of block polyelectrolytes is largely responsible for morphological transitions. From the scaling relation shown in equation 3, we have seen that the degree of stretching of the PS block increases with a decrease in the PAA block length. As chain stretching results in a decrease in the entropy of the system, it will not continue indefinitely as the PAA block length decreases. Below a certain PAA block length, the morphology changes, presumably to reduce the thermodynamic penalty of chain stretching.

In further studies [13b], homopolystyrene was incorporated into the cores of spherical "crew-cut" micelles, by dissolving the homopolystyrene in the copolymer/DMF solutions before the addition of water. During micellization of the copolymers, homopolystyrene was solubilized into the micelle core. The homopolystyrene chain lengths were chosen such that the homopolystyrene had a higher cwc than the block copolymer. In the mixed system PS(500)-*b*-PAA(58)/PS(180), a low weight ratio of homopolystyrene to copolymer (< 10:90), resulted in generally monodisperse micelles (standard deviation in R_{core} was 6%). The core radius increased only slightly with the addition of homopolystyrene. As the ratio increased to 20:80, a more pronounced increase in the core radius was observed, and the core sizes became more polydisperse (standard deviation 12 %). When the ratio was higher than 30 :70, the sizes of the micelles became highly polydisperse and very large particles formed in solution, indicating the accumulation of the homopolystyrene in the core center. No morphological transitions were induced by the addition of homopolystyrene to spherical "crew cuts".

For the non-spherical aggregates described above, it was found that the addition of homopolystyrene induced changes in morphology. Figure 7 shows the morphological transition from vesicles to spheres after the addition of 5 %-wt homopolystyrene; transitions from cylinders to spheres were also observed. It was found that the polydispersity of the resulting spherical micelles was quite low, despite an increase in aggregate dimensions with the addition of homopolystyrene. Although the dimensions of the PS regions are increased as the morphology changes from non-spherical to spherical, it is possible that the stretching of the PS blocks remains constant throughout the transition, as the added homopolymer may accumulate at the core center. Furthermore, even if the stretching is increased, the loss of free energy due to increased stretching could be compensated by the entropy gained in mixing the homopolystyrene with the chains in the core.

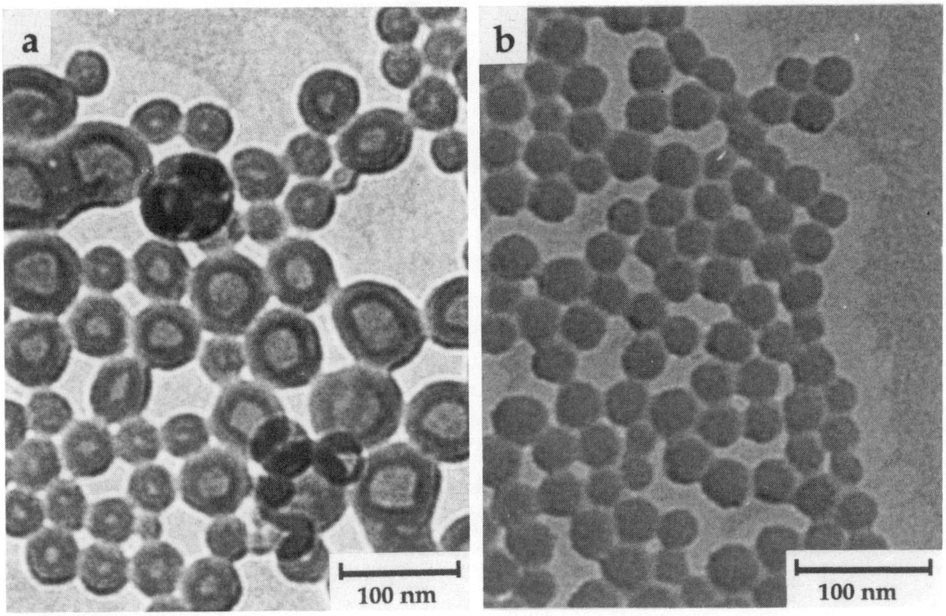

Figure 7. Transmission electron micrographs showing (a) vesicles made from PS(200)-b-PAA(8); and (b) spherical micelles made from 95 %-wt PS(200)-b-PAA(8) and 5 %-wt PS(38).

4. Concluding Remarks

In our survey of ionic block copolymers in organic and aqueous solutions, we have attempted to present the reader with some of the more interesting characteristics of such systems. As evinced by their low cmc's, the driving force for microphase separation of ionic block copolymers is generally much stronger than that of their nonionic counterparts. This feature gives rise to the possibility of studying micellization behavior over a wide range of block copolymer compositions, and microphase separation has been observed even when the ion content is extremely low. In fact, the micellization of block copolymers with very short ionizable PAA blocks results in unprecedented morphological diversity in aqueous solutions. In systems of PS-*b*-PAA in water, transitions from spherical "crew-cut" micelles to cylinders, vesicles, lamellae, and large complex micelles have been observed as the length of the PAA block is decreased. Block polyelectrolyte "star" micelles also exhibit a wide range of colloidal behavior, and micellization parameters such as cmc and aggregation numbers have been shown to be extremely dependent on pH and ionic strength. In organic solvents, block ionomer micelles have been characterized over a range of relative block lengths. Using SAXS, the sizes of ionic cores in reverse micelles have been determined as a function of the ionic block length. This knowledge can be applied to the size-controlled precipitation of nanoparticles within the ionic microdomains, resulting in ionomer-based composites. The diffusion of water into ionic cores has also been studied, opening the door to a range of simple chemistry involving polar species within block ionomer "microreactors".

REFERENCES

1. Tuzar, Z., Kratochvil, P. (1976) Block an Graft Copolymer Micelles in Solution, In *Advances in Colloid and Interface Science,* vol.6, 201-232
2. Price, C.(1982) Colloidal Developments in Block Copolymers, In *Developments in Block Copolymers-- 1*; Goodman, I., Ed.; Elsevier App. Sc. Pub. U.K, 39-79.
3. Tuzar, Z., Kratochvil, P (1993) Micelles of Block and Graft Copolymers in Solution In *Surface and Colloid Science*; Matijevic, E., Ed.; Plenum Press. New York,1-83.
4. Selb, J.and Gallot, Y. (1980) Copolymers with Polyvinylpyridinium Blocks or Grafts: Synthesis and Properties in Solution, In *Polymeric Amines and Ammonium Salts.* Goethals, E. J. Ed., Pergamon Press: New York, 205-218.
5. Selb, J and Gallot, Y. (1985) Ionic Block Copolymers, In *Development in Block Copolymers*; Goodman, I., Ed.; Elsevier Applied Science: London, U.K., vol 2, 27-96.
6. Jalal, N.and Duplessix, R. (1988) Aggregation of Monocarboxylic Chains by Neutralization: Neutron and X-Ray Scattering, J. Phys. France **49**, 1775-1783.
7. **a.** Davidson, N. S., Fetters, L. J., Funk, W. G., Graessley, W. W., Hadjichristidis, N. (1988) Association Behavior in End-Functionalized Polymers. 1. Dilute Solution Properties of Polyisoprenes with Amine and Zwitterion End Groups, Macromolecules **21**, 112-121. **b.** Pispas, S. and Hadjichristidis, N.(1994) Macromolecules **27**, 1891-. **c.** Pispas, S., Hadjichristidis, N., Mays, J.M. (1994) Macromolecules **27**, 6307-.
8. Zhong, X. F. and Eisenberg, A. (1994) Aggregation and Critical Micellization Behavior of Carboxylate-Terminated Monochelic Polystyrene, Macromolecules **27**, 1751-1758.
9. Zhong, X. F. and Eisenberg, A.(1994) Synthesis and Characterization of 1,2-Dicarboxyethyl-Terminated Polystyrene, Macromolecules **27**, 4914-4918.
10. Gao, Z., Varshney, S. K., Wong, S. and Eisenberg, A. (1994) Block Copolymer "Crew-Cut" Micelles in Water, Macromolecules **27**, 7923-7927.
11. Zhang, L., Barlow, R. J. and Eisenberg, A. Scaling Relations and Coronal Dimensions in Aqueous Block Polyelectrolyte Micelles, Macromolecules, accepted for publication.
12. Zhang, L.and Eisenberg, A. Phase Behavior of "Crew-Cut" Micelle Formation in Solutions of Polystyrene-*b*-Poly(acrylic acid), paper in preparation.
13. **a**. Zhang, L. and Eisenberg, A.(1995) Multiple Morphologies of "Crew-Cut" Aggregates of Polystyrene-*b*-Poly(acylic acid) Block Copolymers, *Science* **268**, 1728-1731. **b**. Zhang, L. and Eisenberg, A. Chracteristics and Multiple Morphologies of "Crew-Cut" Micelle-like Aggregates of Polystyrene-*b*-Poly(acylic acid) Block Copolymers in Solution, paper in preparation.
14. Eisenberg, A. and Rinaudo, M. (1990) Polyelectrolytes and Ionomers, Polymer Bulletin **24**, 671.
15. Price, C. and Chan, E. K. M. and Stubbersfield, R. B. (1987) The Effect of Block Length on the Thermodynamic Stability of Micelles Formed by Polystyrene-block-Polyisoprene Copolymers in n-Hexadecane, *Eur. Poly, J.* **23**, 649- 651.
16. Myers, D. (1988) *Surfactant Science and Technology*, VCH publishers; N.Y.

17. Wilhelm, M., Zhao, C. L., Wang, Y., Xu, R., Winnik, M. A., Mura, J. L., Riess, G., and Croucher, M. D. (1991) Poly(styrene ethylene oxide) Block Copolymer Micelle Formation in Water: A Fluorescence Probe Study, *Macromolecules* **24**, 1033 -1040.

18. Nicolas, C. V., Luo, Y. Z., Deng, N. J., Attwood, D., Collet, J. H., Price, C., and Booth, C. (1993) Effect of Chain Length on the Micellization and Gelation of Block Copoly(oxyethylene/oxybutylene/oxyethylene) *E*m*B*n*E*m, *Polymer* **34**, 138-144.

19. Astafieva, I., Zhong, X. F. and Eisenberg, A. (1993) Critical Micellization Phenomena in Block Polyelectrolyte Solutions, *Macromolecules* **26**, 7339-7352.

20. Khougaz, K., Gao, Z., and Eisenberg, A. (1994) Determination of the Critical Micelle Concentration of Block Copolymer Micelles by Static Light Scattering, *Macromolecules* **27**, 6341-6346.

21. Clarke, C., Lennox, B.and Eisenberg, A. (1995) Micellization of Block Copolymers in Two Dimensions, *see subsequent chapter.*

22. Ishizu, K., Kashi, Y., Fukutomi, T. and Kakurai, T. (1982) Synthesis of AB and BAB Poly(styrene-*b*-4-vinylpyridine) and Solution Properties of Their Quaternized Compounds., *Makromol. Chem* **183**, 3099-3107.

23. Zhou, Z., Peiffer, D. J.and Chu, B. (1994) Light Scattering Studies of Block Ionomer Aggregation Characteristics in Nonpolar Sovents, *Macromolecules* **27**, 1428-1433.

24. Desjardins, A. and Eisenberg, A. (1991) Colloidal Properties of Block Ionomers. 1. Characterization of Reverse Micelles of Styrene-*b*-Metal Methacrylate Diblocks by Size Exclusion Chromatography, *Macromolecules* **24**, 5779-5790.

25. Desjardins, A., van de Ven, T. G. M. and Eisenberg, A. (1992) Colloidal Properties of Block Ionomers 2. Characterization of Reverse Micelles of Styrene-*b*-Methacrylic Acid and Styrene-*b*-Metal Methacrylate Diblocks by Dynamic Light Scattering, *Macromolecules* **25**, 2412-2421.

26. Nguyen, D., Williams, C. E. and Eisenberg, A. (1994) Block Ionomer Micelles in Solution 1. Characterization of Ionic Cores by Small-Angle X-ray Scattering, *Macromolecules* **27**, 5090-5093.

27. Halperin, A. (1989) Polymeric Micelles: A Star Model, *Macromolecules* **20**, 2943-2946.

28. Nguyen, D., Williams, C. E. and Eisenberg, A., results to be published.

30. Gao, Z., Desjardins, A. and Eisenberg, A. (1992) Solubilization Equilibria of Water in Nonaqueous Solutions of Block Ionomer Reverse Micelles: An NMR Study, *Macromolecules* **25**, 1300-1303.

29. Gao, Z., Zhong, X. F.and Eisenberg, A. (1994) Chain Dynamics in Coronas of Ionomer Aggregates, *Macromolecules* **27**, 794-802.

31. Khougaz, K., Gao, Z. and Eisenberg, A. Distribution of Water in Reverse Micellar Solutions of AOT and Block Copolymers Ionomers in Toluene-d8, paper in preparation.

32. **a.** Moffitt, M., McMahon, L., Pessel, V. and Eisenberg, A. (1995) Size Control of Nanoparticles in Semiconductor-Polymer Composites. 2. Control via Sizes of Spherical Ionic Microdomains in Styrene-Based Diblock Ionomers, *Chemistry of Materials*, in press **b.** Moffitt, M. and Eisenberg, A. (1995) Size Control of Nanoparticles in Semiconductor-Polymer Composites. 1. Control via Multiplet Aggregation Numbers in Styrene-Based Random Ionomers, *Chemistry of Materials*, in press.

33. Morishima, Y., Hashimoto, T., Itoh, Y., Kamachi, M., and Nozakura, S.-I. (1982) Syntheses of Amphiphilic Block Copolymers. Block Copolymers of Methacrylic Acid and p-N,N-dimethylaminostyrene, *J. Polym. Sci., Polym. Chem. Ed.* **20**, 299-310.

34. Morishima, Y., Itoh, Y., Hashimoto, T., and Nozakura, S.-I. (1982) Amphiphilic Block Copolymer of 9-Vinylphenanthrene and Methacrylic Acid: Fluorescence Quenching of Phenanthrene Residue in Aqueous Media, *J. Polym. Sci., Polym. Chem. Ed.* **20**, 2007-2017.

35. Valint, P. L.and Bock, J. (1988) Synthesis and Characterization of Hydrophobically Associating Block Polymers, *Macromolecules* **21**, 175-179.

36. Cao, T., Munk, P., Ramireddy, C., Tuzar, Z., and Webber, S. E. (1991) Fluorescence Studies of Amphiphilic Poly(methacrylic acid)-block-Polystyrene-block-Poly(methacrylic acid) Micelles, *Macromolecules* **24**, 6300-6305.

37. Kiserov, D., Prochazka, K., Ramireddy, C., Tuzar, Z., Munk, P., and Webber, S. E. (1992) Fluorimetric and Quasi-Elastic Light Sattering Study of the Solubilization of Nonpolar Low-Molar Mass Compounds into Water-Soluble Block-Copolymer Micelles, *Macromolecules* **25**, 461-469.

38. Astafieva., I., Khougaz, K., and Eisenberg, A. Micellization in Block Polyelectrolyte Solutions. II. Fluorescence Study on the cmc as a Function of Soluble Block Length and Salt Concentration, Submitted for publication in *Macromolecules*

39. Khougaz, K., Astafieva., I., and Eisenberg, A. Micellization in Block Polyelectrolyte Solutions. III Static Light Scattering Characterization, Submitted for publication in *Macromolecules*

40. Dan, N. and Tirrell, M. (1993) Self-Assembly of Block Copolymers with a Strongly Charged and a Hydrophobic Block in a Selective, Polar Solvent. Micelles and Adsorbed Layers, *Macromolecules* **26**, 4310.

MICELLIZATION OF BLOCK COPOLYMERS IN TWO DIMENSIONS

C.J. CLARKE, R.B. LENNOX and A. EISENBERG
Department of Chemistry, McGill University
801 Sherbrooke St. West, Montreal, Canada H3A 2K6

M.H. RAFAILOVICH and J. SOKOLOV
Department of Materials Science and Engineering
State University of New York, Stonybrook, New York 11794-2275

1. Introduction

The self-assembly of molecules has been extensively studied in recent years in several different situations, for example low molecular weight amphiphiles in solution and block copolymers both in solution and the melt. In this chapter. we review a systematic study of two dimensional micellization. The system upon which most of the work is based is a family of polystyrene (PS) - poly 4-vinylpyridine (PVP) block copolymers. For example a nearly symmetric copolymer containing 260 hydrophobic styrene units and 240 hydrophilic vinylpyridines has been studied in detail. Each VP unit is quaternized with an alkyl iodide side chain. Micellization at the air-water interface was initially observed for this system by performing transmission electron microscopy on the corresponding Langmuir-Blodgett films and by correlating the observed structures with surface pressure - area (π-A) isotherms obtained at the air-water interface. The effects of a wide range of parameters on the micellization phenomenon have been investigated. We also report the effect of a non-ionic hydrophilic block in the copolymer and the formation of nanofoams. For a more detailed description the reader is referred to the original publications [1-8].

2. Starfish Micelles

To produce the Langmuir and Langmuir-Blodgett films the polymer is dissolved in a 70:30 mixture of chloroform and isopropanol and spread on the water surface. Figure 1 shows a transmission electron micrograph of a LB film produced by dipping a silicon wafer at low surface pressure (2 mN/m). The micelles are seen as round features with diameters and heights of 35 nm and 7.5 nm respectively. The micelles are separated by about 100 nm, and interact with each other to produce a sustainable surface pressure.

S.E. Webber et al. (eds.), Solvents and Self-Organization of Polymers, 73–81.

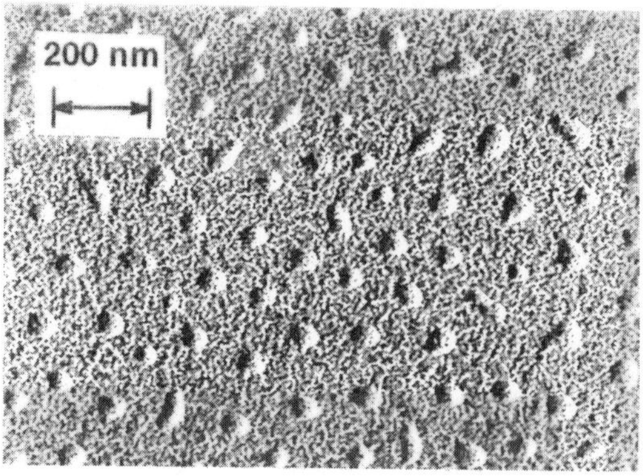

Figure 1 TEM of an LB film of PS-b-PVPC$_{10}$I deposited at 2 mN/m at 25°C.

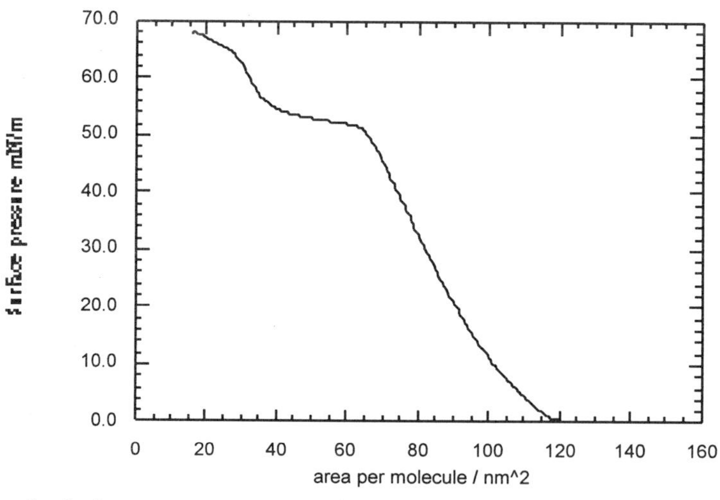

Figure 2 Surface pressure vs. area per molecule at 25°C.

Figure 2 shows a typical π-A curve. Surface pressure appears at 120nm^2/molecule and increases gradually until, at approximately 60nm^2/molecule, a plateau is observed (a first-order transition). The plateau is approximately 20nm^2/molecule wide and is followed by a steep rise in surface pressure until the film collapses. On the basis of this, and much other evidence reviewed below, one can propose that the features observed in the TEM images are micelles with a PS core surrounded by the PVPC$_n$I chains which emanate radially and are located at the air-water interface. The chains must be highly extended since there is a measurable surface pressure when the distance between the closest points of neighbouring cores is 100 nm. A length of 50 nm for a 240 repeat unit chain represents about 80% of the fully stretched length. The name “starfish” micelles is suggested for these because of their many highly extended arms. The aggregation number is found by counting the total number of micelles on a transmission electron micrograph and determining the ratio of the area per micelle to the area per molecule. A typical value is about 120 for the PS_{260}-b-$(PVPC_{10}I)_{240}$ system. This provides no population statistics, though these can be determined by other methods (see the original paper [2]).

2.1. MULTIPLE MORPHOLOGIES

In 3D systems which form micelles, it is well known that multiple morphologies can be produced (see for example, the previous paper). This is also the case in 2D. Figures 1 and 3 show the morphologies for three different $PVPC_nI$ blocks with a fixed PS block length. For very short $PVPC_nI$ blocks (>94% PS) there is a “pancake” structure (figure 3a), at intermediate lengths (between 94% and 86% PS) ribbon-like features are seen (figure 3b) and for longer $PVPC_nI$ chains (<86% PS) the starfish micelles are present (figure 1). Thus the starfish micelles are observed except when the $PVPC_nI$ block is short. The π-A curves are consistent with the observed morphologies. Starting with a symmetric copolymer and reducing the length of the $PVPC_nI$ block first shortens the plateau region of the isotherm. When the ribbon morphology is reached, the first order transition is no longer observed, i.e. the transition occurs for starfish micelles only. The rise in pressure occurs at very low surface area and is extremely steep. This could be due to deformation of the ribbons or pancakes during the compression process.

The phenomena described above were observed initially on ionic block copolymers. More recently it has been shown that identical morphologies and transitions occur in nonionic systems as well. Several different systems have been studied: styrene-b-(t-butyl methacrylate), styrene-b-(n-butyl methacrylate) and styrene-b-(dimethyl siloxane). The boundaries between different morphologies occur at different compositions, but it can be generally stated that ionic character is not a prerequisite for surface micellization: surface micellization is clearly a very general phenomenon which requires only a copolymer with a block of hydrophobic polymer joined to a block of a surface adsorbing one.

2.2. EFFECT OF OTHER PARAMETERS

The effects of a wide range of parameters on the micellization have been studied, including alkyl chain length, temperature and counterion species in the subphase. There is a change in the shape of the isotherm with increasing alkyl chain length [3]. The C_1 system shows no plateau (i.e. no first order transition); the pyridinium methyl iodide is soluble in water. Thus only the sharp increase in pressure at a small value of the molecular area is seen. The C_4 system shows a slight increase in pressure before the sharp rise, which indicates that the butyl-substituted chain is less soluble than the methyl-substituted polymer. The C_6 and C_{10} derivatives show clear evidence of a transition, while the C_{18} chain shows some evidence of a transition but only at a very high surface pressure value. The effect of temperature has been investigated for the C_{10} system [4]. As the temperature increases, the pressure decreases (in contrast to the behaviour of a 3D gas). In its low pressure form the PVP chains are surface adsorbed and the system is highly ordered. As the pressure is increased there is an increase in the entropy (in contrast to a gas). One final parameter which has been investigated is the counterion in the subphase [5]. Various salts were placed in the subphase. Chloride ions give the lowest transition pressure while bromide and iodide increase the pressures considerably. This is because the chloride ion is the least tightly bound to the VP^+ residues and gives the most soluble (i.e. most highly charged and hydrophilic) chain.

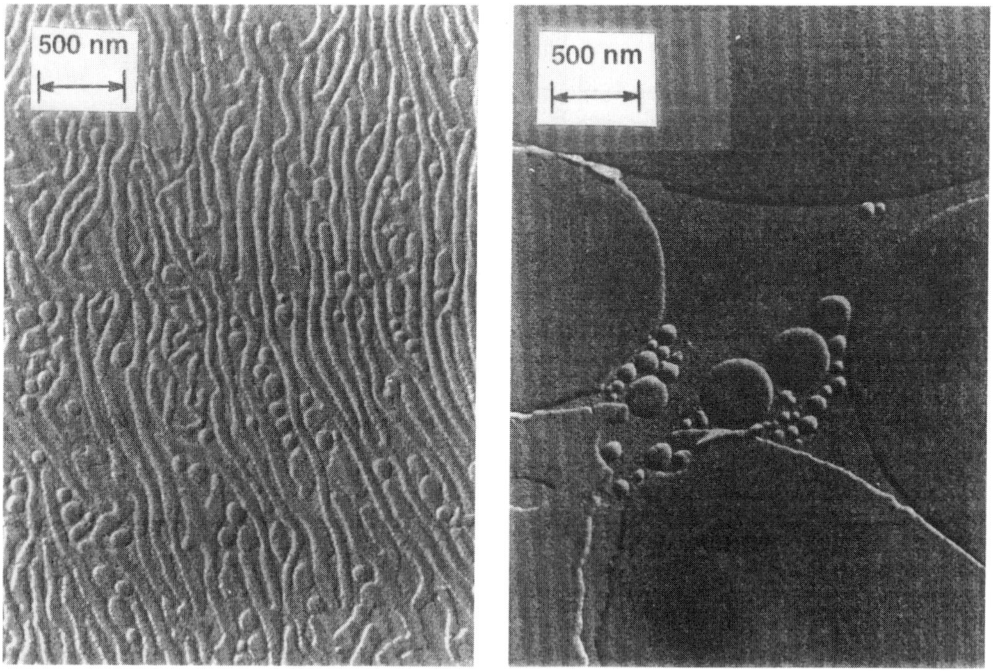

(a) (b)

Figure 3 TEM of PS_{480}-b-$(PVPC_{10}I)_m$ illustrating the different morphologies. (m=200 in figure 1, m=34 in figure 3a and m=13 in figure 3b).

2.3 NANOFOAMS

It has been observed that the morphology of a 3D block copolymer system is influenced by the solvent from which the materials were cast (the block which is the more soluble in a particular solvent becomes the continuous phase). This is also the case for the 2D systems. Most of the Langmuir films were prepared from chloroform/isopropanol, which is a poor solvent for PS. In another study toluene, (a good solvent for PS) was used with block copolymers of PS and polyacrylic or polymethacrylic acid. Regions of these films show features which are best described as 2D nanofoams. These can be regarded as a phase-inverted form of the starfish micelles. The holes have dimensions of ~ 30 nm while the wall heights are ~ 20 nm. The wall thickness depends on the length of the insoluble (PS) block. The % cell area can be varied from about 80% for a short PS block to about 1% for a very long one (1000 units). Their behaviour is similar to macroscopic 2D foams and they may be the smallest 2D foams or honeycombs yet described.

2.4. DISCUSSION

A wide range of experiments have been performed on the transition which we now attempt to interpret in terms of the detailed conformations of the chains at or near the interface. The onset of pressure is due to interactions between surface adsorbed, radial, highly extended $PVPC_nI$ chains from neighbouring micelles. As the pressure is increased, the area per molecule decreases, which implies a change in conformation of the chains at the surface. The area per molecule at the beginning and end of the plateau allow one to conclude that at these two points on the isotherm the pyridine rings are lying parallel and perpendicular to the surface respectively. Thus, at the beginning of the transition, all the space on the surface of the water is completely filled with polymer and any further decrease in surface area would have to be accompanied by disappearance of material from the surface. We have to decide therefore whether the PVP chains are pushed into the air during this transition (as is observed with fatty acids) or whether an alternative mechanism is involved. The studies of the completely soluble $PVPC_1I$ chains show continuity with the $PVPC_{10}I$ material which suggest that the transition involves the movement of the decylated chains into the subphase. The name "starfish to jellyfish" transition has been suggested for this phenomenon.

3. X-ray Reflectivity Study of Surface Micelles

Surface pressure-area isotherms,TEM and atomic force microscopy (AFM) provide only indirect information about the location of the chains. X-ray and neutron reflectivity experiments, however, can give detailed spatial information. Reflectivity experiments measure the ratio of the intensity of a beam after reflection from a sample (usually a thin film) to its incident intensity as a function of the wavevector, q. For x-rays, the contrast arises from heavy elements (e.g. iodine), whereas neutron contrast is produced by deuterium labeling. For example, a particular section of the molecule (e.g. the alkyl chain) is labeled. The experiment is then sensitive to the position of that part of the polymer (see the original paper [8] and its references for details of the technique).

X-ray reflectivity experiments were performed to investigate the film structure in the PS_{260}-b-$(PVPC_nI)_{240}$ system (n = 1, 4, 8, 10), on the Harvard-Brookhaven National Laboratory liquid diffractometer at the BNL National Synchrotron Light Source. Films were deposited on a Langmuir trough as previously described, and the incident beam was perpendicular to the direction of motion of the barrier on the film balance. The isotherms measured in the beam were consistent with previous laboratory measurements. The pressure in the plateau region of the isotherm was found to remain constant for 24 hours. The stability of the films is important in these experiments because data acquisition times ranged from 1 to 3 hours for specular and off-specular scans, respectively.

3.1. SPECULAR REFLECTIVITY

Specular reflectivity measurements provide information about the film structure and its thickness perpendicular to the subphase. Figure 4 shows the model from which the best fits (solid lines) to the reflectivity data (points) shown in figure 5 are calculated. The first layer consists of PS micelle cores whose dimensions are known from TEM and AFM on LB films. The layer thickness and density, and the roughness of the air-PS interface can be obtained from these measurements. The second layer consists of only $PVPC_nI$. Its thickness, density and roughness were varied until the fit best matched the data.

Figure 4 The model: the dark PS micelles and air in between form the top layer; the layer below is PVP. The bottom layer is water. The air-PS, PS-PVP and PVP-water interfaces include roughness.

As the surface pressure increases, the oscillation in the reflectivity curve becomes more pronounced indicating that the $PVPC_nI$ layer thickens as the film is compressed. The thickness increases by less than a factor of 2.5 over the whole compression range (a 4-5 fold increase in surface density). No discontinuity in the thickness is observed at the plateau onset for either n=8 or n=10 (it increases more rapidly only when the limiting surface areas are approached). The PS layer thickness is independent of n as expected since the same PS block length was used for all the experiments. The thickness of the $PVPC_nI$ layer determined from AFM experiments is consistent with the value from x-rays (≈10Å). The other results of the specular reflectivity measurements may be summarized as follows. Firstly, the density of the $PVPC_nI$ layer is close to that of the bulk homopolymer. Secondly, the $PVPC_nI$ - water interface is relatively sharp. Lastly, the roughness of this interface is always greater for n=8 than for n=4.

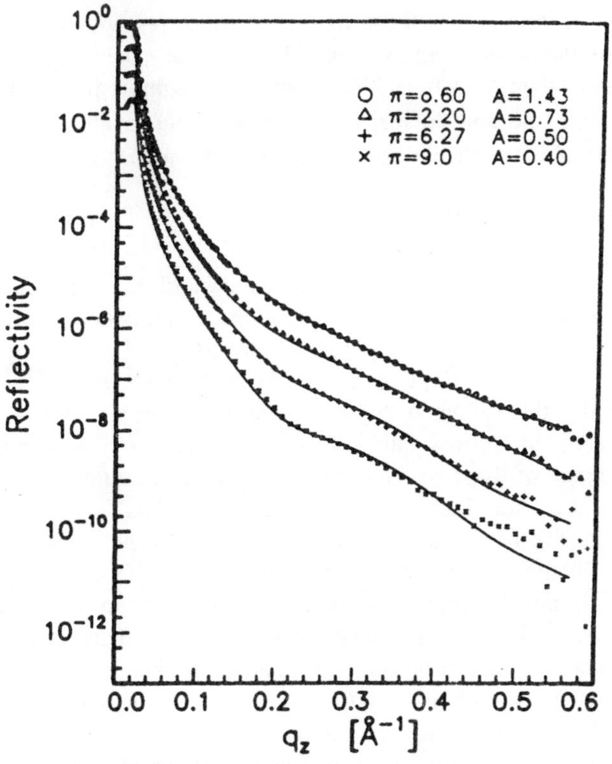

Figure 5 Specular reflectivities (symbols) and best fits (solid lines) for PS_{260}-b-$(PVPC_nI)_{240}$.

We can draw some conclusions about the structure of the film from the above observations. The $PVPC_nI$ must be localized at the air water interface as a dense layer with no significant hydration even for the $n \leq 4$ chains. This is not consistent with the starfish-jellyfish model of the transition in which the chains go from being surface adsorbed to being submerged in the water. In this model, the extent of $PVPC_nI$ submersion at a given area was postulated to depend on the carbon chain length. The longer chains are more hydrophobic and were reasoned to be more effective in tethering the $PVPC_nI$ blocks to the water surface. The x-ray reflectivity results confirm that the chains are entirely surface adsorbed at low pressures, and reveal a gradual submersion with increasing pressure, but the magnitude of the change is small and is not correlated either with the length of the alkyl chain or the existence of a plateau on the surface pressure-area diagram. For example, the reflectivity data for n=1 show that the $PVPC_nI$ has the bulk density and remains as a very thin layer (< 20Å) at the surface even at high pressure. Secondly, the dependence of the roughness of the interface between the $PVPC_nI$ layer and the water on the alkyl chain is consistent with a situation where the tethering alkyl chains protrude from the water surface, with the longer chain producing a rougher surface. In both n=4 and n=8 the roughness increases with increasing surface pressure. This may indicate that some of the side chains are being pushed out of the water into the air, rather than the whole $PVPC_nI$ molecule submerging. Surprisingly the $PVPC_nI$ chains appear to be able to resist osmotic swelling to any significant extent when they exist in this dense, melt-like state at the air-water interface.

3.2. OFF-SPECULAR REFLECTIVITY

For an off-specular reflectivity experiment the angles of incidence and reflection are not equal (unlike specular reflectivity). There is therefore a component of the scattering vector parallel to the surface as well as perpendicular to it and hence structure in the plane of the sample is probed. Experiments were performed with a fixed angle of incidence (0.12°) while the reflected angle was scanned. Peaks in the off-specular reflectivity were observed at approximately 3°. This is indicative of long range order on the surface. The presence of these peaks clearly show the formation of the micelles in the Langmuir *as well as* the Langmuir-Blodgett films. This is proof that micelle formation is not a consequence of the dipping procedure. As the surface pressure was increased from 1 to 36 mN/m, the peak moved to higher angles, corresponding to a decrease in the micelle-micelle spacing from 210nm to 80nm in agreement with the AFM results. The degree of ordering, which can be determined from the peak width, was also found to increase substantially. This establishes that, although the micelles form spontaneously when the polymer solution is spread on the surface, their order is greatly increased by the surface pressure. This is reasonable since the micelles have a distribution of sizes and will therefore adopt different packing arrangements at different surface densities. Also, the contribution of capillary waves decreases at higher pressures as the film becomes stiffer.

4. Conclusions

A review has been presented of the self-assembly of amphiphilic block copolymers at the air-water interface. These systems show many of the same characteristics as ones which form 3D micelles. The inherent driving force (common to block copolymers in which one block is hydrophobic and one hydrophilic) for phase separation and self-assembly is manifest in the formation of micelles. These show multiple morphologies as the relative block size is changed: the starfish, ribbon and pancake phases all have 3D analogues. By appropriate choice of solvents it is possible to prepare nanofoams which are the phase inverted version of the starfish - c.f. complex micelles in 3D. It was also shown that non-ionic species are also capable of self-assembly. X-ray reflectivity experiments of Langmuir films have been performed. Analysis of these data, obtained as a function of film compression, indicate that the block copolymers spontaneously form 2D surface micelles on the water surface which, in turn, order into a micelle superlattice. Once islands of micelles have consolidated, the films are compressible. When the surface density is changed by a factor of 5, the micelle-micelle distance decreases by a factor of about 2.5 while the film thickness increases by a factor of approximately 2. Off-specular data are well fitted by an exact expression for the scattering from a hexagonal array, in excellent agreement with AFM images of the LB films. This confirms that the LB films are representative of the Langmuir films formed at the air-water interface and that film deposition does not greatly perturb the surface micelles. The initial interpretation of the transition as a starfish to jellyfish transition has been modified by the x-ray experiments. It is currently believed that the transition seen as a plateau on the surface pressure - area diagrams is due to compression of the PVP layer. Neutron reflectivity experiments will investigate this further.

References

1. Zhu, J., Eisenberg, A., Lennox, R.B.(1991), Interfacial behavior of block polyelectrolytes. 1. Evidence for novel surface micelle formation. *J. Am. Chem Soc.* **113**, 5583
2. Zhu, J., Lennox, R.B., Eisenberg, A.(1991) Interfacial behavior of block polyelectolytes. 2. Aggregation numbers.*Langmuir* **7**, 1579
3. Zhu, J., Lennox, R.B., Eisenberg, A.(1992) Polymorphism of (quasi) 2-dimensional micelles. *J. Phys. Chem.* **96**, 4727
4. Zhu, J., Eisenberg, A., Lennox, R.B. (1992)), Interfacial behavior of block polyelectrolytes. 5. Effect of varying block length on the properties of surface micelles.*Macromolecules* **25**, 6547
5. Zhu, J., Eisenberg, A., Lennox, R.B. (1992)), Interfacial behavior of block polyelectrolytes. 6. Properties of surface micelles as a function of R and X in P(S260-b-VP240/RX). *Macromolecules* **25**, 6556
6. Li, S., Hanley, S., Khan, I., Eisenberg, A., Lennox, R.B. (1993) Surface micelle formation at the air-water interface from nonionic diblock copolymers.*Langmuir* **9**, 2243
7. Mezaros, M., Lennox, R.B., Eisenberg, A.(1994) Block copolymer self-assembly in two dimensions: nanoscale emulsions and foams *Faraday Discussions* **98**
8. Li, Z., Quinn, J., Rafailovich, M., Sokolov, J., Lennox, R.B., Eisenberg, A., Wu, X., Kim, M.-W., Sinha, S (1995) X-ray reflectivity of diblock copolymer monolayers at the air-water interface. submitted to *Langmuir*

SITE SPECIFIC DRUG-CARRIERS : POLYMERIC MICELLES AS HIGH POTENTIAL VEHICLES FOR BIOLOGICALLY ACTIVE MOLECULES

SANDRINE CAMMAS and KAZUNORI KATAOKA
International Center for Biomaterials Science and Department of Materials Science, Science University of Tokyo, Yamazaki 2641, Noda-shi, Chiba 278, Japan.
Institute for Biomedical Engineering, Tokyo Women's Medical College, Kawada-Cho 8-1, Shinjuku-Ku, Tokyo 162, Japan.

1. Introduction

For centuries, men try to overcome the problems of low efficiency of therapeutic molecules by tremendously increasing the amount of administrated drug. However, since nothing is totally inert with regard to living matter, compromises between side effects, such as toxicity, unwanted disposition and degradation, and therapeutic doses have to be found in order to achieve satisfactorily the treatment of a given disease. Unfortunately, such compromises are not always very easy to obtain mainly due to the excessive toxicity of the drug and/or because of its insufficient solubility [1,2].

To overcome these problems, site-specific drug-carrier conjugates are designed to protect the body from the toxic side effect of a drug as well as to prevent the degradation and the unwanted uptake of the biologically active molecules.

The first proposition of site-specific drug delivery systems was made at the beginning of the 20th century by Paul Ehrlich and is well-known as the "Ehrlich's Magic Bullet" [3]. In Ehrlich's model, the drug is directly bound to the carrier and can ideally exhibit its pharmacology activities only at the targeted site. Since the activity of the drug is only located at the site of action, the toxic side effects may be prevented. Moreover, the efficiency of the drug may be improved thus leading to the possibility of decreasing the administrated doses.

However, even if Ehrlich's concept seems to be very simple and attractive, only few successful examples have been reported with monoclonal antibodies as site-specific carriers [4,5]. The main problems encountered by these kind of drug-carrier conjugates are the antigenicity of the systems, the loss of water solubility, a decrease in targeting efficiency of the carrier, a non-specific capture of the conjugate by mainly the reticuloendothelial system (RES) and also a poor accessibility to the target [6].

For about 30 years, much research has been focused on the use of polymeric materials as site-specific drug-carriers. Indeed, in order to try to overcome the previous

S.E. Webber et al. (eds.), Solvents and Self-Organization of Polymers, 83–113.

cited problems, the drugs have been attached to an antibody via a polymer carrier thus reducing the number of binding points on the antibody. However, problems, such as uptake by the RES and low permeability of the antibody-polymer-drug conjugates from the blood circulation into the cancer cells or tissues, remained [6]. The permeability and thus the tissues accumulation of a conjugate depends on its blood half-life. Indeed, the longer its blood half-life, the higher is its permeability and accumulation [7,8].

Consequently, an important goal is to design systems which can protect the biologically active molecules from degradation and unwanted distribution, decrease the toxic side-effects of the drugs, deliver specifically the chosen drug to the targeted site and also avoid the non-specific uptake.

In 1975, Ringsdorf [9] suggested a model for a polymer-drug conjugate. The drug is attached to the polymer backbone through a spacer. A homing device as well as a solubilizer are also bound to the polymer backbone. As it will be described, several systems were developed based on the Ringsdorf's model. However, the main problem of such systems is the low water solubility of the resulting polymer-drug conjugate. One solution to overcome this problem is to encapsulate the hydrophobic drug into a nano or microscopic vehicles. Indeed, impressive results were obtained with natural carriers such as viruses and lipoproteins [10].

In this chapter, we will first give a short overview on the various site-specific drug delivery systems described in the literature. In a second part, our recent results concerning polymeric micelles as drug carriers will be developed.

2. Site-specific drug delivery : a short overview

2.1 INTRODUCTION

The use of a therapeutic molecule and especially of an anti-cancer drug is rapidly limited by several drawbacks. The first and may be the more important one is the non-selectivity of the drug for the diseased tissues or cells resulting in a high global toxicity. Most of the drugs are low molecular weight compounds and are consequently rapidly excreted from the body. Therefore, large and repeated doses must be administrated in order to maintain a therapeutic action [11]. However, besides the toxic side-effect, this long term treatment with insufficient quantity of drugs at the tumor site leads to the generation of multidrug-resistant tumor cells [12].

To increase the therapeutic index and to protect the normal tissues and cells against the toxic side-effects of the drugs, much effort has been focused on the design of drug-carrier conjugates. Moreover, the possibility to obtain "intelligent" carriers which are able to recognize their objective is of interest.

Since Paul Ehrlich's Magic Bullet model, several systems have been designed and studied for drug transport : macromolecular prodrugs, microcapsules, microspheres, nanoparticles, macromolecular-drug complex, lipophilic microdomain-forming polymeric microphases and polymeric liposomes [1].

To achieve a site-specific delivery of drugs, two kinds of targeting are developed: passive targeting which is based on the physical and/or the physicochemical properties of the carriers and active targeting which results from the introduction of a homing device on the carrier (antibody, sugar residue, etc.) [13].

2.2. MACROMOLECULAR PRODRUGS

As we already mentioned, the first proposition of a polymer-drug conjugate was made by P. Ehrlich who suggested the potential use of antibodies as drug carriers [3].

In 1975, Ringsdorf introduced the idea of tailor-made macromolecular prodrugs [9,14] as shown by scheme 1.

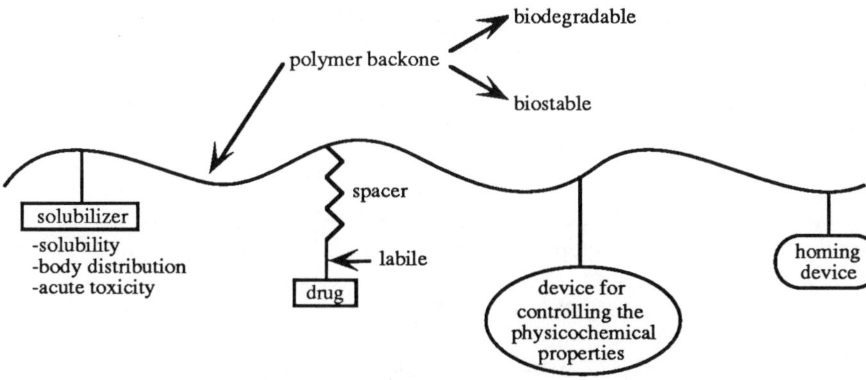

Scheme 1. Ringsdorf's model for pharmacologically active polymer [9,14].

In spite of its simplicity, the Ringsdorf's model was the base of the development of polymeric drug-carriers [7,15].

Kopecek et al. have developed and studied targetable polymeric prodrugs based on N-(2-hydroxypropyl)methacrylamide (HPMA) copolymers [16,17]. The polymeric precursors are copolymers of HPMA containing oligopeptidic side chains with reactive p-nitrophenoxy ester or primary amino groups which allow the binding of therapeutic molecules and/or targeting moieties (scheme 2) [11,18].

The HPMA homopolymer is well-tolerated biologically, non-immunogenic [19] and does not present observable toxicity in vitro and in vivo [20].

Several drugs have been linked covalently to the HPMA copolymers including Doxorubicin (DOX) [21-25], Cyclosporin A (CsA) [26], Daunomycin (DNM) [21,22,25,28], chlorin e_6 [25,29,30]. In addition, targeting moieties such as sugar residues [22,31,32], antibodies [21,26-29,31] or hormones [24] have been bound to the polymeric conjugate.

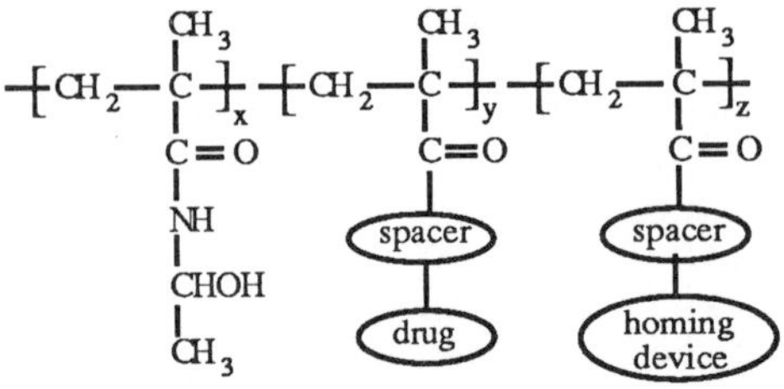

Scheme 2. Structure of targetable polymeric prodrug based on N-(-2-hydroxy propyl)methacrylamide copolymers [11,18].

Targetable polyHPMA copolymers were obtained by the incorporation of antibodies in their backbones : anti-Thy 1-2 antibodies (mice) and anti-CD3 (human). The antibodies were attached to the carriers via a non-specific binding reaction (aminolysis) or via a specific (oriented) binding using an oxidized Fc part of the antibodies [26,29,30]. Studies of the targeting efficiency and of the activity of the drug showed that conjugates containing antibodies bound at a specific location were the most active both in their biological activity and in the targeting efficiency [26,29,30].

Drugs such as DOX, DNM, chlorin e_6 and cyclosporin A were attached to the polymer backbone via non-degradable or biodegradable peptide spacers. Biological studies of the conjugate polyHPMA/DNM showed that a biodegradable spacer (GlyPheLeuGly) has to be used in order to observe good activity for the drug [11,28]. For all the polyHPMA-drug-antibody conjugates, the binding of the drug on the polymer backbone decreases its toxicity and improves the efficiency of the drug. In vitro and in vivo studies of the biological activity and targeting efficiency of these conjugates showed that the activity of the drug is localized at the targeted site [21,26-30]. These results indicate a real possibility of application of the drug-polymer-antibody conjugate in site-specific therapy [11].

Recently, Duncan et al. have studied the possibility of using melanocyte-stimuling hormone (MSH) as a targeting moiety towards malignant melanoma [24]. Doxorubicin (about 8% by weight) was bound to polyHPMA copolymer via a biodegradable peptide spacer (GlyPheLeuGly). The MSH was also linked to the conjugate. In vitro and in vivo experiments using B16F10 cells and mice bearing intraperitoneal B16F10 tumors, respectively, showed that the NSH-conjugate is more active than the free drug [24].

Ulbrich et al. have shown that polyHMPA copolymers containing Doxorubicin and galactosamin can be targeted to the hepatocyte galactose receptor for liver cancer [22]. Moreover, the injection (intraperitonal and subcutaneous) of polyHPMA copolymers bearing a high concentration of galactose residues resulted in a significantly high accumulation in local fatty tissue [31]. These conjugates may be useful for organ-specific chemotherapy of primary and metastic liver cancer as well as for drug-delivery targeted to fat cells [11,21,31].

Recently, preclinical evaluation of polyHMPA-Doxorubicin-galactosamine conjugates was undertaken by Duncan et al. [32]. They showed that this conjugate can be useful for the treatment of liver-associated diseases. However, the reduced number or absence of galactose receptor in some patients may limit the selective targeting to hepatoma of this conjugate [32].

Poly(α-amino acids) have been studied by several authors as possible carriers for anti-tumor drugs. Most of these polymers are water-soluble and, in principle, biodegradable [15]. The rate of the in vivo biodegradation of synthetic poly(α-amino acids) can be controlled by varying the hydrophilicity of the side-chain groups [15,33]. Moreover, the hydrophilic/hydrophobic balance can be adjusted by using appropriate copolymers [15].

For example, Feijen et al. have described the coupling of steroids to a biodegradable poly(hydroxyalkyl-L-glutamine) [33]. Their results show that the controlled delivery of hormone from these polymers by chemical or enzymatic hydrolysis is a promising concept [34]. DOX conjugated to poly(α-L-glutamic acid) was shown to be less toxic than the free DOX [35].

The presence of spacers between the polymer backbone and DOX has an important influence on the activity of the anti-tumor drug [34,36-39].
In vitro and in vivo studies using DOX bound to poly(α-L-glutamic acid) and to poly(N^5-(-2-hydroxyethyl)-L-glutamine) via a cleavable spacer (GlyGlyGlyLeu) have shown an increase in the efficiency of DOX with a decrease in its toxicity [37,38]. However, the use of the poly(α-L-glutamic acid) conjugates is limited because of their RES uptake and because of possible immunogenic reactions. On the other hand, the poly(N^5-(-2-hydroxyethyl)-L-glutamine) conjugates show a slow clearance and a low RES uptake indicating the possibility to use these carriers for anti-tumor drug delivery.
Immunoconjugates of IgM 16.88 and PHEG-DOX have been prepared and biodistribution in mice was studied [37,40]. These immunoconjugates seem to be promising as carriers for the release of drugs in the circulation due to the slow clearance and the low RES uptake of PHEG conjugates [37].

Vert et al. have developed a biodegradable poly(β-hydroxy acid) as potential polyvalent polymeric drug carrier : the poly(β-malic acid) [1].

This polyester presents several advantages for its use in the drug delivery system field : non-toxicity and very low LD_{50}, polymer backbone degradable by hydrolysis of the ester bond, malic acid, a metabolite of the Krebs cycle, as expected final degradation by-product, presence of carboxylic lateral groups for the solubility in water as well as for the drug and/or homing device binding, possibility to modify the carboxylic groups to ester or amide ones turning the PMLA to a hydrophobic polymer, presence of chiral centers in the main chain allowing the adjustment and modification of the physical, physicochemical and conformational properties.

In view of these properties, the poly(β-malic acid) seems to be a good candidate for site-specific delivery of drugs [1].

In the drug delivery system field, an interesting polymer is the poly(N-isopropyl acrylamide) -poly(IPAAm)- which is known to be a thermoreversible water soluble polymer (low critical solution temperature, LCST : 33°C) [41,42].

Recently, Hoffman et al. and Okano et al. investigated the formation of poly(IPAAm)-enzyme conjugates and the properties, especially the enzyme activity, of these conjugates. They showed that the enzyme retains a high percent of its activity and also shows thermal stability over the native enzyme [41,43].

The possibility to target biologically active compounds using this thermosensitive polymer has been also studied by Valuev et al. They concluded that the enzymes immobilized on copolymers having a LCST in aqueous solutions could be used as site specific delivery system because the higher temperature at the targeted site will precipitate the enzyme bound copolymer [44].

All these studies show that poly(IPAAm) and poly(IPAAm) derivatives have a promising future as site-specific drug delivery systems since inflamed areas are characterized by local hyperthemia.

Several natural molecules have been proposed as carriers for biologically active compounds [15].

Dextrans have a well-defined structure, are biologically inert and biodegradable. Moreover, they are available in several different molecular weights with a narrow polydispersity and possess many hydroxy groups allowing chemical modification [15]. Takakura et al. studied drug-dextran conjugates as potential site-specific drug delivery systems [45-50].

Mitomycin C (MMC) was bound to several dextrans having different molecular weights (10,000, 70,000 and 500,000) [51].

In vivo studies of the MMC-dextran conjugates (MMC-D) showed that the activity of MMC towards several tumors was increased and that the size of the dextran influences the efficiency of the conjugates [52]. After intramuscular (i.m.) administration, the MMC-dextran conjugates were retained at the injection site for long period in comparison with the free MMC. The retention time was shown to be increased with an increase of the molecular weight of the dextran. Moreover, the MMC-D conjugates exhibited remarkable accumulation in the regional lymph nodes after i.m. injection for at least 48 hours. These conjugates also suppressed the lymph node metastases introduced by subcutaneous inoculation of L1210 leukemia cells. From these data, Takakura et al. suggested to use MMC-D conjugates in the treatment of lymphatic metastasis of cancer [46].

The tissue distribution of Uricase conjugated with charged dextrans and polyethylene glycol showed that the pharmacokinetic behavior of Uricase can be controlled by conjugation with dextran derivatives [47].

Superoxide dismutase (SOD) was conjugated with several kinds of dextrans ; the SOD derivatives retained 50 to 80% of the original enzymatic activity and were stable during incubation in mouse serum. SOD-dextran conjugates exhibited a long plasma half-life. On the other hand, conjugation with sugar residues leads to SOD derivatives which are taken up by parenchymal and non-parenchymal cells of the liver. Takakura et al. concluded that the in vivo fate of SOD can be controlled by utilizing sugar residues and dextrans [48].

In the following step, these authors modified the carboxymethyl-dextran (CMD) with sugar residues (galactose and mannose) in order to obtain a new carrier for hepatic targeting. The in vivo distribution study of the cytosine β-D-arabinose-CMD-galactose conjugate showed that 80% of the injected dose was selectively delivered to hepatocytes [49]. The same distribution was obtained with several other proteins (bovine γ-globulin, SOD, etc.) modified by the cited sugars [46,53]. On the other hand, mannosylated SOD was shown to be taken up by macrophages in vitro. Takakura et al. proposed a potential use of the mannosilated SOD for inflammatory diseases mediated by superoxide anions generated by macrophages [54].

Succinylated proteins (SOD, BSA, etc.) were shown to be preferentially localized in non-parenchymal cells of the liver suggesting that protein-drug conjugates can be targeted to liver parenchymal cells by direct succinylation through scavenger receptor in vivo [50,54].

Moreover, the conjugation of proteins with sugar residues and polyethylene glycol increases the circulation half-time of the conjugate and allows the hepatic uptake of protein drugs [53,55].

The use of biologically active proteins is sometimes limited because of their immunological characteristics (antigenicity, immunogenicity), their pharmaceutical (stability, solubility, aggregation) and biopharmaceutical (absorption, distribution, metabolism) properties [53].

The chemical modification of proteins with various substrates was studied in order to improve their pharmacokinetic behavior as well as to deliver them specifically to a target site. For example, bovine serum albumin (BSA) was conjugated with galactose or glucose residues. In vivo distribution experiments showed that glycosylated BSA can be utilized to target drugs selectively to liver parenchymal (galactosylated BSA) or non-parenchymal (mannosylated) cells [56]. Finally, Takakura et al. concluded that a polymer-drug conjugate designed for tumor targeting should be polyanionic and should have a molecular weight greater than 70 000 [46].

Low et al. have shown the possibility of targeting in vitro proteins and cytotoxic proteins conjugated with folic acid towards tumor cells bearing folate receptors [57-60]. Moreover, the folate-drug conjugates permitted the selective destruction of tumor cells without modifying the behavior of normal cells in the same culture [57-60]. In view of these results, folate-proteins conjugates can be a useful site-specific carrier towards malignant cells under physiological conditions. However, in vivo studies demonstrating the efficiency of these conjugates have to be done.

All these examples of macromolecular prodrugs give evidence of the usefulness of site-specific polymeric drug-carriers in the treatment of diseases such as cancers. The promising results obtained with some systems showed that the Ehlrich's dream might be achieved. Despite all improvements of this first model, the use of polymeric prodrugs is limited by at least two factors : a rapid urinary clearance of too low molecular weight conjugates and a low capacity for drug binding due to the rapid decrease of water solubility of the conjugate. In order to try to overcome to the previously cited problems, the use of particulate drug-carriers was developed. The drug can be encapsulated inside the core of such carriers and their surface can be modified in order to decrease non-specific interactions with the living milieu and/or to target the carrier.

2.3. PARTICULATE DRUG-CARRIERS

Photoradiative therapy (PRT) utilizing hematoporphyrin derivatives (HPD) and lasers represents a novel treatment method for certain forms of cancer [61]. However, because of the development of cutaneous photosensitization to ambient light in some patients, this technique is limited and the use of some targeting methodology appears important. Thus, the photosensitizer (HPD) was bound to a magnetically responsive matrix (Dynabeads). The magnetic sensitive conjugate can be directed to the target site using external magnetic fields or by insertion of magnetic probe into the target site. In vitro studies showed the high potential of this method. However, the in vivo application presents some problems especially for "deep" organs or sites [62].

Crosslinked poly(N-(2-hydroxypropyl)methacrylamide) are able to form a novel kind of biodegradable hydrogel system [63]. The high potential of poly(HPMA)-drug conjugates was already described in the previous section. The new poly(HPMA) hydrogels are suitable for controlled release of low and high molecular weight drugs or drug conjugates. Moreover, the incorporation of sugar residues (galactose, for example) can allow the targeting of the gel towards the liver [63]. Duncan et al. have shown that the degradation rate depends on the pH of the microenvironment and can be controlled by the hydrogel structure. Moreover, the release rate of the drug is controlled by diffusion, gel degradation or by both processes [63].

The biodistribution of intravenously administrated particles depends to a large extent on their physicochemical properties (nature of the surface, size, etc.) which influences the interactions of these particles with the biological environment [64].

Illum et al. have shown that it is possible to change the biodistribution of particles by modifying their surface [65]. Hydrophobic small particles are coated by plasma components in the bloodstream and are rapidly uptaken by macrophages within the liver. On the other hand, if the surface of the particles is hydrophilic, the circulation time in the bloodstream of these particles is longer [65]. As an example, polystyrene particles with different sizes were coated with a derivative of poly(ethylene glycol), poloxamer 407. The small coated particles (diameter $\leq$ 150 nm) were taken up by the bone narrow (in vivo experiment using rabbits) [65]. Aiming on biodegradable particles, Illum et al. developed the preparation of sub-200 nm microspheres composed by poly(lactide-co-glycolide) copolymers : small microspheres with a diameter of 90 nm can be obtained [64]. However, the behavior of [^{14}C]poly(DL-lactide-co-glycolide) microspheres of 130 nm in diameter after intravenous injection to mice showed that these particles were mainly captured by the RES organs such as liver and spleen [66].

Mehta et al. studied the release of Salmon calcitonin from PLA/PGA microspheres : the release rate depends on the preparation method of the microspheres and on polymer composition, molecular weight and matrix structure [67].

Nanoparticles are solid colloidal particles with a size ranging from 10 nm to 1000 nm (1 μm). They are made of artificial or natural polymers in which the biologically active molecule can be entrapped, dissolved or encapsulated [68]. Different techniques are used to prepare these nanoparticles : emulsion polymerization, polymerization in a continuous aqueous or organic phase, interfacial polymerization, solvent deposition, solvent evaporation, etc. [68]. The size of the nanoparticles is one of the more important parameter which influences the properties and especially the body distribution of the nanoparticles. However, surface characteristics (charge, hydrophobicity, hydrophilicity) will also influence the interactions with the biological environment and thus the resulting body distribution. The release and degradation properties of the nanoparticles are influenced by the density, molecular weight and crystallinity of the polymers constituting the nanoparticles [68].

Lee et al. [69] have described the preparation of pH and thermal-sensitive nanospheres. These nanoparticles are made up by poly(N-isopropylacrylamide-co-methacrylic acid) -poly(NIPAAm/MAA)- copolymers. The poly(IPAAm) blocks are thermosensitive and the poly(MAA) segments are pH sensitive. The size of the particles thus obtained ranges from 114 to 413 nm depending on the polymer composition and temperature. Theophylline was loaded inside the nanoparticles and the loading capacity was shown to depend on the temperature and the drug concentration [69]. Lee et al. did not report any data on either the targeting capacity of their particles or their distribution in vivo.

Magnetically responsive nanoparticles were prepared by the incorporation of magnetic Fe_3O_4 particles (10-20 nm in diameter) into different types of nanoparticles (albumin, polyisobutyl cyanoacrylate, chitosan) [68]. By placing a magnet at the selected site, these magnetically responsive nanoparticles can be targeted and drugs bound to the nanoparticles can be released at the target site. Higher doses of drugs and lower concentration in other part of the body were observed in comparison to the free drug distribution after intravenous administration [68]. However, even if this technique seems to be powerful, its application for human is limited due to the fact that it will be not possible to focus the magnetic field on a restricted area in the interior of the body . In addition, undetected metastases cannot be treated by this method.

Another solution to target nanoparticles is to couple monoclonal antibodies to their surfaces. Several methods were used for the coupling of monoclonal antibodies or normal IgG [68]. However, the binding of antibodies to nanoparticles induced the increase of liver and spleen uptake indicating that the immunogenicity of these nanoparticles may be enhanced [68].

Illum et al. [70] have stabilized polystyrene nanoparticles with a series of synthetic polyoxyethylene(PEO)/polyoxypropylene(POP) block copolymers. These authors have shown that it is possible to control the rate of drainage from the subcutaneous injection site and to manipulate the lymphatic distribution. They found a correlation between the length of the PEO chains of the block copolymer and nanoparticles drainage and passageway across tissue lymph interface in dermal lymphatic capillaries in the rat foodpads. Illum et al. suggested the use of such stabilized nanoparticles for vaccine design, radiation therapy for ablation of metastatic diseases and site-specific delivery of biologically active molecules for the treatment of various lymph node diseases [70].

Couvreur et al. [71,72] have developed biodegradable polyalkylcyanoacrylate nanoparticles as drug carriers. These nanoparticles were synthesized by emulsion polymerization and their diameter was found to be around 150 nm [72]. The drug can be either loaded in the nanoparticles during the polymerization process or adsorbed onto the surface of the formed nanoparticles. The degradation rate of polyalkylcyanoacrylate nanoparticles can be controlled by the nature of the alkyl chain. Moreover, the drug release from these nanoparticles was shown to be a direct result of the polymer erosion [72]. Thus, it seems possible to control the release rate of the drug from the nanoparticles. However, polyalkylcyanoacrylate nanoparticles are rapidly uptaken by the RES ; according to the opinion of these authors, this rapid uptake can represent an opportunity for delivering drugs to certain tissue or cellular sites of the RES and can be defined as passive targeting [72].

The possibility of targeting nanoparticles (polyhexylcyanoacrylate -PHCA- and polymethylmethacrylate -PMMA-) towards the lymphatic system after intraperitoneal administration was studied [73]. The nanoparticles used have a diameter larger than 0.5 μm. One of the nanoparticles properties influencing the lymphatic uptake is the hydrophobicity : an increase of hydrophobicity resulted in an increase of lymphatic uptake [73]. These systems can be efficient for the treatment of cancers with dissemination of metastases in the abdominal cavity and in the lymph nodes [73]. However, the fate of the drugs bound to the nanoparticles after intraperitoneal administration has to be studied.

Couvreur et al. demonstrated, in vitro, that both the uptake and retention by macrophages of ampicillin loaded into polyhexylcyanoacrylate nanoparticles are higher than in the case of the free antibiotic [74,75]. These kind of antibiotic loaded nanoparticles can be useful to kill bacteria which are able to resist intracellular killing by macrophages and to survive and multiply within the cells of the RES [74,75]. In vitro studies showed an intracellular distribution of ampicillin-nanoparticles in vesicular compartments of macrophages [75]. However, no difference was observed between the ampicillin-loaded nanoparticles and the free antibiotic in terms of antibacterial activity against intracellular *S. typhimurin* [74,75].

A problem in cancer therapy is the appearance of multi drug resistant (MDR) cells [76,77]. A strategy to overcome the failure of chemotherapy in case of MDR is to use colloidal particulate carriers for delivering the drugs to the MDR cells. Couvreur et al. studied the effect of the encapsulation of DOX inside polyisohexylcyanoacrylate nanospheres on MDR tumor cells in culture [76,77]. Their in vitro studies with different MDR tumor cells (P388 cells [76] and rat glioblasma cells [77]) showed that encapsulation of DOX inside the nanoparticles provides an useful tool to overcome the MDR of tumor cells.

Liposomes represent a family of particulate drug-carriers which are widely studied in the drug delivery systems field.

Liposomes are made up by phospholipids having either one (unilamellar) or several lipid bilayers (multilamellar) surrounding internal aqueous space(s). Water soluble and hydrophobic drugs can be solubilized in the aqueous interior and in the lipid bilayer(s) of the liposomes, respectively [78,79]. The size, number, position of lamellae and bilayer rigidity of the liposomes can vary widely in function of the selected lipids, the technique and conditions of preparation [79]. The behavior of liposomes in vitro and in vivo depends on the previously cited parameters [79].

The surface of liposomes can be modified in order to increase the stability of the carrier (Stealth liposomes) [80 and references cited therein] and/or to obtain site-specific drug carriers (binding of antibodies on the surface, for example) [79,80].

In 1984, Tirell et al. described the preparation of pH-sensitive liposomes [81]. They showed that the complexation of poly(acrylic acid) derivatives with the membrane of phospholipid vehicles rendered these vehicles becoming sensitive to pH. Moreover, the complexation depended on the pH of the medium and may be controlled by the chemical structure and tacticity of the polymer [81-83]. According to these authors, besides the use of these pH-sensitive vehicles for delivering foreign proteins to the cellular cytoplasmic compartment, pH-sensitive phospholipids vehicles can be used as "microsensors" to determine the kinetics of acidification in the cellular endocytic compartment [82].

Besides their work on polyalkylcyanoacrylate nanoparticles, Couvreur et al. also described site-specific targeting of biologically active molecules (β-lactam, aminoglycosides) entrapped inside liposomes [71]. They concluded that liposomes are more efficient than nanoparticles to deliver β-lactam to the spleens macrophages, whereas the nanoparticles were found to be more efficient than liposomes to deliver this drug to the livers macrophages [71].

Couvreur et al also studied the in vitro efficiency against Friend Retrovirus of oligonucleotides encapsulated inside pH-sensitive liposomes [84,85] which are destabilized at low pH. These liposomes can release their content in the cytoplasm before reaching lysosomes [84]. They showed that the protection of oligonucleotides against degradation and their enhanced intracellular delivery was possible using pH-sensitive liposomes [84]. Moreover, the cellular uptake of such liposomes in fibroblasts was shown to be stimulated by the presence of virions [85]. According to these authors, the fact that retroviral infections stimulate the cellular uptake of oligonucleotides encapsulated inside pH-sensitive liposomes gives new strategies for specific antiviral action [84,85].

Another method to target liposomes is to prepare temperature-sensitive liposomes which can release their content at a temperature slightly higher than their liquid-crystalline phase transition temperature (Tm). It is thus important to prepare thermosensitive liposomes having their Tm a few degrees above the physiological temperature.

Weinstein et al. have given the in vitro properties of temperature-sensitive liposomes and demonstrated the in vivo selective delivery of MTX to tumors [86]. However, they found that heating a tumor does not significantly affected the uptake of liposomes by areas of the reticuloendothelial functions [86]. Another drawback of such thermosensitive liposomes is they cannot be used to treat metastic disease. However, these kind of liposomes can be also useful for the treatment of deep tumor site because selective hyperthermia of deep structure can be realized by using microwaves, radiowave, etc. [86].

Recently, Okada et al. described the preparation of thermosensitive liposomes having a hyper-osmotic internal aqueous phase to release macromolecules. They showed that these liposomes release macromolecules temperature-dependently and efficiently [87]. However, this study is limited to an in vitro evaluation.

Vigneron et al. have studied DOX-loaded thermosensitive liposomes. These liposomes were reported to potentiate the in vitro antitumor activity of DOX when used in combination with 43°C hyperthermia [88]. However, the combination of DOX-loaded thermosensitive liposomes and hyperthermia does not reversed MDR ; but the possibility of obtaining additive cytotoxicity using these conjugates and hyperthermia may, according to the opinion of Vigneron et al., represent an alternative way of increasing toxicity of DOX with the circumvention of MDR, without using MDR reversing agents which have often very toxic side-effects [88].

Low et al. developed a new kind of stabilized site-specific liposomes. They attached the cell-specific ligand to the distal end of few lipid-conjugated PEG molecules on the liposomes surface (66 nm in diameter). The cell-specific ligand used by these authors is a low molecular weight ligand : folic acid. Receptors for folic acid are expressed on certain cancer cells [89]. They studied the possibility of targeting in vitro such systems to KB cells expressing folate receptors. They demonstrated that folate-conjugated liposomes having a 250 Å PEG spacer are uptake by KB cells bearing folate receptors. Following the binding, cells associated folate-PEG-liposomes were internalized by folate-receptor-mediated endocytosis at 37°C. Low et al. concluded that folate-PEG-liposomes may be used to deliver large quantity of low molecular weight compound non-destructively into folate-receptor bearing cells [89].

Low et al. applied the concept of folate-PEG-liposomes to the selective delivery of drugs such as antisene oligodeoxyribonucleodite [90] and DOX [91]. Both studies were accomplished in vitro and showed that the uptake of folate-PEG-liposomal drugs by KB cells was not significantly higher than that of non-targeted liposomes [91]. Testing and optimization of the system in vivo will be the next step in the development of this new technology.

Another strategy was used by Soria et al. in order to deliver specifically liposomes in vivo [92]. They used an indirect targeting method. Indeed, they prepared biotinaled liposomes and showed that it was possible to target these liposomes to avidin or to spectravicin coated cells. Monosiaganglioside-liposomes containing small amounts of N-bioyinbutyl dipalmitoyl phosphatidyl-ethanolamine were shown to have a long circulation life-time in the blood by biodistribution studies in C26 tumor bearing mice [92]. Moreover, these conjugates were detected within the extra-vascular spaces in tumors using a fluorescence microscopic technique [92].

Recently, starburst dendrimers were proposed as possible drug carriers. These dendrimers are spherical polymers that possess known molecular weights, dimensions and number of terminal functional groups depending on the generation of the dendrimers [93]. Several kinds of dendrimers can be prepared in function of the polymers and the method used for their synthesis ; moreover, the end groups can be changed [93,94].

Tomalia et al. reported the possibility to bind synthetic porphyrins on dendrimers for antibody radiolabeling [93]. They observed 100% of the porphyrin bound to the heavy chain of the antibody when dendrimers were used. In contrast, only 69% of porphyrin was attached to the heavy chain by using the random-labeling technique. Moreover, the dendrimer-conjugates were found to contain 90% of the immunoreactivity of the unmodified antibody compared with only 82% in the case of the random-labeling method. In the radiolabeling experiments with copper-67, Tomalia et al. showed that 80% of copper-67 is bound by using the dendrimer-conjugates, whereas, the conjugates obtained by the random-labeling method show an average of 55-60% of radiolabeling yield [93].

Peppas et al. studied the dendrimers and star molecules in order to use them for drug delivery and pharmaceutical applications [95,96].

Barth et al. evaluated boronated starburst dendrimer-monoclonal antibody conjugates as delivery systems for neutron capture therapy [97]. They used starburst dendrimers made up by poly(amidoamino) chains and having reactive amino terminal groups. These dendrimers were boronated and monoclonal antibodies were introduced. The conjugates retained a high level of immunogenicity in vitro but were taken up by the RES (liver and spleen). Barth et al. showed that this uptake is directly related to the molecular weight and number of terminal amino groups of the macromolecule. Studies are going on to determine if the properties of these boronated dendrimer-antibody conjugates can be adjusted in order to decrease the uptake by the liver and spleen [97].

Jansen et al. reported the synthesis of a dendritic box based on the concept of the construction of a chiral shell of protected amino acids onto poly(propyleneimine) dendrimers with amino end groups [98]. The dendritic box is defined by these authors as an internal cavity available for guest molecules. They showed that there is a minimal generation for dendrimers under which the guest molecule can be removed by extraction because of the insufficiently dense shell. Different dye molecules were encapsulated inside the dendritic box [98]. These results show that it might be possible to encapsulate other molecules such as drugs inside the dendritic box. However, studies on the biocompatibility and in vivo behavior of such conjugates have to be carried out.

2.4. CONCLUSION

Since Ehrlich Magic Bullet Model much progress has been made in the design of specific drug carriers. Some of the proposed systems seem to have a promising future in the specific drug delivery field. However, each kind of carrier presents some drawbacks : rapid renal excretion for water-soluble carriers too low in molecular weight , loss of the water solubility after conjugation with hydrophobic drugs, recognition by the RES, too low stability in the physiological fluid.

As we will see in the second part of this chapter, amphiphilic block copolymers are able to form stable polymeric micelles in aqueous medium. One of the advantages of these polymeric micelles is that hydrophobic drugs can be solubilized in their inner core, the hydrophobic core being surrounded by a corona of hydrophilic chains thus preventing the precipitation of the conjugates.

3. Polymeric micelles : high potential drug carriers

3.1. INTRODUCTION

If hydrophobic drugs are introduced on one block of an AB type block copolymer thus creating an amphiphilic drug-block copolymer conjugate, this conjugate is able to form a stable micellar structure in aqueous medium : the hydrophobic inner core contains the drugs and is surrounded by a corona of hydrophilic chains (scheme 3).

Because of their appropriate diameter, the polymeric micelles have a long-term circulation in the blood as renal filtration and RES uptake are avoided ; due to their particular structure, block copolymeric micelles show a high thermodynamic stability (very low critical micellar concentration - cmc) which allows their use in very diluted conditions (body fluid, for example) and for long period of time ; moreover, since the micelles are formed by inter-molecular non-covalent interactions of single polymer chains (unimers), they can be dissociated with very low rate into unimers which are excreted by the renal route so far as the molecular weights of the unimers are designed to be lower than the critical values for renal filtration [99].

In 1984, Ringsdorf et al. reported the idea of using polymeric micelles for the sustained release of drugs [100]. Their micelles were based on block copolymers of poly(ethylene oxide) and poly(L-lysine) which was partially modified with long alkyl chains to increase hydrophobicity and on which an anti-cancer drug was conjugated. Data on dye solubilization suggested the formation of polymeric micelles and retardation in the release of a bound drug showed the advantageous characteristics for drug formulations [101].

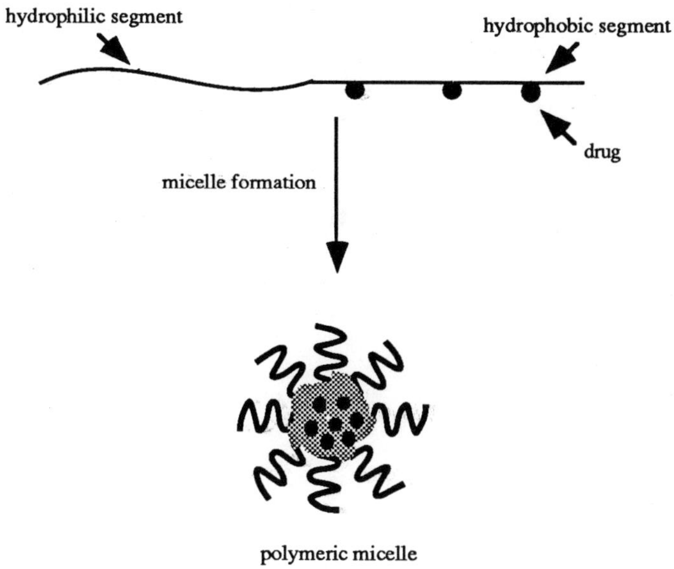

Scheme 3. Representation of micelles-forming block copolymers.

3.2. PEO/P[ASP(DOX)] POLYMERIC MICELLES

In 1987, we first reported the synthesis of a new carrier based on poly(ethylene glycol)-co-poly(aspartic acid) -PEG-P(Aps)- block copolymer [102]. The P(Asp) block is able to retain a large amount of anti-cancer drug and is known to be biodegradable. The PEG block is expected to contribute to the water-solubility as well as to the stability under physiological conditions and to decrease the antigenicity of immunoglobulin-block copolymer-anti-cancer drug conjugate. This AB type block copolymer was prepared to be used as a water-soluble drug carrier.

However, we have found that DOX-conjugated PEG-P(Asp) block copolymer can form stable micelle in a similar approach to the one of Ringsdorf previously cited, leading our group to develop a micelle-forming polymer-drug conjugate starting from this PEG-P(Asp) block copolymer [103-109].

The AB block copolymers of PEG and poly(β-benzyl-L-aspartate) -PBLA- were synthesized by ring opening polymerization of the β-benzyl-L-aspartate-N-carboxy anhydride (BLA-NCA) using α-methoxy-ω-amino PEG as initiator. The block lengths could be easily adjusted by the choice of the PEG molecular weights and the quantity of BLA-NCA. The PEG-PBLA block copolymers were debenzylated under alkaline conditions to form the PEG-P(Asp) block copolymers without detectable chain cleavage. The anti-cancer drug Doxorubicin (DOX) was coupled to the pendent carboxylic groups using 1-ethyl-3-(3-dimethylaminopropyl) carbodiimide (scheme 4) [102,103].

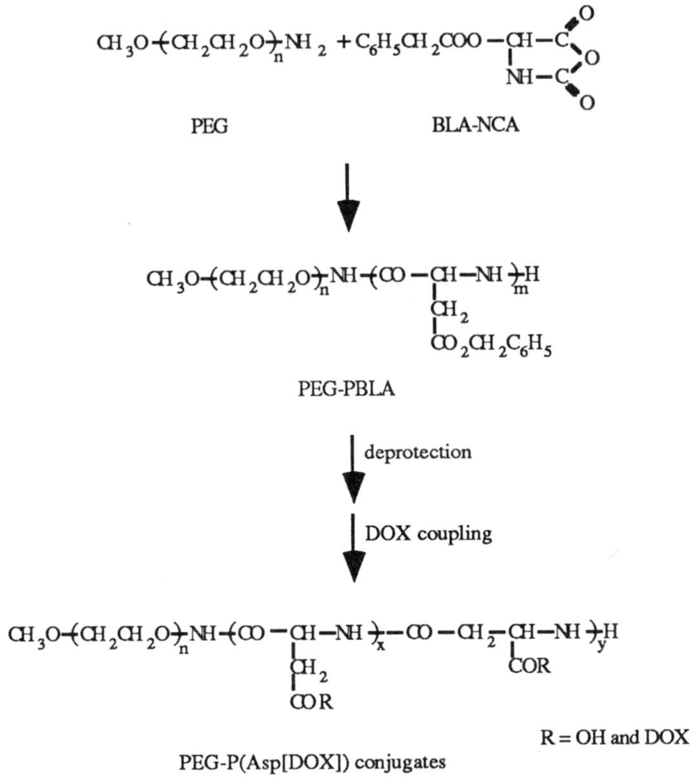

Scheme 4. Synthesis of the PEG-P(Asp[DOX]) conjugates.

The PEG-P(Asp[DOX]) conjugates can form stable micellar structures in aqueous medium because of their amphiphilic character. The functions which are required for drug carriers can be shared by the structurally separated segments of the block copolymers. Indeed, the outer shell made up by the PEG chains is responsible for interactions with biocomponents such as proteins and cells. These interactions determine the pharmacokinetic behavior and the biodistribution of drugs. On the other hand, the inner core made up by DOX-conjugated P(Asp) plays roles in drug binding and release [105,106].

PEG-P(Asp[DOX]) conjugates having different compositions were synthesized by varying the length of both blocks as well as the percent of bound drug (table 1) [104-107,110]. Micelle formation was suggested by fluorescence studies [104,110], by dynamic light scattering (DLS) measurements [99,104-107,110] and by gel permeation chromatography (GPC) [99,107,110,111]. The conjugates are named as followed : 1-40(12) where the first number represents the molecular weight of the PEG chains (here, 1000), the second one the number of aspartic acid units (here, 40 units) and the last one the percent molar substitution ratio of DOX with respect to Asp. residues of PEG-P(Asp) (here, 12%).

The fluorescence of DOX bound to the PEG-P(Asp) block copolymers was found to be strongly quenched in phosphate buffered saline (PBSa) compared to the fluorescence of free DOX in the same solvent. This result suggested intermolecular associations of DOX within the micellar core.

As shown by table 1, DLS measurements give access to the hydrodynamic diameter of the polymeric micelles which is in the 15 to 60 nm range. No micelle formation was observed for the two first composition probably due to short chain lengths of the PEO segments. Unimodal diameter distribution was obtained for the conjugates named 5-20, 12-20 and 12-40, while the other compositions brought about bimodal distribution (table 1).

GPC is a suitable tool to study the micelle properties which depend on the composition. For all the PEG-P(Asp[DOX]) samples, a peak at the gel-exclusion volume (4.2 ml), which corresponds to a molecular weight of over 300 000 on the Pullman standard, was observed. However, the elution profile was found to be dependent on the composition. Indeed, samples with short PEG segments and relatively long P(Asp) blocks have a low tendency to form multimolecular micellar structure and are predominantly as free polymer chains as shown by the presence of a second peak at 6.9 ml or by a very broad peak at 4.2 ml.

TABLE 1. Characterization of the PEG-P(Asp[DOX]) conjugates synthesized.

Run	MW of PEG	Degree of polymerization of Asp. units	Substitution ratio (%)[a]	Diameter (nm)[b]	GPC[d]
1-40	1000	40	12	nd[c]	bimodal
2-10	2000	10	44	nd[c]	bimodal
2-40	2000	40	30	21,123	bimodal
5-10	5000	10	73	24,131	unimodal (broad peak)
5-20	5000	20	30	30	unimodal
5-80	5000	80	46	36,118	unimodal
12-20	12000	20	104	40	unimodal
12-40	12000	40	78	58	unimodal
12-80	12000	80	84	14,91	nm[e]

a) Percent molar substitution of DOX with respect to Asp. residues of PEG-P(Asp). b) Measured by DLS in PBSa, [DOX]=20μg/ml. c) Not detectable, diameter not detectable below 10 nm. d) GPC was measured usingAsahipack CS-520H, eluent : 0.1M PBSa pH 7.4, flow rate : 1 ml/min, temperature = 40°C, detection: UV absorption at 485 nm. e) not measured.

The release of unimers can be followed by GPC [111]. There was a very slow dissociation of unimers from the micelles on a day-scale, indicating a very slow relaxation process of the polymeric micelle structure (figure 1). The dissociation rate was found to be dependent on the composition of the conjugates ; indeed, for the conjugate 12-20(53), no detectable dissociation took place. These results demonstrate the feasibility of controlling the stability of the micelles by varying the composition of the conjugate.

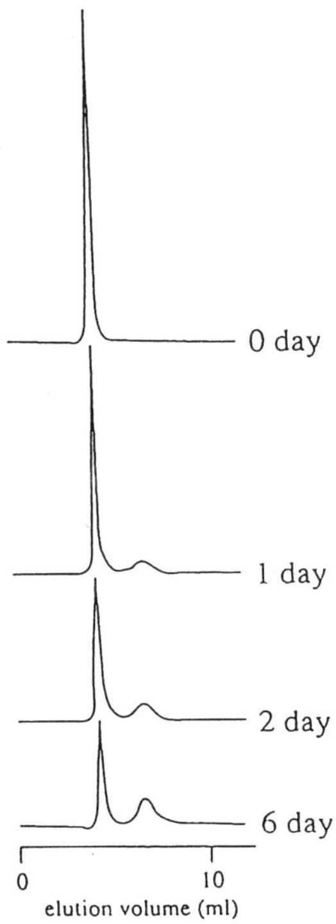

Figure 1. Release of unimers from fractionated micelles [111].

In vivo anti-cancer activity of micelle-forming PEG-P(Asp[DOX]) conjugates was first described for P388 mouse leukemia after intraperitoneal (i.p.) injection[104,105]. The conjugates 4-17(30) and 5-80(37) showed a very high in vivo anti-cancer activity judged from the value of the maximum median life-span over controls (T/C% - figure 2) with a smaller body weight loss than that of free DOX.

The micelle-forming polymeric drug PEG-P(Asp[DOX]) (conjugate 4-19(30)) was observed to express superior in vivo antitumor activity to free DOX against several murine and human solid tumor (C26,C38, M5076 and MX-1) and almost the same activity against MKN-45 cells, all these tumors were advanced stages of development and the conjugate was administrated by intravenous (i.v.) injection [106,108,110]. A complete regression of the tumor was observed in some cases suggesting the possibility of using such conjugates in the treatment of human solid tumors.

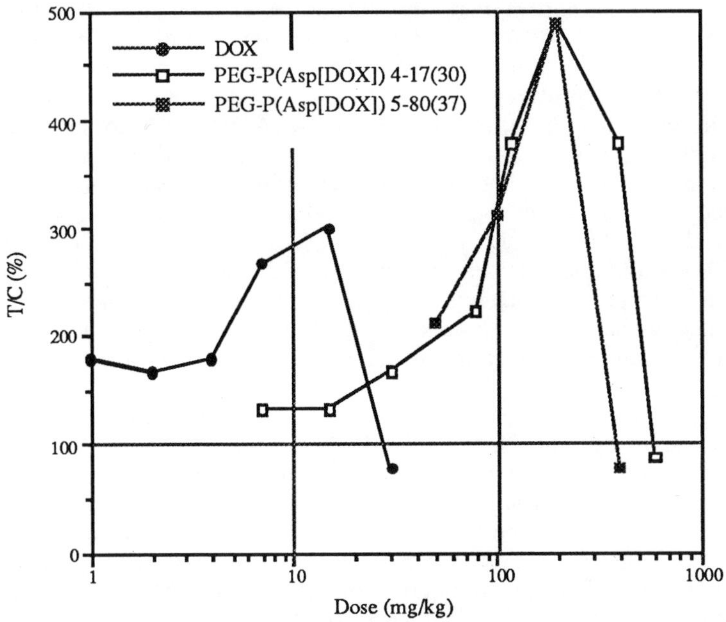

Figure 2. In vivo anti-cancer activity of PEG-P(Asp[DOX]) against P388 mouse leukemia [104,105].

The pharmacokinetic and the biodistribution of the PEG-P(Asp[DOX]) conjugates radiolabeled either with ^{125}I [106,110] or with ^{14}C-benzylamine [112] were studied after i.v. injection in mice. First, these studies showed that the concentration of conjugates in blood was kept significantly higher with longer half-life than those of free DOX. The half-life in blood was determined to be approximately 70 min. The conjugates circulated in blood predominantly under micelles form, even after 4 hours, without rapid excretion, metabolism or adsorption by tissues compared to free DOX which is rapidly uptaken by heart, kidneys, lungs, liver and spleen (table 2).

TABLE 2. Biodistribution of PEG-P(Asp[DOX]) at a dose of 20 mg DOX/HCl equivalent/kg, 1 hour after i.v. injection [106,113].

Organ	Blood[a]	Liver[b]	Spleen[b]	Kidney[b]	Heart[b]	Lung[b]
5-20(30) conjugates	17.1	7.1	11.6	9.6	2.9	4.0
free DOX		12.0	6.0		6.2	8.8

a) % dose/ml. b) % dose/g.

Worthy to notice is that high doses of DOX (ca. 100 mg/kg) can be administrated in the PEG-P(Asp[DOX]) form without significant indication of acute toxicity [106,110].

Following this study, the fate of PEG-P(Asp[DOX]) conjugates in murine colon adenocarcinoma 26 (C26) tumor-bearing mice was investigated and compared with the fate of free DOX after i.v. injection [113]. Two conjugates were studied : 12-20(104) with a diameter of 40 nm and 5-20(50) with a diameter of 30 nm. At 4 and 24 hours, respectively, the 12-20(104) conjugate exhibited 52 and 7% of the original dose in blood and the 5-20(50) conjugate showed 18 and 0.3% of the original dose in blood compared to DOX which is rapidly eliminated from blood circulation. Extended blood circulation times were accompanied by a diminished uptake by the RES. This is in contrast with most colloids which are rapidly and efficiently removed from the circulation within minutes.

As shown by table 3, the micelle-forming PEG-P(Asp[DOX]) conjugates demonstrated an order of magnitude increase in tumor/heart ratio in comparison to free DOX. The levels of the conjugates attained in the solid tumors are higher than free DOX, apparently due to the persistence of the micelles form in blood and low RES uptake of the conjugates. This effect can be described as a passive targeting of micelle-forming PEG-P(Asp[DOX]) conjugates towards tumor cells resulting from the prolonged circulation in blood and low RES uptake.

TABLE 3. Tumor/organ accumulation ratios of PEG-P(Asp[DOX]) conjugates and free DOX at 24 hours [113].

	% dose per g tumor / % dose per g heart	% dose per g tumor / % dose per g muscle
12-20	12	40
5-20	8.1	50
free DOX	0.9	1.5

The mechanism of extravasation of PEG-P(Asp[DOX]) conjugates have not been elucidated ; however, the conjugates may demonstrate enhanced permeability with respect to continuous vascular endothelium and retention at tumors. It can be concluded that PEG-P(Asp[DOX]) conjugates, in a passive manner, can more efficaciously deliver DOX to tumors compared to free DOX.

3.3. PEG-PBLA POLYMERIC MICELLES

We have demonstrated that hydrophobic drugs such as DOX can be conveniently coupled to PEG-P(Asp) block copolymers to form micellar structure. In another approach, hydrophobic drugs can be also physically incorporated within the hydrophobic core of the polymeric micelles (scheme 5) [114-116]. According to this approach, the hydrophobic drugs do not need to have appropriate chemical functional groups for covalent attachment to the block copolymers for incorporation into the polymeric micelles.

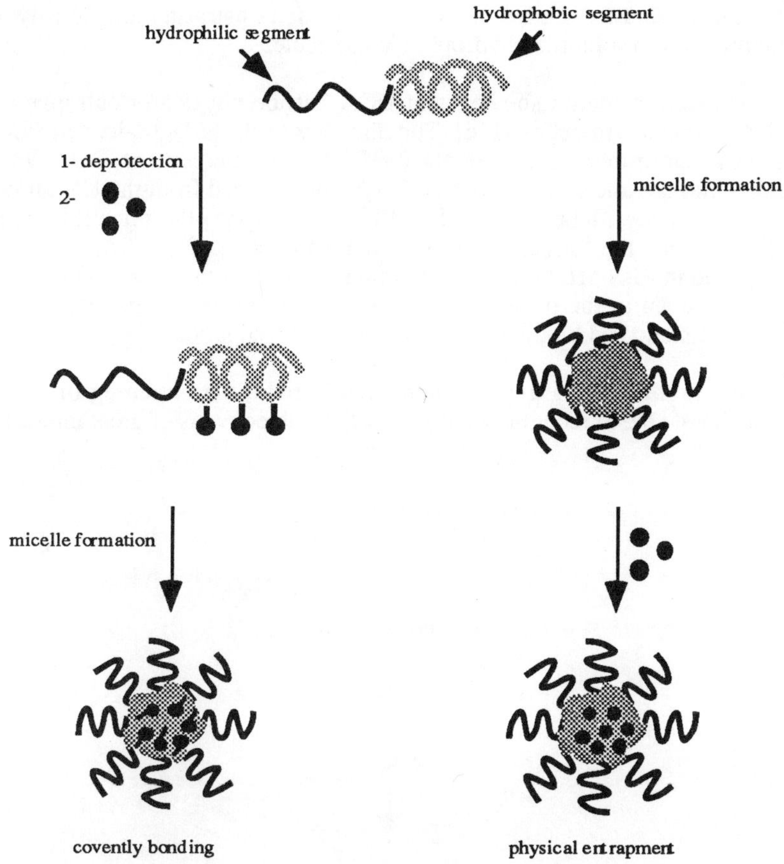

Scheme 5. The chemical and physical incorporation of drugs with PEG-poly(amino acid) micelles [115].

In order to study this possibility, we synthesized several PEG-PBLA block copolymers having different compositions (c.a. 5-10, 5-20 and 12-20) and studied them by DLS and photo physical means [114-115]. The DLS data showed that these micelle-forming PEG-PBLA block copolymers exhibited two subpopulations of particles. The small particles with a diameter of approximately 20 nm are attributed to individual micelles while the larger particles are thought to be micelles that further associated [114]. An association number of 400 was calculated by static light scattering (SLS).

Pyrene was first used as hydrophobic drug model in order to study the properties of these micelles [114,115]. The pyrene will preferentially partition into hydrophobic microdomains with a concurrent change in the molecule's photophysical properties. From these changes, we determined the critical micellar concentration (cmc) of the different block copolymers, c.a. 10 mg/L for 5-10 and 12-20 copolymers and 5 mg/L for 5-20 copolymers. The fluorescence studies showed that pyrene is clearly retained within the PEG-PBLA micelles, thus demonstrating the ability of such polymeric micelles to solubilize hydrophobic molecules.

Our recent research showed that DOX can be physically entrapped into the PEG-PBLA polymeric micelles [116]. The diameter of those DOX-loaded PEG-PBLA micelles was determined to be 30 nm by DLS ; the percent of DOX which was incorporated into the micelles is about 10% w/w, determined from the UV absorption at 485 nm. The very low fluorescence of DOX entrapped into the micelles compared to free DOX supported the formation of DOX-loaded micelles. Data from GPC/HPLC in PBS pH 7.4 and in PBS pH 7.4/serum albumin showed that these DOX-loaded micelles are very stable even in the presence of serum. The in vitro cytotoxicity as well as in vivo pharmacokinetic and biodistribution are under investigation.

Now, in the drug delivery system (DDS) field, the obtaining of "intelligent" carriers which are able to recognize selectively their objective is of great interest.

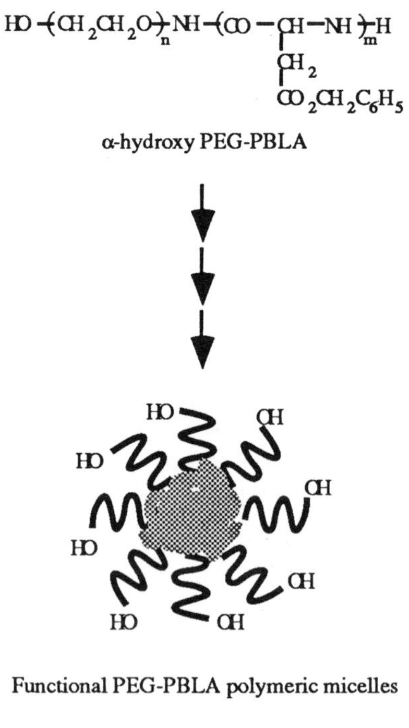

Scheme 6. Functional PEG-PBLA polymeric micelles.

In order to obtain PEG-PBLA polymeric micelles having targeting moieties on their outer shell, we synthesized PEG-PBLA block copolymers with functional groups (c.a. hydroxy groups) at the free end of PEG chains (scheme 6) [117].

We demonstrated that PEG-PBLA polymeric micelles having high density of hydroxy groups on their outer shell can be obtained. Their physico-chemical properties are very close from those of "classical" PEG-PBLA micelles [118]. The diameter of these functional micelles was determined by DLS to be 25 nm (number average distribution). Their cmc in water is 4 mg/L and they are stable in PBS pH 7.4 and in PBS pH 7.4/serum (studied by GPC/HPLC). Moreover, DOX can be physically entrapped inside the hydrophobic inner core of the functional micelles with high loading yield : up to 20% w/w of DOX can be loaded [118]. The in vitro cytotoxicity and in vivo biodistribution of these functional micelles are under investigation. In addition, we are studying the possibility to activate the hydroxyl groups in order to introduce a targeting moiety and consequently to prepare targeted micelles.

3,4. POLYION COMPLEX MICELLES

Recently, we have developed the preparation of stable and monodispersive polyion complex micelles in aqueous milieu (scheme 7) [119].

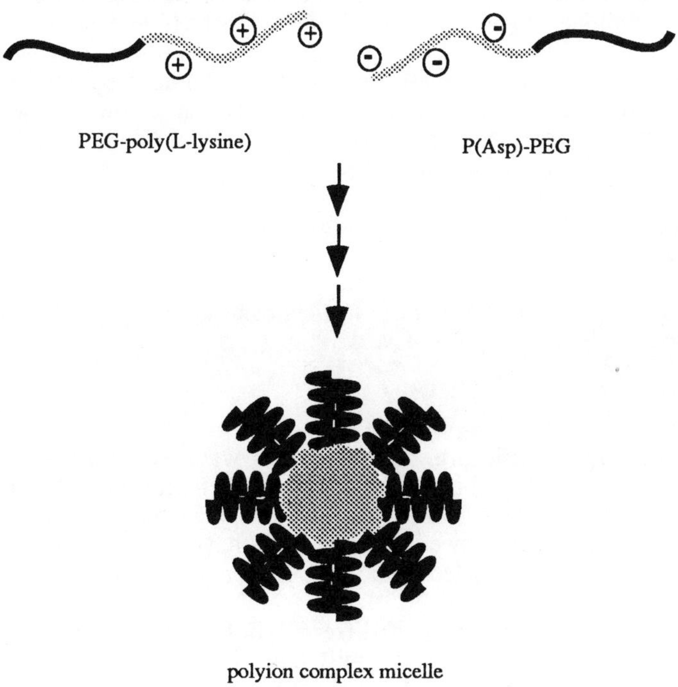

Scheme 7. Polyion complex micelles preparation.

These polyion complex micelles were prepared through electrostatic interaction between a pair of oppositely-charged block copolymers : PEG-poly(L-lysine) and PEG-poly(α,β-aspartic acid) block copolymers.

DLS data showed that these polyion complex micelles were spherical particles without any secondary aggregates. Their hydrodynamic radius was calculated to be 15 nm by the Stokes-Einstein equation. The size of polyion complex micelles was unchanged even after 1 month storing, suggesting that these polyion complex micelles are in thermodynamic equilibrium. These new micelles have potential utility as vehicles for charged compounds such as proteins and nucleic acids.

3.5. CONCLUSION

By using micelle-forming PEG-P(Asp[DOX]) conjugates, a unique delivery system may be constructed. Indeed, the stability, the pharmacokinetic as well as the biodistribution of these polymeric micelles can be adjusted by varying the length of both blocks and by the percent of bound drug.

The PEG-P(Asp[DOX]) polymeric micelles demonstrated high in vivo anti-cancer activity and in some cases a dramatic tumor regression which is one of only few successful example of a strategy of drugs combined with carrier systems. In addition, the toxic score of DOX equivalents in the conjugate was estimated to be less than 1/10 that of free DOX, permitting a higher dose with fewer side effects compared to free DOX.

The possibility to physically entrap hydrophobic drug in the PEG-PBLA polymeric micelles increases the domain of utilization of these new drug carrier systems. In light of our previous experiments with the PEG-P(Asp[DOX]) micelles, we speculate that DOX-loaded PEG-PBLA micelles will show prolonged circulation times in blood and possibly higher uptake at solid tumor sites compared to free DOX. Moreover, the successful preparation of functional micelles opens the way to the obtaining of targeted micelles.

At last, our new polyion complex micelles bring the possibility of entrapment of charged biologically active molecules.

4. Conclusion

Several efficient site-specific drug carriers, from water-soluble macromolecules to liposomes and nanoparticles, have been studied. Their in vivo behavior in animals allows to hope that in a near future such carriers may be utilized to improve the treatment of human diseases. However, carriers will have to be selected and adjusted in function of the disease and the selected biologically active molecule.

Thus, collaboration between different fields such as chemistry, polymer science, biology, pharmacology, etc., will have to develop more intensively for tailoring new site specific carriers with excellent biocompatibility and high selectivity.

Aknowledgements

The authors would like to aknowledge their colleagues for their contribution to the research highlighted in the second part of this chapter and Dr. Carmen Scolz for her critical reading of this manuscript.

References

1. Vert, M. (1985) Polyvalent polymeric drug carriers, in S.D. Bruck (ed.), *Critical Reviews in Therapeutic Drug Carrier Systems*, CRC Press Inc., Florida, **2(3)**, pp. 291-327.

2. Tomlinson, E. (1986) (Patho)physiology and the temporal and spatial aspects of drug delivery, in E. Tomlinson and S.S. Davis (eds.), *Site-Specific Drug Delivery, Cell Biology, Medical and Pharmaceutical Aspects*, John Wiley &Sons Ltd., 1-26.

3. Ehrlich, P. (1906) *Collected Study on Immunology*, John Wiley & Sons, New York, **11**, 442.

4. Trail, P.A., Willner, D., Lasch, S.J., Henderson, A.J., Hofstead, S., Casazza, A.M., Firestone, R.A., Hellström, I. and Hellström, K.E. (1993) Cure of xenografted human carcinomas by BR96-Doxorubicin immunoconjugates, *Science* **261**, 212-215.

5. Uckun, F.M., Evans, W.E., Forsyth, C.J., Waddick, K.G., Ahlgren, L.T., Chelstrom, L. M., Burkhardt, A., Bolen, J. and Myer, D.E. (1995) Biotherapy of B-cell precursor leukemia by targeting genistein to CD19-associated tyrosine kinases, *Science* **267**, 886-891.

6. Okano, T., Yui, N., Yokoyama and M., Yoshida, R. (1994) Advances in Polymeric Systems for Drug Delivery, T. Ikoma, I. Karube and R. Kuroda (eds.), Japanese Technology Reviews (section E : Biotechnology), Gordon and Breach Science Publishers, **4(1)**.

7. Maeda, H., Seymour, L.W. and Miyamoto, Y. (1992) Conjugates of anticancer agents and polymers: advantages of macromolecular therapeutic in vivo, *Bioconj. Chem.* **3(5)**, 128-139.

8. Takakura, Y., Takagi, A., Hashida, M. and Sezaki, H. (1987) Disposition and tumor localization of Mitocyn C-dextran conjugates in mice, *Pharm. Res.* **4(4)**, 293-300.

9. Ringsdorf, H. (1975) Structure and properties of pharmacologically active polymers, *J. Polym. Sci. : Symposium* **51**, 135.

10. Firestone, R.A. (1994) Low-density lipoprotein as a vehicle for targeting antitumor compounds to cancer cells, *Bioconj. Chem.* **5**, 105-113.

11. Ulbrich, K. (1991) Water-soluble polymeric carriers of drugs, *J. Bioactive and Compatible Polymers* **6**, 348-357.

12. Koch, M. (December 7-10, 1994) France-Japan DDS Symposium, *Advances in Novel science and Technology of Drug Delivery and Targeting*.

13. Sezaki, H. and Hashida, M. (1984) Macromolecule-drug conjugates in targeted cancer chemotherapy in *CRC Critical Reviews in Therapeutic Drug Carrier Systems*, CRC press, Inc., **1(1)**, 1-38.

14. Bader, H., Ringsdorf, H. and Schmidt, B. (1984) Water-soluble polymers in medicine, *Angew. Makromol. Chem.* **123/124**, 457-485.

15. Hoes, C.J.T. and Feijen, J. (1989) The application of drug-polymer conjugates in chemotherapy, in F.H.D. Roerdink and A.M. Kroon (eds.), *Drug Carrier Systems* , John Wiley &Sons Ldt., 57-109.

16. Kopecek, J. (1982) Biodegradation of Polymers for Biomedical Use in H Benoît and P. Rempp (eds.), *IUPAC Macromolecules* , Pergamon, Oxford, 305-320.

17. Kopecek, J. and Duncan, R. (1987) Poly[N-(2-hydroxypropyl)methacrylamide] macromolecules as drug carrier systems in I. Illum, S.S. Davies and A. Hilger (eds.), *Controlled Release of Drugs from Polymer Particles and Macromolecules*, Bristol.

18. Kopecek, J. and Duncan, R. (1987) Targetable polymeric prodrugs *J. Control. Rel.*, **5**, 315-327.

19. Rihova, B., Kopecek, J., Ulbrich, K. and Chytry, V. (1985) *Makromol. Chem., Suppl.*, **9**, 13.

20. Duncan,R. (1985) *CRC Critical Reviews in Therapeutic Drug Carrier Systems,.*, **4**, 281.

21. Rihova, B., Strohalm, J., Plocova, D. and Ulbrich, K. (1990) Selectivity of antibody-targeted anthracycline antibiotics on T lymphocytes, *J. Bioactive and Compatible Polymers*, **5**, 249-266.

22. Seymour, L.W., Ulbrich, K., Wedge, S.R., Hume, I.C., Strohalm, J. and Duncan, R. (1991) N-(2-hydroxypropyl)methacrylamide copolymers targeted to the hepatocyte galactose-receptor: pharmacokinetics in DBA_2 mice, *Br. J. Cancer*, **63**, 859-866.

23. Subr, V., Strohalm, J., Ulbrich, K., Duncan, R. and Hume, I.C. (1992) Polymers containing enzymatically degradable bonds, XII. Effect of spacer structure on the rate of release of daunomycin and adriamycin from poly[N-(2-hydroxypropyl)methacrylamide] copolymer drug carriers in vitro and antitumour activity measured in vivo, *J. Control. Rel.*, **8**, 123-132.

24. O'Hare, K.B., Duncan, R., Strohalm, J., Ulbrich, K. and Kopeckova, P. (1993) Polymeric drug-carriers containing Doxorubicin and melanocyte-stimulating hormone: in vitro and in vivo evaluation against murine melanoma, *J. Drug Targeting*, **1**, 217-229.

25. Krinick, N.L., Sun, Y., Joyner, D., Spikes, J.D., Straight, R.C. and Kopecek, J. (1994) A polymeric drug delivery system for simultaneous delivery of drugs activatable by enzymes and/or light, *J. Biomater. Sci. Polymer Edn.*, **5(4)**, 303-324.

26. Rihova, B., Jegorov, A., Strohalm, J., Matha, V., Rossmann, P., Fornusek, L. and Ulbrich, K. (1990) Antibody-targeted cyclosporin A, *J. Control. Rel.*, **19**, 25-40.

27. Rihova, B., Veres, K., Fornusek, L., Ulbrich, K., Strohalm, J., Vetvicka, V., Bilej, M. and Kopecek, J. (1989) Action of polymeric prodrugs based on N-(2-hydroxypropyl)methacrylamide copolymers. II. Body distribution and T-cell accumulation of free and polymer-bound [^{125}I]daunomycin, *J. Control. Rel.*, **10**, 37-49.

28. Flanagan, P.A., Duncan, R., Subr, V., Ulbrich, K., Kopeckova, P. and Kopecek, J. (1992) Evaluation of protein-N-(2-hydroxypropyl)-methacrylamide copolymer conjugates as targetable drug-carriers. 2. Body distribution of conjugates containing transferrin, anti-transferrin receptor antibody or anti-Thy 1.2 antibody and effectiveness of transferrin-containing daunomycin conjugates against mouse L1210 leukaemia in vivo, *J. Control. Rel.*, **18**, 25-38.

29. Krinick, N.L., Rihova, B., Ulbrich, K., Strohalm, J. and Kopecek, J. (1990) Targetable photoactivable drugs, 2[a)] Synthesis of N-(2-hydroxypropyl)methacrylamide copolymer-anti-Thy 1.2 antibody-chlorine e_6 conjugates and a preliminary study of their photodynamic effect on mouse splenocytes in vitro, *Makromol Chem*, **191**, 839-856.

30. Rihova, B., Krinick, N.L. and Kopecek, J. (1993) Targetable photoactivatable drugs. 3. In vitro efficacy of polymer bound chlorin e_6 toward human hepatocarcinoma cell line (PLC/PRF/5) targeted with galactosamine and to mouse splenocytes targeted with anti-Thy 1.2 antibodies, *J. Control. Rel.*, **25**, 71-87.

31. Seymour, L.W., Duncan, R., Chytry, V., Strohalm, J., Ulbrich, K. and Kopecek, J. (1991) Intraperitoneal and subcutaneous retention of a soluble polymeric drug-carrier bearing galactose, *J. Control. Rel.*, **16**, 255-262.

32. Duncan, R., Seymour, L.W., O'Hare, K.B., Flanagan, P.A., Wedge, S., Hume, I.C., Ulbrich, K., Strohalm, J., Subr, V., Spreafico, F, Grandi, M., Ripamonti, M., Farao, M. and Suarato, A. (1992) Preclinical evaluation of polymer-bound doxorubicin, *J. Control. Rel.*, **19**, 331-346.

33. van Heeswijk, W.A.R., Brinks, G.J. and J. Feijen, (1984) Synthesis and characterization of covalently bound polymer-hormone conjugates for the controlled release of hormones, in E. Chiellini and P. Giusti (eds.), *Polymers in Medecine*, Plenun Publishing Corporation, 147-156.

34. Hoes, C.J.T., Potman, W., de Grooth, B.G., Greve, J. and Feijen, J. (1986) Chemical control of drug delivery in A.F. Harms (ed.), *Innovative Approaches in Drug Research*, Elsevier Science Publishers B.V., Amsterdam, 267-283.

35. van Heeswijk, W.A.R., Stoffer, T., Eenink, M.J.D., Potman, W., van der Vijgh, W.J.F., Poort, J.V.D., Pinedo, H.M., Lelieveld, P. and Feijen, J. (1984) Synthesis, characterization and antitumour activity of macromolecular prodrugs of adriamycin in J.M. Anderson and S.W. Kim (eds.), *Recent Advances in Drug Delivery Systems*, Plenum Publishing Corporation, 77-100.

36. van Heewijk, W.A.R., Hoes, C.J.T., Stoffer, T., Eenink, M.J.D., Potman, W. and Feijen, J. (1985) The synthesis and characterization of polypeptide-adriamycin conjugates and its complexes with adriamycin. Part I, *J. Control. Rel.*, **1**, 301-315.

37. Hoes, C.J.T., Boon, P.J., Kaspersen, F., Bos, E.S. and Feijen, J. (1993) Design of soluble conjugate of biodegradable polymeric carriers and adriamycin, *Makromol. Chem, Macromol. Symp.*, **70/71**, 119-136.

38. Hoes, C.J.T., Grootoonk, J., Duncan, R., Hume, I.C., Bhakoo, M., Bouma, J.M.W. and Feijen, J. (1993) Biological properties of adriamycin bound to biodegradable polymeric carriers, *J. Control. Rel.*, **23**, 37-54.

39. Hoes, C.J.T., Grootoonk, J., Feijen, J., Boon, P.J., Kaspersen, F., Loeffen, P., Schlachter, I., Winters, M. and Bos, E.S. (1992) Synthesis and biodistribution of immunoconjugates of a human IgM and polymeric drug carriers, *J. Control. Rel.*, **19**, 59-76.

40. Nukui, M., Hoes, K., van den Berg, H. and Feijen, J. (1991) Association of macromolecular prodrugs consisting of adriamycin bound to poly(L-glutamic acid), *Makromol. Chem.*, **192**, 2925-2942.

41. Chen, G.C. and Hoffman, A.S. (1994) Synthesis of carboxylated poly(NIPAAm) oligomers and their application to form thermo-reversible polymer-enzyme conjugates, *J. Biomater. Sci. Polymer Edn.*, **5(4)**, 371-382.

42. Heskins, M. and Guillet, J.E. (1968) *J. Macromol. Sci.*, **A2**, 1441.

43. Y. Takei, (1994) Temperature-responsive polymers for biomedical modulation systems, pH.D. Thesis, Sophia University, Tokyo (Japan).

44. Valuev, L.I., Zefirova, O.N., Obydennova, I.V. and Platé, N.A. (1994) Targeted delivery of drugs provided by water-soluble polymeric systems with low critical solution temperature (LCST), *J. Bioactive and Compatible Polymers*, **9**, 55-65.

45. Takakura, Y., Matsumoto, S., Hashida, M. and Sezaki, H. (1984) Enhanced lymphatic delivery of mitomycin C conjugated with dextran, *Cancer Res.*, **44**, 2505-2510.

46. Takakura, Y., Fujita, T., Hashida, M. and Sezaki, H. (1990) Disposition characteristics of macromolecules in tumour-bearing mice, *Pharm. Res.*, **7(4)**, 339-346.

47. Fujita, T., Yasuda, Y., Takakura, T., Hashida, M. and Sezaki, H. (1991) Tissue distribution of ^{111}In-labeled uricase conjugated with charged dextrans and polyethylene glycol, *J. Pharmacobio-Dyn.*, **14**, 623-629.

48. Fujita, T., Nishikawa, M., Tamaki, C., Takakura, Y., Hashida, M. and Sezaki, H. (1992) Targeted delivery of human recombinant superoxide dismutase by chemical modification with mono-and polysaccharide derivatives, *J. Pharm. Exp. Therap.*, **263(3)**, 971-978.

49. Nishikawa, M., Kamijo, A., Fujita, T., Takakura, Y., Sezaki, H. and Hashida, M. (1993) Synthesis and pharmacokinetics of a new liver-specific carrier, glycosylated carboxymethyl-dextran, and its application to drug targeting, *Pharm. Res.*, **10(9)**, 1253-1261.

50. Takakura, Y., Fujita, T., Furitsu, H., Nishikawa, M., Sezaki, H. and Hashida, M. (1994) Pharmacokinetics of succinylated proteins and dextran sulfate in mice: Implications for hepatic targeting of protein drugs by direct succinylation via scavenger receptors, *Int. J. Pharm.*, **105**, 19-29.

51. Kojima, T., Hashida, M., Muranishi, S. and Sezaki, H. (1980) Mitomycin C-dextran conjugate: a novel high molecular weight pro-drug of mitomycin C, *J. Pharm. Pharmacol.*, **32**, 30-34.

52. Malek, P., Kolc, J., Herold, M. and Hoffman, J. (1958) Lymphotropic antibiotics-"Antibiolymphins", *Antibiotics Annual*, New York, Medical Encyclopedia Inc., 546-556.

53. Hashida, M., Nishikawa, M., Yamashita, F. and Takakura, T. (1994) Targeting delivery of protein drugs by chemical modification, *Drug Develop. Indust. Pharm.*, **20(4)**, 581-590.

54. Takakura, Y., Masuda, S., Tokuda, H. and Nishikawa, M. (1994) Targeted delivery of superoxide dismutase to macrophages via mannose receptor-mediated mechanism, *Biochem. Pharm.*, **47(5)**, 853-858.

55. Fujita, T., Nishikawa, M., Ohtsubo, Y., Onho, J., Takakura, Y., Sezaki, H. and Hashida, M. (1994) Control of in vivo fate of albumin derivatives utilizing combined chemical modification, *J. Drug Targ.*, **2**, 157-165.

56. Nishikawa, M., Ohtsubo, Y., Ohno, J., Fujita, T., Koyama, Y., Yamashita, F., Hashida, M. and Sezaki, H. (1992) Pharmacokinetics of receptor-mediated hepatic uptake of glycosylated albumine in mice, *Int. J. Pharm.*, **85**, 75-85.

57. Leamon, C.P. and Low, P.S. (1992) Cytotoxicity of momordin-folate conjugates in cultured human cells, *J. Biol. Chem.*, **267(35)**, 24966-24971.

58. Leamon, C.P., Pastan, I. and Low, P.S. (1993) Cytotoxicity of folate-*Pseudomonas* exotoxin conjugates toward tumor cells. Contribution of translocation domain, *J. Biol. Chem.*, **268(33)**, 24847-24854.

59. Leamon, C.P. and Low, P.S. (1993) Membrane folate-binding proteins are responsible for folate-protein conjugate endocytosis into cultured cells, *Biochem. J.*, **291**, 855-860.

60. Leamon, C.P. and Low, P.S. (1994) Selective targeting of malignant cells with cytotoxin-folate conjugates, *J. Drug Targeting*, **2**, 101-112.

61. Dougherty, T.J. (1985) Photodynamic Therapy in D. Kessel (ed.), *Methods in Porphirin Photosensitization*, Plenum press, New York, 313-323.

62. Flynn, G., Hackett, T.J., McHale, L. and McHale, A.P. (1994) Magnetically responsive photosensitizing reagents for possible use in photoradiation therapy, *Cancer Letters*, **78**, 109-114.

63. Ulbrich, K., Surb, V., Seymour, L.W. and Duncan, R. (1993) Novel biodegradable hydrogels prepared using the divinylic crosslinking agent N,O-dimethacryloylhydroxylamine. 1. Synthesis and characterization of rates of gel degradation, and rate of release of model drugs, in vitro and in vivo, *J. Control. Rel.*, **24**, 181-190.

64 Scholes, P.D., Coombes, A.G.A., Illum, L., Davis, S.S., Vert, M. and Davies, M.C. (1993) The preparation of sub-200 nm poly(lactide-co-glycolide) microspheres for site-specific drug delivery, *J. Control. Rel.*, **25**, 145-153.

65. Davis, S.S., Illum, L., Moghimi, S.M., Davies, M.C., Porter, C.J.H., Muir, I.S., Brindley, A., Christy, N.M., Norman, M.E., Williams, P. and Dunn, S.E. (1993) Microspheres for targeting drugs to specific body sites, *J. Control. Rel.*, **24**, 157-163.

66. Le Ray, A.M., Vert, M., Gautier, J.C. and Benoît, J.P. (1994) Fate of [^{14}C]poly(DL-lactide-co-glycolide) nanoparticles after intravenous and oral administration to mice, *Int. J. Pharm.*, **106**, 201-211.

67. Mehta, R.C., Jeyanthi, R., Calis, S., Thanoo, B.C., Burton, K.W. and DeLuca, P.P. (1994) Biodegradable microspheres as depot system for parenteral delivery of peptide drugs, *J. Control. Rel.*, **29**, 375-384.

68. Kreuter, J. (1994) Nanoparticles in J. Swarbrich (ed.), *Colloidal Drug Delivery Systems in Drugs and Pharmaceutical Sciences*, Marcel Dekker, Inc., 219-342.

69. Wu, X.Y. and Lee, P.I. (1993) Preparation and characterization of thermal- and pH-sensitive nanospheres, *Pharm. Res.*, **10(10)**, 1544-1547.

70. Moghimi, S.M., Hawley, A.E., Christy, N.M., Gray, T., Illum, L. and Davis, S.S. (1994) Surface engineered nanospheres with enhanced drainage into lymphatics and uptake by macrophages of the regional lymph nodes, *FEBS Let.*, **344**, 25-30.

71. Couvreur, P., Fattal, E. and Andremont, A. (1991) Liposomes and nanoparticles in the treatment of intracellular bacterial infections, *Pharm. Res.*, **8(9)**, 1079-1086.

72. Couvreur, P. and Vauthier, C. (1991) Polyalkylcyanoacrylate nanoparticles as drug carrier: present state and persectives, *J. Control. Rel.*, **17**, 187-198.

73. Maincent, P., Thouvenot, P., Amicabile, C., Hoffman, M., Kreuter, J., Couvreur, P. and Devissaguet, J.P. (1992) Lymphatic targeting of polymeric nanoparticles after intraperitoneal administration in rats, *Pharm. Res.*, **9(12)**, 1534-1539.

74. Balland, O., Pinto-Alphandary, H., Pecquet, S., Andremont, A. and Couvreur, P. (1994) The uptake of ampicillin-loaded nanoparticles by murine macrophages infected with *Salmonella typhimurium*, *J. Antimicrobial Chemotherapy*, **33**, 509-522.

75. Pinto-Alphandary, H., Ballant, O., Laurent, M., Andremont, A., Puisieux, F. and Couvreur, P. (1994) Intracellular visualization of ampicillin-loaded nanoparticles in peritoneal macrophages infected in vitro with *Salmonella typhimurium*, *Pharm. Res.*, **11(1)**, 38-46.

76. Némati, F., Dubernet, C., Colin de Verdière, A., Poupon, M.F., Trepeul-Acar, L., Puisieux, F. and Couvreur, P. (1994) Some parameters influencing cytotoxicity of free doxorubicin and doxorubicin-loaded nanoparticles in sensitive and multidrug resistant leucemic murine cells: incubation time, number of nanoparticles per cell, *Int. J. Pharm.*, **102**, 55-62.

77. Bennis, S., Chapey, C., Couvreur, P. and Robert, J. (1994) Enhanced cytotoxicity of doxorubicin encapsulated in polyisohexyl-cyanoacrylate nanospheres against multidrug-resistant tumour cells in culture, *European J. Cancer*, **30 A(1)**, 89-93.

78. Poste, G. (1983) Drug targeting in cancer therapy in G. Gregoriadis, G. Poste, J. Senior and A. Trouet (eds.), *Receptor-Mediated Targeting of Drugs*, Published in cooperation with NATO Scientific Affairs Division, 427-474.

79. Crommelin, D.J.A. and Schreier, H. (1994) Liposomes in J. Swarbrick (ed.), *Colloidal Drug Delivery Systems in Drugs and Pharmaceutical Sciences*, Marcel Dekker, Inc., 73-190.

80. Ahmad, I., Longenecker, M., Samuel, J. and Allen, T.M. (1993) Antibody-targeted delivery of doxorubicin entrapped in sterically stabilized liposomes can eradicate lung cancer in mice, *Cancer Res.*, **53**,1484-1488.

81. Seki, K. and Tirrell, D.A. (1984) pH-dependent complexation of poly(acrylic acid) derivatives with phospholipid vesicles membranes, *Macromolecules*, **17**, 1692-1698.

82. Thomas, J.L. and Tirrell, D.A. (1992) Polyelectrolyte-sensitized phospholipid vesicles, *Acc. Chem. Res.*, **25(8)**, 336-342.

83. Tirrell, D.A., Takigawa, D.Y. and Seki, K. (1985) pH sensitization of phospholipid vesicles via complexation with synthetic poly(carboxylic acid)s, in *Macromolecules as Drugs and as Carriers for Biologically Active Materials Vol. 446*, Annals of the New York Academy of Sciences, 237-247.

84. Ropert, C., Lavignon, M., Dubernet, C., Couvreur, P. and Malvy, C. (1992) Oligonucleotides encapsulated in pH sensitive liposomes are efficient toward friend retrovirus, *Biochem. Biophys. Res. Comm.*, **183(2)**, 879-885.

85. Ropert, C., Malvy, C. and Couvreur, P. (1993) Inhibition of the friend retrovirus by antisense oligonucleotides encapsulated in liposomes: mechanism of action, *Pharm. Res.*, **10(10)**, 1427-1433.

86. Magin, R.L. and Weinstein, J.N. (1983) Delivery of drugs in temperature-sensitive liposomes, in G. Gregoriadis, J. Senior and A. Trouet (eds.), *Targeting of drugs*, Plenun Press New York and London, Published in Cooperation with NATO Scientific Affairs Division, 203-221.

87. Oku, N., Naruse, R., Doi, K. and Okada, S. (1994) Potential usage of thermosensitive liposomes for macromolecule delivery, *Biochem. Biophys. Act.*, **1191**, 389-391.

88. Merlin, J.M., Marechal, S., Ramacci, C., Notter, D. and Vigneron, C. (1993) Antiproliferative activity of thermosensitive liposome-encaspulated doxorubicin combined with 43°C hyperthermia in sensitive and multidrug-resistant MCF-7 cells, *European J. cancer*, **29 A(16)**, 2264-2268.

89. Lee, R.J. and Low, P.S. (1994) Delivery of liposomes into cultured KB cells via folate receptor-mediated endocytosis, *J. Biol. Chem.*, **269(5)**, 3198-3204.

90. Wang, S., Lee, R.J., Cauchon, G., Gorenstein, D.G. and Low, P.S. (in press) Delivery of antisense oligodeoxyribonucleotides against the human epidermal growth factor receptor into cultured KB cells using folate-PEG-liposomes, *J. Biol. Chem.*.

91. Lee, R.J. and Low, P.S. (in press) Folate-mediated tumor cell targeting of liposome-entrapped doxorubicin in vitro, *Biochem. Biophys. Act.*.

92. Loughrey, H.C., Ferraretto, A., Cannon, A.M., Masserini, G. and Soria, M.R. (1993) Characterization of biotinylated liposomes for in vivo targeting applications, *FEBS Let.*, **332(1,2)**, 183-188.

93. Tomalia, D.A. (1993) Starburst^{TM}/casdade dendrimers: fundamental building blocks for new nanoscopic chemistry set, *Aldrichimica Acta*, **26(4)**, 91-101.

94. Fréchet, J.M.J. (1994) Functional polymers and dendrimers: reactivity, molecular architecture, and interfacial energy, *Science*, **263**, 1710-1715.

95. Peppas, N.A., Nagai, N. and Miyajima, M. (1994) Prospects of using star polymers and dendrimers in drug delivery and pharmaceutical application, *Pharm. Tech. Japan*, **10(6)**, 611-617.

96. Peppas, N. A. and Langer, R. (1994) New challenges in biomaterials, *Science*, **263**, 1715-1720.

97. Barth, R.F., Adams, D.M., Soloway, A.H., Alam, F. and Darby, M.V. (1994) Boronated starburst dendrimer-monoclonal antibody immunoconjugates: evaluation as a potential delivery system for neutron capture therapy, *Bioconj. Chem.*, **5**, 58-66.

98. Jansen, J.F.G.A., de Bradander-van der Berg, E.M.M. and Meijer, E.W. (1994) Encapsulation of guest molecules into dentritic box, *Science*, **266**, 1226-1229.

99. Kataoka, K. (1994) Design of nanoscopic vehicles for drug targeting based on micellization of amphiphilic block copolymers, in R.A. Guadiana (ed.), *J.M.S. -Pure Appl. Chem.*, **A31(11)**, 1759-1769.

100. Bader, J.H., Ringsdorf, H. and Schmidt, B. (1984) Watersoluble polymers in medicine, *Angew. Makromol. Chem.*, **123**, 457-485.

101. Pratten, M.K., Lloyd, J.B., Hurpel, G. and Ringsdorf, H. (1985) *Makromol. Chem.*, **186**, 725.

102. Yokoyama, M., Inoue, S., Kataoka, K., Yui, N. and Sakurai, Y. (1987) Preparation of adriamycin-conjugated poly(ethylene glycol)-poly(aspartic acid) block copolymer, *Makromol. Chem., Rapid Commun.*, **8**, 431-435.

103. Yokoyama, M., Inoue, S., Kataoka, K., Yui, N., Okano, T. and Sakurai, Y., (1989) Molecular design for missile drug: synthesis of adriamycin conjugated with IgG using poly(ethylene glycol)-poly(aspartic acid) block copolymer as intermediate carrier, *Makromol. Chem.*, **190**, 2041-2054.

104. Yokoyama, M., Miyauchi, M., Yamada, N., Okano, T., Sakurai, Y., Kataoka, K. and Inoue, S. (1990) Characterization and anticancer activity of the micelle-forming polymeric anticancer drug adriamycin-conjugated poly(ethylene glycol)-poly(aspartic acid) block copolymer,*Cancer Res.*, **50**, 1693-1700.

105. Yokoyama, M., Miyauchi, M., Yamada, N., Okano, T., Sakurai, Y., Kataoka, K. and Inoue, S. (1990) Polymer micelles as novel drug carrier: adriamycin-conjugated poly(ethylene glycol)-poly(aspartic acid) block copolymer, *J. Control. Rel.*, **11**, 269-278.

106. Yokoyama, M., Okano, T., Sakurai, Y., Ekimoto, H., Shibazaki, C. and Kataoka, K. (1991) Toxicity and antitumor activity against solid tumors of micelle-forming polymeric anticancer drug and its extremely long circulation in blood, *Cancer Res.*, **51**, 3229-3236.

107. Yokoyama, M., Kwon, G.S., Okano, T., Sakurai, Y., Seto, T. and Kataoka, K. (1992) Preparation of micelle-forming polymer-drug conjugates, *Bioconj. Chem.*, **3**, 295-301.

108. Yokoyama, M., Kwon, G.S., Okano, T., Sakurai, Y., Ekimoto, H., Okamoto, K., Seto, T. and Kataoka, K. (1993) Optimization of in vivo antitumor activity of micelle-forming polymeric drug against murine colon adenocarcinoma 26, *Drug Deliv.*, **1**, 11.

109. Kataoka, K., Kwon, G.S., Yokoyama, M., Okano, T. and Sakurai, Y. (1993) Block copolymer micelles as vehicles for drug delivery, *J. Control. Rel.*, **24**, 119-132.

110. Kataoka, K., Kwon, G.S., Yokoyama, M., Okano, T. and Sakurai, Y. (1992) Polymeric micelles as novel drug carriers and virus mimicking vehicles in J. Kahovec (ed.), *Macromolecules*, 267-276.

111. Yokoyama, M., Sugiyama, T., Okano, T., Sakurai, Y., Naito, M. and Kataoka, K. (1993) Analysis of micelle formation of an adriamycin-conjugated poly(ethylene glycol)-poly(aspartic acid) block copolymer by gel permeation chromatography, *Pharm. Res.*, **10(6)**, 895-899.

112. Kwon, G.S., Yokoyama, M., Okano, T., Sakurai, Y. and Kataoka, K. (1993) Biodistribution of micelle-forming polymer-drug conjugates, *Pharm. Res.*, **10(7)**, 970-974.

113. Kwon, G., Suwa, S., Yokoyama, M., Okano, T., Sakurai, Y. and Kataoka, K. (1994) Enhanced tumor accumulation and prolonged circulation times of micelle-forming poly(ethylene oxide-aspartate) block copolymer-adriamycin conjugates, *J. Control. Rel.*, **29**, 17-23.

114. Kwon, G., Naito, M., Yokoyama, M., Okano, T., Sakurai, Y. and Kataoka, K. (1993) Micelles based on AB block copolymers of poly(ethylene oxide) and poly(β-benzyl L-aspartate), *Langmuir*, **9**, 945-949.

115. Kwon, G.S., Naito, M., Kataoka, K., Yokoyama, M., Sakurai, Y. and Okano, T. (1994) Block copolymer micelles for hydrophobic drugs, *Colloids Surf. B : Biointerfaces*, **2**, 429-434.

116. Kwon, G.S., Naito, M., Yokoyama, M., Okano, T., Sakurai, Y. and Kataoka, K. (1995) Physical entrapment of adriamycin in AB block copolymer micelles, *Pharm. Res.*, **12(2)**, 192-195.

117. Cammas, S., Nagasaki, N. and Kataoka, K. (1995) Heterobifunctional Poly(ethylene oxide) : Synthesis of α-methoxy-ω-amino and α-hydroxy-ω-amino PEOs with the same molecular weights *Bioconj. Chem.*, **6(2)**, 226-230.

118. Cammas, S. and Kataoka, K. (in press) Functional poly(ethylene oxide)-co-poly(β-benzyl-L-aspartate) polymeric micelles : block-copolymer synthesis and micelles formation, *Macromol. Phys. Chem.*

119. Harada, A., Kataoka, K. (in press) Formation of polyion complex micelles in an aqueous milieu from a pair of oppositely-charged block copolymers with poly(ethylene glycol) segments, *Macromolecules*.

EXPERIMENTAL RESULTS ON HYDROPHOBE RELEASE

EMIN ARCA
Hacettepe University
Chemical Engineering Department
06532 Beytepe Ankara Turkey

1. Introduction

Physicochemical properties of self-organizing molecules are of ever increasing experimental interest. This is due to important applications of such materials in coating, adhesives, thin films, micro fabrication of electronic devices, pharmaceutical and photographic technologies, oil recovery, etc. Among the most important self-organizing systems are polymeric micelles. It is well known that block copolymers, when dissolved in selective solvents, i.e., solvents good for one block but poor for the other, assemble into spherical micelles with a dense core formed by the insoluble blocks and a corona consisting of soluble blocks. These micelles differ from the more familiar detergent micelles by their larger size and greater stability. Copolymers with hydrophobic and hydrophilic blocks may form micelles in aqueous media. These aqueous micellar systems are potentially useful for applications in the fields of pharmacology, ecology, and agriculture. They may take up and release organic materials from water solutions or may be utilized as delivery vehicles for hydrophobic drugs or pesticides, as scavengers of hydrophobic pollutants, etc. In this study we are exploring the fundamental aspects of the release of a model compound from the block copolymer micelles with hydrophobic cores and hydrophilic shells into the surrounding media. This kind of self-organized block copolymer micelles are promising candidates as carriers of hydrophobic substances into aqueous media (1). The micelles are formed by block copolymers of styrene and methacrylic acid in a mixture of a good solvent for both blocks (80 vol% of dioxane) and a precipitant for the polystyrene block (20% of water). In this mixture the micelles have a narrow distribution of sizes and the polystyrene core is substantially swollen by dioxane. As a model compound we have selected phenanthrene because it has a good affinity for polystyrene and is strongly hydrophobic. Moreover, its fluorescent properties are very convenient for following the kinetics of its release from the core.

2. Experimental Section

2.1. POLYMER SYNTHESIS.

The copolymers used in this study were prepared by anionic polymerization of styrene followed by addition of tert-butyl methacrylate. The tert-butyl group is then removed by hydrolysis to produce the poly(methacrylic acid). A more detailed description of this synthesis were described elsewhere (2,3).

S.E. Webber et al. (eds.), Solvents and Self-Organization of Polymers, 115–120.

2.2. LOADING OF MICELLES

For preparing chromophore loaded micelles 100 mg of copolymer were dissolved in the mixed solvent (80 vol% of dioxane, 20 vol% water). To the micellar solution were added 5 mg of phenanthrene in the same solvent . The total volume was 10 mL. The mixture was shaken overnight. Presumably, the phenanthrene distributed itself between the core and the outside solvent (including the shell). At this stage the core was swollen and full equilibrium was probably achieved. The mixture was then freeze-dried at -15°C.

2.3. FLUORESCENCE QUENCHING.

Luminescent probe experiments are very powerful techniques in the study of block copolymer micelles (4-8). The objective of the fluorescence quenching studies is to learn how rapidly small molecules can be released from the micelles. The most dramatic experiments are those in which one adds a quencher Q to the aqueous phase and monitors the reduction of fluorescence as a function of time. The general mechanism for the quenching process (9) is summarized in Scheme I.

Scheme I

$$M + h\nu \rightarrow M^* \text{ (Light absorption)} \quad (1)$$

$$M^* \rightarrow M + h\nu \text{ (Fluorescence emission)} \quad (2)$$

$$M^* + Q \rightarrow Q^* + M \text{ (Quenching)} \quad (3)$$

$$Q^* \rightarrow Q + \text{energy} \quad (4)$$

For fluorescence measurements 3 mL of solution was placed 1.0*1.0 cm square quartz cells. All spectra were run on SPEX Fluorolog 2 spectrometer using slit openings of 2.5 mm. This system allow acquisition and integration of standard steady state fluorescence spectra. Spectra were accumulated with an integration time of 1s/1 nm.

3. The Kinetics of the Release

30 μL of 0.5 mg/mL solution of the micelles loaded with phenanthrene in a phosphate buffer pH 7.0 were injected into a fluorescence cell containing 3 mL of the same buffer. Presumably this dilution made the chemical potential of phenanthrene outside the micelles much smaller than its chemical potential inside the micellar cores and started the diffusive release of phenanthrene. The experiment was performed in duplicate: one of the cells contained also 9 mg of thallium nitrate - a contact ionic quencher that cannot penetrate the hydrophobic polystyrene core. No quencher was added to the other cell. In a separate experiment we have found that under these conditions the quencher reduced the fluorescence of phenanthrene dissolved in our buffer by the factor of 0.300. Immediately after the injection of micelles we have started to measure repeatedly (as a function of elapsed time) the spectra of samples in both cells. The excitation wavelength was 293 nm, we have measured the total intensity emitted in the region of 315-450 nm. In order to account for the instrumental drift and possible fluctuation of the lamp intensity we have also measured a third, reference cell containing a dilute solution of phenanthrene in an organic solvent and corrected the data accordingly.

When phenanthrene diffuses out of the core, the quantum yield of its fluorescence changes due to (i) the change of environment and (ii) the presence of the quencher (in one of the cells). By analyzing the time dependent intensities in the two cells and knowing the quenching factor we can evaluate (a) the change of the quantum yield caused by the transfer between the core and the surroundings, (b) the fraction of phenanthrene residing outside the core at the beginning of the experiment, and (c) the time dependence of this fraction. We have performed this experiment using two micellar samples SA-23 and SA-24. The relevant data are presented in Table 1.

TABLE I. Properties of Block Copolymers and Micelles

Sample	M_w (a)	W_{ps}% (b)	$M_w x10^6$ (c)	r_{core} (nm)
SA - 23	66,300	60.0	9.0	25.0
SA - 24	47,700	55.0	3.7	18.0

(a) Molecular weight of block copolymer measured by GPC.
(b) Weight fraction of polystyrene measured by NMR.
(c) Molar mass of micelles in 80D/20W.

3.1. KINETIC ANALYSIS

Block copolymer micelles are usually modeled as containing a spherical core surrounded by a concentric shell. We expect that polystyrene cores in aqueous media are not swollen and that any diffusion in glassy materials will be very slow. However, diffusion of small molecules within the micellar shells will be only slightly slower than their diffusion in the outside solvent. In any case, it will be orders of magnitude faster than their diffusion in the core. We have therefore modeled the release from micelles as a release from a sphere that is uniformly loaded by the probe. The diffusion coefficient of the probe within the sphere is constant and independent of its concentration. Once the probe is released from the sphere it is immediately dispersed uniformly throughout the whole volume of the sample. We are assuming that at the start of the experiment some fraction of the probe was already present outside the sphere. (This fraction corresponds to the probe that is either associated with the shells or simply did not enter the micellar region during the loading procedure). After long time the system will reach an equilibrium governed by the distribution coefficient K_d which is related to the partition coefficient K_p as

$$K_d=K_p(V_s/V_l)=(C_s/C_l)(V_s/V_l) \qquad (5)$$

where V is volume and C is concentration of probe; the subscript s and l refer to the sphere and the outside liquid, respectively.

We have simulated the diffusion/release process using a computer program and the finite difference form of Fick's law in spherical coordinates (10). As the starting outside fraction of the probe we used the values evaluated as described above; we calculated the dependence for a range of values of K_d. It was convenient to plot the results as f_1, the fraction of the probe residing outside of the cores, against ln τ where the reduced time τ is defined as:

$$\tau \equiv Dt/r^2 \qquad (6)$$

where D is the diffusion coefficient, t is time, and r is the radius of the sphere.
In these coordinates the dependencies are S- shaped with the original asymptote corresponding to the starting value of the outside fraction and the asymptote at long times reflecting the equilibrium and its distribution coefficient. Dependencies of f_1 vs lnτ for several values of K_d are presented in Figures 1. and 2. Superimposed on these plots are our experimental curves for the micelles SA23 and SA-24 plotted in f_1 vs. lnτ coordinates. From the shift along the τ axis it is possible to evaluate the quantity D/r^2 for the best fit; the best fit also yields the distribution coefficient. From the known concentration of the micelles and from w_{ps}, the weight fraction of polystyrene in the block copolymers, K_p is easily evaluated. The radius of the core was estimated from the molecular weight of the micelles (measured by an independent experiment) and w_{ps}. Knowing r we easily calculated the diffusion coefficient D. The relevant data are collected in Table II.

TABLE II Diffusion, Distribution and Partition Coefficients of Phenanthrene in Micellar Cores

Micelle	$Dx10^{18}(cm^2/s)$	K_d	$K_p x10^{-5}$
SA - 23	5.32	1.25	4.40
SA - 24	2.19	1.5	3.45

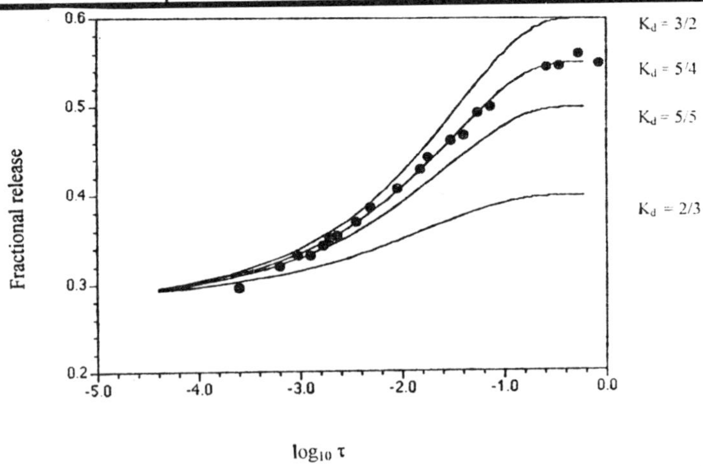

Fig.1. Fractional Release of Phenanthrene from SA-23 micelle in pH=7.0 buffer, as afunction of log τ.

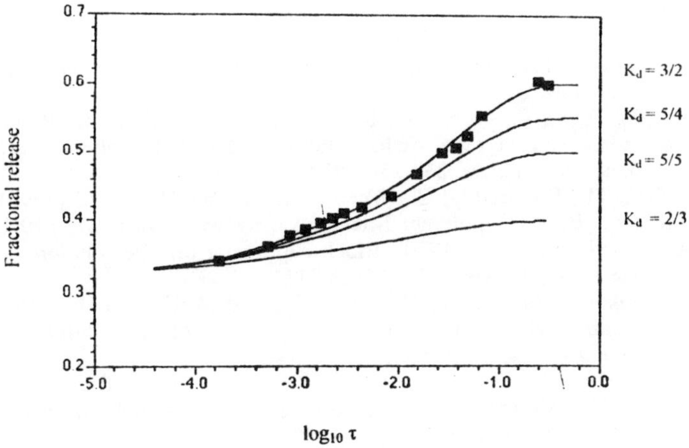

Fig.2. Fractional Release of Phenanthrene from SA-24 micelle in pH=7.0 buffer, as afunction of log τ.

4. Results and Discussion

A fluorescence based technique was developed for following the release of hydrophobic substances from the cores of block copolymer micelles in aqueous media. Using this technique primarily with phenanthrene as a fluorescence probe, we have examined release properties of polymer micelles formed from diblock poly(methacrylic acid) - block - polystyrene. The micelles are very effective in solubilizing hydrophobic aromatics (11,12) , with a partition coefficient on the order of 10^5 with respect to water (5). The release curves were compatible with theoretical dependencies for diffusion controlled release from spherical particles. The diffusion coefficients found are of the order expected for molecules with sizes of our probes and glassy polystyrene. The partition coefficients are also of the expected order of magnitude. The fraction of the probe present outside of the cores at the start of the experiment may serve to characterize the efficiency of various procedures for the loading of various agents into the core.

Acknowledgment

The author would like to express his gratitude to Professor P. Munk and Professor S. E. Webber for providing opportunities to accomplish this study.

5. References

1. Tuzar, Z., Kratochvil, P., in Matijevic, E., Ed., (1993), *Surface and Colloid Science, Plenum* Press: vol. 15, New York.

2. Ramireddy, C., Tuzar, Z., Procházka, K., Webber, S. E., and Munk, P., (1992) Styrene-tert-Butyl Methacrylate and Styrene-Methacrylic Acid Block Copolymers: Synthesis and Characterization, *Macromolecules,* **25**, 2541-2545.

3. Qin, A., Tian, M., Ramireddy, C., Webber, S. E., and Munk, P., (1994) Polystyrene-Poly(methacrylic acid) Block Copolymer Micelles, *Macromolecules,* **27**, 120-126.

4. Yeung, A. S., and Frank, W., (1990), Block copolymer micelle solutions: 2. An intrinsic excimer fluorescence study, *Polymer*, **31**, 2101- 2111.

5. Cao, T., Munk, P., Ramireddy, C., Tuzar, Z., and Webber, S. E., (1991) Studies of Amphiphilic Poly(methacrylic acid) - block - Polystyrene-block - Poly(methacrylic acid) Micelles, *Macromolecules*, **24**, 6300-6305.

6. Wilhelm, M., Zhao, C., Wang, Y., Xu, R., Croucher, M. D., Riess, G., Mura, J., and Winnik, A. M., (1991), Poly(styrene - ethylene oxide) Block Copolymer Micelle Formation in Water : A Probe Study, *Macromolecules*, **24**,1033 - 1040

7. Kiserow, D., Prochazka, K., Ramireddy, C., Tuzar,Z., Munk, P., and Webber, S.E., (1992), Fluorimetric and Quasi - Elastic Light Scattering Study of the Solubilization of Nonpolar Low - Molar Mass Compounds into Water - Soluble Block - Copolymer Micelles, *Macromolecules*, **25**, 461 - 469.

8. Hruska, Z., Piton, M., Yekta, A., Duhamel, J., Riess, G., Croucher, D. M., and Winnik, M. A., (1993), *Macromolecules*, **26**, 1825 - 1828.

9. Guilbault, G. G., (1973), *Practical Fluorescence*, Marcel Dekker, New York.

10. Crank,J., (1975), *The Mathematics of Diffusion*, Clarendon Press, Oxford.

11. Nagarajan, R., Ganesh, K., (1989), Block Copolymer Self - Assembly in Selective Solvents: Theory of Solubilization in Spherical Micelles., *Macromolecules*, **22**, 4312 -

12. Tian, M., Arca, E., Tuzar, Z., Webber, S. E., Munk, P., (1995), Light Scattering Study of Solubilization of Organic Molecules By Block Copolymer Micelles in Aqueous Media, *J. Appl. Poly. Science., part B, Poly. Phys.* **33**, 1713

SOLUBILIZATION OF HYDROPHOBIC SUBSTANCES BY BLOCK COPOLYMER MICELLES IN AQUEOUS SOLUTIONS

R. NAGARAJAN
Department of Chemical Engineering
The Pennsylvania State University
University Park, PA 16802

1. Introduction

Block copolymer molecules consisting of hydrophobic and hydrophilic blocks display the tendency to aggregate in aqueous solutions, forming structures well-known as micelles. In these micelles, the hydrophilic blocks constitute the shell region while the hydrophobic blocks form the micellar core. This phenomenon is entirely analogous to the formation of micelles by the conventional low molecular weight surfactant molecules. The concentration at which the micelles are first detected is known as the critical micelle concentration (CMC). In the case of block copolymers, because of the large size of the hydrophobic block, the CMC is often negligibly small when compared to the CMCs of low molecular weight surfactant systems.

One of the most useful properties of the micellar aggregates is their ability to enhance the aqueous solubility of hydrophobic substances which are otherwise only sparingly soluble in water. The enhancement in the solubility arises from the fact that the micellar cores can serve as a compatible micro environment for the water-insoluble solute molecules. This phenomenon of enhanced solubility is referred to as solubilization. The solubilization characteristic of block copolymer micelles holds great potential for the development of aqueous block copolymer solutions as environmentally benign substitutes for the organic solvents currently being used in applications such as reaction media in chemical process industry, as chemical extractants in separation processes, and as industrial cleaning agents. Other potential applications such as tissue-specific drug delivery are also based on the exploitation of the solubilization process.

Extensive experimental studies of solubilization in aqueous solutions of conventional surfactants have appeared in the literature [1]. In contrast, there have been very few studies of the solubilization tendencies of low molecular weight hydrophobic substances in block copolymer micelles. In general, one is interested in knowing the solubilization capacity of the micelles for different hydrophobic substances (also referred to as solubilizates), the changes in the size and shape of the micelles as a consequence of solubilization and the change in the CMC due to the presence of the solubilizates. In our earlier paper [2], experimental results on the

S.E. Webber et al. (eds.), Solvents and Self-Organization of Polymers, 121–165.

solubilization of aliphatic and aromatic hydrocarbons in micelles formed of poly(ethylene oxide)-poly(propylene oxide) (PEO-PPO) and poly(vinyl pyrrolidone)-polystyrene (PVP-PS) block copolymers were presented. Further, the selectivity in the solubilization was explored via measurements on binary mixtures of hydrocarbons. In a later paper [3a], we presented experimental results on the solubility of a number of alcohols, ethyl esters, ketones and aldehydes (all of which are typical components present in food flavors) in PEO-PPO block copolymer micelles. Also, the selectivity in solubilization of components of orange oil in the block copolymer micelles was investigated [3b]. Hurter and Hatton have conducted a systematic study of the solubilization of three polycyclic aromatics in solutions of PEO-PPO-PEO triblock copolymers and four-armed PEO-PPO star block copolymers of varying molecular weights and compositions [4a]. The location of solubilized benzene in micelles formed of the above copolymers were identified by Nivaggioli et al. [4b] as the PPO core region of the aggregates. The solubilization of xylene in PEO-PPO-PEO triblock copolymer micelles and the changes in aggregation number induced by the solubilizate were measured by Chu and coworkers [5a,b]. They observed a growth in the aggregation number and the hydrodynamic radius of the micelle on solubilization and also concluded that xylene is present in the PPO core of the micelle. The solubilization of fluorescent probes in PEO-PPO-PEO triblock copolymer solutions has been investigated by Kabanov et al. [6].

From a theoretical point of view, the solubilization in PEO-PPO diblock copolymer micelles and PEO-PPO-PEO triblock copolymer micelles were treated using a mean-field approach in our earlier work [7a,b] considering spherical as well as non-spherical shapes for the micellar aggregates. Subsequently, a treatment based on the star polymer model was presented using the scaling approach and focusing on spherical micelles [8]. A scaling analysis of solubilization in diblock copolymer aggregates has been presented by Dan and Tirrell [9] who have investigated the formation of spherical, cylindrical and lamellar structures. The solubilization phenomenon has been treated using a self-consistent mean-field approach taking into account composition inhomogeneities inside the aggregate structure by Cogan et al. [10], Hurter and Hatton [11a,b], Linse [12] and Leermakers et al. [13].

This paper emphasizes our own work and is organized as follows. In section 2, we briefly mention the methods used for the measurement of the solubilization capacity of the micelles and present results for many hydrophobic solutes. This is followed by a thermodynamic analysis of solubilization in section 3 using the mean-field approach. The free energy expressions are presented here in a general manner so that they can be applied to both diblock and triblock copolymers (with the middle block being hydrophobic) and also to spherical, rodlike and lamellar aggregates. Section 4 explores the influence of solubilization on shape transitions of the aggregates. The final section presents some general conclusions.

The solubilizates in block copolymer micelles need not be limited to low molecular weight hydrophobic substances. Homopolymers have been solubilized in dilute solutions of block copolymer micelles [14a,b]. Also, high molecular weight membrane integral proteins with hydrophobic characteristics have been solubilized in block copolymer systems [15]. However, this paper will be confined to a study of the solubilization of only low molecular weight hydrophobic substances in dilute aqueous solutions of block copolymer micelles.

2. Experimental Studies

2.1 DETERMINATION OF SOLUBILIZATION CAPACITY OF MICELLES

The solubilization capacity of micelles has been experimentally determined using simple, classical methods [2,3]. The turbidity method involves sequential addition of the solute to a block copolymer solution containing a specified amount of the block copolymer and measuring the turbidity of the solution. For concentrations of the solute below the solubilization capacity, the micellar solution is optically clear and no turbidity is observed. As the solubilization limit is approached a change in turbidity follows. The turbidity vs concentration relation is linear both below the solubilization limit and also immediately above the solubilization limit. Therefore, from an intersection of the two straight lines on such a plot, the solubilization limit is estimated.

Another method involves contacting the aqueous micellar solution with the pure solubilizate phase until equilibration is achieved. The phases are then separated and the concentration of the solute in the aqueous phase is determined by a direct method such as gas chromatography. Such direct analysis is the only method applicable for the measurement of solubilization of mixtures of solutes. For solutes whose aqueous solubility is very small as in the case of the polycyclic aromatics investigated by Hurter and Hatton [4a], special consideration needs to be given to achieving equilibration. Kabanov et al. used the fluorescence emission data to estimate the partition coefficient for the two probe molecules [6].

The solubilization characteristics of the block copolymer micelles have been presented in the literature in terms of micelle-water partition coefficients defined as the ratio between the concentration of the solubilizate inside the micelle and the concentration of the solubilizate that is molecularly dispersed in the aqueous phase. The concentrations have been expressed either as mole fractions or as molarities. The solubilization characteristics have has also been specified in terms of the maximum amount of solubilization in the micelles. This limiting solubilization occurs when the aqueous micellar solution is allowed to coexist with an excess solubilizate phase. The solubilization capacity can also be expressed in the form of the volume fraction of the solubilizate in the micellar core. This is defined as the ratio between the volume of the solubilizate in the micelle and the sum of the volumes of both the solubilizate and the hydrophobic blocks of the copolymer present in the micelle. In estimating the amount of the solubilizate in the micelle, the amount of molecularly dissolved solubilizate in the aqueous solution is subtracted from the total amount of the solubilizate present in the solution (for very hydrophobic solubilizates, the amount of molecularly dissolved solubilizate is negligibly small). Similarly, the amount of micellized block copolymer is taken to be the difference between the total block copolymer concentration and the CMC (which is often too small and hence, neglected).

2.2. SOLUBILIZATION OF HYDROCARBONS

The solubilization capacity of block copolymer micelles for aromatic and aliphatic hydrocarbons have been determined using the gas chromatographic method mentioned above. Two block copolymer molecules were employed: PEO-PPO having a molecular weight of 12,500 and containing 30 weight percent PPO (Polysciences Inc.) and PVP-PS of unspecified molecular weight having 60 weight percent PS (Scientific Polymer Products Inc.). The copolymers were not well characterized in terms of the molecular polydispersity and there is some uncertainty as to whether the PEO-PPO copolymer is a diblock or triblock copolymer [4a]. The extent of solubilization of a number of aromatic and aliphatic hydrocarbons in aqueous solutions of 10 weight percent PEO-PPO block copolymer or 20 weight percent PVP-PS block copolymer are listed in Table 1. Also listed for comparison are the results obtained in solutions of a conventional low molecular weight surfactant, sodium dodecyl sulfate (SDS).

TABLE 1. Solubilization capacity of block copolymer micelles for aromatic and aliphatic hydrocarbons at 25°C and some molecular properties of the solubilizates

Solubilizate	Molecular properties of solubilizates			mmoles solubilized per gram of hydrophobic block		
	v_J Å^3	σ_{JW} dyne/cm	δ_J MPa$^{1/2}$	PEO-PPO	PVP-PS	SDS
Benzene	146	33.93	18.80	11.67	30	9.9
Toluene	176	36.1	18.19	6.33	14.8	8
O-Xylene	200	36.1	18.40	4.0	31.6	4.43
Ethyl benzene	204	38.4	17.99	5.67	26.7	-
Cyclohexane	179	50.2	16.76	1.97	3	4.6
Hexane	217	50.7	14.92	0.667	0.77	2.39
Heptane	243	51.2	15.13	0.567	0.47	-
Octane	270	51.5	15.53	0.5	0.18	-
Decane	323	52	15.74	0.387	0.072	1.18

In Table 1, the limiting amounts solubilized are expressed as mmoles solubilized per gram of the hydrophobic block of the copolymer (or of the hydrophobic surfactant tail). Also listed in the Table are some important molecular characteristics of the solubilizates including their molecular volume (v_J), their interfacial tension against water (σ_{JW}), and their Hildebrand-Scatchard solubility parameter (δ_J) values. The subscript J is used to denote the solubilizate while W refers to the solvent water. Subscripts A and B denote the hydrophobic and hydrophilic blocks of the copolymer, respectively.

The most notable feature of the results is the large difference in the amounts solubilized of aromatic hydrocarbons compared to aliphatic hydrocarbons. The 4-fold difference between the solubilized moles of benzene and hexane in the low molecular weight surfactant SDS is replaced by a 17-fold difference in the PEO-PPO block copolymer and by a 40-fold difference in the PVP-PS block copolymer.

The amounts solubilized in the low molecular weight surfactant micelles have been correlated with sufficient accuracy to the molecular volume of the solubilizates (v_J), and also to a non-dimensional volume-polarity parameter defined as ($\sigma_{JW} v_J^{2/3}/kT$) where k is the Boltzmann constant and T is the absolute temperature [16]. The solubility of various solubilizates in water needed for the calculation of the solubilization capacities is obtained from ref. [17]. In defining the volume-polarity parameter, the solubilizate-water interfacial tension is taken as a measure of the solubilizate's polarity and the consequent interfacial activity. The solubilization data for the block copolymers and for the low molecular weight SDS listed in Table 1 are correlated to these molecular properties in Figures 1 and 2. The correlations show that the amount solubilized for a homologous family of solubilizates decreases with increasing molecular size of the solubilizate. The difference between the aromatic and aliphatic solubilizates of similar molecular volumes (toluene and cyclohexane, for example) has been explained in terms of the polarity and the consequent interfacial activity of the aromatics and the correlation based on the polarity-volume parameter has been found to be quite satisfactory for many surfactants [16].

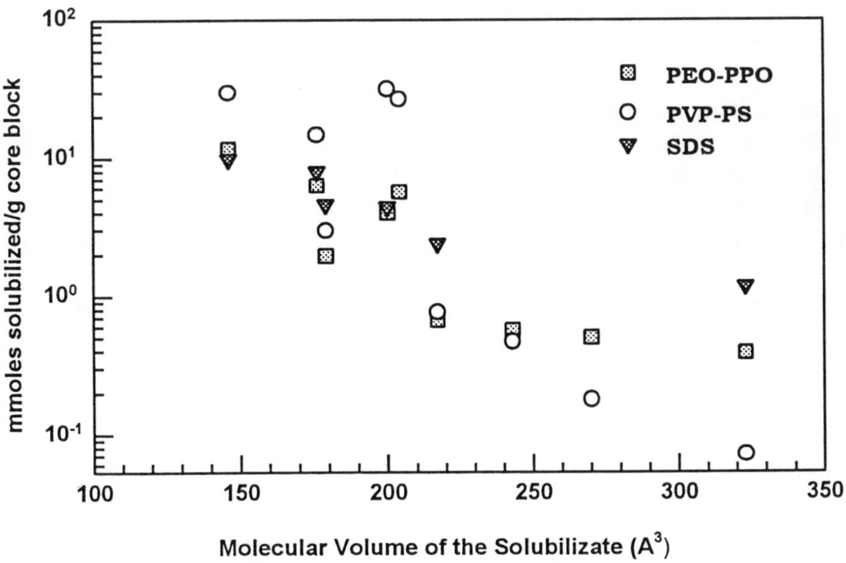

Figure 1. Dependence of the amount of hydrocarbons solubilized on the molecular volume of the hydrocarbons.

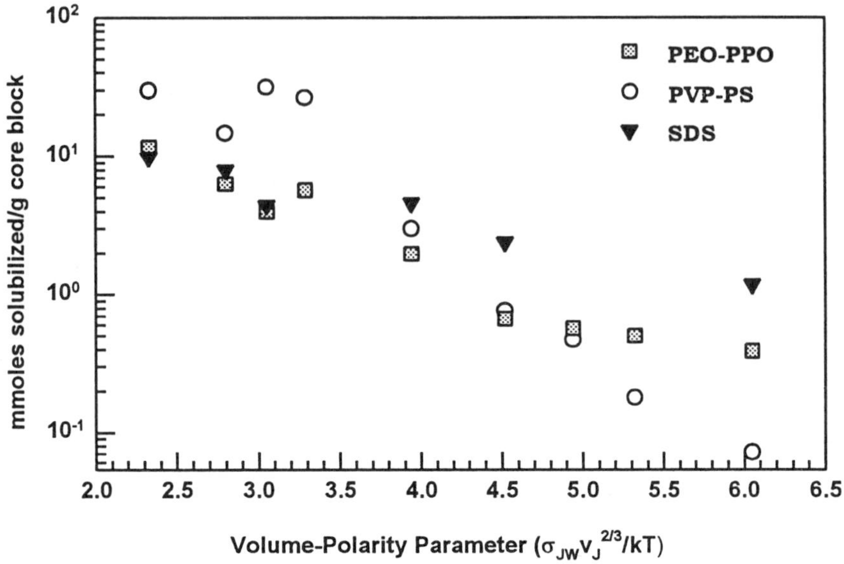

Figure 2. Dependence of the amount of hydrocarbon solubilized on the polarity-volume parameter of the hydrocarbons.

Besides the above correlations which have been applied to low molecular weight surfactants, one can expect in the case of polymeric surfactants, that the interactions between the core block A of the copolymer and the solubilizate J may be directly related to the solubilization capacity. These interactions are represented by the Flory interaction parameter χ_{AJ} which can be estimated [2] from knowledge of the Hildebrand solubility parameters of both A and J via the relation $\chi_{AJ} = (\delta_A - \delta_J)^2 v_J/kT$, where δ_A is the solubility parameter for the core block. Taking δ_A to be 19 $MPa^{1/2}$ for PPO, 18.6 $MPa^{1/2}$ for PS, and 15.94 $MPa^{1/2}$ for the dodecyl tail of the SDS, and using the molecular properties of solubilizates listed in Table 1, we can calculate the Flory interaction parameters for all the solubilizates [2]. The measured solubilization capacity is correlated in Figure 3 with the Flory interaction parameter. One can see that the aromatic solubilizates with the small χ_{AJ} values would constitute good solvents for the PPO and the PS blocks. In contrast, the aliphatic solubilizates are poorer solvents for these blocks. A reasonable correlation between the solubilization capacity and the Flory parameter is seen for both block copolymers. In contrast, for the low molecular weight surfactant SDS, one observes the counter-intuitive result of an increasing amount of solubilization with increasing χ_{AJ}. Obviously, the Flory parameter bears no correlation in this case. This behavior can be traced to the smaller aggregate sizes and the relatively more prominent role for the interfacial interactions in the case of SDS micelles. Consequently, the volume-polarity parameter exhibits a better correlation in SDS micelles. When compared to the SDS micelles, the block copolymer micelles are larger in size and one therefore, expects to find a more important role for the interactions in the core of the aggregate.

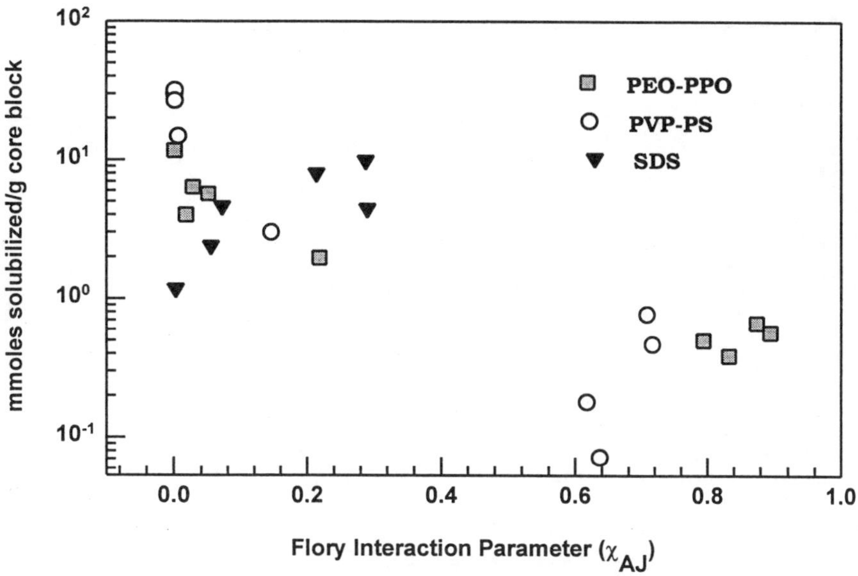

Figure 3. Dependence of the amount of hydrocarbon solubilized on the Flory interaction parameter between the hydrocarbon and the core block.

2.3. SELECTIVE SOLUBILIZATION FROM MIXTURES

The single component solubilization results in Table 1 suggest that solubilization will be selective for aromatic molecules when mixtures of aromatic and aliphatic molecules are solubilized. This is confirmed by the experimental results obtained in the PEO-PPO and PVP-PS block copolymer systems for binary mixtures of benzene and hexane. The amounts of each hydrocarbon solubilized as a function of the composition of the bulk solubilizate phase that coexists with the micellar solution are shown in Figure 4 for the PEO-PPO block copolymer and in Figure 5 for the PVP-PS block copolymer. One can observe that over the entire composition range, the amount of hexane solubilized is very small. Such selectivity was seen also in the case of low molecular weight surfactant micelles. However, in SDS micelles the selectivity diminishes when the size of the aromatic solubilizate increases from benzene to xylene as shown by the data in Table 1. In contrast, the block copolymer systems display large selectivity for aromatic hydrocarbons of differing sizes as can be seen from the results for the PVP-PS copolymer in Table 1. In this case, the compatibility of the various aromatic solubilizates with the PS block of the micellar core results in high selectivity for these molecules irrespective of their molecular size variations. The selectivity in the solubilization behavior is of practical importance to chemical separations.

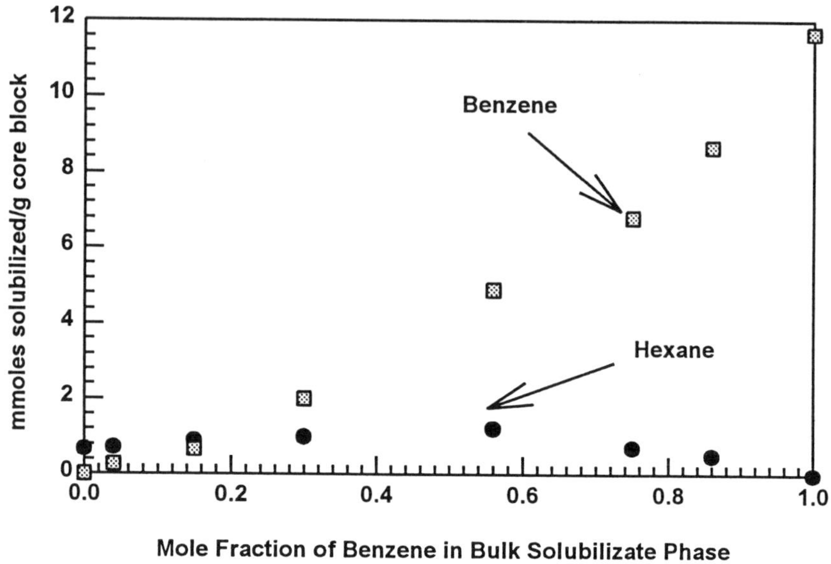

Figure 4. Amounts of benzene and hexane solubilized in the aqueous PEO-PPO micellar solution as a function of the composition of the hydrocarbon phase coexisting with it.

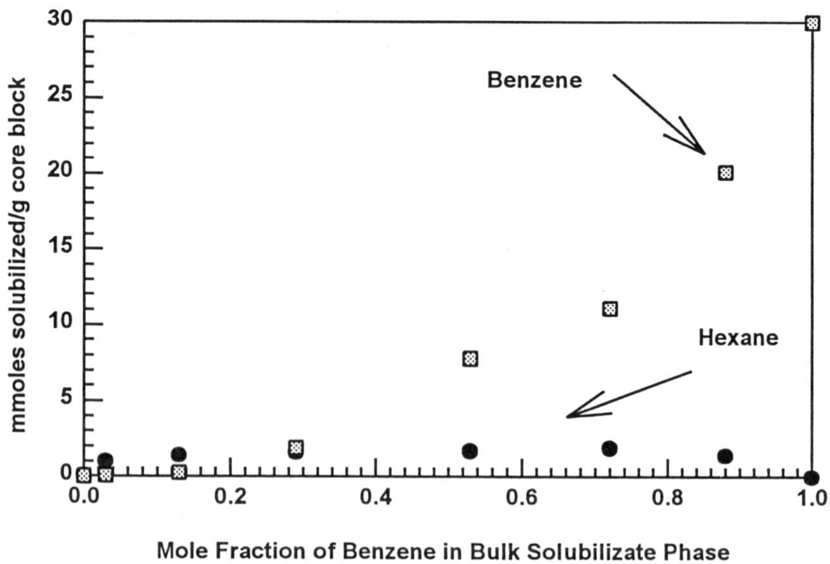

Figure 5. Amounts of benzene and hexane solubilized in the aqueous PVP-PS micellar solution as a function of the composition of the hydrocarbon phase coexisting with it.

3. Thermodynamics of Solubilization

A theory of solubilization should allow the a priori prediction of the micellar core dimension R, the thickness of the micellar shell D, the aggregation number of the micelle g, and the volume fraction of the solubilizate inside the micellar core region η as a function of the molecular properties of the copolymer, the solvent, and the solubilizate. Such a theory was developed in our earlier work [7] by adapting the theoretical approaches to the formation of block copolymer micelles pioneered by de Gennes [18], Leibler, Orland and Wheeler [19], and Noolandi and coworkers [20]. The thermodynamic treatment is outlined below for both diblock and triblock copolymer molecules and for aggregates having spherical, cylindrical or lamellar structures.

3.1. GENERAL THERMODYNAMIC RELATIONS

3.1.1. *Size and Composition Distribution of Aggregates*

We consider a solution consisting of solvent molecules, singly dispersed copolymer and solubilizate molecules and micellar aggregates. The aggregates are assumed to be made up of g block copolymer and j solubilizate molecules where g and j are allowed to take all possible values. Each of the above species, including micelles of different sizes, shapes and compositions, is treated as a distinct chemical component. The standard state of the solvent is defined to be that of pure solvent whereas the standard states of all the other components are taken to be those corresponding to infinitely dilute solution conditions. The standard chemical potentials of the solvent (water, designated by W), the singly dispersed copolymer, the singly dispersed solubilizate (designated by J) and micelles of aggregation number g containing j solubilizate molecules, are denoted by μ_W^o, μ_1^o, μ_J^o and μ_g^o, respectively. The distribution of micelle size and composition at equilibrium can be obtained by minimizing the total free energy of the system [7]. Denoting the mole fraction of species i by X_i, the micelle size and composition distribution equation can be written in the form

$$X_g = X_1^g \ X_{1J}^j \exp - \left(\frac{\mu_g^o - g\,\mu_1^o - j\,\mu_{1J}^o}{kT} \right) \tag{1}$$

In writing eq. (1), it is assumed that either the inter-micelle interactions are not present or that they do not affect the size distribution expression. Also the system entropy of the multicomponent solution is written as for an ideal solution. In general, the expression for the aggregate size and composition distribution depends on the assumptions pertaining to the system entropy as well as the nature of inter-aggregate interactions. In dilute solutions such as those of interest here, the inter-micellar interactions are not important. Therefore, one may neglect the free energy contributions associated with such interactions. The consequences of using a few plausible models of system entropy in the theory of micellization have been analyzed in detail in reference [21]. It was found that for nonionic systems such as those considered here, the choice of a model for the system entropy affects only the magnitude of the CMC and has no influence on the micellar size parameters. Also, the value for the CMC predicted on the basis of one of the entropy models can be simply related to those predicted on the basis of alternate entropy models.

From the size distribution equation, various size dependent properties of the solution can be computed. The number, weight and z-average aggregation numbers of the micelles are defined respectively by the relations

$$g_n = \frac{\sum g X_g}{\sum X_g} \quad , \quad g_w = \frac{\sum g^2 X_g}{\sum g X_g} \quad , \quad g_z = \frac{\sum g^3 X_g}{\sum g^2 X_g} \tag{2}$$

The average molar ratio of the solubilizate to the block copolymer molecules in the micelle is calculated from

$$\left(\frac{j}{g}\right)_{avg} = \frac{\sum j\, X_g}{\sum g\, X_g} \tag{3}$$

In the above two equations, the summation refers to that over the aggregation number g as well as the number of solubilizate molecules j contained within a micelle.

The CMC can be estimated by plotting any one of the above size-dependent solution properties as a function of the total concentration of the block copolymer. The CMC can be identified as the total concentration at which one observes a sharp change in slope in any of the plotted solution properties.

If one wants to calculate the maximum amount of solubilization of a compound that is possible inside the micelle, then the concentration of the singly dispersed solubilizate X_{1J} is assigned a value equal to the saturation concentration X_{1J}^* of the solubilizate in water.

$$\mu_J^H = \mu_{1J}^o + kT \ln X_{1J}^* \tag{4}$$

The standard state of the solubilizate μ_J^H appearing in the above equation refers to a pure solubilizate phase. Denoting by f, the ratio between the molecularly dissolved solubilizate concentration X_{1J} and the saturation concentration X_{1J}^*, the micelle size and composition distribution equation simplifies to

$$X_g = X_1^g \, f^{j/g} \exp - \left(\frac{g\, \Delta\mu_g^o}{kT} \right) , \text{ where } \Delta\mu_g^o = \left(\frac{\mu_g^o}{g} - \mu_1^o - \frac{j}{g}\, \mu_{1J}^H \right) \tag{5}$$

The condition f=1, corresponding to the solubilization limit is achieved when the aqueous micellar phase coexists with the pure solubilizate phase.

3.1.2. *Pseudophase Description of Micelles*

The micelle containing the solubilizates can also be represented as a pseudophase in equilibrium with the singly dispersed solubilizate and copolymer molecules in solution. This is a convenient simplification generally used in the theory of micellization. For narrowly dispersed systems, this representation provides results practically identical to those obtained from the detailed size distribution calculations. The equilibrium characteristics of the micelle in the pseudophase approximation are obtainable from the condition

$$\frac{\partial}{\partial g} \left[\frac{\Delta\mu_g^o}{kT} \right] = 0 \ , \quad \frac{\partial}{\partial j} \left[\frac{\Delta\mu_g^o}{kT} \right] = 0 \quad \text{at} \quad g = g_{opt} \ , \quad j = j_{opt} \tag{6}$$

where g_{opt} and j_{opt} refer to the numbers of block copolymer and solubilizate molecules, respectively, constituting the optimal or equilibrium aggregate. The critical micelle concentration in the pseudophase approximation is calculated from eq. (5) to be

$$X_{CMC} = \exp \left(\frac{\Delta\mu_g^o}{kT} \right) \quad \text{at} \quad g = g_{opt} \ , \quad j = j_{opt} \tag{7}$$

The quantity $\Delta\mu_g^o$ appearing in eq. (5) and in various subsequent expressions represents the change in the reference state free energy when a singly dispersed block copolymer molecule from the aqueous solution and j/g solubilizate molecules from their pure phase are transferred into an isolated micelle in the aqueous solution. The magnitude of this free energy difference controls the CMC as shown by eq.(7). In contrast, the equilibrium structural features of the micelle are determined by how this free energy difference depends on the variables g and j as is evident from eq. (6). In order to formulate an expression for this free energy difference, the geometrical features of the micelle should be specified.

3.1.3 *Geometrical Properties of Aggregates*

We consider the formation of spherical, cylindrical and lamellar aggregates. Corresponding to each shape, one can visualize two kinds of aggregates depending upon how the solubilizate is contained inside the aggregates. The two types are illustrated in Figure 6 for aggregates having spherical shapes. These structures resemble the models employed for simple solubilization and microemulsification, respectively, in systems involving low molecular weight surfactants.

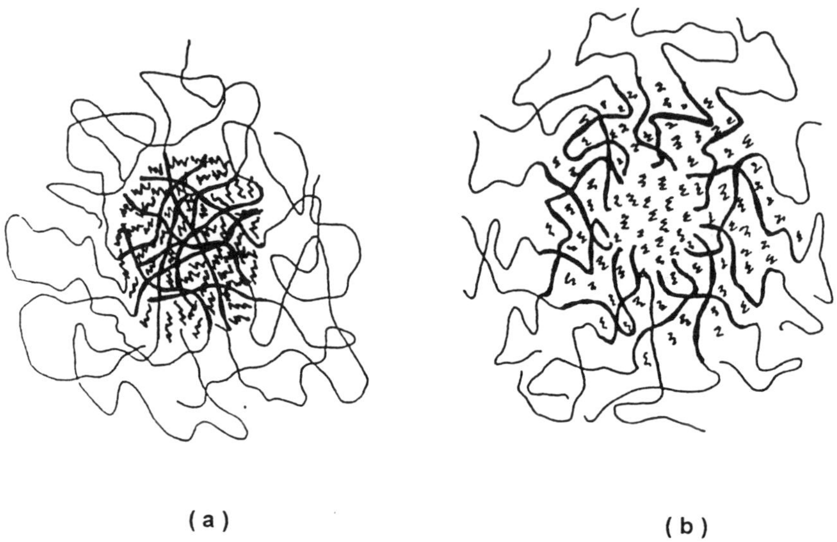

Figure 6. Schematic representation of a spherical micelle containing the solubilizate. The darker lines denote the hydrophobic block and the lighter lines, the hydrophilic block. In (a) all of the solubilizate molecules interact with the core block. In (b), a part of the solubilizate molecules are present in a separate domain while the remaining interact with the core block.

In the first type of structure, the micellar core is made up of the solvent incompatible A blocks and the solubilizate J. The solvent compatible B blocks and solvent W are present in the spherical shell region of the micelle. The second type of structure shown is analogous to that used for droplet microemulsions. Here, a region of pure solubilizate J is allowed to exist within the micellar core. This core is surrounded by an inner shell region consisting of the solvent incompatible block A and the solubilizate J while the outer shell of the micelle contains the solvent compatible block B and the solvent S. This structural model differs in two important respects from the first model. The solubilizate J is present as a pure fluid in the core and as a solution in the inner shell; the solvent incompatible block A stretches over only the inner shell region and does not extend to the center of the micelle. In the absence of a pool of pure solubilizate J in the core, the second structure reduces identically to the first type. Free energy calculations of the kind described below showed that the condition of minimum free energy always occurred corresponding to a zero size for the pure solubilizate pool. Thus, the thermodynamic equilibrium criterion always favored the occurrence of the first structure. Consequently, the free energy expressions corresponding to the first type of structure are discussed here.

The shape of the aggregates along with the assumption of incompressibility allow one to write geometrical relations describing the structure of the aggregates. These relations are summarized in Table 2.

TABLE 2. Geometrical properties of spherical, cylindrical and lamellar aggregates

Property	Sphere	Cylinder	Lamella
V_c	$4\pi R^3/3$	πR^2	$2R$
V_s	$V_c\,[(1+D/R)^3-1]$	$V_c\,[(1+D/R)^2-1]$	$V_c\,[(1+D/R)-1]$
g	$V_c\,(\phi_A/v_A)$	$V_c\,(\phi_A/v_A)$	$V_c\,(\phi_A/v_A)$
a	$3\,v_A/\,(R\phi_A)$	$2\,v_A/\,(R\phi_A)$	$v_A/\,(R\phi_A)$
ϕ_B	$(v_B/v_A)\,\phi_A\,(V_c/V_s)$	$(v_B/v_A)\,\phi_A(V_c/V_s)$	$(v_B/v_A)\,\phi_A(V_c/V_s)$

We will use the variable R to denote the hydrophobic core radius in the case of the spherical and the cylindrical micelles, and the half-thickness of the bilayer in the case of the lamellar aggregates. The variable D denotes the shell thickness in all three types of aggregates. The surface area of the aggregate core per constituent block copolymer molecule will be denoted by a. The number of block copolymer molecules in a micelle are denoted by g while the number of solubilizate molecules are denoted by j, with the understanding that both g and j refer to the total numbers of molecules in the case of spherical aggregates, numbers per unit length in the case of cylindrical aggregates and those per unit area in the case of lamellar aggregates. The molecular volumes of the A and the B blocks, the solubilizate and the solvent are denoted by v_A, v_B, v_J and v_W, respectively. The variables m_A, m_B and m_J refer to the ratios of the molecular volumes of block A, block B and the solubilizate J to that of the solvent W, respectively. The volume of the hydrophobic core of the aggregate V_c is calculated from the numbers of block copolymer and solubilizate molecules present in an aggregate as $V_c = g\,v_A + j\,v_J$. The volume of the hydrophilic shell of the aggregate is denoted by V_s. Both the core volume and the shell volume refer to that per unit length in the case of cylindrical aggregates and that per unit area in the case of lamellar aggregates. The volume fraction of the solubilizate molecules within the core of the micelle is denoted by η, where $\eta = j\,v_J\,/\,(g\,v_A + j\,v_J)$. We denote by ϕ_A, the volume fraction of the A block in the micellar core ($\phi_A = 1-\eta$), and by

ϕ_B, the volume fraction of the B block in the micellar shell. The polymer concentrations are thus assumed to be uniform both in the core of the micelle as well as in the shell. One may note that, if any three structural variables are specified, then all the remaining geometrical variables can be calculated through the relations given in Table 2. For convenience, we have chosen the variables R, D and η (or ϕ_A) as the independent variables, and use them to express other remaining variables.

3.2. MODEL FOR FREE ENERGY OF SOLUBILIZATION

An expression for the free energy of solubilization $\Delta\mu_g^o$ as defined in eq.(5) can be developed by considering all the physicochemical changes accompanying the transfer of the solubilizate molecules from its pure phase and the singly dispersed copolymer molecules from the solution to an isolated micelle also in the solution. Firstly, the transfer of the solubilizate and the singly dispersed diblock copolymer to the micellar core is associated with changes in the state of dilution and in the state of deformation of the A block. There is also the swelling of the A blocks inside the micellar core by the solubilizate J. Secondly, the B block of the singly dispersed copolymer is transferred to the solvent penetrated shell region of the micelle. This transfer process also involves changes in the states of dilution and deformation of the B block. Thirdly, the formation of the micelle localizes the copolymer such that the solvent incompatible A block is confined to the core while the solvent compatible B block is confined to the shell. Further, the formation of the micelle is accompanied by the generation of an interface between the micelle core made up of A blocks and the solubilizate J and the micelle shell consisting of the solvent W and the B blocks. Finally, in the case of a BAB triblock copolymer, folding or loop formation of the A block occurs ensuring that the B blocks at the two ends are in the aqueous domain while the folded A block is within the hydrophobic core of the micelle. The free energy of solubilization can be written as the sum of the above contributions.

$$(\Delta\mu_g^o) = (\Delta\mu_g^o)_{A,dil} + (\Delta\mu_g^o)_{A,def} + (\Delta\mu_g^o)_{B,dil} + (\Delta\mu_g^o)_{B,def} + (\Delta\mu_g^o)_{loc} + (\Delta\mu_g^o)_{int} + (\Delta\mu_g^o)_{loop} \tag{8}$$

Expressions for each of these contributions are formulated below. In developing the free energy expressions, m_A and m_B are used to denote the block sizes of the hydrophobic and hydrophilic blocks of both the AB diblock copolymer and the BAB triblock copolymer. Therefore, in the triblock copolymer, we are assuming a size of $m_B/2$ for both hydrophilic end blocks.

3.2.1. *Change in State of Dilution of Block A*

In the singly dispersed copolymer molecule, the A block is in a collapsed state minimizing its interactions with the solvent. We consider the region consisting of the collapsed A block with some solvent entrapped in it to be a spherical globule, whose diameter $2R_{-A}$ is equal to the end to end distance of block A in the solvent. The volume of this spherical region is denoted by $v_{\infty A}$.

$$v_{\infty A} = \frac{4\pi R_{\infty A}^3}{3} \quad , \quad 2\, R_{\infty A} = \alpha_A\, m_A^{1/2}\, L \tag{9}$$

Here, L is the characteristic segment length which is calculated from the volume of the solvent molecule as $L = v_W^{1/3}$. The chain expansion parameter α_A describes the swelling of the polymer block A by the solvent W. From eq.(9), one can write

$$\alpha_A = \left(\frac{6}{\pi}\right)^{1/3} m_A^{-1/6}\, \phi_{A1}^{-1/3} \tag{10}$$

where, ϕ_{A1} $(=v_A/v_{\infty A})$ is the volume fraction of A within the monomolecular globule. It is calculated from the condition of osmotic equilibrium between the monomolecular globule treated as a distinct phase and the solvent surrounding it, as suggested by de Gennes [22]. ϕ_{A1} is obtained [23] as the solution of

$$\ln\,(1 - \phi_{A1}) + \phi_{A1} + \chi_{AW}\, \phi_{A1}^2 = 0 \tag{11}$$

χ_{AW} in eq. (11) refers to the Flory-Huggins interaction parameter between the pure A polymer and the solvent W. The state of the A block within the singly dispersed copolymer is thus specified. The state of A block within the micelle is defined by the A block being confined to the core region where it is swollen by the solubilizate J. We consider this region to be uniform in concentration and employ a mean field description to calculate the free energy of this region.

One can thus estimate how the difference in the state of dilution of block A contributes to the free energy of solubilization from the following relation

$$\begin{aligned} \frac{(\Delta\mu_g^o)_{A,dil}}{kT} = {} & m_A \left[\frac{1}{m_J}\, \frac{1 - \phi_A}{\phi_A} \ln\,(1 - \phi_A) + \frac{1}{m_J}\,(1 - \phi_A)\, \chi_{AJ}\right] \\ & - m_A \left[\frac{1 - \phi_{A1}}{\phi_{A1}} \ln\,(1 - \phi_{A1}) + (1 - \phi_{A1})\, \chi_{AW} + \left(\frac{\sigma_{AW} L^2}{kT}\right) \frac{6}{\alpha_A\, m_A^{1/2}}\right] \end{aligned} \tag{12}$$

In the above equation, the first two terms account for the entropic and enthalpic contributions arising from the mixing of pure A block and the pure solubilizate J within the micellar core. They are written in the form of the Flory expression for the swelling of a network [24] by a solvent. The third and the fourth terms account for the entropic and enthalpic changes associated with the removal of A block from its infinitely dilute condition to a pure A state. These terms are written in the framework of the Flory expression [24] for an isolated polymer molecule. The last term accounts for the fact that the interface of the monomolecular globule disappears on micellization. This term is written as the product of the surface area of the monomolecular globule ($4\pi R_{A}^2$) and an effective interfacial tension, ($\sigma_{AW}\,\phi_{A1}$). Here, σ_{AW} is the interfacial tension between pure A and solvent W. The additional factor ϕ_{A1} which is the volume fraction of the polymer A in the monomolecular globule takes into account the reduction in this interfacial tension caused by the presence of some solvent molecules inside the monomolecular globule. If the interfacial tension σ_{AW} is not available from direct experimental measurements, it can be approximately estimated from knowledge of the Flory interaction parameter χ_{AW} using the relation $\sigma_{AW} = (\chi_{AW}/6)^{1/2}\,(kT/L^2)$, which is usually employed for the calculation of polymer-polymer interfacial tensions.

3.2.2. *Change in State of Deformation of Block A*

The A block is stretched within the micelle over a length equal to the radius R of the micellar core. The free energy of deformation compared to the unperturbed end to end distance of block A can be estimated using the Flory model [24] for the deformation of a polymer chain along one direction keeping the volume of the molecule constant. In the singly dispersed state of the copolymer, the conformation of the A block is characterized by the chain expansion parameter α_A which is the ratio between the actual end to end distance and the unperturbed end to end distance of the polymer block. The free energy of this deformation is written using the Flory expression [24] derived for an isolated polymer molecule. In the case of a BAB triblock copolymer, the A block deformation is calculated by considering the folded A block of size m_A to be equivalent to two A blocks of size $m_A/2$, each stretched over the micellar core. On this basis, one obtains

$$\frac{(\Delta\mu_g^o)_{A,def}}{kT} = \left[q\,\frac{1}{2}\left(\frac{R^2}{(m_A/q)\,L^2} + \frac{2\,(m_A/q)^{1/2}\,L}{R} - 3\right)\right] - \left[\frac{3}{2}\left(\alpha_A^2 - 1\right) - \ln \alpha_A^3\right] \tag{13a}$$

where q = 1 for a AB diblock copolymer and q = 2 for a BAB triblock copolymer having a middle hydrophobic block.

As mentioned earlier, in representing the free energy of deformation of the A block within the micelle, the above equation assumes that the chain is uniformly stretched along the length of the chain. It is known that in order to maintain an uniform concentration inside the micellar core, it is necessary for the A block to stretch non-uniformly. The free energy contribution allowing for non-uniform chain deformation can be calculated using the results from the analysis of chain packing pioneered by Semenov [25]. Such an alternate model allowing for non-uniform chain deformation yields [7,26],

$$\frac{(\Delta\mu_g^o)_{A,def}}{kT} = [\, q\, (\frac{p\pi^2}{80})\, \frac{R^2}{(m_A/q)\, L^2}\,] - [\frac{3}{2}\, (\alpha_A^2 - 1) - \ln\, \alpha_A^3\,] \qquad (13b)$$

where the parameter p has the values of 3, 5, and 10 respectively, for spherical, cylindrical and lamellar aggregates. In eq.(13a) and (13b), the first term defines the A block deformation free energy in the micelle while the second term defines the deformation free energy in the singly dispersed copolymer.

3.2.3. *Change in State of Dilution of Block B*

In the singly dispersed state of the copolymer, the polymer block B is swollen with the solvent. As mentioned before, m_B denotes the size of the B block for the AB diblock copolymer while for a BAB triblock copolymer, the end blocks are taken to be of equal size $m_B/2$. We consider this swollen B block to be a sphere, whose diameter 2 $R_{\infty B}$ is equal to the end to end distance of isolated block B in the solvent. The volume of this spherical region $v_{\infty B}$ is

$$v_{\infty B} = \frac{4\pi R_{\infty B}^3}{3}\ , \quad 2\, R_{\infty B} = \alpha_B\, (m_B/q)^{1/2}\, L \qquad (14)$$

where the chain expansion parameter α_B is estimated using the expression developed by Flory [24]. In the Flory expression for α_B, Stockmayer [27] has suggested decreasing the numerical coefficient by approximately a factor of two to ensure consistency with the results calculated from perturbation theories of excluded volume. Consequently, one can obtain [23] α_B as the solution of

$$\alpha_B^5 - \alpha_B^3 = 0.88\, (1/2 - \chi_{BW})\, (m_B/q)^{1/2} \qquad (15)$$

where χ_{BW} is the Flory interaction parameter between the hydrophilic B block and the solvent water. The state of the B block within the singly dispersed copolymer is thus specified. Within the micelle, the B blocks are present in the solvent penetrated shell region of volume V_s. The shell region is assumed to be uniform in concentration and its free energy can be written using the Flory expression [24] for a network swollen by the solvent.

Based on the definitions of the states of dilution of the B block in the micelle and in the singly dispersed state, one can write an expression for the corresponding contribution to the free energy of micellization.

$$\frac{(\Delta\mu_g^o)_{B,dil}}{kT} = m_B \left[\frac{1-\phi_B}{\phi_B} \ln(1-\phi_B) + (1-\phi_B)\chi_{BW} \right]$$
$$- m_B \left[\frac{1-\phi_{B1}}{\phi_{B1}} \ln(1-\phi_{B1}) + (1-\phi_{B1})\chi_{BW} \right] \qquad (16)$$

The first two terms in eq. (16) describe the entropic and enthalpic contributions to the free energy of swelling of the B block by the solvent in the shell region of the micelle while the last two terms refer to the corresponding contributions in the singly dispersed copolymer molecule.

3.2.4. *Change in State of Deformation of Block B*

In the singly dispersed state, the B block has a chain conformation characterized by the chain expansion parameter α_B which is the ratio between the actual end to end distance of the chain and the unperturbed end to end distance of the chain. Within the micelle, the B block is stretched over a length equal to the thickness D of the micellar shell. The difference between these deformation states provides a contribution to the free energy of micellization which can be written analogous to that for A blocks. One obtains

$$\frac{(\Delta\mu_g^o)_{B,def}}{kT} = q \left[\frac{1}{2} \left(\frac{D^2}{(m_B/q)\, L^2} + \frac{2\,(m_B/q)^{1/2}\, L}{D} - 3 \right) \right]$$
$$- q \left[\frac{3}{2} (\alpha_B^2 - 1) - \ln \alpha_B^3 \right] \qquad (17a)$$

The first term in eq. (17a) represents the free energy of deformation of the B block in the micellar shell while the second term denotes the corresponding free energy in the singly dispersed copolymer molecule. Both terms are written with respect to the unperturbed dimension of the B block.

As in eq. (13a), here also the chain deformation free energy within the micelle is written assuming uniform chain stretching. An alternate expression to describe non-uniform chain stretching in the micellar shell can be written based on Semenov's [25] analysis assuming for simplicity that the termini of all B blocks lie at the distance D from the core surface. One gets [7,26]

$$\frac{(\Delta\mu_g^o)_{B,def}}{kT} = \left[q\, \frac{3}{2}\, \frac{L\, R}{(a/q)\, \phi_B}\, P \right] - q \left[\frac{3}{2} (\alpha_B^2 - 1) - \ln \alpha_B^3 \right] \qquad (17b)$$

where a is the surface area per molecule of the micelle core, $q=1$ for AB diblock and $q=2$ for BAB triblock, as before and P is a shape-dependent parameter. For spheres, $P=(D/R)/[1+(D/R)]$, for cylinders, $P=\ln [1+(D/R)]$ and for lamellae, $P=(D/R)$. The first term in eq.(17b) represents the free energy of deformation of the B block in the micellar shell while the second term denotes the corresponding free energy in the singly dispersed copolymer molecule.

3.2.5. *Localization of the Copolymer Molecule*

As a result of micellization, the copolymer becomes localized in the sense that the joint linking blocks A and B in the copolymer is constrained to remain in the interfacial region rather than occupying all the positions available in the entire volume of the micelle. The entropic reduction associated with localization is modelled using the concept of configurational volume restriction. Thus, the localization free energy is calculated on the basis of the ratio between the volume available to the A-B joint in the interfacial shell of the micelle (surrounding the core and having a thickness L) and the total volume of the micelle.

$$\frac{(\Delta\mu_g^o)_{loc}}{kT} = -q \ln\left[\frac{p\,L}{R\,(\,1+D/R\,)^p}\right] \qquad (18)$$

The parameter p accounts for the aggregate geometry and assumes values of 3,2, and 1, respectively, for spherical, cylindrical and lamellar aggregates.

3.2.6. *Formation of Micellar Core-Solvent Interface*

When micelle forms, an interface is generated between a core region consisting of the A block and the solubilizate J and a shell region consisting of the solvent W and the B block. The free energy of formation of this interface can be estimated as the product of the surface area of the micellar core and an interfacial tension characteristic of this interface. The appropriate interfacial tension is that between a solution of block A in solubilizate J within the micelle core and a solution of block B in solvent W within the micellar shell. Since the shell region is often very dilute in block B, the interfacial tension can be approximated to be that between the solvent W and a solution of the A block with the solubilizate J within the core. Noting that polymer A-solvent W interfacial tension is σ_{AW} and solubilizate J- solvent W interfacial tension is σ_{JW}, the free energy of generation of the micellar core-solvent interface is calculated from

$$\frac{(\Delta\mu_g^o)_{int}}{kT} = \frac{\sigma_{agg}}{kT}\,a\;, \quad \sigma_{agg} = \sigma_{AW}\,\phi_A + \sigma_{JW}\,(1-\phi_A) \qquad (19)$$

In writing eq. (19), the interfacial tension of a polymer solution of block A in solubilizate J against another liquid W is approximated to be the composition averaged interfacial tensions of pure polymer A and pure solubilizate J against the solvent W. The volume fraction is used as the composition variable. Such a simple dependence of the interfacial tension on bulk solution composition is not generally obeyed in case of free solutions of polymers or of low molecular weight components. The origin of the deviation from linearity lies in the preferential adsorption or depletion of one of the components at the interface, which causes the surface composition to differ from the bulk composition [28]. However, the micellar interface is somewhat different from the interface of a free polymer solution. Specifically, because of the localization of the A-B link at the interface, the segments of the A block are forced to be at the interface independent of any selective adsorption or depletion. It is very likely that the difference between the surface and overall compositions in the micellar core is smaller when compared to a free polymer solution. Consequently, the composition averaging of interfacial tension as expressed by eq. (19) is assumed in the present calculations. An alternate approach to calculating the interfacial tension as that between two free solutions has been explored in our study of solubilization in low molecular weight surfactants [26].

3.2.7. *Backfolding or Looping in Triblock Copolymer*
The backfolding or looping of the middle block in a BAB triblock copolymer contributes an entropic term to the free energy of solubilization. This contribution is absent for the case of a diblock copolymer. Jacobson and Stockmayer [29] showed that the reduction in entropy for the condition that the ends of a linear chain of m segments are to lie in the same plane or on one side of a plane is proportional to ln m. Therefore, the assumption that the backfolding of the middle block in the micelle follows the same functional form is made. Hence, the backfolding or looping entropy makes the following contribution in the case of a BAB copolymer.

$$\frac{(\Delta\mu_g^o)_{loop}}{kT} = -\frac{3}{2}\,\beta\,\ln\,[\,m_A\,] \tag{20}$$

where β is an excluded volume parameter and is taken to be unity as long as the excluded volume effects can be neglected. It can take on values larger than unity when these effects become important. For the case of the BAB copolymer, the core is usually concentrated in polymer (not very large solubilization) as a result of which β is taken to be unity in the present calculations.

3.2.8. *Estimation of Model Parameters*
To perform quantitative calculations, the values of molecular constants appearing in eqs. (9) to (20) are needed. Illustrative calculations have been carried out here for PEO-PPO diblock and PEO-PPO-PEO triblock copolymers with water as the solvent and hydrocarbons as the solubilizates. The molecular volumes of the solubilizates (v_J), their solubility parameters (δ_J), and the interfacial tensions between the solvent and the solubilizates (σ_{JW}) are all listed in Table 1. Values for the Flory-Huggins

interaction parameters χ_{AJ} between the solubilizates J and the hydrophobic A block have been estimated utilizing the solubility parameters as indicated in Section 2.2. The interfacial tension σ_{AW} between the solvent and the hydrophobic block A is estimated using the expression provided in Section 3.2.1 involving the Flory interaction parameter χ_{AW}. For PPO-water, the interaction parameter χ_{AW} has been estimated to be 2.1 [23]. For calculating the values of m_A and m_B, the molecular volumes of the repeating units are taken to be 96.5 Å^3 for propylene oxide and 64.6Å^3 for ethylene oxide while the molecular volume of water is taken to be 30 Å^3. For all the systems, the numerical computations were carried out for m_A/m_B ranging from 0.1 to 10, and $M=m_A+m_B$ in the range 50 to 2500.

3.3. MODEL PREDICTIONS FOR DIBLOCK COPOLYMERS

All the calculations described in Section 3.3 for diblock copolymer micelles and in Section 3.4 for triblock copolymer micelles have been carried out for spherical aggregates.

3.3.1. *Physicochemical Interpretation of Solubilization Capacity*

The various free energy contributions calculated using the expressions developed in Section 3.2 (and based on uniform chain deformation model) are plotted against the aggregation number g in Figure 7. The remaining two independent variables D and η are not kept constant but are chosen to be those values which minimize the free energy of solubilization per molecule $\Delta\mu_g^o$ for each given value of g.

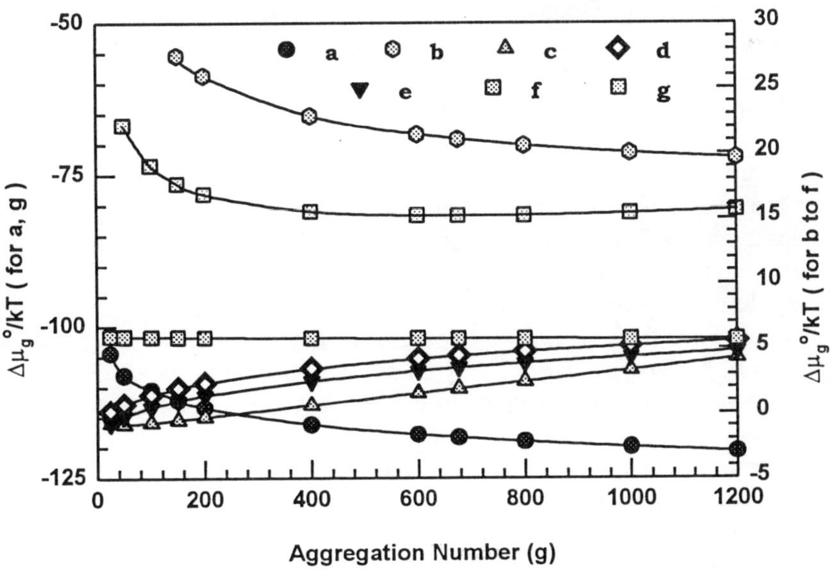

Figure 7. Dependence of various free energy contributions on the aggregation number g for PEO-PPO diblock copolymer (30% PPO, MW=12,500). The other independent variables D and η assume values that minimize the free energy of solubilization of benzene for each value of g.

The formation of micelles in preference to the singly dispersed state of the copolymer occurs because of the large negative free energy contribution arising from a change in the state of dilution of the solvent incompatible A block following solubilization (curve a). In the absence of the solubilizates, this free energy contribution is a constant independent of the size of the micelle and hence does not govern the aggregation number of micelle [23]. However, in the presence of the solubilizate, this free energy also accounts for the swelling of the A blocks by the solubilizate J. This free energy decreases with an increase in the aggregation number g and is thus favorable to the growth of the aggregates. A second contribution favorable to the growth of the aggregates is provided by the interfacial energy (curve b). In general, the surface area per molecule of the micelle decreases with an increase in the aggregation number. Consequently, the positive interfacial free energy between the micellar core and the solvent decreases with increasing aggregation number of the micelle and thus this contribution promotes the growth of the micelle. The changes in the state of deformation of the A and the B blocks (curves c and d) and the change in the state of dilution of the B block (curve e) provide positive free energy contributions that increase with increasing aggregation number of the micelle. Therefore, these factors are responsible for limiting the growth of the micelle. More interestingly, when solvent W is a very good solvent for the B block, the positive free energy contributions resulting from changes in the states of dilution and of deformation of the B block outweigh the contribution resulting from the changes in the state of deformation of the A block. Under such conditions, the B block related free energy contributions strongly influence the structural properties of the equilibrium micelles. The free energy of localization (curve f) is found to be practically independent of g and thus has no influence over the determination of the equilibrium aggregation number. The net free energy of the micelle per molecule is shown by curve g. As one would expect, this free energy is negative and shows a minimum at the equilibrium aggregation number.

In Figure 8, the various free energy contributions are plotted as a function of the volume fraction of the solubilizate η in the micellar core. The remaining independent variables g and D/R are kept constant at the values corresponding to the global minimum of the free energy (i.e. with respect to g, D and η) per molecule of the micelle. Here also, the expressions for uniform chain deformation are used in the calculations. While Figure 7 helps illustrate how the equilibrium aggregation number is influenced by various free energy contributions, Figure 8 reveals the importance of various contributions to determining the equilibrium uptake of the solubilizate.

Curve a shows the large negative free energy contribution provided by the change in state of dilution of block A. This free energy decreases with an increase in the volume fraction of the solubilizate. This factor thus favors the uptake of the solubilizate within the micelle. One may note that the magnitude of this contribution is larger if the solubilizate-core block interaction parameter χ_{AJ} is smaller and if the molecular size of the solubilizate m_J is smaller. Correspondingly, the capacity of the micelles for such solubilizates (i.e. the equilibrium value of η) will be larger. When micelles incorporate solubilizates, the radius of the micellar core and the interfacial area per molecule of the micellar core both increase. The former increases the positive free energy contribution arising from the increased deformation of the A block (curve c). The latter increases the positive free energy contribution associated with the micellar core-solvent interfacial free energy (curve b). Thus both these factors serve to restrict the swelling of the micellar core by the solubilizate and consequently, the extent of solubilization. One may note that the increase in the positive interfacial free energy accompanying the uptake of solubilizates by the micelles (shown by curve b) is also dependent on the solubilizate-solvent interfacial tension σ_{JW}. Therefore, given two solubilizates, the micellar capacity η will be larger for the solubilizate associated with a lower solubilizate-solvent interfacial tension σ_{JW}. Further, the increase in the amount of solubilizate within a micelle of specified aggregation number also changes the state of deformation (curve d) as well as the state of dilution (curve e) of the B block in the shell region of the micelle. Of these two positive free energy contributions, the former increases with η, thus disfavoring solubilization while the latter decreases with increasing η, thus favoring increased solubilization. The free energy of localization (curve f) is practically independent of η and thus has little influence over the nature and extent of solubilization. The net free energy of the micelle per molecule is represented by curve g. The minimum in this net free energy occurs at the equilibrium value for η. All the results described in the following sections can be interpreted in terms of the above-described free energy variations accompanying solubilization.

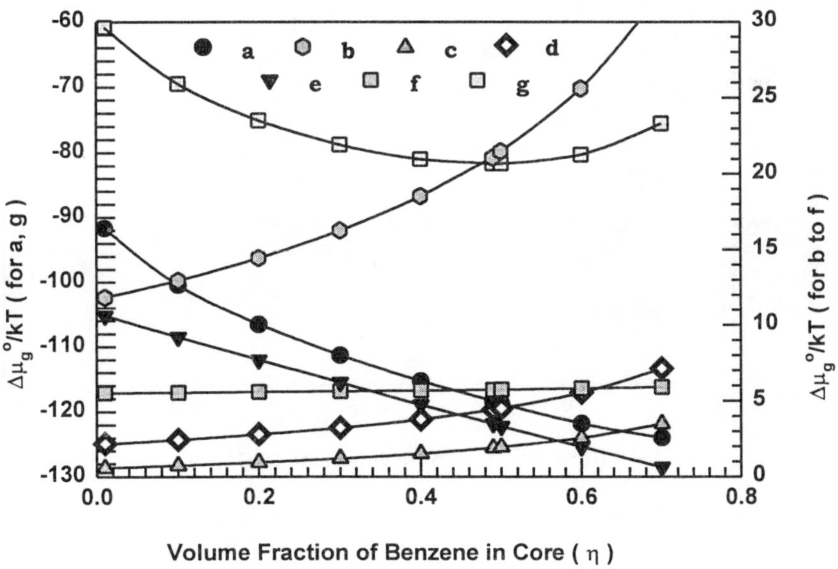

Figure 8. Free energy contributions as a function of the volume fraction η of benzene in the micelle core for PEO-PPO diblock copolymer (30% PPO, MW=12,500). The other independent variables D and g are assigned values corresponding to the global optimum of the free energy of solubilization of benzene for this block copolymer molecule.

3.3.2. *Validity of the Pseudophase Approximation*

As mentioned earlier, the visualization of micelles as a pseudophase is quite satisfactory and very convenient when the aggregates are narrowly dispersed in their sizes. Here, the size and composition distributions (namely, the dispersion in the values of variables g and j around their most probable values) have been calculated for a number of micellar systems containing the solubilizates. A typical result is shown in Figure 9 for benzene solubilized within PEO-PPO micelles in aqueous solutions.

In constructing this figure, the volume fraction of the solubilizate within the micellar core η is chosen as the independent variable in place of j. The point enclosed by the closed curves corresponds to the most populous micelles, that is those aggregates whose concentration in the solution is the largest. The three closed curves surrounding this point are the loci of micellar sizes and compositions corresponding to which the micellar concentrations are respectively, 1, 2 and 3 orders of magnitude smaller compared to the concentration of the most populous micelles. One can observe that the aggregate concentrations fall off very rapidly when the values of g and η deviate from those of the most populous micelles represented by the point in Figure 9.

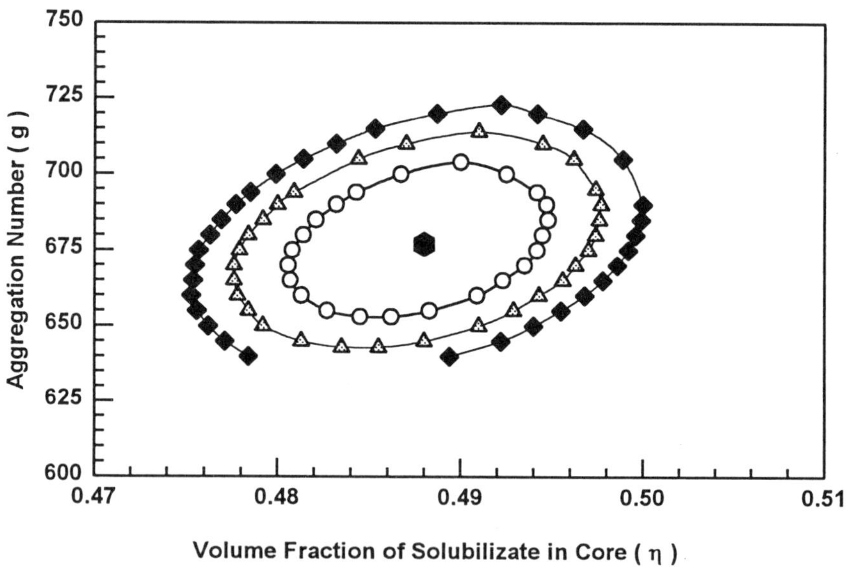

Figure 9. Distribution of the aggregation number g and the solubilization capacity η in the equilibrium micelles. The closed curves are loci of sizes and compositions of aggregates present at constant concentrations. The calculations are for benzene solubilized in PEO-PPO diblock copolymer (30% PPO, MW=12,500).

Similar numerical calculations of the aggregate size and composition distributions for many systems considered here indicate that the micelles are virtually monodispersed both in relation to the number of constituent block copolymer molecules as well as the number of solubilizate molecules. Therefore, it is quite satisfactory to simplify the calculations by invoking the pseudophase approximation and then estimating the micellar characteristics by the minimization of the free energy per molecule $\Delta\mu_g^o$ of an isolated micelle with respect to the three independent variables R, D and η.

3.3.3. *Comparison with Experimental Data*

The predictions of the present theory employing the equations valid for non-uniform chain deformations are summarized in Table 3. Also shown within the parenthesis in Table 3 are the measured solubilization capacities from Table 1 expressed as volume fractions within the micelle core. In general, the agreement between the experimental and measured values of η are reasonably satisfactory for all the solubilizates. Whereas the theory permits the prediction of all the structural features of the micelles such as g, R, D and the CMC, experimental data for these variables are currently not available and hence the corresponding comparisons have not been possible.

TABLE 3. Predicted solubilization capacity, core radius, shell thickness, aggregation number and CMC of PEO-PPO (30% PPO, MW=12,500) diblock copolymer micelles containing aromatic and aliphatic hydrocarbon solubilizates at 25°C

Solubilizate	R (Å)	D/R	η	$-\ln X_{CMC}$	g	D (Å)
Benzene	113	1.53	0.52 (0.51)	72.5	470	173
Toluene	102	1.67	0.44 (0.40)	66.7	398	170
O-Xylene	97	1.73	0.41 (0.33)	64.2	365	168
Ethyl benzene	93	1.80	0.36 (0.41)	62.7	342	167
Cyclohexane	83	1.97	0.25 (0.18)	60.3	293	164
Hexane	69	2.24	0.08 (0.08)	55.7	202	155
Heptane	68	2.26	0.07 (0.08)	55.2	196	154
Octane	68	2.27	0.07 (0.08)	55.0	194	154
Decane	66	2.31	0.05 (0.06)	54.4	185	152
None	62	2.39	-	53.6	163	149

The predicted results show that the solubilization capacity is larger if the core block is very compatible with the solubilizates (small χ_{AJ}), if the solubilizate -solvent interfacial tension is lower and if the molecular volume of the solubilizate is smaller. Consequently, the aromatic molecules are found to display a larger solubilization limit compared to the aliphatic molecules. Solubilization is found to increase the micellar core radius and decrease the critical micelle concentration. The larger the solubilization capacity, the more significant are the changes in R and the CMC. The increase in the core radius R results not only from the incorporation of the solubilizate but also because of the increasing number of block copolymer molecules that are accommodated within a micelle. This increase in g is more dramatic for solubilizates whose uptake by the micelles is large. The dimensionless shell thickness D/R is found to decrease with increasing solubilization capacity of the micelles while the shell thickness D is not very much affected by solubilization.

For solubilizates such as benzene which are also good solvents for PEO and for the more polar solubilizates such as the oxygenated compounds discussed in ref. 3a, there is the likelihood that in addition to the solubilizate being present in the micellar core, it could also be present in the micellar shell. The model presented here does not describe such a situation. Obviously, the presence of the solubilizate in the micellar shell will alter the micellar dimensions and the predicted solubilization capacity due to the fact that its presence in the shell influences the various free energy contributions.

The predictions based on the uniform chain deformation model have been presented before [7a] for the solubilization of various hydrocarbons. The results reported in that paper differ significantly from those listed in Table 3 for the predicted values of the aggregation number g, the core radius R and the shell thickness D, but only marginally for the volume fraction of the solubilizate η.

3.3.4. *Generalized Scaling Relations*

The dependence of the solubilization capacity and the other structural features of the micelles on the block copolymer size and composition have been examined through numerical simulations taking benzene as the solubilizate in PEO-PPO micelles present in aqueous solutions. Figure 10 shows the dependence of R, D and η on the molecular weight of the PPO block keeping the molecular weight of the PEO block constant at 8750. It is found that the core radius R and the solubilization capacity η change significantly whereas the shell thickness D remains more or less unaffected. The CMC is lower compared to that calculated for solubilizate-free solutions and the deviation caused by the presence of the solubilizate increases with increasing size of the core block PPO. It is possible to interpret these variations recognizing that the core block is considered to constitute the site of solubilization. Therefore, a change in the size of the core block directly influences the values of R, η and the CMC. Since the shell thickness is primarily dependent on the solvent compatible B block (PEO), the value of D is only very weakly dependent on m_A.

Figures 11 presents similar calculated results, but as a function of the size of the shell block PEO, keeping the molecular weight of the core block PPO constant at 3750. The calculations show that the variation in the magnitude of m_B affects the core radius R and η in addition to having a more significant influence on the shell thickness D. An increase in the value of m_B causes a reduction in both R and η. In solubilizate-free systems, the shell block B influences the micellar radius R (or equivalently, the aggregation number g) significantly, especially if the solvent W is a good solvent for the B block, as is the case for PEO-water [23]. The present calculations show that in the presence of solubilizates, the magnitude of the influence

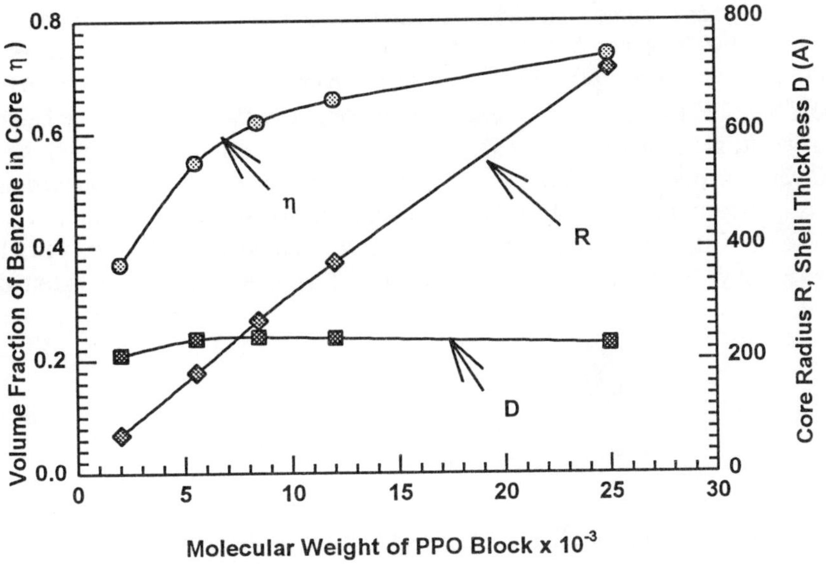

Figure 10. The core radius R, shell thickness D and volume fraction in core η for benzene solubilized in PEO-PPO micelles as a function of the molecular weight of the PPO block. For the PEO block, MW=8750.

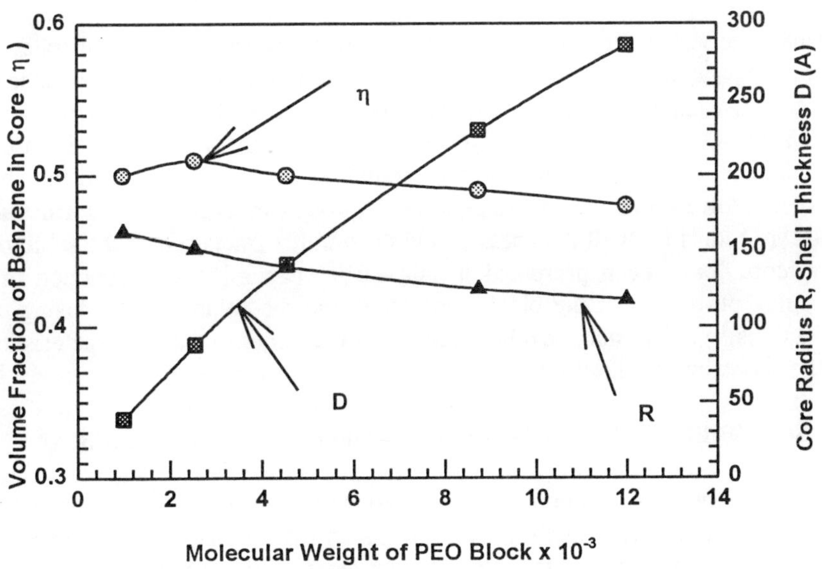

Figure 11. The core radius R, shell thickness D and volume fraction in core η for benzene solubilized in PEO-PPO micelles as a function of the molecular weight of the PEO block. For the PPO block, MW=3750.

of the B blocks is somewhat reduced while being qualitatively similar to that in solubilizate-free systems. The larger the solubilization capacity of the micelles, the less prominent is the role played by the B block-solvent W interactions. The variation in the CMC due to a change in the molecular weight of PEO parallels the behavior observed in the absence of the solubilizate [23]. The difference between the two CMC values is practically uninfluenced by the magnitude of m_B. By correlating the results from numerical simulations, scaling relations between micellar size characteristics and the block sizes m_A and m_B have been obtained. For the PEO-PPO diblock copolymer micelles in water, with benzene as the solubilizate, it is found [7] that

$$R \propto m_A^{0.92}\, m_B^{-0.13}, \quad D \propto m_A^{-0.08 \sim -0.125}\, m_B^{0.79}$$

$$g \propto m_A^{1.42}\, m_B^{-0.39}, \quad \eta \propto m_A^{0.17}\, m_B^{-0.017}$$

One may compare these scaling relations against those obtained [23] in the absence of the solubilizate:

$$R \propto m_A^{0.73}\, m_B^{-0.17}, \quad D \propto m_A^{0.06}\, m_B^{0.74}, \quad g \propto m_A^{1.19}\, m_B^{-0.51}$$

Evidently, the scaling relations are modified in the presence of the solubilizate. The solvent compatible B block continues to influence the magnitude of the micellar core parameters R and g, but the influence is diminished by the presence of the solubilizate. One may note that these scaling relations are specific in the sense that they depend on the nature of the solubilizate as well as on the block copolymer-solvent system. More general algebraic expressions for the calculation of the core radius R, shell thickness D and the volume fraction η of the solubilizate in the core, have been presented in our earlier paper [7] as a function of the molecular properties of the block copolymer and the solubilizates. Using these expressions, one can make explicit numerical predictions of the characteristics of micelles containing solubilizates.

3.4. SOLUBILIZATION IN TRIBLOCK COPOLYMER MICELLES

We will confine our attention to BAB triblock copolymers where B refers to the hydrophilic and A to the hydrophobic blocks as before. Due to the sequence of blocks in the copolymer, the A blocks have to be subjected to some degree of backfolding so as to exclude the B blocks from the micellar shell. Two extreme possibilities for the configuration of the A block within the micellar core can be visualized. In the first visualization, the A blocks can be assumed to be stretched across the micellar core along the micellar diameter so that the B blocks connected to the two ends of the A block are diametrically opposite each other. In the second visualization, the A block in the micellar core loops back on itself so that the two B blocks connected to it are present next to each other in the micellar shell. In general, all possible configurations (i.e. different degrees of backfolding of the A blocks) are possible. This looping provides a free energy contribution given by eq.(20) to the free energy of solubilization. In qualitative terms, the only difference between the AB diblock copolymer systems and the BAB triblock copolymer systems is in the presence of this backfolding or looping entropic contribution. All other features are essentially retained. Micellization of triblock copolymers taking into account such a looping contribution has been modelled before by ten Brinke and Hadziioannou [30a], by Balsara et al. [30b] and by Prochazka et al. [30c].

3.4.1. *Model Predictions of Solubilization in PEO-PPO-PEO Micelles*

Table 4 lists the calculated results for PEO-PPO-PEO triblock copolymer (30% PPO, MW=12,500) in water at 20°C. The calculations have been performed using the non-uniform chain deformation equations. A comparison with the results for diblock copolymers of the same molecular weight and composition (presented in Table 3) shows that the micellar core radius R, aggregation number g, and the shell thickness D are all considerably smaller for the case of triblock copolymers. The solubilization capacities are also somewhat reduced. The CMC's are significantly larger for the triblock copolymer compared to the diblock copolymer. The values for η shown in parenthesis are the measured values that were already presented in Table 3. As mentioned earlier, there is some uncertainty as to whether the block copolymer used in the experiments was a diblock or a triblock copolymer.

TABLE 4. Predicted solubilization capacity, core radius, shell thickness, aggregation number and CMC of PEO-PPO-PEO (30% PPO, MW=12,500) triblock copolymer micelles containing aromatic and aliphatic hydrocarbon solubilizates at 25°C

Solubilizate	R (Å)	D/R	η	$-\ln X_{CMC}$	g	D (Å)
Benzene	67	1.40	0.41 (0.51)	41.8	120	94
Toluene	61	1.52	0.32 (0.40)	36.7	101	92
O-Xylene	58	1.57	0.29 (0.33)	34.6	94	92
Ethyl benzene	56	1.62	0.25 (0.41)	33.4	89	91
Cyclohexane	51	1.74	0.15 (0.18)	31.1	77	89
Hexane	46	1.87	0.05 (0.08)	28.5	63	87
Heptane	46	1.89	0.04 (0.08)	28.1	61	86
Octane	45	1.89	0.04 (0.08)	28.0	61	86
Decane	45	1.92	0.02 (0.06)	27.6	59	86
None	43	1.95	-	27.2	55	85

3.4.2. *Physicochemical Interpretation of Solubilization*

The aggregation behavior of triblock copolymer systems is in all qualitative aspects identical to that of diblock copolymer systems. The looping energy for the triblock copolymer, in the form given by eq.(20), has no effect on micelle characteristics. It is a positive constant dependent on the block copolymer molecular weight and composition and its only effect is on altering the CMC. As a result, the CMC is significantly larger for the triblock copolymer compared to that for a diblock copolymer of identical molecular weight and composition. In other words, it is more difficult to form micelles with triblock copolymers when compared to their diblock counterparts. Solubilizates which have greater compatibility with the core blocks (i.e. smaller χ_{AJ}), which are smaller molecules, and which have a lower interfacial tension against water are in general solubilized to greater extents. Again, the looping effect has no influence on the equilibrium solubilizate uptake since it is just a constant term. The aggregation numbers of triblock copolymer micelles are in general smaller than the aggregation numbers of diblock copolymer micelles for identical copolymer molecular weight and composition. This is because the A block in the micellar core is treated as being equivalent to two A chains of half the molecular weight each. So is the case with the B block in the micellar shell. In other words, a triblock copolymer behavior is approximately equivalent to that of a diblock copolymer of half its molecular weight in many respects.

4. Aggregate Shape Transitions Induced by Solubilization

Only a few studies in the literature refer to the formation of non-spherical structures in dilute solutions of block copolymers. Price et al. [31a,b] observed a transition in micellar shape from spheres to worm-like micelles with increasing temperature in dilute solutions of polystyrene-polybutadiene-polystyrene (PS-PB-PS) triblock copolymer (molecular weight of the blocks being 12,900, 66,000, and 13,700, respectively) in ethyl acetate. Electron micrographs showed rod-like micelles which were polydisperse in lengths, but monodisperse in the radial dimension. Price et al. [31c] also investigated micelle formation in a dilute solution of polystyrene-polyisoprene (PS-PI) diblock copolymer in N,N'- dimethylacetamide, which is a selective solvent for PS. Electron micrographs indicated the presence of stable worm-like micelles at 20°C. Shape transitions have also been observed by Mandema et al. [32a,b] in solutions of PS-PI diblock copolymer in a mixed solvent of *trans*-decalin and decane. Decane is a good solvent for PI, whereas *trans*-decalin is a good solvent for both blocks. Transitions from spherical to rod-like micelles were observed with increasing *trans*-decalin concentrations, before complete dissolution of the copolymer takes place. The rod-like micelles were polydispersed with large molecular weights. Large non-spherical aggregates have also been observed in dilute solutions of polystyrene-poly methyl methacrylate (PS-PMMA) diblock copolymer in 1-chloro n-hexane (a selective solvent for PS) in the vicinity of the theta temperature of PMMA [32c,d]. At temperatures well below the theta temperature of the PMMA, spherical micelles were found in the system. Cylindrical micelles have been reported also from PEO-PPO-PEO block copolymers in aqueous solutions [32e]. A very recent paper reports the formation of bilayer vesicles [33] which are spherical analogs of lamellar aggregates in dilute block copolymer solutions. No experimental studies have appeared in the literature examining the influence of solubilization on shape transitions of micellar aggregates.

4.1 FORMATION OF CYLINDERS AND LAMELLAE

In this section, we examine the formation of cylindrical and lamellar aggregates in dilute solutions of diblock copolymers. Model independent thermodynamic analysis [21,26] suggests that when cylindrical aggregates form, they will be highly polydispersed. This is because the free energy per molecule of an optimal rod does not change by addition of molecules along the length of the rod. Therefore, rods of all possible lengths can exist in solution. The thermodynamic analysis also suggests that the average size of the polydispersed rodlike aggregates will increase with increasing concentration of the block copolymer in solution. It is possible that these conclusions may be modified when factors such as inter-rod interactions and the flexibility of rods are taken into account in deriving the aggregate size distribution. For cylindrical micelles that are monodispersed in their radial dimension, the pseudophase approximation can be conveniently applied to determine the optimal radial dimension and the degree of solubilization within the rods. This approach is used in this paper. As mentioned above, experimental observations of bilayer structures in dilute diblock copolymer solutions have appeared only recently in the literature [33]. Bilayer aggregates can be visualized as being disc-like with a planar middle surrounded by a hemi-cylindrical rim or they can be spherical vesicles.

The three independent variables R, D, and η characterize all the aggregate structures. The free energy of solubilization $\Delta\mu_g^\circ$ is minimized (within the framework of the pseudophase approximation) with respect to these variables in order to determine the optimal values for R, D, and η characterizing the optimal sphere, rod or planar bilayer. The preferred structure is taken to be that associated with the lowest free energy. In the absence of solubilizates, only two independent variables (R and D) characterize each of these structures.

Results from numerical simulations are presented in this section illustrating the effects of the copolymer molecular weight, the copolymer composition, and polymer block-solvent interaction parameters on shape transitions in systems consisting of PEO-PPO diblock copolymers in water. The molecular constants discussed in Section 3 are used in these simulations as well. Specifically, the calculations have been carried out for $M=m_A+m_B=750$, $\chi_{AW}=2.1$, $\chi_{BW}=0.2$, $\chi_{AJ}=0.2$, $v_J=180Å^3$, $v_W=30Å^3$, $\sigma_{AW}=25.2$ dyne/cm, and $\sigma_{AJ}=50.2$ dyne/cm, The solubilizate properties listed above correspond to those for cyclohexane.

4.2. SHAPE TRANSITIONS IN THE ABSENCE OF SOLUBILIZATE

4.2.1 *Effect of block composition*

Simulations have been performed using both the uniform and the non-uniform deformation equations for the A and the B blocks. The optimal free energy of micellization per molecule ($\Delta\mu_g^\circ/kT$) is calculated as a function of m_A/M using the non-uniform deformation model and the results are plotted in Figure 12 in the form of a difference in the free energy between the optimal sphere and the optimal rod or the optimal bilayer.

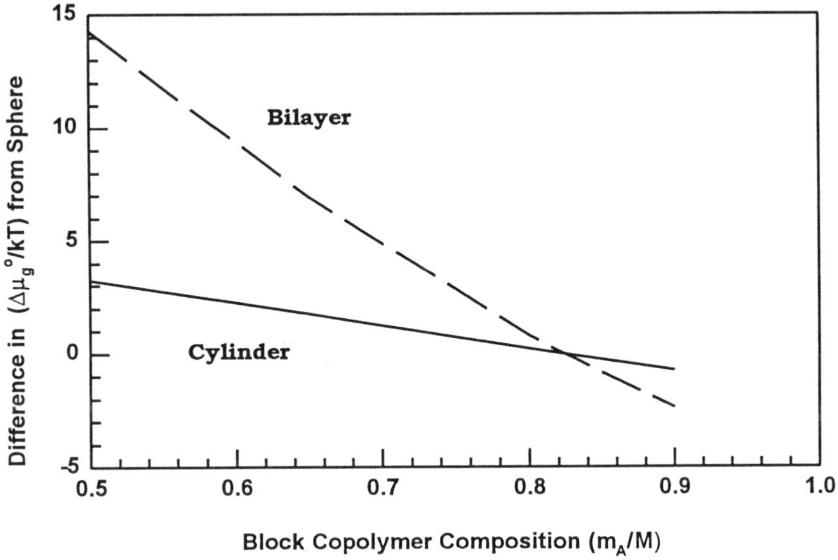

Figure 12. The difference in free energy between optimal spheres and optimal rods or optimal bilayers as a function of the block composition calculated using the non-uniform deformation model for $M=750$, $\chi_{AW}=2.1$, and $\chi_{BW}=0.2$.

The optimal free energy of micellization is lowest for spheres in the region $m_A/M < 0.82$. When $m_A/M > 0.82$, bilayers have the lowest free energy and become the preferred structure. As can be seen from the figure, the free energy of micellization associated with the optimal rods is never the lowest of the three structures. Significantly differing results are predicted when free energy expressions valid for uniform chain deformations are used in the calculations. Indeed, for the same conditions corresponding to those in Figure 12, the uniform deformation model predicts the formation of rodlike aggregates over a finite m_A/M domain. The use of non-uniform deformation equations widens the range over which spheres are the preferred structure compared to the predictetions based on uniform chain deformation.

In Figures 13 and 14, the optimal R and D (in Å) are plotted for the preferred structures as a function of the composition of the block copolymer. The optimal R and D values for the non-uniform deformation model (Figure 13) are smaller than those corresponding to the uniform deformation model (Figure 14). Whereas the values of R are decreased by a small magnitude, the values of D are significantly diminished when the non-uniform deformation model is employed. Since only

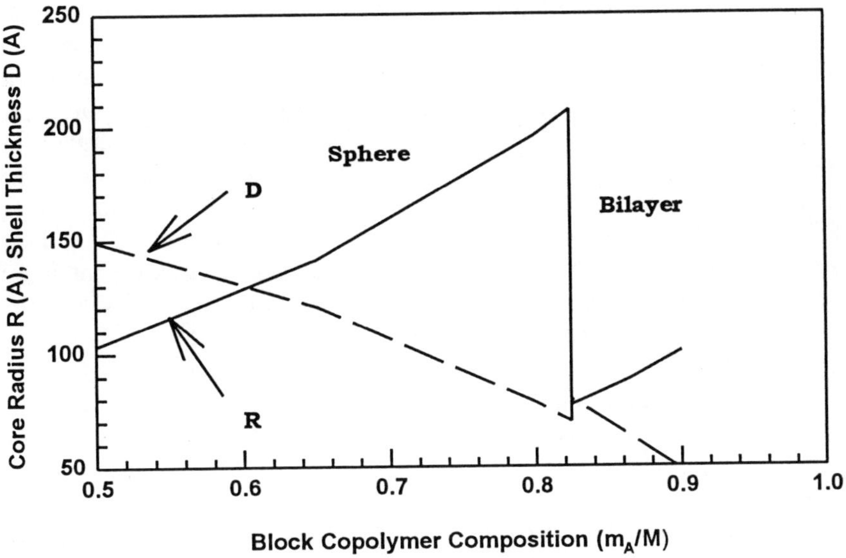

Figure 13. The optimal core dimension R of the preferred geometry as a function of block composition calculated using the non-uniform deformation model for M=750, χ_{AW}=2.1, and χ_{BW}=0.2.

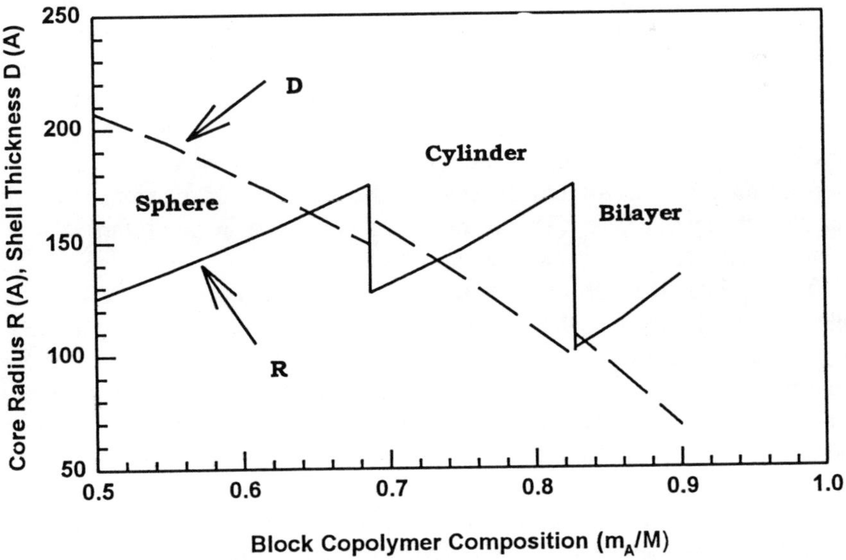

Figure 14. The optimal core dimension R of the preferred geometry as a function of block composition calculated using the uniform deformation model for M=750, χ_{AW}=2.1, and χ_{BW}=0.2.

spheres and bilayers are predicted according to the non-uniform deformation model, the dimensions of only these structures are shown in that case. The reasons for rods not being preferred when the non-uniform deformation model is used can be explored in terms of the various free energy contributions [7b]. For $m_A/M < 0.7$, spheres have the lowest optimal interfacial energies. While the interfacial energies of rods decreases more rapidly compared to that of spheres as m_A/M is increased, bilayers show an even more rapid rate of decrease. The rate of decrease for the rods is not sufficient to make it a favorable structure in any composition domain. On the basis of the interfacial energy alone, transitions from spherical structures are not preferred for $m_A/M < 0.7$ and for $m_A/M > 0.7$, transition to bilayers becomes favorable. In contrast, when the uniform deformation model is used, the rods show a lower optimal interfacial energy than spheres over the entire composition range.

The optimal A block deformation free energies (based on non-uniform chain stretching) increase as m_A/M is increased for all three structures, in a similar rate. Spheres show the largest A block deformation energies. Hence, this contribution favors transition to non-spherical shapes. The difference in the optimal A block non-uniform deformation energies between spheres and rods is smaller compared to the corresponding difference in the uniform deformation case . This is another reason why the tendency to form rods from spheres is greatly reduced for the non-uniform deformation case. Bilayers, show the lowest optimal deformation energies and hence have a tendency to become the favored structure in some block copolymer composition domain at least.

The optimal B block non-uniform deformation energy variations are similar to the uniform deformation case, but the energies are of larger magnitude. Spheres are associated with the lowest optimal B block deformation energy. Hence, from the point of view of B block deformation (non-uniform stretching model), transitions from spherical structures are not favored. Also, the difference in the optimal non-uniform B deformation energies of rods and bilayers is smaller than the corresponding values in the uniform deformation case. As a result, this contribution does not have as unfavorable an effect on the formation of bilayers as it is in the uniform deformation case. This results again in bilayers being preferred over rods when non-uniform deformation model is used.

The optimal B block dilution energies (for non-uniform deformation model) follow a trend identical to that in the case of the uniform deformation model. The magnitudes are also very similar. The optimal localization free energy follows the same pattern for the two deformation models but the energies are lower for the non-uniform case compared to the uniform case. The difference between the energies of rods and spheres, and rods and bilayers are similar to that in the case of the uniform deformation model. Hence, their role in the shape transitions is not changed by the way the deformations are accounted for. The energies remain favorable to non-spherical structures. Overall, the interfacial free energy and the A block deformation energy contributions decisively influence the formation of non-spherical aggregates. They are also responsible for the preference for bilayers over rods when the non-uniform deformation model is used.

4.2.2 *Shape transition diagram in molecular size-composition space*

To calculate the domains of preferred aggregate shape with respect to the molecular size and the composition of the block copolymer, the same model system as before is chosen. Given M and m_A/M, the transition points are generated by comparing the optimal free energies of spheres, rods, and bilayers. The preferred structure is taken to be that associated with the lowest optimal free energy of micellization. Figure 15 is generated using the non-uniform deformation equations, while Figure 16 is generated using the uniform deformation equations. The solid lines are drawn as the loci of points denoting the sphere-rod, sphere-bilayer and rod-bilayer transitions. The shape transition points in Figures 15 and 16 show a dependence on the molecular weight of the copolymer. Spheres are in general preferred over larger composition domains with increasing molecular weights of the copolymer. The spherical domain size increases sharply at low molecular weights and is not strongly influenced at high molecular weights.

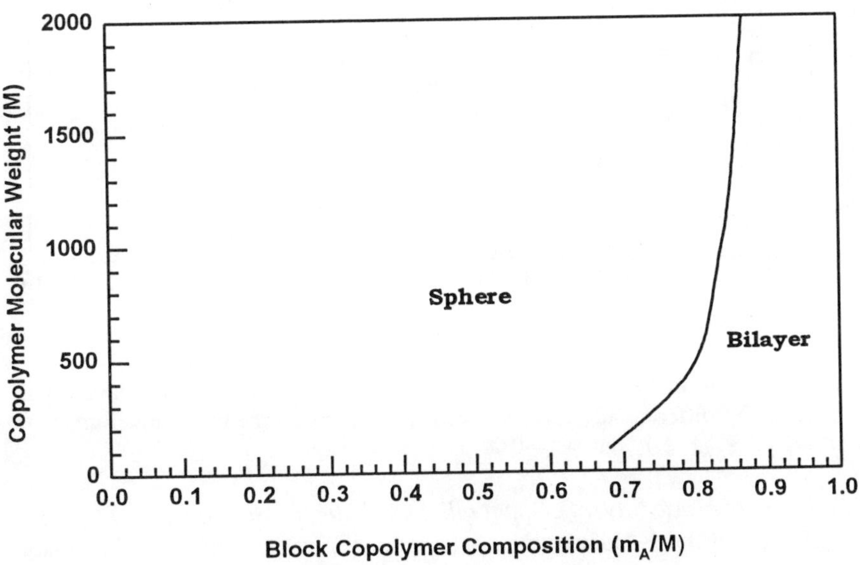

Figure 15. The molecular size-block composition diagram showing the domains of existence of different aggregate shapes calculated using the non-uniform deformation model for $\chi_{AW}=2.1$, and $\chi_{BW}=0.2$.

For very low values of m_A/M, one can expect the copolymer to be molecularly dispersed in the solvent rather than aggregate since the large B block can virtually shield the small A block without necessarily having to form multi-molecular aggregates (i.e. monomolecular globules may be the energetically favored structure). This molecularly dispersed domain is not shown in the figures. It can be expected to occur at low m_A/M, for all values of M and for very low molecular weights M over a range of block compositions. In the other limit of large m_A/M, there is the possibility that macrophase separation may be preferred energetically to the formation of bilayers. This essentially reduces the domain of existence of the bilayer. The macrophase separated region is also not shown in either plot. The reason for phase separation to be preferred over aggregation is intuitive. With large A blocks and small B blocks, the B blocks lack sufficient capacity to keep the A blocks in solution by shielding them in aggregates.

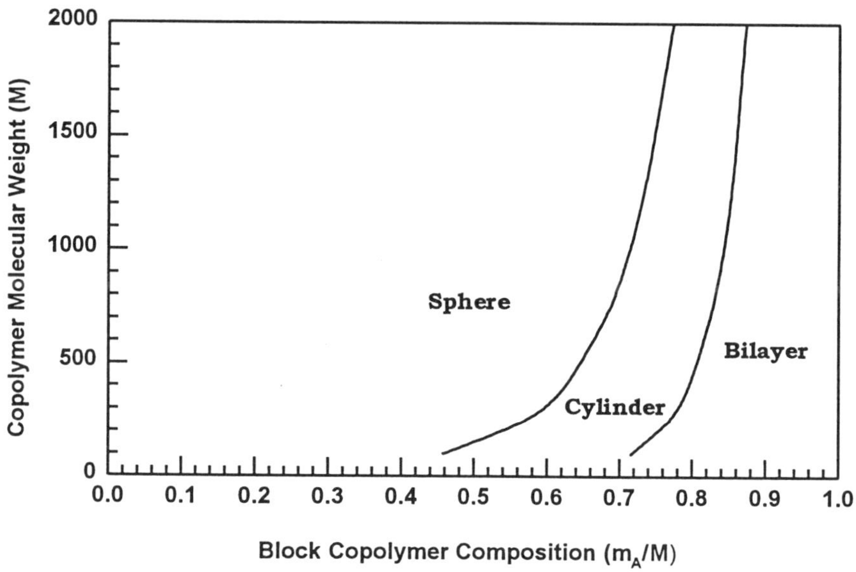

Figure 16. The molecular size-block composition diagram showing the domains of existence of different aggregate shapes calculated using the uniform deformation model for $\chi_{AW}=2.1$, and $\chi_{BW}=0.2$.

4.2.3 *Effect of block-solvent compatibility on shape transitions*
The influence of the B block-solvent interaction parameter and the A block-solvent interaction parameter on shape transition is examined in this section. For determining the effect of χ_{BW} on transitions, a value of 0.45 is chosen. An increase in this interaction parameter in PEO-water systems can occur as a result of an increase in the temperature or by the addition of electrolytes. All other parameters are fixed at the values previously used in the simulations. This permits comparison with the results obtained before for $\chi_{BW}=0.2$. A comparison of calculated optimal free energy of micellization as a function of m_A/M for the case of uniform deformation shows that the optimal free energy of micellization for spheres is the lowest of the three shapes in the interval $m_A/M < 0.47$. The transition from spheres to rods takes place at $m_A/M > 0.47$ and rods become the energetically preferred geometry in the range $0.47 < m_A/M < 0.72$. The transition from rods to discs is

made at $m_A/M = 0.72$. For $\chi_{BW} = 0.2$, the sphere to rod transition occurred at a composition of 0.69 and a rod to disc transition took place at 0.82. Therefore, one can conclude that increasing χ_{BW} makes the transition to non-spherical shapes a more favorable phenomenon. Increasing χ_{BW} (i.e. making the solvent a poorer solvent for the B block) decreases the m_A/M domain space over which spheres exist and therefore increases the non-spherical regimes. The dimensions of the optimal aggregates are of course functions of χ_{BW}. Increasing χ_{BW} in general yields aggregates larger in core sizes than the corresponding aggregates obtained with lower χ_{BW} values. The shell thickness is reduced when χ_{BW} is increased, all other parameters remaining the same. The reason for this decrease rests in the fact that as the solvent quality diminishes with respect to the B block, the B block will be stretched to lesser extents in the shell since the B block prefers to expose lesser extents of itself to the solvent. This results in smaller values of D.

When the non-uniform deformation equations are used in the simulations with $\chi_{BW}=0.45$, the transition from spheres to bilayers occurs at $m_A/M=0.76$ (as against 0.82 for the case of $\chi_{BW}=0.2$). Again, as in the case with $\chi_{BW}=0.2$, rods never have the lowest optimal free energy of the three structures at any composition. As in the uniform deformation case, the sphere regime is decreased with increasing χ_{BW}.

The effect of the interaction parameter between the A block and solvent χ_{AW} on transitions has also been examined. One may note that the interfacial tension between the solvent and the A block (σ_{AW}) is also dependent on χ_{AW} and hence varies as the interaction parameter is altered. Thus, the simulation results reflect the consequences of changes in both these molecular constants. The simulations are carried out with $\chi_{AW} = 1.25$, $\chi_{BW}=0.2$, and $M=750$, so that a comparison with the earlier results can be made. The optimal free energy of micellization as a function of composition calculated for the uniform deformation case shows that the transitions from spheres to rods take place at $m_A/M=0.7$, and rods to discs at $m_A/M=0.84$. The spherical aggregate domain is therefore marginally increased with decreasing χ_{AW} (the transition from sphere to rod occurs at 0.69 for $\chi_{AW}=2.1$). The dimensions of the optimal cores and the shells are diminished with decreasing χ_{AW}. This is because decreasing χ_{AW} and decreasing σ_{AW} imply that the solvency for the A block and thus for the copolymer is improved. This, in turn, is reflected in larger CMC values and smaller micelle sizes. Decreasing χ_{AW} decreases the magnitude of the optimal interfacial energy for all three structures. However, the interfacial energies for spheres are larger than for rods only by a smaller amount with decreasing values of χ_{AW}. Hence, the tendency to form rods is diminished as a result of which spheres may exist over a wider composition range.

The non-uniform deformation model calculations with $\chi_{AW}=1.25$ shows that the transition from spheres to discs occurs around 0.84, which is a marginal increase over that for $\chi_{AW}=2.1$. Rods are again never a preferred shape. The marginal increase in the domain of existence of spheres is a consequence of the effects discussed in the previous paragraph.

4.3 SHAPE TRANSITIONS IN THE PRESENCE OF SOLUBILIZATES

Solubilization calculations involve the three independent variables η, R and D. For a given copolymer molecular weight and composition, shape transitions have been examined with respect to the degree of solubilization in the micelles. The simulations discussed below have been carried out for T=25°C, M=750, χ_{AW}=2.1, χ_{BW}= 0.2, v_J=180Å^3, χ_{AJ}=0.2, σ_{JW}=50.2 dyne/cm and σ_{AW}=25.2 dyne/cm. In the first set of calculations, the block composition m_A/M is fixed at 0.6 and the effect of solubilization on the transition phenomenon is investigated. Both uniform and non-uniform deformations are examined in this context. For the second set of simulations, m_A/M is fixed at 0.75 and the changes in the composition domains of occurrence of aggregates of different shapes are investigated. Given a copolymer composition, the structure associated with the lowest optimal free energy of solubilization is taken to the preferred structure at the η under consideration.

4.3.1 *Block Composition m_A/M=0.6*

The global optimal sphere corresponds to an η value of 0.46. The global optimal rod and the global optimal bilayer have free energies larger than that associated with the global optimal sphere. As a result, aggregates with these geometries will never form according to the non-uniform deformation model. For solubilizate concentrations in excess of η=0.46, the system will macrophase separate into two phases, one phase being the pure excess solubilizate and the other being the micellar phase of the global optimal sphere (η=0.46). Figure 17 is a plot of the dimensions of the micelle (in Å) as a function of η in the range $0<\eta<0.46$. The calculations based on the uniform chain deformation equations show very different results. Figure 18 is a plot of the dimensions of the aggregates (in Å) as a function of η for the case of the uniform deformation model. In this case, spherical, rodlike and lamellar aggregates are all favored over some compositions domains. An interesting feature is the predicted coexistence of rodlike and lamellar aggregates in the region $0.51<\eta<0.61$. In general, the non-uniform deformation model yields smaller micelles compared to the corresponding uniform deformation model.

The transition behavior with respect to the two types of deformation equations is once again compared on the basis of the various free energy contributions with respect to η. For any given η in the uniform deformation model, all three structures are associated with the same A block dilution energy. The optimal interfacial energy per molecule associated with spheres is the largest of the three geometries for a given η. This quantity therefore favors transitions to non-spherical structures. Also, as η increases, this quantity increases due to the fact that the micellar cores are being swollen to increasing extents causing the interfacial area to increase.

For the non-uniform deformation model, the interfacial energy per molecule associated with spheres is the smallest of the three structures at low η, but it does increase more rapidly than the interfacial energies of non-spherical structures. Due to the small interfacial energy values associated with spheres, the tendency to undergo transitions from spheres is greatly reduced.

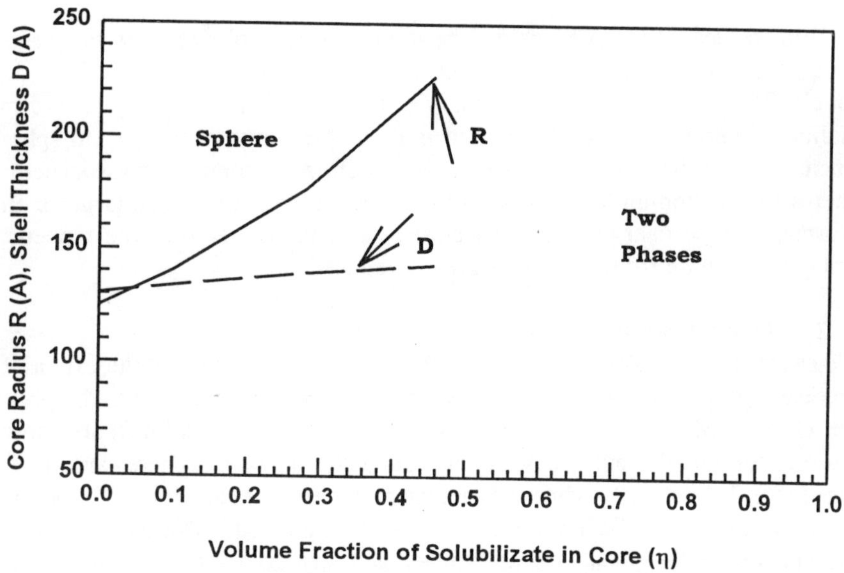

Figure 17. The dimensions of the optimal aggregates as a function of the volume fraction η of benzene in the hydrophobic core. The calculations are based on the non-uniform deformation model for $m_A/M=0.6$.

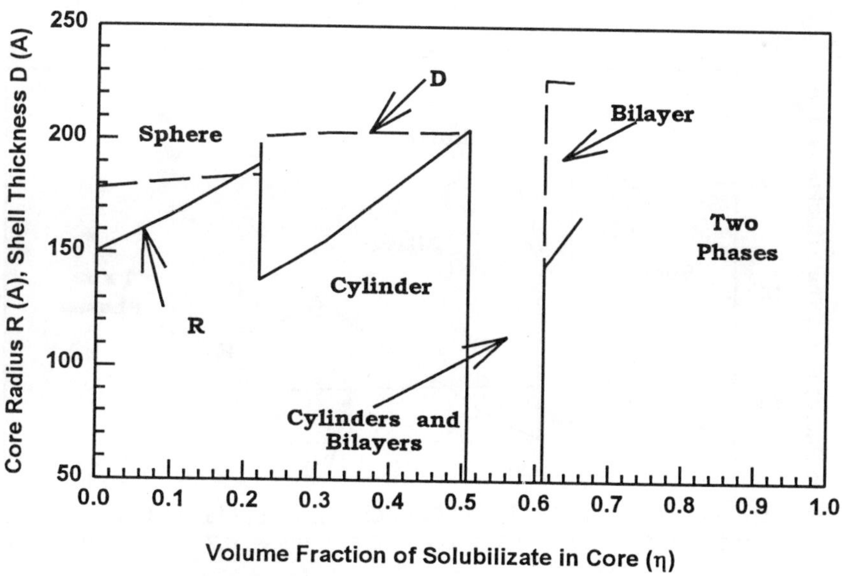

Figure 18. The dimensions of the optimal aggregates as a function of the volume fraction η of benzene in the hydrophobic core. The calculations are based on the uniform deformation model for $m_A/M=0.6$.

With respect to the A block deformation energies, spheres show the largest A block deformation energies for a given η in the case of the uniform deformation model. The A block deformation energies for non-spherical aggregates are much smaller. Therefore, this contribution is favorable for transitions from spherical structures. On the other hand, the A block deformation energies associated with spheres for the non-uniform deformation model are not very much larger than the deformation free energies of non-spherical structures. Hence, the tendency to display a transition from spheres is greatly reduced.

4.3.2. *Block composition $m_A/M=0.75$*

Calculations based on the non-uniform deformation model show that the extent of the spherical aggregate domain is reduced when $m_A/M=0.75$ compared to the previous case when $m_A/M=0.6$. The dimensions of the optimal micelles (in Å) are shown in Figure 19 for the calculations based on the non-uniform model. Spheres are preferred in the range $0<\eta<0.22$, while rods are the preferred geometry for $0.22<\eta<0.27$, and in the region $0.27<\eta<0.61$, bilayers are preferred. The value of $\eta=0.61$ corresponds to the global optimal bilayer and any excess solubilizate will be in equilibrium with this phase. The interesting feature in this calculation is the appearance of a domain of stable rods which were absent in the absence of the solubilizate as discussed before (see Figure 15). Also, for $m_A/M=0.75$, the presence of the solubilizate has facilitated the transition to rods over some range of values for η while for $m_A/M=0.6$ only spheres were predicted (see Figure 17).

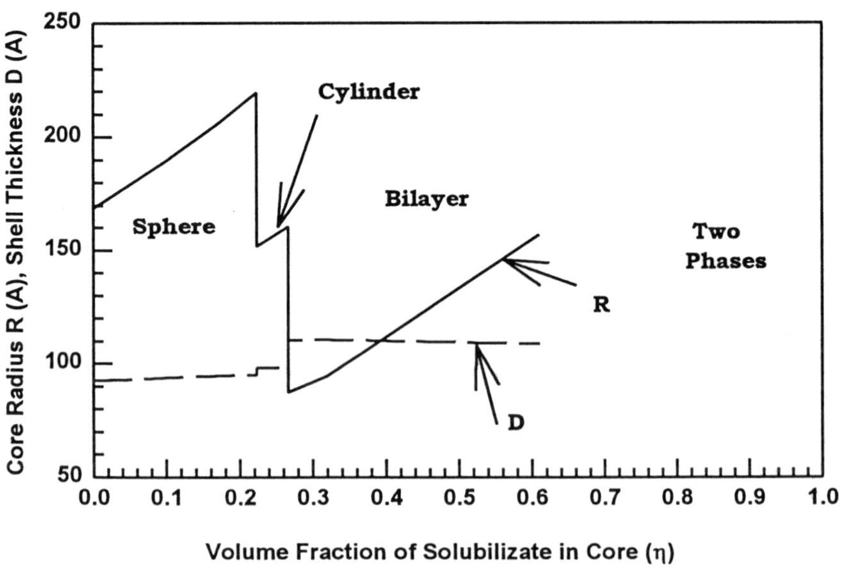

Figure 19. The dimensions of the optimal aggregates as a function of the volume fraction η of benzene in the hydrophobic core. The calculations are based on the non-uniform deformation model for $m_A/M=0.75$.

Similar results obtained assuming the uniform deformation model are plotted in Figure 20. In this case, the spheres are never preferred and rods exist for $0<\eta<0.28$, and bilayers are formed over the range $0.28<\eta<0.66$. The global optimal bilayer exists at $\eta=0.66$. One may note that the increase in the A block size (compared to when $m_A/M=0.6$) has eliminated the existence of spherical aggregates at all solubilizate concentrations.

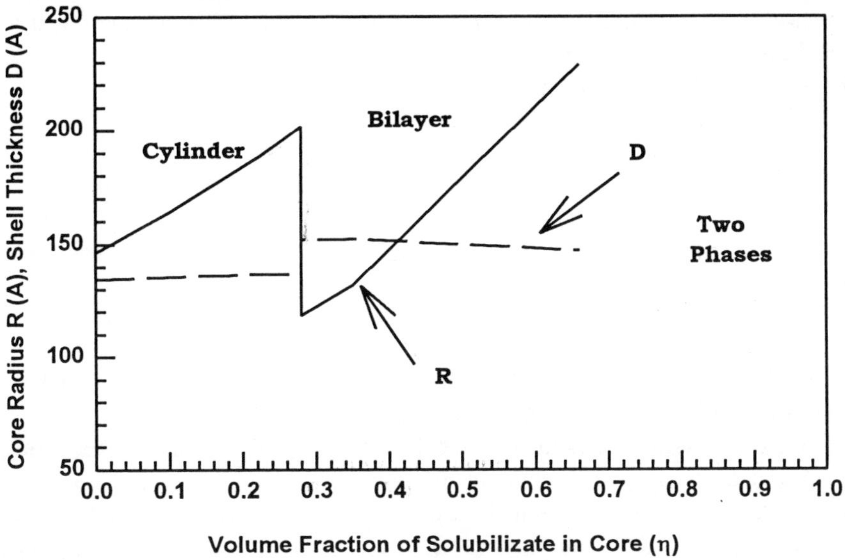

Figure 20. The dimensions of the optimal aggregates as a function of the volume fraction η of benzene in the hydrophobic core. The calculations are based on the uniform deformation model for $m_A/M=0.75$.

4.3.3. *General Conclusions on Shape Transitions*

The transition boundaries predicted here are not exact since the free energy contributions from the ends of rods and hemi-cylindrical rims of discs are ignored. Also, the system entropy that influences the phase boundary has been ignored since the present calculations provide results in the pseudophase representation of the aggregation. However, when these features are included, only marginal variations in phase boundaries are expected. In general, one can expect non-spherical geometries when the copolymer has a very large hydrophobic block compared to the size of the hydrophilic block. For the case of micellization, the non-uniform deformation model disfavors the formation of rods, unlike the uniform deformation model. Also, the copolymer molecular weight-composition space over which spheres are the preferred geometry is larger for the non-uniform deformation model. On making the solvent a poorer solvent for the B block, transitions to non-spherical shapes become more favorable. This is the case for both deformation models. On the other hand, improving the solvent quality for the A block (the solvophobic block) tends to disfavor transitions to non-spherical shapes and eventually, the aggregation itself.

The tendency for a transition to non-spherical shapes with increasing solubilizate concentration is less for the non-uniform stretching model compared to the uniform stretching model. Also, increasing the fraction of the A block in the copolymer keeping the copolymer molecular weight constant increases the tendency to form non-spherical aggregates with increasing solubilizate concentrations. Another interesting result is the appearance of rods over a finite solubilizate concentration range when the non-uniform deformation model is used. This is in contrast to the micellization case (absence of solubilizate) where for the same model, rods are disfavored in the copolymer molecular weight-composition space.

5. Concluding Remarks

Experimental and theoretical studies of solubilization of hydrophobic substances in aqueous solutions of block copolymer micelles have been described here. The paper emphasizes our work but important contributions by many other authors are cited. Experimental results show that large solubilization capacities and high selectivities can be derived through the use of block copolymer micelles when compared to micelles formed of conventional low molecular weight surfactants.

Thermodynamic treatment of solubilization has been presented here for both diblock and triblock copolymer molecules that allows the explicit calculation of the solubilization capacity of the micelles, the dimensions of the hydrophobic core and the hydrophilic shell of the aggregates, the aggregation number and the critical micelle concentration. The analysis is based on the mean-field approach and utilizes the assumption of uniform concentrations in the micellar core and shell regions. Two descriptions of chain packing energies based on uniform and non-uniform stretching have been employed in the calculations. Reasonable agreement is found between the experimental values of the solubilization capacity and the predicted values. Further, the formation of non-spherical structures of cylinders and lamellae are explored for diblock copolymer micelles using the mean-field approach.

Quantitative predictions of aggregate characteristics, based on the thermodynamic analysis presented here, are strongly influenced by the estimates for the interfacial tension of the hydrophobic core-hydrophilic shell interface. Only approximate methods to estimate this interfacial tension are employed in the present calculations and a more fundamental analysis for its calculation is necessary.

The thermodynamic analysis presented here assumes a sharp interface and no penetration of water or the hydrophilic block into the hydrophobic core. Contrary conclusions have been obtained from calculations based on the self-consistent- field theories. This needs further examination since the SCF theories make similar predictions in the case of conventional low molecular weight surfactants where many experiments have ruled out the possibility of water penetration into the hydrophobic core.

An alternate approach that allows for concentration inhomogeneities in the micellar core and shell regions has been described in an earlier paper [8] utilizing the star polymer model [34] and adopting a scaling analysis [35-37].

The mean-field approach, the SCF theory and the star polymer model all employ the Flory interaction parameter for describing the interactions between the hydrophilic B block and water. It is known that the thermodynamic properties of the aqueous solutions are not described well through the use of such an interaction parameter. Therefore, a more fundamental treatment for describing the shell region is desirable.

Finally, experimental data on solubilization in block copolymer micelles are currently very sparse and inadequate for the purposes of testing available theories especially with respect to scaling relations.

Acknowledgements: It is a pleasure to acknowledge the contributions of my former students Mark Chaiko, Maureen Barry, Sherri Slocum and Ganesh Kailasam. This paper is based on the experimental studies on solubilization conducted by Mark Chaiko, Maureen Barry and Sherri Slocum and the thermodynamic modelling of various aspects of block copolymer self-assembly carried out by Ganesh Kailasam.

6. References

1. Attwood, D. and Florence, A. T. (1983) *Surfactant Systems. Their Chemistry, Pharmacy and Biology*, Chapman and Hall, London.
2. Nagarajan, R., Maureen Barry and Ruckenstein, E. (1986) Unusual selectivity in solubilization by block copolymer micelles, *Langmuir* **1**, 210-215.
3. (a) Slocum, S. A., Kilara, A. and Nagarajan, R. (1990) Solubilization of food flavor compounds, in George Charalambous, Ed. *Flavors and Off-Flavors*, Elsevier, Amsterdam, pp. 233-251 . (b) Slocum, S. (1990) Solubilization of essential oil components and orange oil in microemulsions, Ph.D. Thesis, The Pennsylvania State University.
4. (a) Hurter, P. N. and Hatton, T. A. (1992) Solubilization of polycyclic aromatic hydrocarbons by poly(ethylene oxide-propylene oxide) block copolymer micelles: Effects of polymers structure, *Langmuir* **8**, 1291-1299. (b) Nivaggioili, T., Tsao, B., Alexandridis, P. and Hatton, T.A. (1995) Microviscosity in pluronic and tetronic poly(ethylene oxide)-poly(propylene oxide) block copolymer micelles, *Langmuir* **11**, 119-126.
5. (a) Wu, G., Chu, B. and Schneider, D. K. (1995) SANS study of the micellar structure of PEO/PPO/PEO aqueous solution, *J. Phys. Chem.* **99**, 5094-5101. (b) Tontisakis, A., Hilfeker, R. and Chu, B. (1990) Effect of xylene on micellar solutions of block-copoly(oxyethylene/oxypropylene) in water, *J. Colloid Interface Sci.* **135**, 427-434.
6. Kabanov, A. V., Nazarova, I. R., Astafieva, I. V., Batrakova, E. V., Alakov, V. Yu., Yaroslavov, A.A. and Kabanov, V. A. (1995) Micelle formation and solubilization of fluorescent probes in poly(oxyethylene-b-oxypropylene-b-oxyethylene) solutions, *Macromolecules* **28**, 2303-2314.
7. (a) Nagarajan, R. and Ganesh, K. (1989) Block copolymer self-assembly in selective solvents: Theory of solubilization in spherical micelles, *Macromolecules* **22**, 4312-4325. (b) Ganesh, K. (1991) Molecular assembly of block copolymers, Ph.D. Thesis, The Pennsylvania State University.
8. Nagarajan, R. and Ganesh, K. (1993) Solubilization in spherical block copolymer micelles: Scaling analysis based on star model, *J. Chem. Phys.* **98**, 7440-7450.
9. Dan, N. and Tirrell, M. (1993) Diblock copolymer microemulsions. A scaling model, *Macromolecules* **26**, 637-642.

10. Cogan, K. A., Leermakers, F. A. M. and Gast, A. P. (1992) Predictions of copolymer micelle behavior in immiscible solvents, *Langmuir* **8**, 429-436.
11. (a) Hurter, P. N., Scheutjens, J. M. H. M. and Hatton, T. A. (1993) Molecular modelling of micelle formation and solubilization in block copolymer micelles. 1. A self-consistent mean-field lattice theory. *Macromolecules* **26** 5592-5601. (b) Hurter, P. N., Scheutjens, J. M. H. M. and Hatton, T. A. (1993) Molecular modelling of micelle formation and solubilization in block copolymer micelles. 1. Lattice theory for monomers with internal degrees of freedom, *Macromolecules* **26** 5030-5040.
12. Linse, P. (1994) Micellization in poly(ethylene oxide)-poly(propylene oxide) block copolymer in aqueous solution: Effect of polymer impurities. *Macromolecules* **27**, 2685-2693.
13. Leermakers, F. A. M., Wijmans, C. M. and Fleer, G. J. (1995) On the structure of polymeric micelles: Self-consistent-field theory and universal properties for volume fraction profiles, *Macromolecules* **28**, 3434-3443.
14. (a) Tuzar, Z. and Kratochvil, P. (1993) Micelles of block and graft copolymers in solutions, *Surface and Colloid Science* **15**, 1-82. (b) Tuzar, Z., Bahadur, P. and Kratochvil, P. (1981) Solubilization of homopolymers and copolymers by block copolymer micelles in dilute solutions, *Makromol. Chem.* **182**, 1751-1760.
15. Gupte, A., Nagarajan, R. and Kilara, A. (1991) Block copolymer microdomains: A novel medium for enzymatic reactions, *Biotechnol. Prog.* **7**, 348-354.
16. Chaiko, M. A., Nagarajan, R. and Ruckenstein, E. (1984) Solubilization of single component and binary mixtures of hydrocarbons in aqueous micellar solutions, *J. Colloid Interface Sci.* **99**, 168-182.
17. Abraham, M. H. (1984) Thermodynamics of solution of homologous series of solutes in water, *J. Chem. Soc. Faraday Trans. I*, **80**, 153-181.
18. de Gennes, P. G. (1978) Macromolecules and liquid crystals: Reflections on certain lines of research, in J. Liebert, ed., *Solid State Physics*, Academic Press, New York, Suppl.14, pp. 1-18.
19. Leibler, L., Orland, H. and Wheeler, J. C. (1983) Theory of critical micelle concentration of solutions of block copolymers, *J. Chem. Phys.* **79**, 3550-3557.
20. (a) Noolandi, J. and Hong, K. M. (1983) Theory of block copolymer micelles in solution, *Macromolecules* **16**, 1443-1448. (b) Whitmore, D. and Noolandi, J. (1985) Theory of micelle formation in block copolymer-homopolymer blends, *Macromolecules* **18**, 657-665.
21. Nagarajan, R. (1993) Modelling solution entropy in the theory of micellization. *Colloids and Surfaces A. Physicochem. and Eng. Aspects* **71**, 39-64.
22. de Gennes, P. G. (1979) *Scaling Concepts in Polymer Physics,* Cornell University Press, Ithaca.
23. Nagarajan, R. and Ganesh, K. (1989) Block copolymer self-assembly in selective solvents: Spherical micelles with segregated cores, *J. Chem. Phys.* **90**, 5843-5856.
24. Flory, P.J. (1962) *Principles of Polymer Chemistry*, Cornell University Press, Ithaca.
25. Semenov, A. N. (1985) Contributions to the theory of microphase layering in block copolymer melts, *Sov. Phys. JETP* **61**, 733-742.
26. Nagarajan, R. and Ruckenstein, E. (1991) Theory of surfactant self-assembly: A predictive molecular thermodynamic approach, *Langmuir* **7**, 2934-2969.
27. Stockmayer, W. H. (1955) Chain dimensions near the Flory temperature, *J. Polym. Sci.* **15**, 595-598.
28. Siow, K. S. and Patterson, D. (1973) Surface thermodynamics of polymer solutions, *J. Phys. Chem.* **77**, 356-365.
29. Jacobsen, H. and Stockmayer, W. H. (1950) Intramolecular reaction in polycondensations. I. The theory of linear systems, *J. Chem. Phys.* **18**, 1600-1606.
30. (a) ten Brinke. G. and Hadziioannou, G. (1987) Topological constraints and their influence on the properties of synthetic macromolecular systems. 2. Micelle formation of triblock copolymers, *Macromolecules* **20**, 486-489. (b) Balsara, N. P., Tirrell, M. and Lodge, T. P. (1991) Micelle formation of BAB triblock copolymers in solvents that preferentially dissolve the A block, *Macromolecules* **24**, 1975-1986. (c) Prochazka, O., Tuzar, Z. and Kratochvil, P.

(1991) Association of a three-block copolymer BAB in selective solvents for blocks B: Spherical micelles, *Polymer* **32**, 3038-3044.

31. (a) Cohnheim, P.C., Laxly, T.V., Price, C. and Stubbersfield, R.B. (1980) Formation of worm-like micelles from a polystyrene-polybutadiene-polystyrene block copolymer in ethyl acetate, *J. Chem. Soc., Faraday I* **76**, 1857-1867. (b) Laxly, P. and Price, C. (1974) Some observations on the colloidal behavior of block copolymers, *Polymer* **15**, 325-326. (c) Price, C., Chan, E.K.M., Hudd, A.L. and Stubbersfield, R.B. (1986) Worm-like micelle formation by a polystyrene-b-polyisoprene block copolymer in N,N'- dimethylacetamide, *Polym. Commun.* **27**, 196-198.
32. (a) Mandema, W., Zeldenrust, H. and Emeis, C.A. (1979) Association of block copolymers in selective solvents, 1. Measurements on hydrogenated poly(styrene-isoprene) in decane and in trans-decalin, *Makromol. Chem.* **180**, 1521-1538. (b) Mandema, W., Zeldenrust, H. and Emeis, C.A. (1979) Association of block copolymers in selective solvents, 2. Viscosity, diffusion and light-scattering measurements on hydrogenated poly(styrene-isoprene) in trans-decalin/decane mixtures, *Makromol. Chem.* **180**, 2163-2174. (c) Duval, M. and Picot, C. (1987) Hydrogenated polystyrene-poly(methyl methacrylate) diblock copolymer micelles in selective solvent: 1. Single solvent system-effect of temperature, *Polymer* **28**, 793-797. (d) Duval, M. and Picot, C. (1987) Hydrogenated polystyrene-poly(methyl methacrylate) diblock copolymer micelles in selective solvent: 2. Binary solvent-precipitant mixtures-effect of mixture composition and temperature, *Polymer* **28**, 798-803. (e) Schillen, K., Brown, W. and Johnsen, R. M. (1994) Micellar sphere-to-rod transition in aqueous triblock copolymer system. A dynamic light scattering study of translational and rotational diffusion. *Macromolecules* **27**, 4825-4832.
33. Zhang, L. and Eisenberg, A. (1995) Multiple morphologies of "crew-cut" aggregates of polystyrene-b-poly(acrylic acid) block copolymers. *Science* **268**, 1728-1731.
34. Daoud, M. and Cotton, J. P. (1982) Star shaped polymers: A model for the conformation and its concentration dependence, *J. Physique* **43**, 531-538.
35. Zhulina, Ye. B. and Birshtein, T. M. (1985) Conformations of block copolymer molecules in selective solvents (micellar structures) *Polym. Sci. USSR* **27**, 570-578.
36. Halperin, A. (1987) Polymeric micelles: A star model. *Macromolecules* **20**, 2943-2946.
37. Wang, Z. and Safran, S. A. (1990) Equilibrium emulsification of polymer blends by diblock copolymers, *J. Phys. France* **51**, 185-200.

MONTE CARLO SIMULATION OF SELF-ASSEMBLY IN MACROMOLECULAR SYSTEMS

TÜRKAN HALILOĞLU
TUBITAK - IPOM and Polymer Research Center
Boğaziçi University
80815 Bebek, Istanbul, Turkey

WAYNE L. MATTICE
The Maurice Morton Institute of Polymer Science
The University of Akron
Akron, Ohio 44325-3909, USA

Many arbitrarily chosen pairs of polymer A and polymer B are immiscible. When placed together at bulk density, they separate into two distinct macroscopic phases. If the pair of homopolymers is joined end-to-end by a covalent bond to make an AB diblock copolymer, macroscopic separation of A from B is no longer possible. Instead, the immiscibility of A and B drives the self-assembly of the block copolymer into organized microstructures. Self-assembled structures are also obtained when the block copolymer is dispersed in dilute solution in a selective solvent. The selective solvent is simultaneously a poor (and good) solvent for the A (and B) blocks.

The simulation of the structure and dynamics of block copolymers is described in the three parts of this chapter. The concepts and methods used in the simulation of self-assembly in macromolecular systems are introduced first. Then we review briefly several recent simulations of self-assembly of AB diblock and ABA triblock copolymers. The final section examines in more detail the simulation of the exchange of chains between micelles formed by diblock copolymers in dilute solution in a selective solvent.

1. General Considerations in the Simulation of Self-Assembly in Macromolecular Systems

Analytical theory and computational simulation are complimentary methods for the study of the self-assembly of macromolecular systems. Theory offers the important advantages of the insight that can be gained by inspection of the form of an equation, and the speed with which numerical results can often be calculated

S.E. Webber et al. (eds.), Solvents and Self-Organization of Polymers, 167–196.

from an equation in closed form. Sometimes, however, the theory is limited by the nature of the approximations that must be introduced in order to make the problem tractable. These approximations can be severe in the complex systems of interest here. They can have an important influence on the extent to which the final equation describes the physical behavior of real macromolecular systems. In contrast, the simulations require relatively few approximations, thereby providing access to the physical behavior of systems that are more complicated than those accessible to analytical theory. For example, in systems where there are extensive fluctuations about an idealized structure, these fluctuations are easily incorporated in a simulation, but may present an enormous complication for the analytical approach. The important disadvantage of the simulations is the requirement for extensive computation.

This section presents fundamental considerations that are important for the simulation of the self-assembly of macromolecular systems. There are trade-offs between atomistic and coarse-grained methods. For these large systems, the coarse grained methods are more important. Lattices provide a popular framework for the simulation of the self-assembly of coarse-grained chains.

1.1. ATOMISTIC VS. COARSE GRAINED METHODS

Time and distance scales determine the types of techniques that can be employed for simulating the physical properties of molecules [1]. These issues become more restrictive as the size of the individual molecule increases. They are of extreme importance in the simulation of macromolecules. They require trade-offs between the scientific merit of performing simulations in which fully atomistic models move in real time in response to reasonable descriptions of interatomic forces, and the practical consideration of the computational time required for completion of the simulation to the desired degree of accuracy. Often many physical properties common to families of macromolecules, such as nonionic diblock copolymers, can be deduced from simulations that do not incorporate the details of the local covalent structure.

1.1.1. *Distance Scale*

Fully atomistic simulations at bulk density are currently limited to systems with no more than $\sim 10^4$ atoms. The largest fully atomistic systems for which we have computed molecular dynamics trajectories are inclusion compounds of polymeric hydrocarbons (poly(1,4-*trans*-butadiene) [2], poly(1,4-*trans*-isoprene) [3], and polyethylene [4]) in perhydrotriphenylene. The number of carbon and hydrogen atoms in these systems is in the range 4422–4452. This number of atoms is found in a single polyethylene chain with a molecular weight near 20×10^3. Phenomena that are unique to much larger chains, or that arise from the cooperative interaction of several such chains, are too large for practical simulations in the fully atomistic representation. Simulation of these larger systems becomes feasible using coarse-grained methods, where each fundamental unit in the simulation represents a segment comprised of several (perhaps many) atoms.

If our interest is in the simulation of a single micelle formed by diblock copolymers in a selective solvent, each of the individual chains might have a molecular weight of $\sim 20 \times 10^3$. The micelles might contain, on average, 10–10^2 such chains. Therefore the problem is at least an order of magnitude too large for atomistic methods. If we want to extend the problem to a study of the interactions between micelles, the shortcomings of the atomistic methods are even worse. So we conclude, on the basis of the size of the systems required for simulation, that coarse grained methods must usually be used for simulations of the self-assembly of macromolecules.

1.1.2. *Time Scale*

Molecular dynamics trajectories with fully atomistic models are computationally intensive, due to the necessity for performing an enormous series of consecutive numerical integrations with a very small time step, typically on the order of 10^{-15} s [5]. The lengths of the trajectories are limited to 1–10×10^{-9} s. When the nanosecond time scale of the molecular dynamics trajectory is compared with the time scale of important dynamic phenomena that occur in self-assembled systems of block copolymers, the gap can be enormous. The time scale for the exchange of chains between micelles of block copolymers in dilute solution in a selective solvent can easily be thirteen or more orders of magnitude longer than the nanosecond time scale of a molecular dynamics trajectory for a fully atomistic model [6]. So we conclude, on the basis of time scales, that coarse grained methods must usually be used in the simulation of the dynamics of the chains in self-assembled macromolecular systems.

1.2. COARSE GRAINED METHODS

Coarse grained methods are of two types, depending on whether they employ a continuous space for the conformations of the chain and the configurations of the system, or whether this space is discretized on a lattice. The "conformation" of a chain with N_b bonds is determined by the internal degrees of freedom, which are the lengths of the N_b bonds, the $N_b - 1$ angles between successive bonds and the $N_b - 2$ torsions about internal bonds. The "configuration" of the system includes the conformation of all of the chains, as well as information about the locations of their centers of mass, and the orientation of the chains about their centers of mass.

1.2.1. *Continuum*

Several of the simplest models for ideal polymer chains use a continuum. Such ideal models include the *Freely Jointed Chain* and the *Worm-Like (Porod-Kratky) Chain* [7,8]. Both of these models have an internal distance scale, which for the Freely Jointed Chain is the bond length, L_b, and for the Worm-Like chain is the persistence length, a. The mean square unperturbed end-to-end distances, $\langle r^2 \rangle_0$, of the two models are

$$\langle r^2 \rangle_0 = N_b L_b^2 \tag{1}$$

$$\langle r^2\rangle_0 = 2ar_{\max}\left\{1 - \frac{a}{r_{\max}}\left[1 - \exp\left(\frac{-r_{\max}}{a}\right)\right]\right\} \tag{2}$$

where $r_{\max}$ denotes the contour length of the Worm-Like Chain.

The two parameters in the Freely Jointed Chain can be assigned so that the model reproduces $\langle r^2\rangle_0$ for a real unperturbed chain with n real covalent bonds of length l. This ***Equivalent Chain*** is designed so that it has the same $\langle r^2\rangle_0$ and $r_{\max}$ as the real chain, which implies the assignments

$$L_b = \frac{\langle r^2\rangle_0}{r_{\max}} \tag{3}$$

$$N_b = \frac{r_{\max}^2}{\langle r^2\rangle_0} \tag{4}$$

for the Equivalent Chain. For the example of polyethylene, $r_{\max}$ is the length of the planar zig-zag conformation in which each internal bond is in a ***trans*** rotational isomeric state. Therefore $r_{\max} = nl\sin(\theta/2)$, where θ is the C—C—C bond angle. Measurements in a Θ solvent at 140° C yield $\langle r^2\rangle_0 = 6.7nl^2$ in the limit of large n [9]. When this structural and conformational information is placed in Eq. 3, $L_b = 6.7l/\sin(\theta/2)$, or 1.23 nm, in the Equivalent Chain for polyethylene. The Equivalent Chain requires an L_b that is about 8 times longer than the C—C bond length of 0.153 nm. Thus L_b for the Equivalent Chain is not the length of a real covalent bond, but is instead the length of a ***segment*** that is several times longer than a covalent bond.

Neither the Freely Jointed Chain nor the Worm-Like Chain, as described in Eq. 1 and 2, takes specific account of long-range interactions or intermolecular interactions. Such interactions could be included by, for example, incorporating a potential that describes the interaction of any pair of segments in the Freely Jointed Chain. A simple example of such a potential is produced in the representation of each junction between bonds by a hard sphere. Continuum models become computationally less efficient when systems with long-range and intermolecular interactions are examined at high density [10]. Discretization of space on a lattice may be more efficient in dealing with these dense systems.

1.2.2. *Lattice*

Simulations of multichain macromolecular systems in three dimensions often employ either a cubic lattice or a diamond lattice. Simple avoidance of double occupancy of any site on the lattice is the counterpart of the assignment of a hard sphere to each junction between bonds for the Equivalent Chain in the continuum. The notion of a ***segment*** still applies. Individual steps on the lattice should not be equated to individual covalent bonds, but instead represent a collection of such bonds.

Various schemes are currently being explored by several groups for the incorporation of some aspects of the local structure of the chain into the lattice models [11–14]. One approach, known as the ***Bond Fluctuation Model*** [11], allows different lengths for the individual steps on the lattice, and different angles between two successive steps on the lattice. These lengths and angles can be selected to mimic the local structure of a particular chain, such as the polycarbonate of bisphenol A [15]. Another approach uses lattices of high coordination, which are based on an underlying cubic [12] or diamond [14] lattice. Models for specific globular proteins on high coordination lattices based on an underlying cubic lattice can be folded specifically, and reversibly, into structures that are nearly indistinguishable from the native structures obtained by experiment [12]. This impressive accomplishment demonstrates that lattice methods can incorporate important features of the local structure of copolymers with a specific sequence of monomer units. When applied to commercial polymers, the discretized chains in the *Rotational Isomeric State Model* [16,17] can be accurately mapped onto the 2nnd lattice, which is a high-coordination lattice based on an underlying diamond lattice (Rapold and Mattice, work in progress).

The simpler lattices, such as the cubic lattice, have been used to date in most of the simulations of the self-assembly of block copolymers. On these simple lattices, one learns the general rules of behavior for diblock copolymers as a family. In the near future it is likely that simulations on high-coordination lattices will permit study of the self-assembly of specific copolymers, with incorporation of information about the local structure.

1.3. MONTE CARLO SIMULATIONS ON LATTICES

When conformational space is discretized on a lattice, a short chain can be studied by enumeration of all of its conformations. If we ignore those aspects of the configuration that arise from the placement of the center of mass and the direction adopted for the first step, and focus entirely on the internal degrees of freedom, a chain of N beads ($N-1$ steps) has 5^{N-2} internal conformations in a non-reversal walk on a cubic lattice. The number of internal conformations is 1.2×10^9 at $N = 15$, which is accessible by discrete enumeration. But at $N = 30$, this number has risen to 3.7×10^{19}, and discrete enumeration of all internal conformations is no longer a viable option. Since the simulation of the self-assembly of block copolymers requires the use of systems containing many more than 30 beads, discrete enumeration cannot be employed. Instead a representative subset of the allowed configurations is generated by Monte Carlo techniques [18]. Special care must be exercised in the construction of the subset. If $\langle f \rangle_s$ denotes the average of a property for a subset of s configurations, and $\langle f \rangle_\infty$ denotes the average of that same property in the ensemble that contains all configurations, we seek a means of generating the subset such that

$$\langle f \rangle_\infty = \lim_{s \to s_m} \langle f \rangle_s \tag{5}$$

when s_m is very much smaller than the number of configurations accessable to the system. The smaller we can make s_m, the more efficient the simulation will become.

For the systems of interest here, the best approach has been found to be one where consecutive configurations in the subset are not generated independently from one another, but instead where the proposed new configuration is generated by a small modification in the current configuration. The acceptance of the proposed new configuration is conditional, being governed by an appropriately constructed acceptance criteria. Some of the criteria can be "all or none", such as the requirement that no site be doubly occupied. Other criteria dictate a probability for acceptance of the new configuration that is between 0 and 1. Comparison of the probability for acceptance and a random number (hence the name "Monte Carlo") determines the fate of the proposed new configuration. If the acceptance criteria are fulfilled, the proposed new configuration is accepted; if they are not fulfilled, the proposed new configuration is rejected and the current configuration is counted again as the new configuration.

1.3.1. *Large-Scale vs. Local Moves*

The moves that convert one configuration into another are of two types, depending on the number of beads affected by the move.

Large-Scale Moves. Large-scale moves affect the coordinates of many beads. Translation of a chain as a rigid body is a large-scale moves, even if the center of mass moves only one step on the lattice. The translation alters the coordinates of N beads, independent of the size of the translation. Rotation about a bond as a rigid body alters the coordinates of all beads that are not colinear with the bond selected for the rotation. The number of beads that is affected can range from 0 (if the chain is a rod with all beads in a line) to $N-2$. Rotation as a rigid body is a large-scale move, because for the flexible chains of interest the number of beads affected is often nearly as large as $N-2$. The related pivot algorithm rotates the subchain between a randomly selected bead and one end of the chain [19,20].

Reptation can be accomplished conceptually by removal of a bead at one end of the chain, and attachment of this bead at the other end [21]. Expressed in this way, it might appear that the coordinates of only a single bead have been affected. However, if we index the beads in the chain, it is apparent that the coordinates of all beads have been changed. If the chain were a diblock copolymer, the step at the junction between the two blocks would move upon reptation of the chain, clearly demonstrating that the coordinates of the internal beads are changed upon reptation. Therefore reptation is a large-scale move. Large-scale moves are often useful in the efficient initial approach to equilibration of a system far from equilibrium, because each successful move alters the coordinates of approximately N beads, rather than just one or two beads, as in the local moves described below.

The large-scale moves have potential drawbacks. They may produce a system that is not ergodic. In an ergodic system, there are readily accessable pathways by which every configuration can be converted into any other configuration. Ergodic systems should be studied if thermodynamic properties are desired. As a very simple example of a non-ergodic simulation, consider an attempt to study flexible chains using translation of the center-of-mass as the only move. Then no chain could change from one internal conformation to another, and the simulation might produce quite bad estimates of the actually behavior of flexible chains. The absence

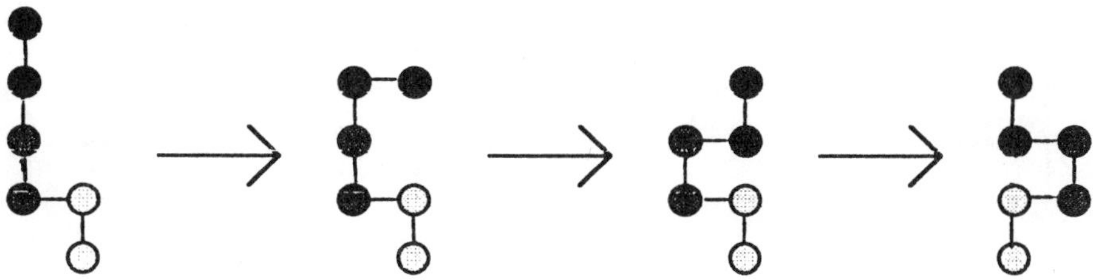

Figure 1. **Three consecutive local moves. The first two moves affect one bead only, and the last move affects two beads.**

of a pathway by which a chain in internal conformation i could be converted to internal conformation j causes the simulation to be non-ergodic.

Repation will produce a change in the internal conformation of the chain. But reptation of a self- and mutually avoiding chain on a cubic lattice is possible only if there is a vacant site adjacent to one of its ends. If the chain occupies a conformation that is severely congested at both ends, such that beads from that chain occupy all of the nearest-neighbor sites for both of its ends, reptation is impossible. Neither translation nor rotation as a rigid body will enable reptation. Therefore the chain is locked in one internal conformation, unless we introduce local moves that can relieve the congestion at the ends. Hence we must be aware of the possibility of the introduction of a bias into the subset if only large-scale moves are used in the simulation.

Another drawback of the large-scale moves comes when we seek to understand not only the static properties of the system at equilibrium, but also the dynamics of the equilibrated system. If only large-scale moves are used, the dynamics evaluated in the simulation will necessarily be determined by the types of moves selected. If a simulation were constructed using reptation as the only move for converting one configuration into another, we shouldn't read very much significance into the observation that the dynamics of the equilibrated system appears to fit the reptation model very well! When the interest is in understanding the dynamics, it becomes imperative to employ local moves with activation energies that are small compared to the large activation energies for the slow processes of interest (see Section 1.3.6.). Individual local moves usually affect the coordinates of one or two beads only.

Local Moves. Local moves that affect one or two beads are easily incorporated into simulations of chains on lattices [22]. The ***end-flip*** detaches the bead at one end, and then re-attaches it to the same end, but in a different position, as shown by the change in position of the top bead in the first move in Figure 1. A comparable move can be performed by an internal bead on a square or cubic lattice, provided the two steps to this bead make an angle of 90°. A square can conceptually be constructed, using these two steps as two of its sides. The internal bead at the junction of these two steps is then moved diagonally across the square, which

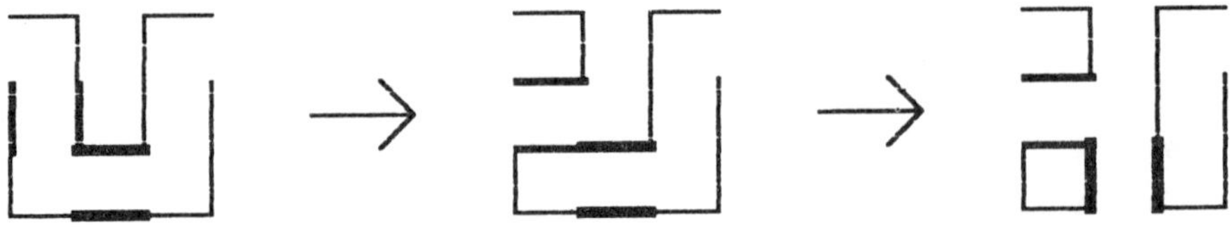

Figure 2. Two consecutive bond flips in a small system on a square lattice that has no vacant sites.

changes the position of that one internal bead and the two steps to it, as shown by the second move in Figure 1. Another local move, the ***kink-jump***, affects the position of two successive beads. It becomes possible on a cubic lattice whenever three successive steps form three sides of a square. These three steps are rotated by 90° or 180° about the axis defined by the remaining side of the square, as in the 180° flip of the middle two beads in the last move in Figure 1.

These local moves are not efficient in rapidly equilibrating a system, because each accepted move alters the coordinates of only one or two beads. However, when the system is close to equilibrium, the introduction of local moves may reduce the potentially adverse consequences of locked-in configurations. If one wishes to investigate the dynamics of an equilibrated system, it is desirable to use local moves only, in order to avoid a bias toward that type of dynamics defined by the large-scale moves used. Sometimes computational expediency will require the less desirable use of a mix of local and large-scale moves in the study of the dynamics of large system, when the dynamics of interest occurs on a slow time scale.

The individual moves that have been described here affect the coordinates of beads in one chain only. Simulations can be devised where the individual moves affect the coordinates of beads in more than one chain [23,24]. Sometimes this objective is obtained by simultaneous coordination of multiple local changes in the system. The number of beads that is affected grows accordingly, so while each local change may affect only 2 beads, all such local changes that contribute to the proposed new configuration may alter the coordinates of substantially more beads. Therefore this approach can have some of the character of a large-scale move.

Moves have been constructed for use in systems so dense that every site is occupied. These moves obviously cannot make use of voids, but instead utilize the simultaneous destruction and reformation of bonds, which may reside on different chains [23]. In such cases, there may be changes in the topology and polydispersity of the system, as illustrated in Figure 2. Here each move flips two parallel vertical (or horizontal) bonds to two horizontal (or vertical) bonds. The polydispersity of the system changes in the first bond flip, and the second bond flip changes the number-average molecular weight and creates a cyclic chain. The polydispersity, number-average molecular weight, and topology can be confined to narrow limits by appropriately designed selection criteria.

1.3.2. *Metropolis Monte Carlo*

Simulations are sometimes performed with systems in which all allowed configurations have the same probability. For example, when the chains are described by a random walk on a lattice, every conformation is allowed, and all conformations are weighted equally. Other simulations discriminate between conformations, but do so on an "all-or-none" basis, as in the non-reversal random walk (which prohibits simultaneous occupancy of any site by beads i and $i+2$), the self-avoiding walk (which prohibits simultaneous occupancy of any site by any two beads from the same chain), and the self- and mutually-avoiding walk (which prohibits simultaneous occupancy of any site by two beads from any chains). In none of these cases are the results dependent on temperature. Some of the conformations may be disallowed, but the allowed conformations are all weighted equally.

Incorporation of finite interaction energies into the simulation brings with it the concept of temperature, T.

Types of Energies Incorporated. Short-range intrachain interactions can be incorporated into lattice simulations. The stiffness of individual chains on a cubic lattice can be manipulated in a simple way by the introduction of an energetic contribution that is determined by the angle between each pair of successive steps. This description might assign bending energies of 0, E_θ, and ∞ when two successive bonds make angles of 180°, 90°, and 0°, respectively. The chain becomes stiffer as E_θ increases. This increase in stiffness causes a decrease in the critical micelle concentration and an increase in the average micelle size, when it is incorporated into the insoluble block of an AB diblock copolymer in dilute solution in a selective solvent [25]. In comparison, there is little affect when the stiffness of the soluble block is modified. Local structure can also be manipulated by assigning a torsion potential energy function to each internal bond.

Longer range intramolecular interactions, as well as intermolecular interactions, are usually introduced by the assignment of energies to pairs of non-bonded beads that are nearest neighbors on the lattice. For a two component system, such as homopolymer chains (A) in a one-component solvent (S), a single parameter will suffice,

$$\Delta E_{AS} = E_{AS} - \frac{1}{2}(E_{AA} + E_{SS}) \tag{6}$$

where ΔE_{AS} is related to the Flory-Huggin χ and the coordination number, z, for the lattice.

$$\chi = \frac{\Delta E_{AS}}{kT}\left[\frac{N_A(z-2)+2}{N_A}\right] \tag{7}$$

Here N_A denotes the number of beads in the chain of A, and k is Boltzmann's constant. In systems of AB diblock copolymers in a one-component solvent, there are three quasi-binary energetic parameters that can be assigned independently,

$$\Delta E_{AB} = E_{AB} - \frac{1}{2}(E_{AA} + E_{BB}) \tag{8}$$

$$\Delta E_{BS} = E_{BS} - \frac{1}{2}(E_{BB} + E_{SS}) \tag{9}$$

and the ΔE_{AS} from Eq. 6. The energies employed in the simulation are used in developing the probabilities for acceptance of proposed new configurations for the system.

Selection/Rejection of Proposed New Configurations. The most common method for generating equilibrated systems in a Monte Carlo simulation uses the Metropolis criterian for the acceptance of new configurations [26]. This criterian states that a proposed new configuration is accepted if it is of lower energy than the current configuration. Otherwise the proposed new configuration is accepted conditionally, with the acceptance probability, p, being given by the Boltzmann factor constructed from the ratio of the increase in energy, ΔE, to kT.

$$p = \min\left[1,\ \exp\left(\frac{\Delta E}{kT}\right)\right] \tag{10}$$

Temperature enters the simulation via the exponential in Eq. 10. The configurations accepted in the simulation are determined by dimensionless reduced energies of the form $\epsilon_{AB} = E_{AB}/kT$.

1.3.3. *Periodic Boundaries*

Simulations of the self-assembly of macromolecules are rarely performed with a sufficiently large system so that edge effects can be ignored. Therefore it is desirable to treat the edges of the system in a manner that minimizes their influence on the properties that are evaluated in the simulation. Periodic boundary conditions are commonly used for this purpose. The essence of the method can be illustrated with a very simple simulation performed in one dimension, with a periodic one-dimensional box located between 0 and L_x. At any instant, every bead within this one-dimensional box is counted as such, and all beads outside this one-dimensional box generate an image inside the box and located at

$$x_{\text{box}} = x - L_x \ \text{INT}\left(\frac{x}{L_x}\right) \tag{11}$$

where x denotes the coordinate of the parent bead (which may be outside the box), x_{box} is the coordinate of its image inside the periodic box, and the function INT returns the integer part of its argument. Extension to a three-dimensional cubic box is obvious, requiring only the counterparts of Eq. 11 for computing y_{box} and z_{box} from y and z.

The minimum image convention is employed to avoid double counting of the interactions between beads i and j. If a particular image of bead i is in the central box of a $3 \times 3 \times 3$ array of periodic boxes, this array contains 27 copies of bead j with which this image of bead i might interact. The interaction that is counted is the one between the image of bead i and the nearest of the 27 copies of bead j. This nearest copy may be in a different periodic box. Figure 3 depicts an example

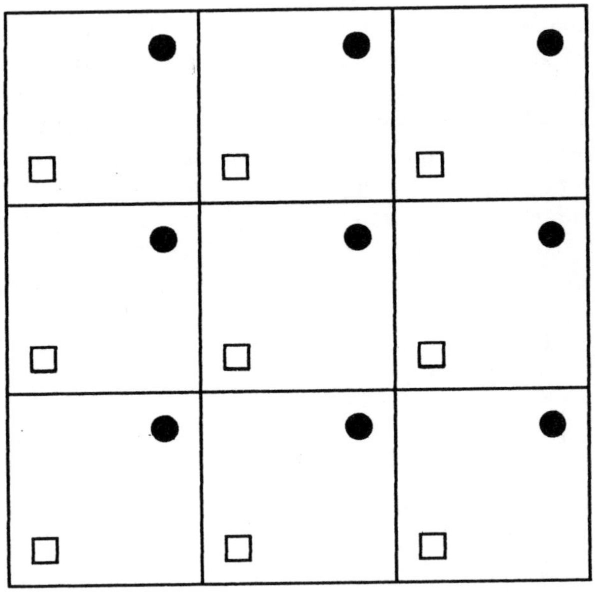

Figure 3. Beads i (filled circles) and j (open squares) in nine periodic boxes.

in two dimensions of a case where the shortest distance between beads i and j crosses the boundary of a periodic box.

The smallest dimension of the periodic box should be selected so that it is at least twice the size of the correlations in the system. If this requirement is not fulfilled, the simulation may be subject to size effects generated by the periodic boundaries [27]. The size effects are easily illustrated by considering the perfect lamellar phases of a symmetric AB diblock copolymer in a periodic cubic box of volume L^3 on a cubic lattice. If the lamellae are perfect, their repeat spacings, w, are restricted to

$$w = \frac{L}{i}, i > 0 \tag{12}$$

$$w = \frac{\sqrt{2}L}{i}, i > 1 \tag{13}$$

$$w = \frac{\sqrt{3}L}{i}, i > 1 \tag{14}$$

depending on whether the normal to the perfect lamellae connects two parallel surfaces, two opposite parallel sides, or two opposite corners, respectively, of the periodic box [28]. The best method for detecting such size effects is to repeat the simulation in periodic boxes of different sizes. One seeks evidence that the results are independent of box size for the largest periodic boxes used.

In special cases, where the formation of a macroscopic self-assembled structure is highly cooperative, the simulation may provide useful information about the formation of structures much larger than the periodic box itself. Thus the scaling behavior of the critical concentration for the formation of infinite transitory networks by ABA triblock copolymers with sticky ends has been determined in periodic boxes of finite size, even though these periodic boxes are much smaller than an infinite network [29]. This procedure was successful because the periodic boxes used in the simulation were large enough to capture the cooperative unit for gelation.

The periodic boundary conditions can be manipulated to produce special effects. If one of the lengths, such as L_z, is longer than the range of the interactions plus the sum of the lengths of the fully extended chains, the chains in the box cannot feel their periodic images along the z axis. If L_x and L_y are much smaller than the length of the fully extended chain, short-range attractions between the beads can cause the system to collapse into a thin film [30]. Alternatively, the two parallel surfaces at $z = 0$ and $z = L_z$ may be treated as reflecting or adsorbing plates [31], in order to investigate interactions of block copolymers with surfaces that are repulsive [32] or attractive [33,34] for one of the blocks.

1.3.4. *Generation of Initial Configurations*

Static properties for an equilibrated system must be independent of the selection of the initial configuration if the system is ergodic. Often it is easiest to construct a dense system by initially producing an ordered array, perhaps using parallel extended chains. This ordered array can be randomized by running the simulation with all energies turned off. Then the desired energies can be turned on, either gradually or in a single step, in order to equilibrate the system at the desired conditions.

Of course, a non-equilibrium initial state must be selected if the objective is to determine the kinetics of the approach to equilibrium. For example, if one wants to study the kinetics of the adsorption of AB diblock copolymers onto a surface that has an affinity for A, but not for B, one can initially equilibrate the simulation with the surface treated as a reflecting wall for both A and B. Then, at the start of the run from which the kinetics will be evaluated, the interaction between A and the surface is turned on in one step. The kinetics of the development of the adsorbed layer on the surface is then produced in the simulation [33].

The dynamics of the system at equilibrium is assessed by starting with a configuration selected from an equilibrated system.

1.3.5. *Equilibrium vs. Metastable States*

A necessary, but not sufficient, condition for equilibration in a simulation is that the average properties become independent of simulation time. When strong interactions are present, this behavior may be observed because the simulation is stuck in a local minimum, surrounded by energy barriers that cannot be surmounted by the individual moves employed. Then the results obtained in the steady state will be determined by the prior history of the simulation, in much the same way that properties of some real macromolecular systems depend on their prior history. While this non-equilibrium behavior in a simulation may be "real", in the sense that it might mimic the behavior of a real system, it is rarely the objective of the simulation. Much more common is the objective of simulating the behavior of a system that is in an equilibrium state, independent of the path used for its generation.

If the system is truly equilibrated, the same average properties should be obtained in a series of runs, each of which starts with an entirely different configuration. Also, if a simulation is performed by gradually turning on the desired energies in small steps, and running the simulation until the average properties achieve a steady state before incrementing the energies, then the same set of average properties should be obtained upon reversal of the path, where the energies are gradually turned off. The observation of hysteresis in the simulations demonstrates the absence of equilibration.

1.3.6. *Dynamic Monte Carlo*

If the objective of a Monte Carlo simulation is confined to the evaluation of static properties for equilibrated systems, the types of moves employed need not be related to the types of moves accessible to real chains. The moves need only produce an equilibrated system. We are not concerned if they do so by using non-physical moves, provided we seek static information only. Two successive accepted configurations of the system need not be separated by a single transition state of relatively low energy, if our interest is in static properties.

But if our interest is in the dynamics, the results are intimately related to the mechanisms by which the chains can move. Therefore the moves selected must not produce a bias in the dynamics. Ideally, this objective is obtained by using local moves only. The overall dynamics in the simulation is then the collective result of the random motions of individual beads, just as random thermal motion produces the dynamics in real systems that are at dynamic equilibrium.

Monte Carlo Time. Dynamic Monte Carlo simulations have their own version of time, which is measured in Monte Carlo time steps. One Monte Carlo time step is that number of iterations that would have moved NM beads, if every move were accepted. Here M denotes the number of parent chains, each with N beads. Therefore, on average, each bead has an opportunity to execute one move in each Monte Carlo time step. The number of accepted new configurations in a Monte Carlo time step is not constant, because the acceptance rate need not be exactly the same in each time step. The trajectory of sequential configurations is analyzed in terms of Monte Carlo time steps, or multiples of these time steps.

"Dynamics" in Monte Carlo Simulations. The notion that dynamics might be studied in Metropolis Monte Carlo simulations may initially seem bizarre, because the acceptance of a new configuration is determined solely by the difference in energy of the proposed new configuration and the current configuration. There is no consideration of the size of the activation energy required to step from the

Figure 4. Schematic depiction of the energy of the system (vertical) with reaction coordinate (horizontal). The fast fluctuations in energy are depicted by the thinner line, and the thicker line is a smoothe curve that averages out the fast fluctuations. The bar denotes the distance that might be moved along the reaction coordinate in one MC time step. Several time steps are required to pass over the barrier.

current configuration to the proposed new configuration. How can dynamics be inferred from a simulation that may appear to ignore activation energies?

The rationale for the dynamic Monte Carlo approach supposes that the energy of the entire system in the simulation in the vicinity of a transition state varies in the manner depicted schematically in Figure 4. The thinner line shows not only the slower long-term evolution, but also all of the faster (and smaller) local fluctuations in energy. The thicker line is a smoothe curve that averages out the rapid short-term fluctuations, but retains the long-term trends. Now imagine stepping through the time-evolution of the system, using a step along the reaction coordinate with a length represented by the horizontal bar in Figure 4. This length is meant to be large compared to the scale of the rapid short-term fluctuations, but small compared to the scale of the slower processes with large activation energies. On the scale of this step, the thicker curve provides a reasonable approximation to the evolution of the energy along the reaction coordinate, since we do not see the rapid small-scale fluctuations. This step corresponds to the one used in dynamic Monte Carlo simulations. Our interest is in obtaining the dynamics of the slower processes, and we average out the dynamics of the faster local processes.

According to the concepts in transition state theory, the ratio of the populations at any two positions along the approach to the transition state is given by equilibrium concepts. Thus the probability of a move along the reaction coordinate from a to b (or from b to a) in Figure 4 can be evaluated using the Metropolis rules. The passage over the barrier is the result of many small steps. Most attempts at crossing the barrier will be unsuccessful, because a configuration at b always has a higher probability of falling back to a than of continuing to climb toward the transition state. A simulation that uses large-scale moves (which might take the system directly from a to c, without feeling the barrier) may give a very unrealistic picture of the dynamics.

If the simulation is performed with local moves only, one hopes that Monte Carlo time is proportional to real time. If the simulation permits evaluation of a relaxation time (in Monte Carlo time) for a process for which a reliable relaxation time is available from experiment, the proportionality constant relating Monte Carlo time and real time can be established for that simulation. Better still, if the simulation can measure the entire relaxation for a process that is known to behave as a single exponential in real time, one can determine whether that relaxation is also a single exponential in Monte Carlo time (which it must be, if Monte Carlo time is linear in real time for that simulation).

2. Brief Review of Selected Recent Simulations of the Self-Assembly of Block Copolymers in a Selective Solvent

Here we briefly review several simulations of self-assembly of block copolymers that have recently been performed in our group. Following this review, Section 3 presents current work on the simulation of the dynamics of the exchange of chains between micelles formed by AB diblock copolymers in dilute solution in a selective solvent.

2.1. INTRAMOLECULAR INTERACTION IN FREE CHAINS

Individual ABA triblock copolymers at infinite dilution in a medium that is a good solvent for the internal block, but a nonsolvent for the middle block, can undergo intramolecular association of the end blocks, resulting in the formation of “hair-pins”, as depicted in Figure 5. This association is opposed by a reduction in conformational entropy of the middle block. The size of the reduction in entropy is important, because it has a bearing on the ability of these triblock copolymers to form micelles at higher concentration, with both end blocks placed in the same micellar core. When a family of simulations is performed with a varying number of beads, N_{B}, in the middle block, but with constant energetics and size of the end blocks, the ratio of the number of isolated chains that are intramolecularly associated, and unassociated, denoted $n_{\mathrm{a}}/n_{\mathrm{u}}$, is given by

$$\ln\left(\frac{n_{\mathrm{a}}}{n_{\mathrm{u}}}\right) = \frac{3}{2}\beta \ln N_{\mathrm{B}} + \text{ constant} \qquad (15)$$

where $\beta = 1$ if the loop entropy for the internal aggregation is described by the Jacobson and Stockmayer approach to macrocyclization equilibria in long unperturbed homopolymers [35]. Simulations of a family of ABA chains with constant N_{B}, but varying N_{A}, find that β decreases as N_A decreases, being near one only when N_A is very small [36]. This result shows that the internal aggregation

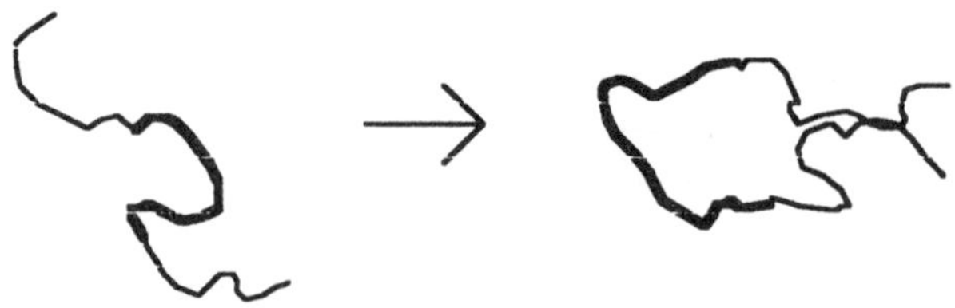

Figure 5. Intramolecular aggregation of a single ABA triblock copolymer.

(and hence also micellization at higher concentration) will be easier for ABA with $N_A > 1$ than might have been expected based on the loop entropy considered by Jacobson and Stockmayer. As the end blocks become larger, there is an increase in the number of conformations of the middle block that can support the formation of "hair-pins" (or micelles).

2.2. MICELLES

Several aspects of the aggregation of block copolymers in dilute solution in a selective solvent have been studied with Monte Carlo simulations. In these simulations, two chains are defined as forming an aggregate when at least one A bead from one chain has an A bead from the other chain as a nearest neighbor.

2.2.1. *The Critical Micelle Concentration*

The critical micelle concentration (cmc) can be determined by monitoring either the free chains or the aggregates. Let ϕ_{free} and ϕ_{total} denote the volume fractions of free chains and all chains, respectively. The value of ϕ_{total} is determined by the initial composition of the periodic box, but ϕ_{free} is evaluated after equilibration. A series of simulations is performed with a particular set of N_A, N_B, and ϵ's, but with different ϕ_{total}. At very low concentration, all chains are free chains, and $\phi_{\text{free}} = \phi_{\text{total}}$. Aggregates form at higher concentrations, causing ϕ_{free} to become less than ϕ_{total}. Eventually ϕ_{free} reaches a plateau, which is finally followed by a gradual decline (ϕ_{free} must return to 0 in the limit where $\phi_{\text{total}} \rightarrow 1$, because no chain is free when the system reaches bulk density). The cmc can be identified with the maximum value of ϕ_{free} or with the intersection between a linear horizontal extension of the plateau and the line $\phi_{\text{free}} = \phi_{\text{total}}$ [36]. Alternatively, one can evaluate the volume fraction of the aggregates, ϕ_{micelle}, as a function of ϕ_{total}, and extrapolate it to the total concentration at which ϕ_{micelle} becomes zero [29].

For AB diblock copolymers in media where $\epsilon = \epsilon_{AB} = \epsilon_{AS} > 0$, the cmc is exponentially dependent on ϵN_A [36]. This behavior was predicted symmetric AB copolymers at strong segregation [37], and the simulations are consistent with that prediction. The simulations cannot be extended indefinitely into the region of strong segregation because of the onset of nonequilibrium behavior, which manifests itself by hysteresis. When extended in the other direction, the simulations show that the prediction by Leibler *et al.* [37] reaches further into the region of weak segregation than might have been expected. It also holds for diblock copolymers in which the ratio of the number of beads in the blocks is as large as three, provided the cmc is expressed in terms of the volume fraction of the insoluble block, ϕ_A, and not the volume fraction of the entire copolymer. The cmc is less sensitive to N_B than to N_A. When N_B is very much larger than N_A, the system may no longer form micelles.

Symmetric ABA triblock copolymers can also form micelles, but at a cmc that is larger than the one observed with the corresponding diblock copolymers [36]. Individual chains of ABA experience a decrease in conformational entropy for the soluble block when both terminal blocks are placed in the same micellar core. For this reason ABA triblock copolymers have a higher cmc than the corresponding AB diblock copolymers. The cmc of the triblock copolymers in the simulation can be described by a scaling law with $b \sim 0.4$ [29].

$$\text{cmc} \propto (N_A \epsilon)^{-b} \tag{16}$$

2.2.2. *Size and Structure of the Micelles*

In the simulations, the micelles of AB diblock copolymers approach the classic core-shell model as ϵN_A increases [36]. This model has a spherical core in which the volume fraction of A is one, surrounded by a corona composed of B blocks swollen by solvent. The micelles of the corresponding ABA triblock copolymers are smaller and less well organized, with more solvent in their cores [29,36]. Often the micelles of ABA triblock copolymers contain only 3–6 chains, which is an order of magnitude smaller than micelles for the corresponding AB copolymers under comparable conditions. The polydispersity of the ABA micelles is often large [38]. The scattering functions can be computed for micelles in selective solvents that are isorefractive with one of the blocks [39].

The ABA chains in a micelle can adopt three distinct types of conformations: ***dangling ends***, in which one of the A blocks is in the core, but the A block at the other end of the chain has no interactions with other A blocks, ***loops***, in which both A blocks are in the core of the same micelle, and ***bridges***, in which the two A blocks are in cores of different micelles, as depicted in Figure 6. Bridges are favored at large ϕ_{total} and large ϵN_A [29]. In contrast, ***dangling ends*** tend to decrease as ϵN_A increases, because they are converted to loops (particularly at small ϕ_{total}) or to bridges (particularly at large ϕ_{total}). Neither loops nor bridges are observed in the corresponding AB block copolymers. The bridges in the ABA system are of special interest because they can produce assembly of the micelles on a much larger scale, as described in the next section.

2.3. TRANSITORY NETWORKS FROM ABA TRIBLOCK COPOLYMERS

Bridging of micelles by ABA triblock copolymers produces clusters of micelles that are bonded together by transitory links. Eventually the amount of bridging will become sufficient to produce a three-dimensional transitory network [40,41].

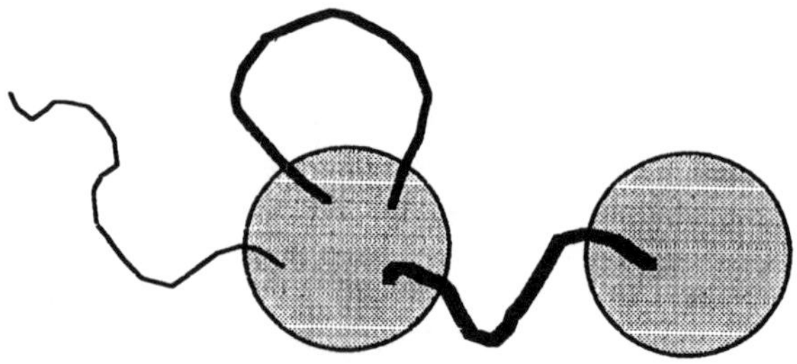

Figure 6. Dangling end (thin line), loop (medium line), and bridge (thick line). The shaded circles denote the cores of two micelles.

This network is transitory because the bridges between junctions are not formed by covalent bonds, but are formed instead by the noncovalent association of the A blocks of a bridge with two different micellar cores. This network spans the periodic box in all three directions. If the periodic box is large enough so that box size effects have been eliminated, the concentration at which this transitory network forms can be identified with the concentration for macroscopic gelation.

2.3.1. *The Critical Gel Concentration*

The critical concentration for gelation (cgc) can be evaluated from the simulations by several methods. The most direct method uses visualization of the periodic box, and the search for bridging connections between a subset of the micelles that provide a continuous path between all three pairs of parallel faces of the box. Numerical analysis can also be employed, using criteria for gelation that have their counterparts in theory and in experiment [42,43]. Thus one can evaluate the weight-average cluster aggregation number, and seek the ϕ_{total} at which it diverges [29]. Alternatively, one can evaluate the functionality of the largest cluster [29]. Flory demonstrated that gelation occurs in a macroscopic system when this functionality rises above two [43]. These two analytical methods produce nearly the same values for the cgc for the systems in the simulations. These two closely spaced values typically bracket the value of the cgc that is deduced by visual inspection of the periodic box [29]. The functionality based critical gel concentration tends to be slightly higher than the cluster size based cgc. Both critical gel concentrations can be summarized as

$$\text{cgc} \propto (N_{\text{A}}\epsilon)^{-a} \tag{17}$$

with values of a in the range 0.14 – 0.17 [29].

2.3.2. *Lifetime Spectrum of the Bridges, and Their Contribution to the Modulus*

The network formed at concentrations above cgc is transitory in nature. The "cross-link" sites between network chains are the micelles, and the network chains are the B blocks of the bridges. In contrast with typical covalent networks, in

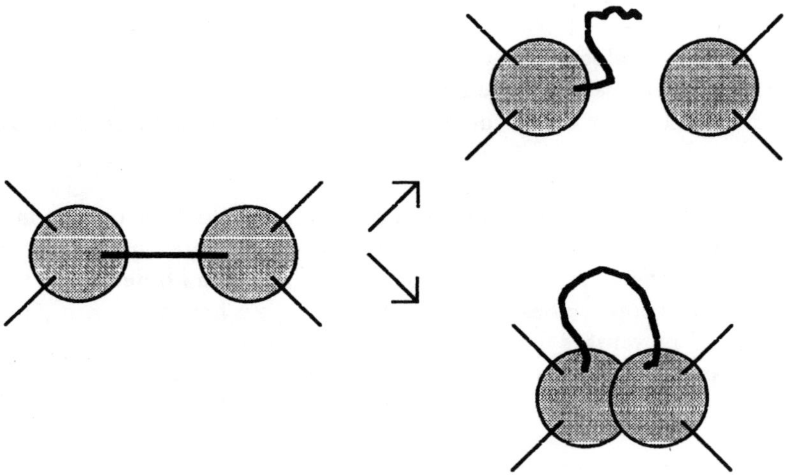

Figure 7. Block expulsion (top) and micellar fusion (bottom).

which the network chains can be destroyed only by breaking covalent bonds, the network chains in these transitory networks have lifetimes that are limited by the dynamics of non-covalent interactions.

A network chain becomes inactive by either of the two mechanisms depicted in Figure 7. In *block expulsion*, one of the A blocks of the bridging chain is expelled from the core of a micelle, thereby converting the bridging chain to a dangling chain. Block expulsion in a micelle of AB diblock copolymers would produce a free chain instead [44,45]. The bridging attraction [46–49] causes *micellar fusion* by a merger of the cores of the two micelles at the ends of the bridging chain, thereby converting the bridge to a loop. Both mechanisms are observed in the simulations, and either one can be dominant, depending on the conditions employed [38]. Block expulsion is favored over micellar fusion at small ϵN_A and low ϕ_{total}, but micellar fusion becomes dominant as either of these variables rises.

The spectrum of the lifetimes of the bridging chains, evaluated in a dynamic Monte Carlo simulation and expressed in Monte Carlo time steps, can be used to determine the contribution of the bridging chains to the modulus upon a small shear of the transitory network [38]. The shear must be assumed to be small, because the analysis is based on the lifetime distribution evaluated for the bridges in the undeformed network at dynamic equilibrium. The fraction of the stress remaining after a time interval $\Delta\tau$ is equated to the fraction of the bridges that continuously remained active (were not converted to dangling ends or loops) during this period. The lifetime distribution for the bridges cannot be described by a single exponential. When ϵN_A increases, the width of the rubbery plateau region broadens, and the transition to flow is slower. The shape of the stress relaxation function is much more sensitive to N_A that to either ϕ_{total} or N_B.

2.4. DIBLOCK COPOLYMERS AT HIGH CONCENTRATION

The micelles of diblock copolymers cannot form transitory networks by bridging chains. However, fusion of their cores at high concentration does produce interesting structures that can be detected in the simulations [27,28,39,50,51]

2.4.1. *Types of Structures*

As ϕ_{total} increases above the cmc for a symmetric diblock copolymer in a selective solvent, simulations detect a series of interesting structures [52]. As the concentration increases, these structures appear sequentially as an ordered array of micelles with spherical cores, coalescence of the cores into a hexagonal array of cylinders, formation of perforated lamellae by lateral extensions from one cylinder to its neighbor, and ultimately the formation of lamellae by fusion of parallel cylinders. These microstructures are expected for diblock copolymers [53].

2.4.2. *Dynamics in Lamellar Structures*

The dynamics of the AB diblock copolymers in a lamellar system is expected to be anisotropic, with the translational diffusion coefficient being different in the plane of the lamellae and perpendicular to this plane. This expection is realized in dynamic Monte Carlo simulations, which find that the diffusion coefficients in these two directions differ by at least an order of magnitude, being larger in the plane of the lamellae [54]. The translational diffusion coefficient, D, can be obtained from a dynamic Monte Carlo simulation as

$$D = \frac{1}{6} \lim_{\Delta t \to \infty} \frac{\langle (\Delta r)^2 \rangle}{\Delta t} \tag{18}$$

where Δr denotes the displacement of the center of mass in time interval Δt, and the angle brackets denote the average of all $(\Delta r)^2$ in the trajectory that are consistent with a time interval of the size Δt. The mean square displacement should become linear in Δt as this time interval increases. When this condition is fulfilled, D is obtained from the slope. The largest Δt that can be used is typically no greater than half the length of the trajectory, because the statistical uncertainty in $\langle (\Delta r)^2 \rangle$ increases at large Δt, due to fewer observations.

2.4.3. *Changes in Dynamics at the Transitions between Different Structures*

The anisotropy of the translational diffusion coefficient in these self-assembled structures can be exploited to provide a sensitive indicator of the point at which one structure converts into another. Consider, for example, the anisotropy of the translational diffusion coefficient of symmetric AB diblock copolymers as ϕ_{total} is varied to produce the morphological transitions hexagonal phase $\rightarrow$ performated lamellae $\rightarrow$ lamellae. In the hexagonal phase, there is rapid diffusion in one direction (along the axis of the cylinders), and slow diffusion in the two directions perpendicular to this axis. As mentioned in Section 2.4.2, the AB diblock copolymers in the lamellar phase have rapid diffusion in two directions (in the plane of the lamellae) and slow diffusion in the remaining direction, perpendicular to this plane. The onset of the transition from the hexagonal phase to the performated lamellae is accompanied in the simulations by an easily detectable change in the anisotropy of the translation diffusion coefficient of the diblock copolymers, from one fast and two slow components in the hexagonal phase to two fast and one slow components [55].

2.5. INTERACTIONS WITH SURFACES AND INTERFACES

Manipulation of the periodic boundary conditions permits the simulation of block copolymers in the presence of impenetrable surfaces and penetrable interfaces.

2.5.1. *Adsorption onto a Planar Surface*

Interaction with a planar surface is achieved by placing impenetrable walls at the two parallel surfaces of the periodic box that are spaced by L_z. If the two surfaces are to be independent, L_z must be large enough so that chains cannot interact with both surfaces simultaneously, or interact simultaneously with any adsorbed layers that develop at these surfaces.

Adsorption from a nonselective solvent can be studied when all ϵ's are zero except for the one that describes the interaction of A beads with the surface. The system is pre-equilibrated with all energies turned off. If the attractive interaction of A with the surface is then turned on instantaneously, the kinetics of the development of the adsorbed layer in the simulations shows an initial rapid growth that is diffusion controlled, followed by a slower second growth phase when there must be an adjustment of chains already on the surface before new chains can be adsorbed [33], which is the behavior expected from experiment [56,57]. The adsorption isotherms exhibit Langmuir behavior only when N_A and the interaction energy are both small. The fraction of the A beads in contact with the surface decreases as N_A or ϕ_{total} increases [58]. Beads in the middle of the A block have a higher probability for adsorption than does the bead at the free end of the A block, which in turn has a higher probability for adsorption than the A bead at the AB junction.

When the solvent is selective, there will be a competition between micellization of the diblock copolymers in solution and adsorption of the copolymer onto the surface. At concentrations slightly above the cmc, the adsorption takes place primarily via the pool of free chains, and the structures of the adsorbed layer are similar to those obtained in the nonselective solvent [59]. At higher concentrations, the micelles themselves participate in the adsorption, and the structure of the adsorbed layer changes as a result.

2.5.2. *Adsorption between Two Parallel Plates*

If L_z is made small enough so that a single chain can interact simultaneously with both opposing surfaces, the simulation addresses the issue of the behavior of a system with two closely spaced parallel plates. The chains might be telechelic chains, in which only the two end beads have an affinity for the plates [60], or they might be symmetric ABA triblock copolymers, in which all A beads have an affinity for the plates (Nguyen-Misra *et al.*, submitted). The spectrum of the lifetime of the bridging chains has a simple exponential form only in the telechelic chains, where $N_A = 1$. This spectrum becomes more complicated at larger N_A (Nguyen-Misra *et al.*, submitted).

2.5.3. *Adsorption at an Interface*

The periodic boundary conditions are easily manipulated to reveal some aspects of the behavior of diblock copolymers in a mixture of two immisible liquids. For this purpose, the voids are assigned one set of energies in the half of the box where z is less than $L_z/2$, and another set of energies in the remainder of the box. Reflecting walls are placed at $z = 0, L_z$. When the voids in one half of the box are treated as energetically equivalent to A, and those in the other half are energetically equivalent to B, the AB diblock copolymers can form micelles in a phase where the medium is comprised of very small oligomers of either A or B, or they can adsorbed at the sharp interface between the two phases. The micelles are of different structure in the two halves of the box. They have cores rich in A blocks in the half of the box where the voids are energetically equivalent to B, but in the other half of the box the cores of the micelles are rich in B blocks instead. A chain with its center of mass near $z = L_z/2$ can place its A block in the A-rich phase, and simultaneously place its B-block in the B-rich phase. The adsorption of symmetric AB diblock copolymers at the interface is very close to Langmuir in character, both when the concentration of AB diblock copolymer is low enough so that micelles are not present in the two homopolymer phases, and at higher concentrations, where micelles are formed [61].

When the diblock copolymers are asymmetric, they have different cmc's in the two halves of the box. The concentrations of free chains are then different in these two regions. The ratio of the concentrations of the free chains provides a sensitive test of how well the equilibrium distributions of the chains in these simulations is described by Flory-Huggins theory. No major deviation from the prediction of Flory-Huggins theory is detected [62].

3. Exchange of Chains Between Micelles

When simulations of the formation of micelles by AB diblock copolymers in a selective solvent can be successfully equilibrated, there must be a dynamic equilibrium with pathways by which the micelles can exchange chains with one another. This phenomenon is detected in the dynamic Monte Carlo simulations as a small fluctuation in the fraction of the chains that exist as free chains [36]. It can be detected experimentally as a time dependence of the efficiency of nonradiative singlet energy transfer in an equilibrated system where initially some of the micelles are comprised of chains that are covalently labelled with a donor, while the remainder of the micelles are initially comprised of chains that are covalently labelled with an acceptor [6,63,64]. Dynamic Monte Carlo simulations have the potential for showing how the variables in the system affect the time scale and heterogeneity of the relaxation, as well as the mechanism by which the micelles exchange chains with one another [65,66].

3.1. DIRECT EVALUATION OF CORRELATION FUNCTIONS

A variety of correlation functions can be used to study different aspects of the dynamics of the exchange of chains between micelles. The tracer correlation function, $M(\tau)$, is defined using the number of chains, N_c, in the aggregate that contains the tracer chain at times t and $t+\tau$.

$$M(\tau) = \frac{\langle N_c(t)N_c(t+\tau)\rangle - \langle N_c(t)\rangle^2}{\langle N_c(t)^2\rangle - \langle N_c(t)\rangle^2} \tag{19}$$

The averaging uses all possible time origins and all possible chains as the tracer chain. The nature of the relaxation deduced from $M(\tau)$ is dependent upon the polydispersity of the micelles [65]. Very little change in $M(\tau)$ will be apparent if the micelles are all of nearly the same size, with the same value of N_c.

Correlation functions can also be defined for the extraction of a chain from a micelle, and for the addition of a chain to a micelle [65]. The rate at which they decay is strongly dependent on ϵ and N_A, being more sensitive to changes in ϵ than to changes in N_A [66].

3.2. TRANSITIONS DEDUCED DIRECTLY FROM THE SIMULATIONS

The decay of the correlation functions may be very different in form from a single exponential, and it often exhibits a very long tail. Therefore evaluation of an average lifetime by the conventional method of measuring the area under the correlation function is difficult. For this reason, and also in order to attempt to gain insight into the mechanism, we attempt another type of analysis of the dynamic Monte Carlo simulations, based on direct analysis of the numbers of transition of three types. The long-term dynamics is deduced from a long series of snapshots of the system. Successive snapshots are separated by 100–120 Monte Carlo time steps, depending on the trajectory.

3.2.1. *Classification of Processes*

Inspection of the dynamic equilibrium for several systems of diblock copolymers in dilute solution in a selective solvent shows the occurrence of three distinct types of exchange mechanisms. The transition states for these exchange mechanisms are sketched in Figure 8.

- *Chain Expulsion*: A chain is expelled by a micelle, exists for at least one (or two) snapshots as a free chain, and in some later snapshot is captured again by a different (or the same) micelle. A transition state, showing the free chain between two micelles, is depicted in the top portion of Figure 8. The requirements stated in parenthesis in the opening sentence of this definition exclude the process in which a chain is expelled from a micelle, and then immediately recaptured by the same micelle.

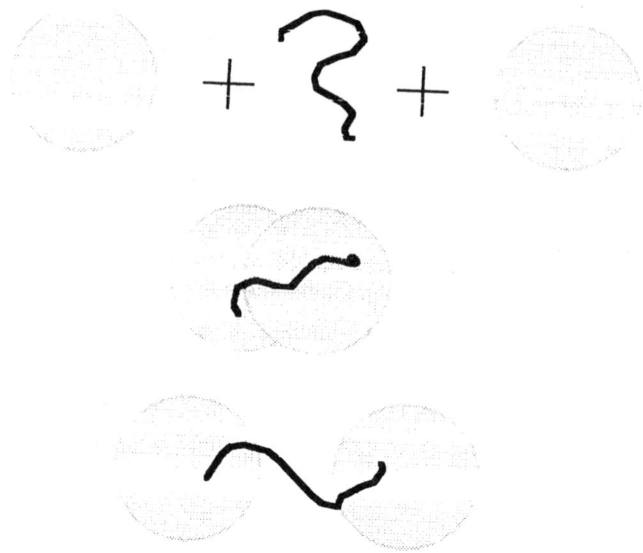

Figure 8. **Chain expulsion (top), micellar merger (middle), and micellar spanning (bottom)**

- *Micellar Merger/Splitting*: The number of chains in a micelle changes from m to $m + \Delta m$, and remains at $m + \Delta m$ for a least two successive snapshots. The value of Δm cannot be -1, 0, or 1. Two micelles merge if $\Delta m > 1$, and a micelle splits if $\Delta m < -1$. This mechanism is depicted in the middle portion of Figure 8, showing two micellar cores in contact, and a chain that is part of both cores. The chain may migrate from one core to the other, without exposure of its A block to the solvent, when the two cores are in contact.
- *Micellar Spanning*: When the exteriors of the cores of two independent micelles are separated by less than the fully extended length of an A block, a chain can transfer from one micellar core to the other micellar core without ever becoming a free chain. Of course, the A block of the migrating chain is exposed to the solvent-swollen coronas. The mechanism is sketched in the bottom part of Figure 8. A chain is migrating from one micellar core to another, without becoming a free chain, but without contact of the cores.

As will shall see, Micellar Spanning is much less important that the other two processes.

3.2.2. *Transition Rates*

The transition rates are inversely proportional to the number of events per unit time. The rates for Chain Expulsion, Micellar Merger/Splitting, and Micellar Spanning were evaluated directly from the dynamic Monte Carlo simulations, by

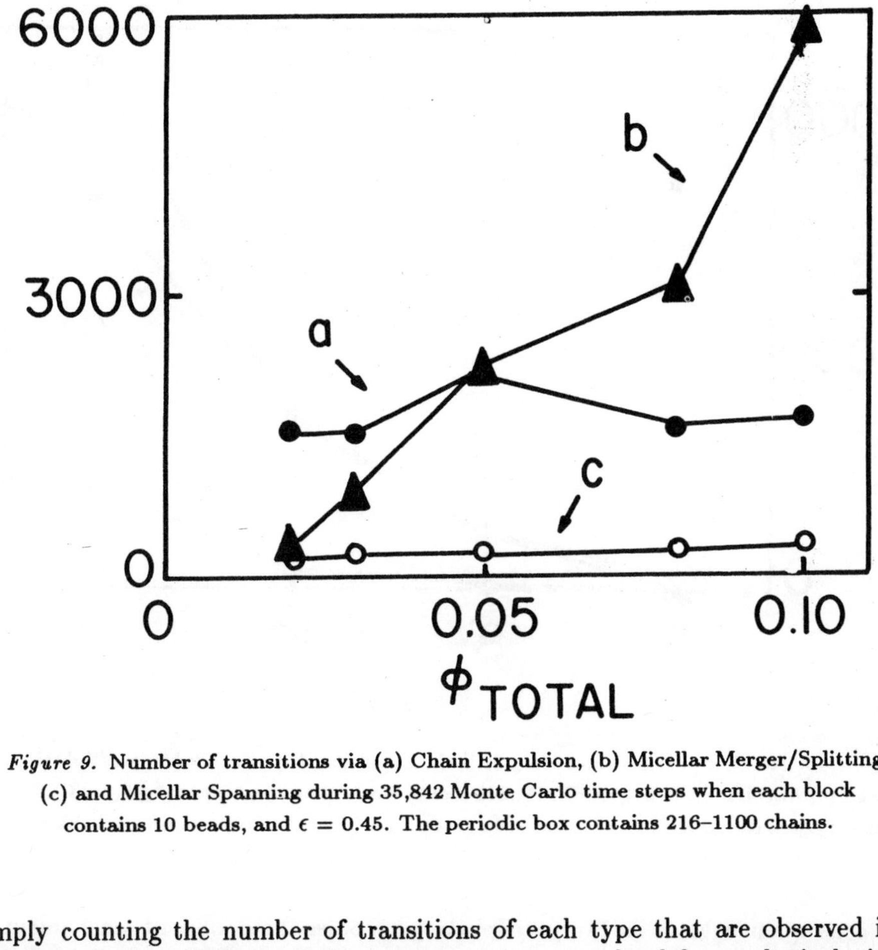

Figure 9. **Number of transitions via (a) Chain Expulsion, (b) Micellar Merger/Splitting, (c) and Micellar Spanning during 35,842 Monte Carlo time steps when each block contains 10 beads, and $\epsilon = 0.45$. The periodic box contains 216–1100 chains.**

simply counting the number of transitions of each type that are observed in a specified number of Monte Carlo time steps. An example of the results is depicted in Figure 9 for a simulation under conditions where the cmc has previously been reported to be 0.008 [36]. The number of chains in the simulations varies from 216 to 1100. The trajectories have lengths that provide an average of 7–9 transitions per chain. Chain Expulsion is the dominant mechanism when ϕ_{total} is nearest the cmc. At these concentrations, a large fraction of the diblock copolymer is present as free chains, and the exchange mechanism that utilizes the free chains is naturally the one that is dominant. The number of transitions that occur by Chain Expulsion is nearly independent of ϕ_{total} for the range covered in Figure 9, perhaps because the concentration of free chains is nearly constant in this range of ϕ_{total}.

In contrast, the number of transitions in Figure 9 that occur via Micellar Merger/Splitting increases strongly as ϕ_{total} becomes larger. The number density of micelles becomes larger as ϕ_{total} rises above the cmc, and the likelihood of close encounters between pairs of micelles increases as the square of the number density.

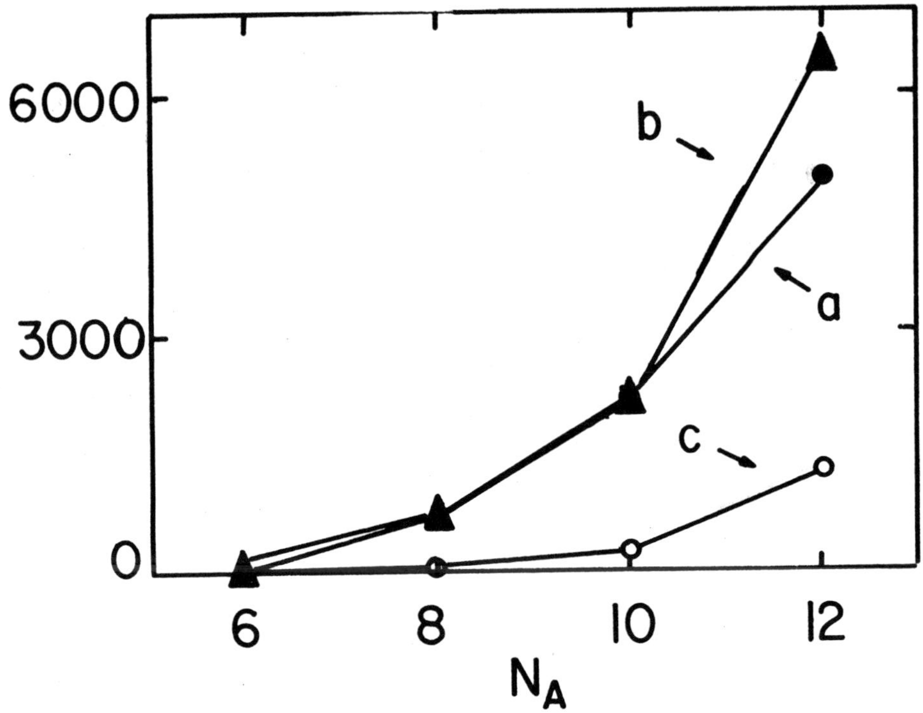

Figure 10. Number of transitions via (a) Chain Expulsion, (b) Micellar Merger/Splitting, and (c) Micellar Spanning during 35,842 Monte Carlo time steps when $\phi_{\text{total}} = 0.05$, the B block contains 10 beads, and $\epsilon N_A = 4.5$. The periodic box contains 550 chains.

As a consequence, Micellar Merger/Splitting becomes the dominant mechanism of exchange at higher concentrations. In the conditions used for the simulations, there are comparable numbers of exchanges via Chain Expulsion and Micellar Merger/Splitting when the concentration is about six times larger than cmc.

If the rate of exchange of chains between micelles were controlled by the composite variable ϵN_A, the transitions rates should be unaffected by an increase in N_A by a factor of x, provided ϵ is decreased by the same factor, $\epsilon N_A = (\epsilon/x)(xN_A)$. The actual behavior observed in the dynamic Monte Carlo simulations is depicted in Figure 10. The transition rates for all three mechanisms are slower at small N_A and large ϵ than they are at large N_A and small ϵ, even though the produce ϵN_A has been held constant at 4.5. Clearly ϵN_A is not a useful variable in summarizing the transition rates.

Figure 10 also shows that the dynamics of the exchange of chains between micelles will shut down as ϵ increases. There were 550 chains in the simulations reported in this figure. At the largest ϵ used, the average number of transitions per chain has fallen below 1. It would become very much less than 1 at larger ϵ.

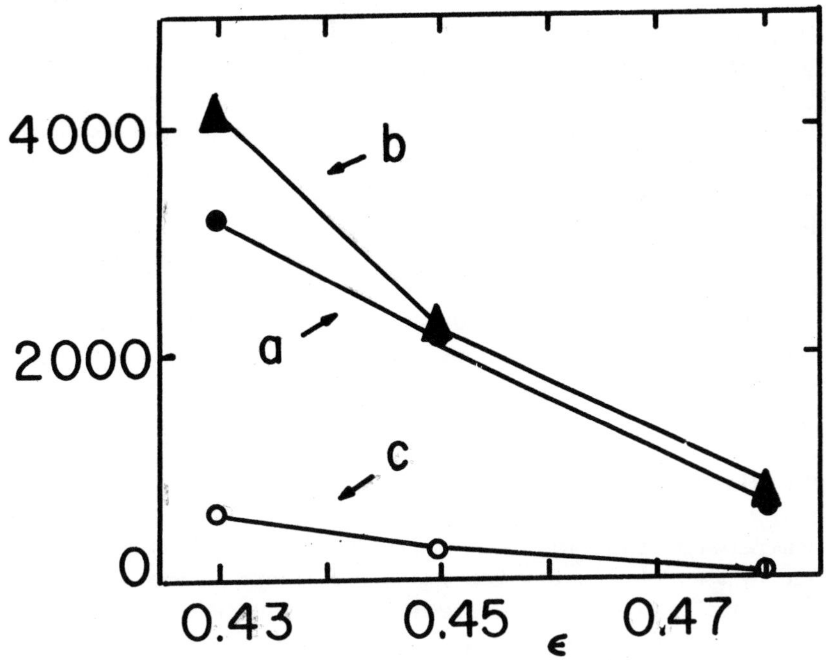

Figure 11. **Number of transitions via (a) Chain Expulsion, (b) Micellar Merger/Splitting, and (c) Micellar Spanning during 35,842 Monte Carlo time steps when $\phi_{total} = 0.05$, and both blocks contain 10 beads. The periodic box contains 550 chains.**

The influence of ϵ itself on $A_{10}B_{10}$, at $\phi_{total} = 0.05$, is depicted in Figure 11. The transition rates for all three mechanisms decrease as the interaction energy becomes stronger, as expected. This system will become frozen at ϵ slightly larger than 0.5. The larger energy makes chain expulsion more difficult because the A block feels a stronger energetic penalty when it is surrounded by solvent. The larger energy improves the organization of the cores of the micelles, causing them to approach more closely a classic core-shell structure. This improvement in the organization of the micelles makes direct micelle-micelle interaction more difficult, due to the repulsions of the corona of the two micelles. The corona in the classic core-shell model is formed by blocks of B that are swollen by the solvent.

Micellar Spanning accounts for only a small fraction of the transitions under all conditions examined, as is apparent in Figures 9–11. The simulation suggest that little harm might follow from ignoring Micellar Spanning entirely, and constructing a theoretical analysis based on the transition rates for Chain Expulsion and Micellar Merger/Splitting only. The detailed analysis, based on the Master Equation and using transitions rates deduced from the dynamic Monte Carlo simulations, will be reported elsewhere (Haliloğlu *et al.*, submitted).

4. Conclusion

The self-assembly of block copolymers occurs on distance scales and time scales that are too large for study by fully atomistic simulations. Important information on the structure and dynamics of these systems is accessable from Monte Carlo simulations, which provide access to the necessary sizes and time scales. These simulations can study several successive levels of self-assembly: (1) internal aggregation of single chains, (2) formation of micelles by 10–100 chains in dilute solution in a selective solvent, (3) formation of very large transitory networks by bridging between micelles, (4) self-assembly into cylinders, perforated lamellae, and lamellae at high concentration, and (5) adsorption at surfaces and at interfaces. Dynamic Monte Carlo simulations of previously equilibrated systems provide information on the mechanisms by which the chains in the self-assembled structures equilibrate with one another.

5. Acknowledgments

Our recent research on the simulation of the self-assembly of large systems has been supported by the Polymers and International Programs at the National Science Foundation (grants DMR 90-14502, DMR 92-20369 and INT 93-12285) and by a gift from the BF Goodrich Company.

6. References

1. Allen, M.P. and Tildesley, D.J. (1987) *Computer Simulation of Liquids*, Oxford University Press, Oxford.
2. Dodge, R. and Mattice, W.L. (1991) *Macromolecules* **24**, 2709–2714.
3. Zhan, Y. and Mattice, W.L. (1992) *Macromolecules* **25**, 3439–3442.
4. Zhan, Y. and Mattice, W.L. (1992) *Macromolecules* **25**, 4078–4083.
5. Brooks, C.L., III, Karplus, M., and Pettitt, B.M. (1988) *Adv. Chem. Phys.* **71**, 1–259.
6. Wang, Y., Kausch, C.M., Chun, M., Quirk, R.P., and Mattice, W.L. (1995) *Macromolecules* **28**, 904–911.
7. Porod, G. (1949) *Monatsh. Chem.* **80**, 251–255.
8. Kratky, O. and Porod, G. (1949) *Rec. Trav. Chim.* **68**, 1106–1122.
9. Chiang, R. (1965) *J. Phys. Chem.* **69**, 1645–1653.
10. Gerroff, I., Milchev, A., Binder, K., and Paul, W. (1993) *J. Chem. Phys.* **98**, 6525–6539.
11. Carmesin, I. and Kremer, K. (1988) *Macromolecules* **21**, 2819–2823.
12. Skolnick, J. and Kolinski, A. (1990) *Science* **250**, 1121–1125.
13. Paul, W. and Pistoor, N. (1994) *Macromolecules* **27**, 1249–1255.
14. Rapold, R.F. and Mattice, W. L. (1995) *Faraday Trans.*, in press.
15. Paul, W., Binder, K., Kremer, K., and Heermann, D.W. (1991) *Macromolecules* **24**, 6332–6334.
16. Flory, P.J. (1969) *Statistical Mechanics of Macromolecules*, Wiley, New York.

17. Mattice, W.L. and Suter, U.W. (1994) *Conformational Theory of Large Molecules. The Rotational Isomeric State Model in Macromolecular Systems*, Wiley, New York.
18. Binder, K. and Heermann, D.W. (1992) *Monte Carlo Simulation in Statistical Physics*, Springer-Verlag, New York.
19. Madras, N. and Sokal, A.D. (1988) *J. Stat. Phys.* **50**, 109–186.
20. Clancy, T.C. and Webber, S.E. (1993) *Macromolecules* **26**, 628–636.
21. Wall, F.T. and Mandel, F. (1975) *J. Chem. Phys.* **63**, 4592-4595.
22. Verdier, P.H. and Stockmayer, W.H. (1962) *J. Chem. Phys.* **36**, 227–235.
23. Mansfield, M. (1982) *J. Chem. Phys.* **77**, 1554–1559.
24. Pakula, T. (1987) *Macromolecules* **20**, 679–682.
25. Adriani, P., Wang, Y., and Mattice, W.L. (1994) *J. Chem. Phys.* **100**, 7718–7721.
26. Metropolis, N., Rosenbluth, A.W., Rosenbluth, M.N., Teller, A.H., and Teller, E. (1953) *J. Chem. Phys.* **21**, 1087–1092.
27. Larson, R.G. (1994) *Macromolecules* **27**, 4198–4203.
28. Balaji, R., Wang, Y., Foster, M.D., and Mattice, W.L. (1993) *Comput. Polym. Sci.* **3**, 15–22.
29. Nguyen-Misra, M. and Mattice, W.L. (1995) *Macromolecules* **28**, 1444–1457.
30. Zhan, Y. and Mattice, W.L. (1994) *Macromolecules* **27**, 7056–7062.
31. Binder, K. (1993) *Makromol. Chem., Macromol. Symp.* **65**, 175–187.
32. Kikuchi, M. and Binder, K. (1993) *Europhys. Lett.* **21**, 427–432.
33. Zhan, Y., Mattice, W.L., and Napper, D.H. (1993) *J. Chem. Phys.* **98**, 7502–7507.
34. Misra, S. and Mattice, W.L. (1994) *Macromolecules* **27**, 2058–2065.
35. Jacobson, H. and Stockmayer, W.H. (1950) *J. Chem. Phys.* **18**, 1600–1606.
36. Wang, Y., Mattice, W.L., and Napper, D.H. (1993) *Langmuir* **9**, 66–70.
37. Leibler, L., Orland, H., and Wheeler, J.C. (1983) *J. Chem. Phys.* **79**, 3550–3557.
38. Nguyen-Misra, M. and Mattice, W.L. (1995) *Macromolecules*, in press.
39. Molina, L.A. and Freire, J.J. (1995) *Macromolecules* **28**, 2705–2713.
40. Stockmayer, W.H. (1991) *Macromolecules* **24**, 6367–6368.
41. Tanaka, F. and Stockmayer, W.H. (1994) *Macromolecules* **27**, 3943–3954.
42. Flory, P.J. (1946) *Chem. Rev.* **39**, 137–197.
43. Flory, P.J. (1941) *J. Am. Chem. Soc.* **63**, 3083–3090.
44. Anniason, E.A.G., Wall, S.N., Almgren, M., Hoffmann, H., Kielmann, I., Ulbricht, W., Zana, R., Lang, J., and Tondre, C. (1976) *J. Phys. Chem.* **80**, 905–922.
45. Halperin, A. and Alexander, S. (1989) *Macromolecules* **22**, 2403–2412.
46. Johner, A. and Joanny, J.-F. (1991) *Europhys. Lett.* **15**, 265–270.
47. Dai, L. and Toprakcioglu, C. (1991) *Europhys. Lett.* **16**, 331–335.
48. Milner, S.T. and Witten, T.A. (1992) *Macromolecules* **25**, 5495–5503.
49. Wijmans, C.M., Leermakers, F.A.M., and Fleer, G.J. (1994) *J. Coll. Interf. Sci.* **167**, 124–134.
50. Fried, H. and Binder, K. (1991) *J. Chem. Phys.* **94**, 8349–8366.
51. Binder, K. and Fried, H. (1993) *Macromolecules* **26**, 6878–6883.
52. Larson, R.G. (1994) *Chem. Eng. Res.* **49**, 2833–2850.
53. Bates, F.S. and Fredrickson, G.F. (1990) *Ann. Rev. Phys. Chem.* **41**, 525–557.
54. Haliloğlu, T., Balaji, R., and Mattice, W.L. (1994) *Macromolecules* **27**, 1473–1476.
55. Ko, M.B. and Mattice, W.L. (1995) *Macromolecules*, in press.

56. Tassin, J.F., Siemens, R.L., Tang, W.T., Hadziioannou, G., Swalen, J.D., and Smith, B.A. (1989) *J. Phys. Chem.* **93**, 2106–2111.
57. Motschmann, H., Stamm, M., and Toprakcioglu, C. (1991) *Macromolecules* **24**, 3681–3688.
58. Zhan, Y., Mattice, W.L., and Napper, D.H. (1993) *J. Chem. Phys.* **98**, 7508–7514.
59. Zhan, Y. and Mattice, W.L. (1994) *Macromolecules* **27**, 683–688.
60. Misra, S., Nguyen-Misra, M., and Mattice, W.L. (1994) *Macromolecules* **27**, 5037–5042.
61. Wang, Y. and Mattice, W.L. (1993) *J. Chem. Phys.* **98**, 9881–9887.
62. Wang, Y., Li, Y., and Mattice, W.L. (1993) *J. Chem. Phys.* **99**, 4068–4075.
63. Procházka, K., Bednář, B., Mukhtar, E., Svoboda, P., Trěná, J., and Almgren, M. (1991) *J. Phys. Chem.* **95**, 4563–4568.
64. Wang, Y., Balaji, R., Quirk, R.P., and Mattice, W.L. (1992) *Polym. Bull.* **28**, 333–338.
65. Haliloğlu, T. and Mattice, W.L. (1994) *Chem. Eng. Res.* **49**, 2851–2857.
66. Haliloğlu, T. and Mattice, W.L. (1995) *Comput. Polym. Sci.* **5**, 65–70.

SELF-ASSEMBLIES IN ION-CONTAINING POLYMERS

A.R. KHOKHLOV AND O.E. PHILIPPOVA
Physics Department, Moscow State University,
117234 Moscow, Russia

1. Polyelectrolyte Macromolecules and Gels

Polyelectrolytes are macromolecules containing electrically charged monomer units. Ionization is normally a result of dissociation of electrically neutral monomer unit into charged unit and counterion. Therefore, counterions are always present in polyelectrolyte systems, and the number of charged monomer links is equal to the number of counterions.

Polyelectrolyte macromolecules can be subdivided into strongly charged polyelectrolytes and weakly charged ones. In strongly charged polyelectrolytes all monomer links (or at least large fraction of links) are charged, while in weakly charged polyelectrolytes only small fraction of monomer units is ionized. Thus, in strongly charged polyelectrolytes the long-range Coulomb interactions usually dominate over all other kinds of interaction of links. On the other hand, for weakly charged polyelectrolytes the effective interplay between Coulomb interactions and usual short-range forces acting between the uncharged links (e.g. of Van-der-Waals type) is possible.

Among the examples of strongly charged polyelectrolytes we will mention here polyanion macromolecules of poly(sodium methacrylate) in water

$$\begin{array}{lll} & CH_3 & \\ & \vert & \\ -[CH_2- & C-]_n & \\ & \vert & \\ & COO^- & Na^+ \end{array} \qquad (1)$$

and polycation macromolecules of poly(diallyldimethylammonium bromide) in water

S.E. Webber et al. (eds.), Solvents and Self-Organization of Polymers, 197–225.

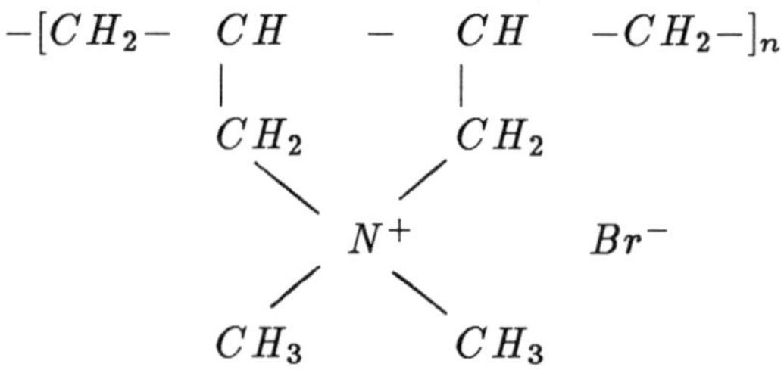

(2)

because we will frequently refer to these types of macromolecules in the subsequent discussion. Other strongly charged macromolecules which are abundantly studied in the literature are DNA and poly(styrene sulfonate) in water.

As to the weakly charged polyelectrolytes, they can be further subdivided into two classes. The first of these classes is formed by the polymers like poly(acrylic acid) (PAA) or poly(methacrylic acid) (PMAA)

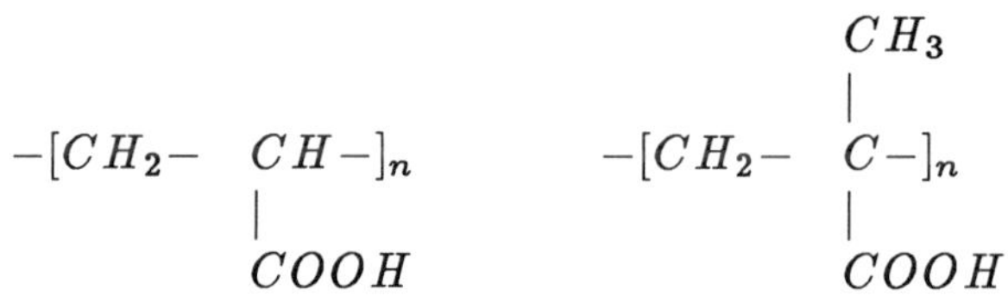

(3)

PAA *PMAA*

in water where the ionization (i.e. the dissociation of groups) is induced by the addition of a base (e.g. $NaOH$), i.e. by changing of the pH. For such polymers the places of the ionized COO^- units are not fixed, because H^+ ions can migrate along the chain. The second class of weakly charged polyelectrolytes is formed by copolymers containing charged units (like (1) or (2)) and neutral monomer units (e.g. acrylamide unit (4)).

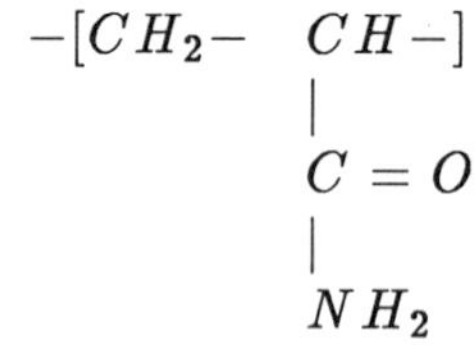

(4)

If the fraction of neutral links is large, the polyelectrolyte macromolecule is weakly charged, with the ionized units on predetermined places in the chain.

In the present chapter we will deal mainly with various effects for weakly charged polyelectrolytes.

In addition to linear polyelectrolytes we will consider also polyelectrolyte gels. They can be obtained, for example, by copolymerization of neutral and charged monomers with multifunctional crosslinker (e.g. N, N'-methylenebisacrylamide). Depending on the type of polymerizing units, polyelectrolyte gels can be weakly or strongly charged; with charges on predetermined places or migrating charges.

Polyelectrolyte gel can be regarded as a giant three-dimensional macroscopic macromolecule. Any conformational changes occuring in linear chains constituting a gel result in the changes of macroscopic dimensions of a gel sample, therefore these changes can be directly detected in macroscopic experiments. This advantage of studying polyelectrolyte gels vs polyelectrolyte solutions is, however, counterbalanced by the disadvantage: too high equilibration times, τ, for macroscopic gel samples (days or even weeks for $1cm$ gels; the value of τ scales with linear size of a gel, L, $\tau \sim L^2$).

2. Basic Physical Factors for Polyelectrolyte Systems

From our experience in analyzing various effects in polyelectrolyte systems we can underline the following main physical factors which should be taken into account for these systems.

(1) Translational entropy of counterions.

If monomer units of the type (1) or (2) are immersed in aqueous medium they dissociate, and counterions can travel through the whole system independently of a monomer unit, i.e. they acquire some translational entropy. This translational entropy contribution leading to a special osmotic pressure of counterions can play a very essential (often dominant) role in the behavior of polyelectrolyte systems.

(2) Coulomb interactions.

Direct Coulomb interactions between the charged species should be, of course, taken into account as well. It is, however, important to understand that most of the systems are macroscopically electrically neutral, i.e. the number of negatively charged units (e.g. monomer links) is equal to the number of positively charged units (e.g. counterions). Coulomb interactions between all the species (both positively and negatively charged) should be considered, and then it turns out that the net effect of electrostatic interactions is in many cases attraction, not repulsion (cf. Coulomb interaction term in electroneutral plasma which is negative [1]).

For strongly charged polyelectrolytes direct Coulomb interactions are essential and lead to the effects of counterion condensation [2-4] and additional electrostatic chain stiffening (formation of electrostatic persistent length) [4-6]. However, for weakly charged polyelectrolytes (which will be mainly considered in the present chapter) the role of Coulomb interaction term is normally much less significant than that of counterion translational entropy (cf.[4,7]).

(3) Interactions of uncharged links.

Since for weakly charged polyelectrolytes the majority of links are not charged, the interactions of these links constitute an important factor. In the physically most interesting situations translational entropy of counterions and non-Coulomb interactions are responsible for opposing tendencies, e.g. osmotic pressure of counterions tend to expand the system while uncharged links attract each other and force the system to collapse. The interplay between these tendencies leads to interesting microstructures and self-assemblies which will be discussed below.

(4) Formation of ion pairs and ionomer multiplets.

We have already mentioned the effect of counterion condensation for strongly charged polyelectrolytes [2-4]. For weakly charged polymer chains this effect in its classical form does not exist. Instead, the notion of the formation of ion pairs is usually introduced: it is assumed that the pair ionic monomer unit / counterion (e.g. typical examples (1) and (2)) can under given conditions be in the bound form, i.e. the counterion is no longer free and form a dipole with the corresponding co-ion.

What are the conditions for the ion pair formation? The simplest arguments are the following. Each counterion can be either free or form ion pair with the co-ion. In the latter case there is some gain in the energy of Coulomb attraction between the ions in the pair and some loss in the translational entropy of the counterion. If a is the distance between the ions in the ion pair, ϵ is the dielectric constant of the medium and e is an elementary charge, the Coulomb energy gain resulting from the formation of an ion pair is $e^2/\epsilon a$, while the loss of the translational entropy of counterion is kT times some concentration dependent logarithmic factor (T is temperature and k is Boltzmann constant). Neglecting logarithmic factors one can say that the dimensionless parameter governing ion pairing is

$$u = \frac{e^2}{\epsilon a k T}. \tag{5}$$

If $u < 1$ the translational entropy is more important, and the ion pairs are practically absent. At $u > 1$, on the contrary, Coulomb energy dominates, and most of the charges form ion pairs.

For room temperatures ($T \sim 300K$), usual characteristic distances between the charges in an ion pair ($a \sim 7\AA$) and for highly polar medium (such as water, $\epsilon \sim 81$), the parameter u is of order unity, so a significant fraction of counterions are in the mobile form. This is the reason why in aqueous medium monomer units (1) and (2) dissociate. Polymer chains are indeed charged in this case, and it is natural to call this situation a "polyelectrolyte regime".

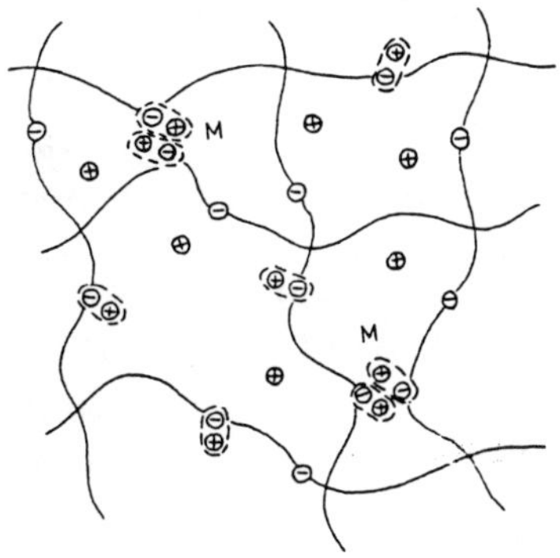

Figure 1. Polyelectrolyte network: Some counterions are free (mobile), others form ion pairs (marked schematically by dotted lines). Some of the ion pairs aggregate and form multiplets (marked by letters M).

However, in the medium of lower polarity ($\epsilon \sim 2 \div 5$), e.g. for polymer melt or solution in organic solvent, the value of u is $u \gg 1$, hence practically all the counterions form ion pairs. Polymer chain is uncharged, and the character of its behavior is completely different from that of polyelectrolytes [8-11]. This is the "ionomer regime", i.e. ion-containing polymer exists in the form of ionomer.

Intermediate cases (between "pure polyelectrolyte" and "pure ionomer" behavior) can be realized as well. For these cases $u \sim 1$, and some part of counterions are free, while the other form ion pairs. The ion pairs can essentially influence the behavior of otherwise polyelectrolyte system, and this factor should be taken into account in many physical situations (see, for example, [12-18]).

The reason for the strong influence of the formation of even a small number of ion pairs is the strong dipole-dipole attraction between these pairs (the characteristic energy is $\sim 10 \div 25kT$, depending on the type of the ion pair and the dielectric constant of the medium). Usually, this leads to the formation of aggregates of ion pairs (Fig.1). The average size of the aggregate is determined by the interplay of the attraction energy and the steric restrictions which increase progressively as more and more ion pairs are included into the aggregate. These aggregates are called multiplets in the theory of ionomers [8-10]. The self-assembled microheterogeneities caused by multiplet formation play an important role in a number of physical situations, for the modern approaches in the theory of multiplet structure in ionomers see [9,11,19,20].

In the remaining part of this chapter we will illustrate the importance of the basic physical factors for ion-containing polymers outlined of in the present section by considering several typical effects. We will pay most of attention to the situations when the interplay of different factors leads to the formation of various self-assembled structures.

3. Compatibility Enhancement in Mixtures of Polyelectrolytes

We first describe two most spectacular examples to illustrate the importance of the translational entropy of counterions for the systems containing weakly charged polyelectrolytes. The first of these examples deals with the effect of compatibility enhancement in polyelectrolyte mixtures caused by counterions.

If we have a mixture of two neutral polymers A and B (Fig.2a), it is well known that even small excess positive energy of $A-B$ contact (Flory-Huggins parameter χ_{AB} of order $1/N$, where N is the number of monomer units in the chain) gives rise to a macroscopic phase demixtion [21,22]. The reason for this is the relatively small translational entropy loss in this process: the entropy cost of demixtion is roughly speaking one entropy unit per whole long chain, because monomer units are connected in the chain and do not possess the entropy of independent translational motion.

The situation changes significantly if one of the polymers in the mixture is slightly ionized (see e.g. Fig.2b where the mixture of weakly charged polymer A / neutral polymer B is presented). If macroscopic phase separation occurs in this case, counterions should follow the corresponding polyions, because different phases should obey the condition of total macroscopic electroneutrality (Fig.2b). On the other hand, free counterions have the entropy of independent translational motion, so the entropy cost of demixing is roughly one entropy unit per counterion. Since normally there are many counterions per polymer chain, this is much higher entropy loss than for Fig.2a. Therefore, we clearly have the effect of compatibility enhancement due to the presence of a small fraction of charged monomer units in the mixing component. This effect was first theoretically described and experimentally verified in [23] (see also [24-28]).

In particular, it was shown that the characteristic Flory-Huggins parameter, χ_{AB}, needed for the macroscopic phase separation is now $\chi_{AB} \sim f/N$, where f is the number of counterions per chain (instead of $\chi_{AB} \sim 1/N$ for neutral chains). So the effect of the presence of even slight number of counterions is indeed very pronounced.

In Fig.3 reproduced from [23] one can see the dependence of the volume fraction of the components $\Phi_A = \Phi_B = \Phi_1$, corresponding to the phase separation, on the number of charges on the polyelectrolyte chain for the system: weakly charged poly(4-vinylpyridine) (polyelectrolyte A) - poly(ethylene glycol) (polymer B) - solvent (water - methanol mixture). The experimental points fall well on the theoretical curve. The effect of

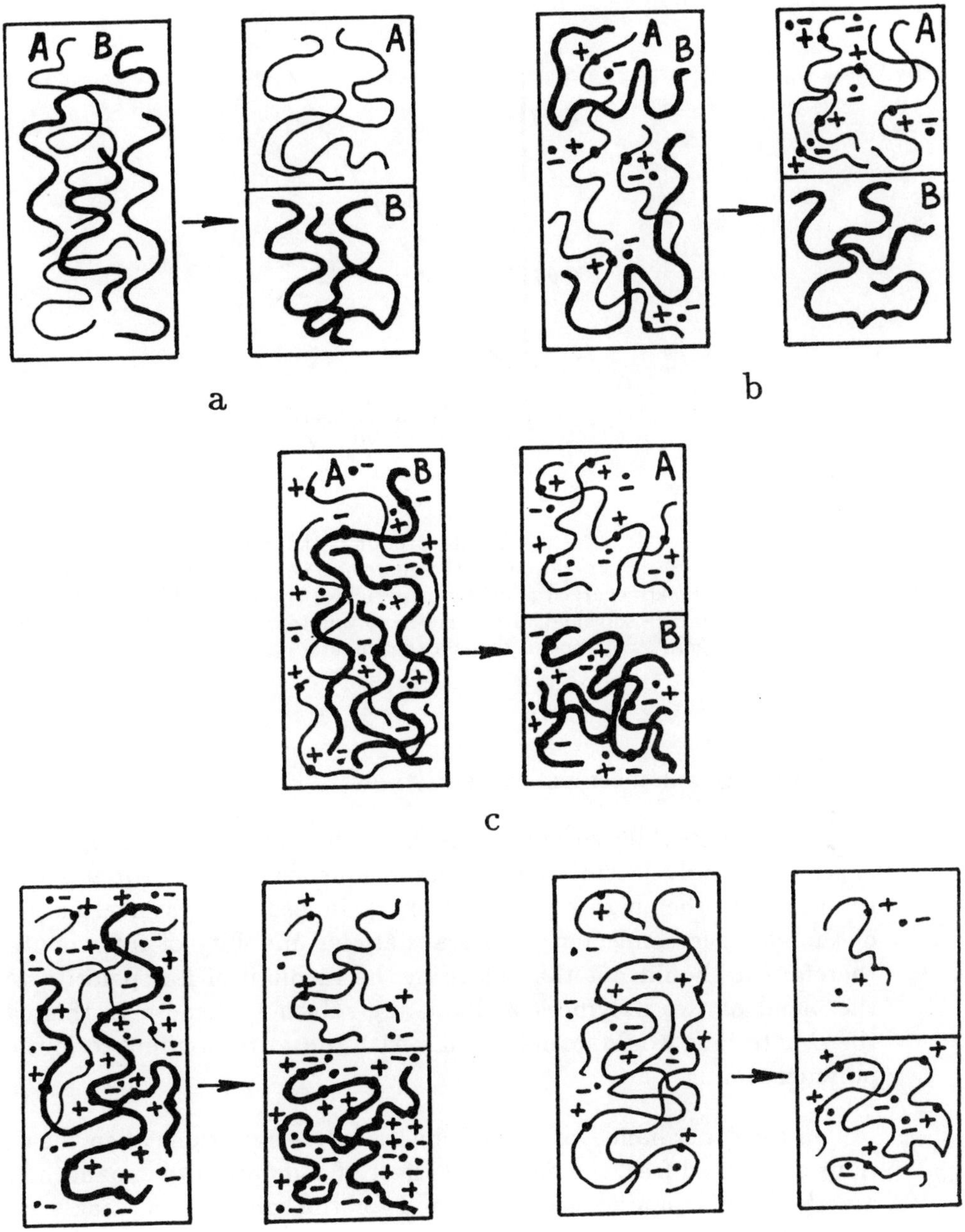

Figure 2. Schematic representation of phase separation in the systems:
a - neutral polymer A / neutral polymer B mixture,
b - charged polymer A / neutral polymer B mixture,
c - weakly charged polycation A / weakly charged polyanion B mixture,
d - mixture of two polymers charged with the same sign, but with essentially different fraction of monomer units,
e - poor solvent solution of weakly charged polyelectrolyte.

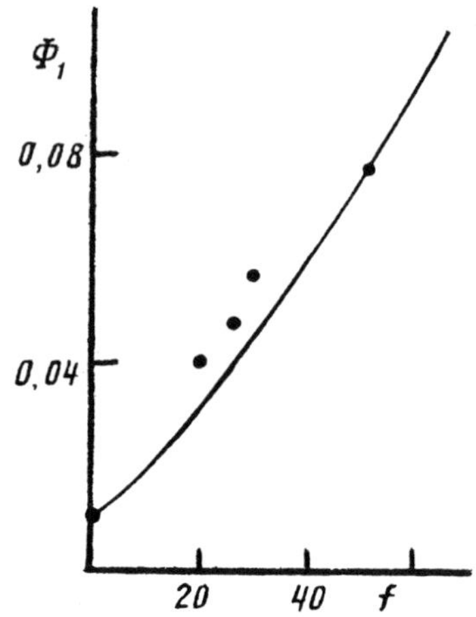

Figure 3. The dependence of the volume fraction of the components $\Phi_A = \Phi_B = \Phi_1$, corresponding to the phase separation, on the number of charges on the polyelectrolyte chain for the system: weakly charged poly(4-vinylpyridine) (polyelectrolyte A) - poly(ethylene glycol) (polymer B) - solvent (water - methanol mixture). The curve depicts the calculated data, the points correspond to the experimental data.

compatibility enhancement is clearly seen.

It is worth-while to emphasize here that this effect normally can be observed for polyelectrolyte mixtures in aqueous solutions (or at least in the solutions of other highly polar solvents). Indeed, the physical background of the effect is connected with translational mobility of free counterions, therefore it should not take place in the medium of low polarity (e.g. in the blend of two polymers without any solvent), because in this case all the counterions would form ion pairs with the corresponding co-ions on the polymer chains.

On the other hand, for truly polyelectrolyte situation, when the counterions are indeed free, the described effect of compatibility enhancement has a very general significance. It is valid for the mixtures of weakly charged polycation and polyanion (Fig.2c), mixtures of the chains charged with the same sign, but which have essentially different fraction of charged monomer units (Fig.2d), and for poor solvent solutions of weakly charged polyelectrolytes (Fig.2e). In all these cases macroscopic phase separation leads to the essentially inhomogeneous distribution of counterions in both phases (due to the macroscopic electroneutrality condition), which is unfavorable from the viewpoint of their translational entropy. For the experimental manifestations of this effect see [27,28].

4. Collapse of Polyelectrolyte Gels

Another example illustrating the importance of translational entropy of counterions for polyelectrolyte systems is connected with the phenomenon of collapse of polyelectrolyte gels. This effect was first theoretically predicted in [29] and experimentally observed by Tanaka in 1978 [30].

Tanaka studied the swelling behavior of polyacrylamide gels in water-acetone mixtures. Water is a good solvent for polyacrylamide, while acetone is a poor solvent. It was observed that upon the addition of a certain amount of acetone the swollen gel undergoes an abrupt shrinking; the corresponding change in volume can be very significant (several orders of magnitude). It was immediately realized that the gel collapse is just the manifestation of the coil-globule transition in the chains constituting the gel: each chain undergoes this transition when the solvent becomes poorer and as a result the gel sample collapses as a whole [30,31].

Afterwards it was observed that the character of the collapse depends essentially on the presence of even a small fraction of charges on the gel chains: the more is this fraction the larger is the amplitude of the change in volume at the point of the collapse and the abruptness of the collapse transition (for neutral gel the collapse is normally continuous, and it takes place in a jumpwise manner only for slightly charged gels) [32,33]. To explain these facts it is necessary to take into account the translational entropy of counterions for charged gels [32,34].

Indeed, suppose that we have a slightly charged gel swollen in a large volume of solvent (Fig.4). Counterions are also present in the gel, and they cannot leave it because of the condition of macroscopic electroneutrality. Therefore, the "gas of counterions" inside the gel is responsible for the osmotic pressure leading to the additional expansion of the gel. As a result of this exerting osmotic pressure of counterions the gel swells extensively in a good solvent (i.e. in water). On the other hand, the phase of collapsed gel which is stabilized by the attraction of the uncharged monomer units in poor solvent is relatively unaffected by the presence of counterions (only the region of stability of this phase should become smaller with the increase of the number of charges on the gel). Thus, the difference between swollen and collapsed phase for charged gels becomes larger leading to the increase of the amplitude of the collapse.

Below we present the simplest theory of the collapse transition in polyelectrolyte gels [7,34], mainly to illustrate the point mentioned in section 2 that the translational entropy of counterions for this case is much more important than the direct Coulomb interactions.

The free energy of a polyelectrolyte gel swollen in a solvent can be

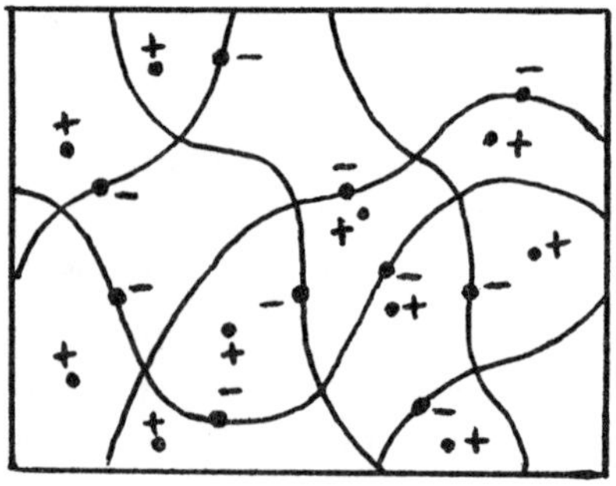

Figure 4. Schematic representation of slightly charged gel swollen in a large volume of solvent.

written as a sum of four contributions:

$$F = F_{int} + F_{el} + F_{trans} + F_{Coulomb}, \tag{6}$$

where F_{int} is the free energy of non-Coulomb interactions of monomer units inside the gel, F_{el} is the free energy of elastic deformation of gel chains, F_{trans} is the contribution connected with the translational entropy of counterions, while $F_{Coulomb}$ is the free energy of Coulomb interaction of charges inside the gel.

From the theory of the collapse transition in individual macromolecules [4,35-39], it is known that to describe this transition, it is enough to take into account the first two terms of the expansion of free energy of non-Coulomb interaction of monomer units in the powers of polymer concentration (second and third virial terms). Therefore, for F_{int} we write

$$F_{int} = NkT\left(Bn + Cn^2\right), \tag{7}$$

where N is the total number of monomer units in the gel, n the concentration of monomer units, and B and C second and third virial coefficients of their non-Coulomb interaction.

The elastic contribution to the free energy, F_{el} , can be written in the form [31,40]

$$F_{el} = \frac{3}{2}\nu kT\left(\alpha^2 - 1\right), \tag{8}$$

where ν is the number of elastically active subchains between two crosslink points in the gel, and α is the swelling factor of the gel with respect to some reference state which depends on the conditions of gel synthesis [31]. For ν and α we can write

$$\nu = N/m, \qquad \alpha = (n_o/n)^{1/3}, \tag{9}$$

where m is the average number of monomer units in one elastically active subchain and n_o is the concentration of links in the reference state. The value of n_o depends on m, the normal scaling dependence is $n_o \sim 1/a^3 m^{1/2}$, where a is the characteristic microscopic size of a monomer unit [31].

The contribution F_{trans} can be written as the free energy of ideal gas of counterions:

$$F_{trans} = \frac{N}{\sigma} kT \ln \frac{n}{\sigma}, \tag{10}$$

where σ is the average number of neutral monomer units between two charged ones ($\sigma \gg 1$, since we are considering weakly charged polyelectrolyte gels). Eq.(10) is just due to the fact that the concentration of counterions inside the gel is n/σ.

Finally, the Coulomb interaction contribution can be written in the form of the Debye-Huckel term in electroneutral plasma with concentration of charged species of one sign n/σ [1]:

$$F_{Coulomb} = -\frac{N}{\sigma} kT \left(\frac{na^3}{\sigma} \right)^{1/2} u^{3/2}, \tag{11}$$

where $u \equiv e^2/\epsilon akT$ is the characteristic dimensionless parameter connected with electrostatic interaction (cf. eq.(5)). It should be emphasized that this contribution to the free energy is negative, i.e. it corresponds to the effective attraction.

Now let us estimate the relative importance of the contributions (10) and (11). It was already mentioned that for aqueous systems the parameter u is of order unity. Thus, apart from logarithmic factors

$$F_{Coulomb}/F_{trans} \sim \left(na^3/\sigma \right)^{1/2} \ll 1, \tag{12}$$

since $na^3 \ll 1$ and $\sigma \gg 1$. We see that Coulomb interaction contribution is indeed negligible in comparison with the translational entropy of counterions.

The free energy defined by eqs.(6)-(11) should be minimized with respect to n in order to obtain the equilibrium concentration of monomer units in the gel and hence the swelling ratio of the gel α (see eq.(9)). The result of these calculations [7] is presented in Fig.5 where α is plotted as a function of the parameter $x = Bn_o m$ for different degrees of charging of the gel. Variation of x determines the variation of the solvent quality: $x > 0$ corresponds to the good solvent region, while $x < 0$ defines poor solvent region. One can see that in accordance with the experimental observations the amplitude of the collapse transition and the degree of its abruptness increase significantly with the increase of the number of charged units in the gel.

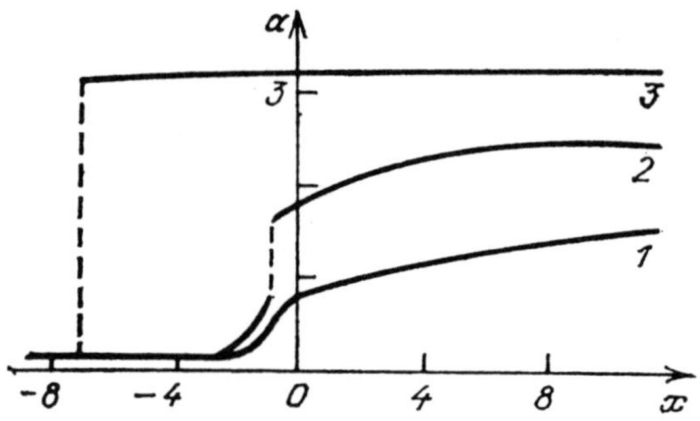

Figure 5. The swelling ratio of the gel α as a function of the parameter x for different degrees of charging of the gel: 1 - uncharged gel, 2,3 - charged gels with six (2) or twelve (3) charges between two cross-linking points.

5. Microphase Separation in Polyelectrolyte Systems

Having illustrated the importance of translational entropy of counterions for polyelectrolyte systems, we now turn to the situations where the interplay of this factor with other effects leads to the formation of self-assembled microstructures. The first example of this kind is the microphase separation in weakly charged polyelectrolytes.

This effect was first theoretically predicted in [41] for the poor solvent solutions of weakly charged polyelectrolytes (see also [42,43]). Later it turned out that the same effect can take place in the mixtures of polyelectrolytes [24,25,43-45] and in polyelectrolyte gels in a poor solvent [46-51].

The first experimental observations of the tendency for microphase separation in polyelectrolyte systems were reported in [46-51]. In the papers by Strasbourg group [46-50] the weakly ionized gels of PAA in deuterium oxide were studied using the method of small angle neutron scattering (SANS), while in [51] the same method was applied for the gels obtained by copolymerization of N-isopropylacrylamide (NIPA) and acrylic acid in the ionized form. In Fig.6 one can see the SANS intensity curves as a function of the scattering vector q for the latter system for different temperatures. As the temperature increases the peak on these curves becomes more and more pronounced. The existence of the maximum on SANS intensity curves at finite values of q is a manifestation of the tendency to microphase separation (cf. the analogous behavior for block copolymers [52,53]).

What is the physical reason for this behavior? Poly(N-isopropylacrylamide) is a weakly hydrophobic polymer, and its degree of hydrophobicity increases with the increase of temperature. Upon the increase of temperature from 30^oC to 50^oC we are passing from the good solvent to poor

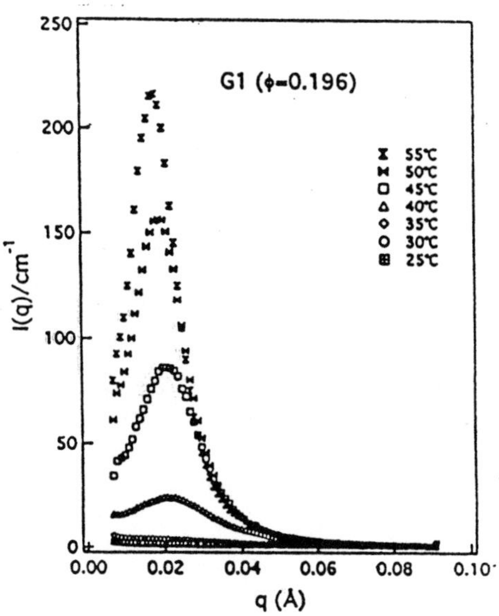

Figure 6. Temperature dependence of SANS profiles for poly(N-isopropylacrylamide-co-acrylic acid) gel.

solvent behavior for neutral links of NIPA (this can be also seen from the swelling curve of the gel, Fig.7). Therefore, with the increase of temperature the gel should rearrange in such a manner that the number of contacts of NIPA monomer units with water becomes as low as possible. One way of achieving this is just to make the gel to collapse (Fig.8). However, in this case the osmotic pressure of counterions would increase dramatically (see previous section). To avoid this it is sometimes more thermodynamically advantageous to keep the gel volume practically unchanged and to form a kind of hydrophobic micelles located periodically inside the gel (Fig.8). In this way, on the one hand, the number of contacts of hydrophobic links with water is diminished and, on the other hand, the loss in the translational entropy of counterions is not so dramatic as in the case of the collapse of the gel sample as a whole. The above arguments are just a qualitative description of the exact theory first proposed in [43].

The detailed quantitative theory-experiment comparison performed in [51] has demonstrated good agreement. The form of SANS intensity curves and the position of the peak coincide with the theoretical predictions. From the comparison of Figs.6 and 7 one can see that the pronounced peak on SANS curves can be observed not only for collapsed gels, but also for swollen gels at somewhat lower temperatures. This is in agreement with the mechanism of hydrophobic micelles formation explained in Fig.8.

It is important to emphasize that the same mechanism is valid for a

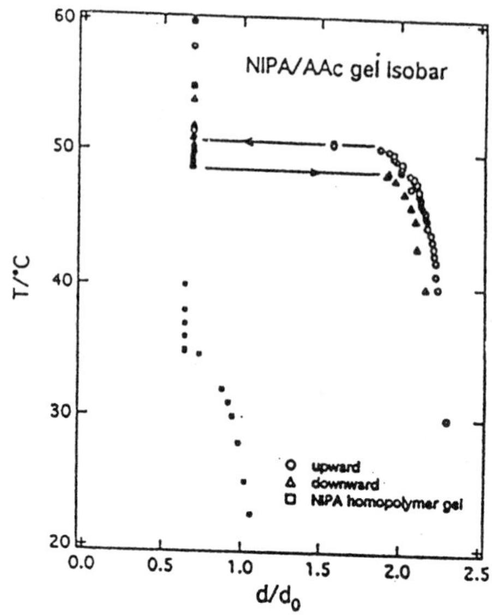

Figure 7. Swelling curves of the poly(N-isopropylacrylamide-co-acrylic acid) gels in deuterium oxide. Open circles and triangles denote the data points taken in the heating, and cooling processes, respectively.

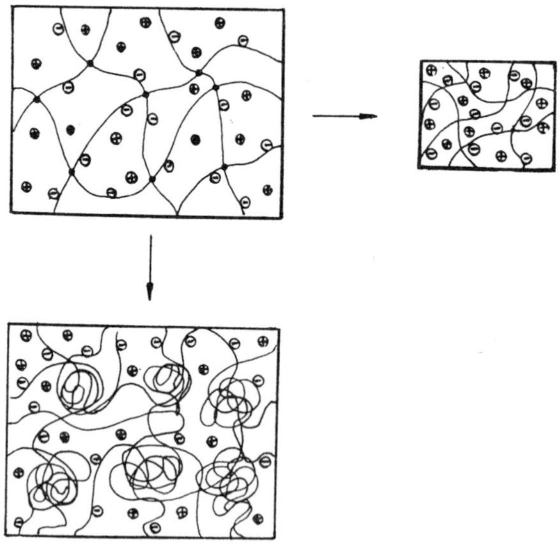

Figure 8. Schematic representation of possible conformational changes in a polyelectrolyte gel arising from polyelectrolyte/hydrophobic competition.

much wider class of polyelectrolyte systems, for example for all the systems shown in Figs.2b-e [24,25,43,44]. Let us consider, for example, a mixture of weakly charged polycations and polyanions in water or other highly polar solvent (Fig.2c). With the increase of the degree of immiscibility of neutral monomer units of polymers A and B the tendency for macroscopic phase separation will increase. However, as it was explained before, the phase demixtion may be impossible because of a too high loss in the translational entropy of counterions. Instead, in analogy with Fig.8 the thermodynamic equilibrium may correspond to a periodic array of alternating microregions rich with $A-$ and $B-$ components, respectively. The exact calculations performed in [25,44] show that this is indeed the case.

To understand the character of the calculations giving the theoretical justification to the existence of microphase separation transition in polyelectrolyte systems, below we write down the expression for the free energy for the mixture of weakly charged polymer A, neutral polymer B and solvent S (see Fig.2c) which was used in [26]. To calculate the phase diagram of this system including the possibility of microphase separation, it is necessary to write down its free energy for an inhomogeneous state in which concentrations of all the components depend on spatial coordinates. Thus, in the framework of the Flory-Huggins lattice model we have

$$\begin{aligned}\frac{F}{kT} &= \int \frac{d^3r}{v}\left(\frac{\Phi_A}{N_A}\ln\Phi_A + \frac{\Phi_B}{N_B}\ln\Phi_B + \Phi_S\ln\Phi_S + \chi_{AB}\Phi_A\Phi_B + \right.\\ &+ \chi_{AS}\Phi_A\Phi_S + \chi_{SB}\Phi_S\Phi_B + l_A b\left[\nabla\left(\Phi_A(r)\right)^{1/2}\right]^2 + \\ &+ \left. l_B b\left[\nabla\left(\Phi_A(r)\right)^{1/2}\right]^2\right) + \int d^3r\sum_i n_i\ln n_i + \\ &+ \frac{1}{2\epsilon kT}\int\int d^3r\, d^3r'\frac{\rho(\vec{r})\rho(\vec{r'})}{|\vec{r}-\vec{r'}|},\end{aligned} \tag{13}$$

where $\Phi_A\left(\vec{r}\right)$, $\Phi_B\left(\vec{r'}\right)$ and $\Phi_S\left(\vec{r}\right)$ are volume fractions of the components A, B and S at the point $\vec{r}$, $v = b^3$ is the volume of elementary lattice site, b is the lattice spacing, N_A and N_B are the numbers of monomer units in $A-$ and $B-$chains, l_A and l_B are their Kuhn segment lengths, $\chi_{\alpha\beta}$ are Flory-Huggins interaction parameters between the species α and β ($\alpha, \beta = A, B$ or S), $n_i\left(\vec{r}\right)$ is the concentration of the small ions of type i at the point $\vec{r}$ (we allow for the presence of low-molecular salt in the system), and $\rho\left(\vec{r}\right)$ is the charge density at the point $\vec{r}$:

$$\rho\left(\vec{r}\right) = e\left[\sum_i z_i n_i\left(\vec{r}\right) + z_A f_A \Phi_A\left(\vec{r}\right)/v\right] \tag{14}$$

In eq.(14) f_A is the fraction of charged monomer units in $A-$chains, and z_i and z_A are valences of the ions of type i and of charged units of $A-$chains, respectively.

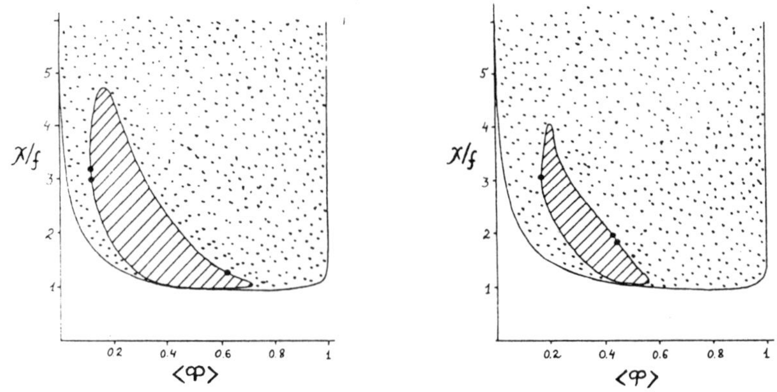

Figure 9. Typical phase diagrams for polyelectrolyte A / neutral polymer B mixture. Regions of existence of microdomains are shaded by inclined lines, phase separation regions are shaded by disperse points, regions corresponding to the homogeneous phase are left white. Left phase diagram corresponds to higher dielectric constant of the solvent.

The first three terms in the integrand of eq.(13) represent the contribution to the free energy from the translational entropies of the species A, B and S. The next three terms describe the energy of nonelectrostatic interaction of the components. The terms with gradients represent a specific polymer contribution to the free energy which is connected with entropy loss due to the inhomogeneous concentration profiles of polymer components (Lifshitz entropy, see [38,54]). The second integral in eq.(13) is due to the translational entropies of small ions, while the third integral describes the free energy of Coulomb interactions.

To obtain the phase diagram of the system the free energy (13) should be minimized with respect to all possible distributions $\Phi_A\left(\vec{r}\right)$, $\Phi_B\left(\vec{r}\right)$, $\Phi_S\left(\vec{r}\right)$ and $n_i\left(\vec{r}\right)$ at given average values of $<\Phi_A>$, $<\Phi_B>$, $<\Phi_S>$ and $<n_i>$. The additional conditions for the minimization are the incompressibility condition and Poisson-Boltzmann equation (for more details see [26,55,56]). This program was accomplished for several particular cases in [26,56], and in Figs.9 and 10 one can see the resulting phase diagrams obtained in these papers.

The phase diagram of Fig.9 was calculated in [26] via the direct numerical minimization of the free energy of the type (13) (without any simplifying assumptions) for a two-component system: polyelectrolyte A / neutral polymer B. Because of the fact that numerical calculations are rather bulky, only lamellar microdomain phase was taken into account.

Phase diagram of Fig.10 was obtained in [56] for poor solvent solutions of weakly charged polyelectrolytes (Fig.2e) in the so-called weak crystal-

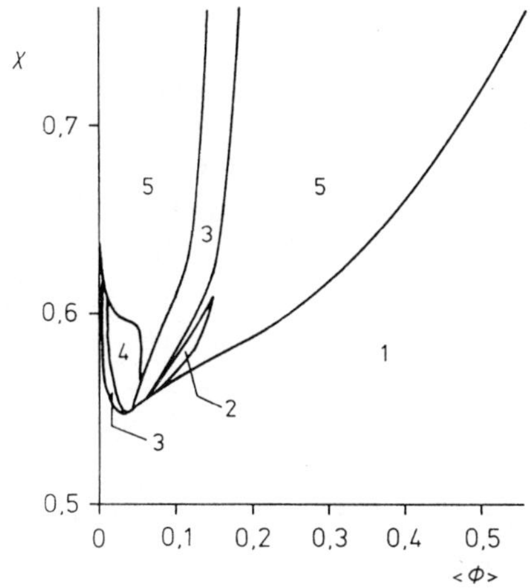

Figure 10. Phase diagram for poor solvent solution of weakly charged polyelectrolytes. The phase diagram contains the following regions: disordered phase (1), body-centered-cubic microdomain phase (2), triangular microdomain phase (3), lamellar microdomain phase region (4) and the regions of phase separation (5).

lization approximation which is valid only near the critical point [45]. On the other hand, the use of this approximation allowed to study all possible microdomain phases (in particular, hexagonal lattice of cylinders and body-centered cubic lattice of spherical micelles), not only lamellar microphase.

In spite of the differences in details of the phase diagrams in Figs.9 and 10 which are mainly due to the different approximation applied, one can notice several common features. 1. The region of stability of microdomain phases lies always within the region of macroscopic phase separation. 2. The regions of macrophase separation are rather wide. 3. If one increases the uncompatibility parameter χ starting from the homogeneous phase, one normally comes to the region of macrophase separation between the microdomain and homogeneous phase. This gives the strategy for the search for true microdomain phases in such systems: at first it is necessary to induce a macroscopic phase separation, then one of the coexisting phases will probably contain microdomains.

As to the period of the emerging microdomain structure, the calculations in [26] have shown that its characteristic value is about 100 monomer unit sizes. Therefore, the best experimental methods to detect such microdomains are SANS and small angle X-ray scattering (SAXS).

6. Mixed Polyelectrolyte/Ionomer Behavior. Collapse Induced by Extra Ionization

Another example of self-assemblies in polyelectrolyte systems is connected with the polyelectrolyte/ionomer interplay briefly described in section 2. As it was shown theoretically in [13], this effect should be of primary importance for the collapse of charged gels. Indeed, even for collapse in pure water the volume fraction of nonpolar polymer in the collapsed gel can be significant, leading to the much lower dielectric constant and to the higher fraction of ion pairs. Thus, ionomeric multiplets can be formed in the collapsed gels [13]. This effect is even more probable for less polar solvents or for water/organic solvent mixtures.

Below we consider three experimental situations illustrating the existence of ionomer-type microstructures (multiplets) in the collapsed gels.

6.1. COLLAPSE OF CATIONIC GEL IN WATER INDUCED BY THE LOW MOLECULAR SALT - SODIUM IODIDE

The study of the swelling behavior of cationic poly(diallyldimethylammonium chloride) (PDADMACl) gel in the presence of sodium salts of chloride, bromide, iodide and acetate in aqueous media demonstrated the striking effect of the nature of salt anion [16]. Among the studied low molecular salts only sodium iodide can induce a jump-wise collapse of the gel (Fig.11). This effect was explained by the formation of ion pairs between the positively charged network units and salt anions. The I^- having a higher polarizability among the other salt anions under study form a larger number of ion pairs. The ion pairs formed by I^- have a great dipole moment which favors the clustering of the ion pairs in the multiplets. Both effects (ion pairs formation and their clustering) cause the gel collapse.

A decrease in the charge density of the cationic network resulted in the disappearance of the phase transition. Therefore, the collapsed ionomeric state of the network becomes competitive with the swollen polyelectrolyte state only at high enough fractions of the ion-containing segments because of the tendency of the ion pairs to associate.

The transition from polyelectrolyte to ionomer regime has also affected the kinetics of the gel collapse. It was shown that in the presence of NaI the gel collapse proceeds in two steps (Fig.12). The first one (also observed for the other salts) corresponds to a fast polyelectrolyte contraction. The last step corresponds to slower processes of ion pairs formation. It was suggested [16] that the two-step kinetics of the collapse reflects a potential barrier which separates the two states of the collapsed gel: with and without ionomeric multiplets. The existence of these two states (called collapsed and supercollapsed) was theoretically predicted in [13].

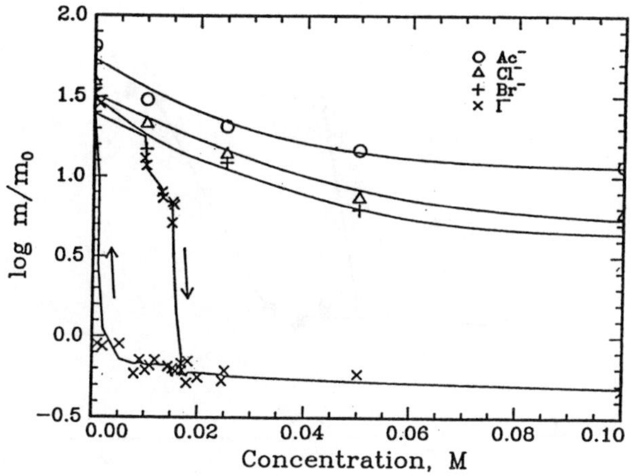

Figure 11. Plot of the degree of swelling versus salt concentration of sodium acetate (o), chloride (△), bromide (+) and iodide (X) at 21^0C. Symbols x show the direct and reverse transition in the degree of swelling versus the sodium iodide concentration.

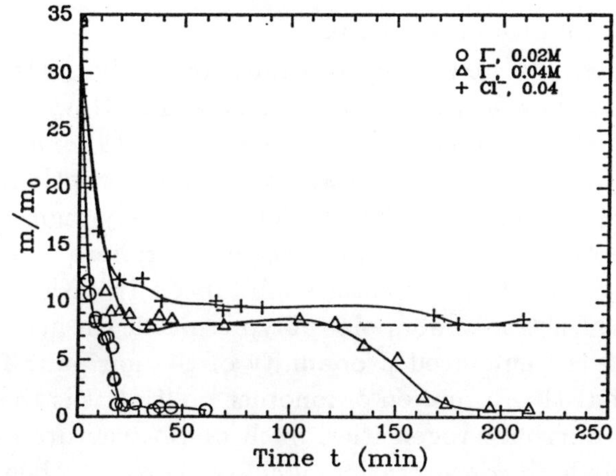

Figure 12. Plots of the degree of swelling versus time for PDADMACl gels in 0.04 and 0.02 M sodium iodide and 0.04 M sodium chloride. The dashed line indicates a volume of the gel sample after being immersed in 0.04 M sodium chloride solution for 2 days.

6.2. IONOMER-TYPE MULTIPLET STRUCTURE IN COLLAPSED POLYELECTROLYTE GELS IN POOR ORGANIC SOLVENT

When the gel collapse is induced by the addition of a poor organic solvent of low polarity, one can expect the microsegregation connected with ionomer-type multiplet structure. The formation of such microstructure was observed experimentally by SAXS for polyacrylamide gels containing 10 mol-% of cationic diallyldimethylammonium bromide units (PADADMAB-

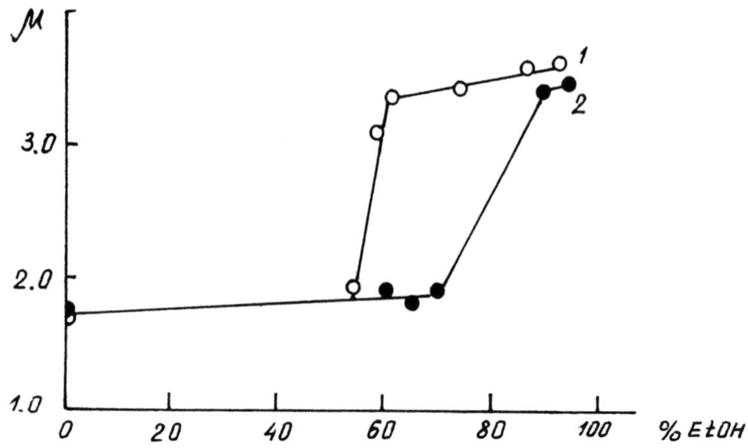

Figure 13. Scattering exponent, μ, as a function of ethanol concentration in water-ethanol mixture for PADADMAB gel, containing 10 mol-% (1) and 5 mol-% (2) of charged units.

10) in water-ethanol mixtures [14,15].

The SAXS intensity curve $I(q)$ as a function of the scattering vector q for the gel did not show any scattering maxima [14]. However, the scattering exponents, μ, evaluated according to the fitting $I(q) \sim q^{-\mu}$ in the range $0.1\, nm^{-1} < q < 0.47\, nm^{-1}$, were shown to increase abruptly from 1.7 to 3.3 together with the gel collapse (Fig.13). The relatively high values of scattering exponents may indicate the presence of interphase boundaries in the system. Since the value of μ increases simultaneously with the gel collapse (when the volume fraction of nonpolar polymer inside the gel becomes much higher leading to the enhanced probability of the ion pairs formation), it was concluded that the formation of ionomer multiplets is the main reason for the observed microheterogeneities. Such microstructure can be formed only at sufficiently high concentration of charged groups when the ion pairs easily aggregate in multiplets. Indeed, in contrast to PADADMAB-10 gel, the collapse of the PADADMAB-5 gel containing 5 mol-% of charged units (at 52 vol-% of ethanol) has no effect on μ (Fig.13) [14,15].

6.3. COLLAPSE OF POLYELECTROLYTE GELS IN LOW POLAR MEDIA INDUCED BY EXTRA IONIZATION

The transition from the swollen polyelectrolyte regime to the supercollapsed ionomer regime of the gel can be further illustrated by the swelling behavior of PMAA gel with different degrees of ionization z in methanol [18]. The ionization of PMAA was performed by the addition of a strong base sodium

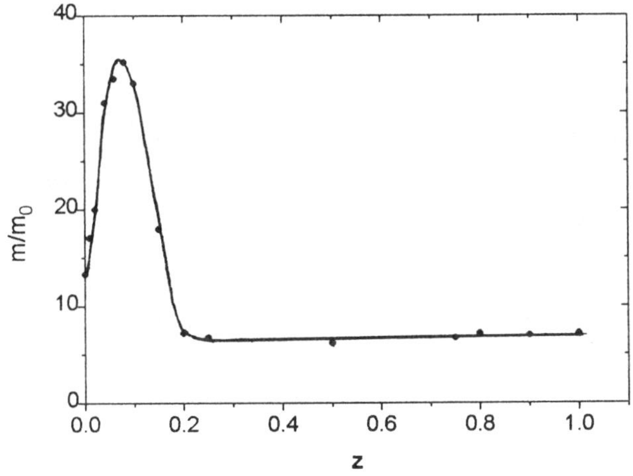

Figure 14. The dependence of the swelling ratio of PMAA gel in methanol on the degree of ionization z.

methoxide. It was found that the initial increase of the degree of ionization leads to an usual polyelectrolyte swelling of the gel due to the exerting osmotic pressure of counterions neutralizing the network charges. After a certain degree of ionization was reached ($z \approx 0.1$) a pronounced gel collapse was observed. At a further increase of z the swelling ratio of the collapsed gel remains constant (Fig.14). The gel collapse was assigned to the formation of ion pairs. This was supported by the results of conductivity measurements. It was shown that the gel swelling at low z correlates well with the increase of the reduced conductivity, while the gel collapse is accompanied by a significant drop of the reduced conductivity. The last fact indicates that the ion pairs are formed.

The more is the number of charged monomer units, the more thermodynamically advantageous is the state of the collapsed gel with ionomer multiplet structure, because of the great energy gain due to the formation of ion pairs and their subsequent aggregation into multiplets. This leads to the sudden collapse of the gel at some definite degree of ionization. At higher ionization degrees the gel is always in the supercollapsed state with developed ionomer multiplet microstructure.

To study the aggregation of ion pairs in the multiplets the dielectric method was used. It was found that the dielectric constant increases with the decrease of the frequency of the applied electric field. The very high values of the dielectric constant at low frequencies evidence the formation of large aggregates with high polarizability, which can be regarded as ionomer multiplets.

7. Partial Vitrification. Microheterogeneities in the Collapsed Gels

The polymer gels may exhibit supramolecular order due to the formation of glassy kinetically frozen polymer-rich regions. The most common gels with high swelling capacity are based on polymers which are in the glassy form at room temperatures, for example, PAA, PMAA, polyacrylamide. In the course of the gel collapse dense polymer nuclei can be formed which undergo the transition to the glassy state. After that the structure becomes kinetically frozen, and no further collapse is possible.

The formation of kinetically frozen structures was observed for polyacrylamide gels containing 2, 5 or 10 % of positively (diallyldimethylammonium bromide) or negatively (sodium or cesium methacrylate) charged units in water - ethanol mixtures [14,15].

If the nonequilibrium kinetically frozen structures exist, the swelling ratio of the gel should depend on the manner by which the given state of the gel was reached: by swelling of the dry gel or by shrinking of the swollen gel. It turns out that this is just the case for the gels immersed in the solvent mixtures with the ethanol content higher than 70 vol-% (see Fig.15). The water-swollen gels immersed in a poor solvent with more than 70 vol-% of ethanol do not reach the equilibrium collapsed state because of the vitrification of polymer-rich regions in the gel emerging during the contraction of the gel. The vitrification prevents further removing of solvent from the gel and stabilizes the nonequilibrium gel structure. The SAXS data show that the nonequilibrium structures emerge simultaneously with the rise of the scattering exponent from 1.7 to 3.5 in Fig.13 which indicates the formation of the microstructure with pronounced phase boundaries [14]. The nonequilibrium gels were found to have the modulus of elasticity by two orders of magnitude higher than that of the equilibrium collapsed gels. Such high values of the modulus of elasticity are reasonable if one assumes the formation of the glassy structure.

8. Interaction of Polyelectrolyte Gels with Oppositely Charged Surfactants. X-ray Analysis

When polyelectrolyte gel is immersed in the solution of oppositely charged surfactant the surfactant molecules penetrate the network and replace counterions inside the gel via the ion exchange reaction. The concentration of the surfactants inside the gel can easily exceed the critical micelle concentration (CMC) [57-59]. Moreover, CMC inside the gel is lower than in the solution, because in the gel the charge of the surfactant micelles is neutralized by the immobilized charges of the gel chains (contrary to the case of the solution where the compensation of the micellar charge by the mobile

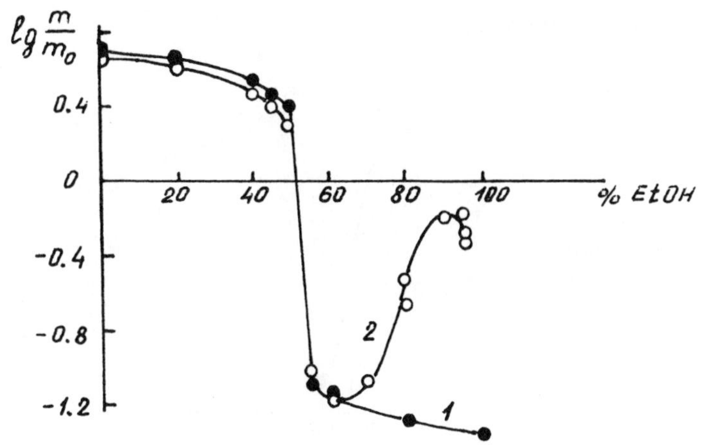

Figure 15. The swelling ratio as a function of ethanol concentration in water-ethanol mixtures for PADADMAB gel, containing 5 mol-% of charged units (1 - swelling of the dry gel, 2 - collapse of the swollen gel).

counterions leads to a significant loss of their translational entropy) [59].

The formation of micelles inside the network results in the gel collapse. The reason is the significant decrease of the exerting osmotic pressure of counterions inside the gels since counterions-surfactants aggregated in micelles loose the mobility. The attraction between the charged surfactant aggregates and oppositely charged network chains also contributes to the gel shrinking. These effects predicted theoretically in [59,60] were observed experimentally.

Experimental investigations were performed for both cationic network - anionic surfactant and anionic network - cationic surfactant systems [57,58, 61-71].

It was shown that polyelectrolyte gels absorb effectively the oppositely charged surfactants. The equilibrium concentration of surfactant ions in the gel phase was found to be by several orders of magnitude higher than in the external solution [61]. Absorption of a considerable amount of surfactant ions results in a phase transition of the swollen network in the collapsed state (Fig.16). As was mentioned above, the gel collapse is caused by the aggregation of surfactant ions within the network. If the micelles inside the gel are destroyed by the addition of low molecular salt or of organic solvent which weakens the hydrophobic interactions between surfactant hydrocarbon chains, the gel reswells [61,65].

The micelles in the gel phase possess the ability to solubilize various organic substances [61,63,67]. In Table 1 the effective distribution constants of different organic compounds between the gel-surfactant complex and

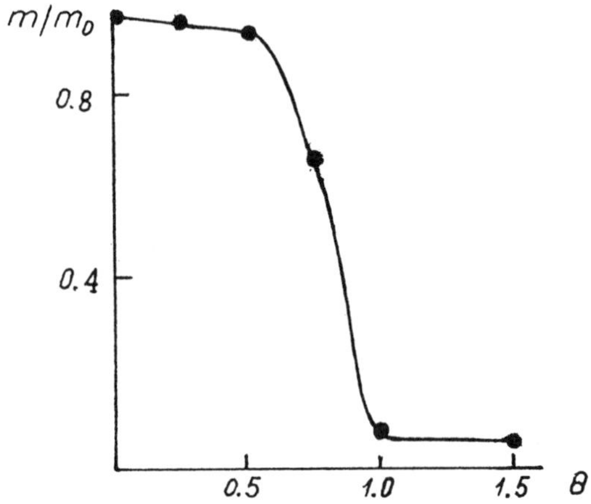

Figure 16. The dependence of the swelling ratio of cationic PDADMAB gel on the fraction θ of the absorbed anionic surfactant $C_{13}H_{27}COOK$.

aqueous solution are presented. It is seen that the gel-surfactant complexes are very effective absorbents for the organic substances. This fact can have obvious applications for the problem of purification of water from organic impurities.

TABLE 1. Fraction of the absorbed organic substance Q and effective distribution constants K of the organic substances between the gel phase and exterior solution for equimolar complexes cationic PDADMAB gel / anionic surfactant $C_nH_{2n+1}COOK$. ($K = C_{in}/C_{out}$, C_{in} being the concentration of organic substance inside the gel, C_{out} the concentration of organic substance in the external solution; the volume of water was 400 L per 1 mol of charged network units).

organic substance	Q,%			K		
	n=11	n=13	n=17	n=11	n=13	n=17
pyrogallol	93	92	96	1040	450	800
gallic acid	99	99	99	13000	3200	2800
tetrachloroguaiacol	86	69	80	220	50	130
α−naphthol	-	72	75	-	210	80

It was shown that the surfactant ions can form micelles in the network, even if the concentration of surfactants in the gel is below the CMC for the external solution. The aggregation of anionic surfactant sodium dodecyl sulfate (SDS) in cationic PDADMAB network just before the gel collapse was studied by fluorescence probe method [63]. Pyrene was used as a probe.

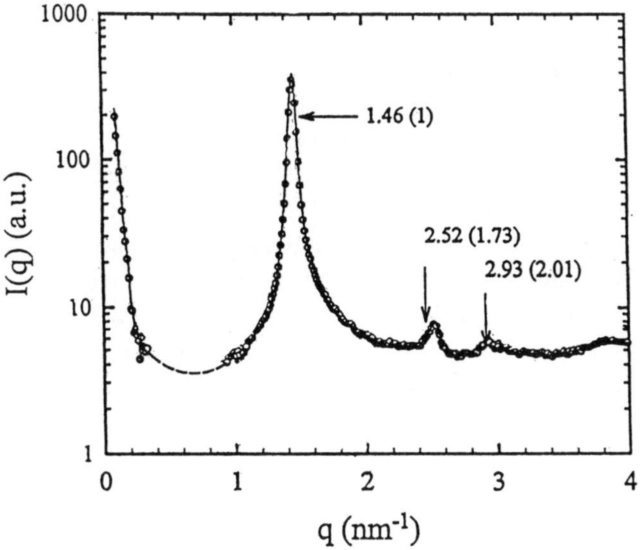

Figure 17. Integrated SAXS profile of PDADMACl gel / sodium tetradecyl sulfate complex.

The formation of micelles results in two effects: (1) the sharp decrease of the polarity of local environment of pyrene molecules which incorporate in micellar core, (2) the appearance of fluorescence band of pyrene excimers. It was found that the values of the CMC of SDS in the oppositely charged gel ($5 \times 10^{-4} - 8 \times 10^{-4}$ mol/l) are 10-15-fold lower than in aqueous solution of SDS which is consistent with the theoretical predictions.

At the same time the microenvironment of the probe molecules included in SDS micelles in the swollen PDADMAB network is more polar than that in SDS micelles in water [63]. Thus it was concluded that the structure of micelles in the gel phase differs strongly from their structure in water.

The structure of surfactant micelles in the gel phase was studied by means of SAXS [72,73]. The SAXS investigation of the self-assemblies of anionic sodium alkylsulfate surfactants in the collapsed cationic PDADMACl gels revealed a surprisingly perfect quasicrystalline ordering of the micelles inside the gels [72]. The observed SAXS peaks are very narrow and correspond to regions of practically perfect spatial organization of micelles of the size of at least 10 intermicellar distances (Fig.17). The position of the peaks corresponds to hexagonal type of spatial arrangement of the micelles. The surprising feature is that the mesh size of the gel is much smaller than the distance over which the micelles are practically perfectly ordered, i.e. the gel chains do not disturb the order.

The analogous results (narrow SAXS peaks corresponding to a high level of structural organization of surfactant aggregates in the gels) were also obtained for anionic poly(sodium acrylate) gels interacting with various cationic surfactants (alkyltrimethylammonium or alkylpyridinium halides) [73]. In this case the emerging supramolecular order was associated with the existence of lamellar surfactant phase.

For both cationic gel / anionic surfactant and anionic gel / cationic surfactant complexes the obtained values of the characteristic length correlated with the length of the surfactant molecules (were somewhat larger than the double length of the surfactant molecule).

Thus, polyelectrolyte gels can form the appropriate medium for the self-assembling of surfactants. This fact can have interesting consequences for biological systems.

References

1. Landau, L.D. and Lifshits, E.M. (1976) ***Statisticheskaya Fizika (Statistical Physics)***, Nauka, Moscow, v.5.
2. Manning, G.S. (1972) Polyelectrolytes, ***Annu. Rev. Phys. Chem.*** **23,** 117-140.
3. Oosawa, F. (1971) ***Polyelectrolytes,*** Marcel Dekker, New York.
4. Grosberg, A.Yu. and Khokhlov, A.R. (1994) ***Statistical Physics of Macromolecules,*** AIP Press, New York.
5. Odijk, T. (1977) Polyelectrolytes near the rod limit, ***J. Polym. Sci., Part B: Polym. Phys.*** **15,** 477-483.
6. Skolnick, J. and Fixman, M. (1977) Electrostatic persistence length of a wormlike polyelectrolyte ***Macromolecules*** **10,** 944-948.
7. Khokhlov, A.R., Starodubtzev, S.G., and Vasilevskaya, V.V. (1993) Conformational transitions in polymer gels: theory and experiment, ***Adv. Polym. Sci.*** **109,** 123-171.
8. Eisenberg, A. (1970) Clustering of ions in organic polymers. A theoretical approach, ***Macromolecules*** **3,** 147-154.
9. Eisenberg, A., Hird, B., and Moore, M. (1990) A new multiplet-cluster model for the morphology of random ionomers, ***Macromolecules*** **23,** 4098-4107.
10. Mauritz, K.A. (1988) Review and critical analyses of theories of aggregation in ionomers, ***J. Macromol. Sci. - Rev. Macromol. Chem. Phys.*** **C28,** 65-98.
11. Nyrkova, I.A., Khokhlov, A.R., and Doi, M. (1993) Microdomains in block-copolymers and multiplets in ionomers: parallels in behavior, ***Macromolecules*** **26,** 3601-3610.
12. Starodubtsev, S.G., Khokhlov, A.R., and Vasilevskaya V.V. (1985) Collapse of polyacrylamide networks: influence of mechanical deformation of a sample and of type of solvent, ***Dokl. Akad. Nauk SSSR*** **282,** 392-395.
13. Khokhlov, A.R. and Kramarenko, E.Yu. (1994) Polyelectrolyte/ionomer behavior in polymer gel collapse, ***Makromol. Chem., Theory Simul.*** **3,** 45-59.
14. Khokhlov, A.R., Makhaeva, E.E., Philippova, O.E., and Starodubtzev, S.G. (1994) Supramolecular structures and conformational transitions in polyelectrolyte gels, ***Makromol. Chem., Macromol. Symp.*** **87,** 69-91.
15. Philippova, O.E., Pieper, T.G., Sitnikova, N.L., Starodoubtsev, S.G., Khokhlov, A.R., and Kilian, H.G. (in press) Conformational transitions in polyelectrolyte networks in binary solvents: microheterogeneities in the collapsed state, ***Macromolecules.***
16. Khokhlov, A.R. and Kramarenko, E.Yu. (in press) Weakly charged polyelectrolytes: Collapse induced by extra ionization, ***Macromolecules.***
17. Starodoubtsev, S.G., Khokhlov, A.R., Sokolov, E.L., and Chu, B. (in press) Evidence for polyelectrolyte/ionomer behavior in the collapse of poly-

cationic gels, ***Macromolecules.***
18. Philippova, O.E. and Sitnikova, N.L. (in press) Conformational behavior of ionic gels in nonpolar media, in ***Abstracts of the First International Symposium on Polyelectrolytes "Polyelectrolytes in Solution and at Interfaces"***, September 18-22, 1995, Potsdam.
19. Semenov, A.N., Joanny, J.F., and Khokhlov, A.R. (1994) Associating polymers: equilibrium and linear viscoelasticity, ***Macromolecules* 28,** 1066-1075.
20. Semenov, A.N., Nyrkova, I.A., and Khokhlov, A.R. (in press) Polymers with strongly interacting groups: theory for non-spherical multiplets, ***Macromolecules.***
21. Paul, D.R. and Newman, S. (1978) ***Polymer Blends,*** Academic Press, New York.
22. Olabisi, D., Robeson, L.M., and Shaw, M.T. (1979) ***Polymer-Polymer Miscibility,*** Academic Press, New York.
23. Vasilevskaya, V.V., Starodubtzev, S.G., and Khokhlov, A.R. (1987) Miscibility enhancement of polymer mixtures at charging of one of the components, ***Vysokomol. Soedin., Ser. B*** 29, 930-933.
24. Nyrkova, I.A., Khokhlov, A.R., and Kramarenko. E.Yu. (1990) Possibility of microphase separation in polyelectrolyte systems, ***Polym. Sci. (USSR)* 32,** 852-861.
25. Khokhlov, A.R. and Nyrkova, I.A. (1992) Compatibility enhancement and microdomain structuring in weakly charged polyelectrolyte mixtures, ***Macromolecules* 25,** 1493-1502.
26. Nyrkova, I.A., Khokhlov, A.R., and Doi, M. (1994) Microdomain structures in polyelectrolyte systems: calculation of the phase diagrams by direct minimization of the free energy, ***Macromolecules* 27,** 4220-4230.
27. Piculell, L. and Lindman, B. (1992) ***Adv. Colloid Interface Sci.* 41,** 149-178.
28. Philippova, O.E. and Starodubtzev, S.G. (1993) Miscibility enhancement of polymers in polar media by incorporation of a small amount of charged groups into the polymer chains, ***Makromol. Chem., Rapid Commun.* 14,** 421-425.
29. Dusek, K. and Patterson, D. (1968) Transitions in swollen polymer networks induced by intramolecular condensation, ***J. Polym. Sci., Part A-2* 6,** 1209-1216.
30. Tanaka, T. (1978) Collapse of gels and the critical endpoint, ***Phys. Rev. Lett.* 40,** 820-823.
31. Khokhlov, A.R. (1980) Swelling and collapse of polymer networks, ***Polymer* 21,** 376-380.
32. Tanaka, T., Fillmore, D.J., Sun, S.-T., Nishio, J., Swislow, G., and Shah, A. (1980) Phase transitions in ionic gels, ***Phys. Rev. Lett.* 45,** 1636-1639.
33. Stejskal, J., Gordon, M., and Torkington, J.A. (1980) Collapse of polyacrylamide gels, ***Polym. Bull.* 3,** 621-625.
34. Vasilevskaya, V.V. and Khokhlov, A.R. (1986) Influence of low molecular weight salt on the collapse of polymer networks, ***Vysokomolek. Soed., Ser.A* 28,** 316-320.
35. Ptytsyn, O.B. and Eizner, Yu.E. (1965) Theory of coil-globule transitions in macromolecules, ***Biofizika* 10,** 3-9.
36. de Gennes, P.G. (1975) Collapse of a polymer chain in poor solvents, ***J. Phys. France Lett.* 36,** 55-57.
37. Lifshitz, I.M., Grosberg, A.Yu., and Khokhlov, A.R. (1976) Structure of polymer globule formed by saturated bonds, ***Sov. Phys. JETP* 71,** 1634-1643.
38. Lifshitz, I.M., Grosberg, A.Yu., and Khokhlov, A.R. (1978) Some problems of the statistical physics of polymer chains with volume interactions, ***Rev. Mod. Phys.* 50,** 683-713.
39. Lifshitz, I.M., Grosberg, A.Yu., and Khokhlov, A.R. (1979) Volume interactions in the statistical physics of a polymer macromolecule, ***Usp. Fiz. Nauk* 127,** 353-389.
40. Flory, P.J. (1953) ***Principles of Polymer Chemistry,*** Cornell University Press, Ithaca, New York.
41. Borue, V.Yu. and Erukhimovich, I.Ya. (1986) Structural phase transitions in solutions of weakly charged polyelectrolytes, ***Dokl. Akad. Nauk USSR* 286,** 1373-1376.
42. Borue, V.Yu. and Erukhimovich, I.Ya. (1988) A statistical theory of weakly charged

polyelectrolytes: fluctuations, equation of state, and microphase separation, ***Macromolecules* 21,** 3240-3249.
43. Joanny, J.F. and Leibler, L. (1990) Weakly charged polyelectrolytes in a poor solvent, ***J. Phys. France* 51,** 545-557.
44. Brereton, M.G. and Vilgis, T.A. (1990) Compatibility and phase behavior in charged polymer systems and ionomers, ***Macromolecules* 23,** 2044-2049.
45. Dobrynin, A.V. and Erukhimovich, I.Ya. (1991) Weak crystallization and structural phase transitions in weakly charged polyelectrolyte systems, ***Zh. Eksp. Teor. Fiz.* 99,** 1344-1359.
46. Moussaid, A., Munch, J.P., Schosseler, F., and Candau, S.J. (1991) Light scattering study of partially ionized poly(acrylic acid) systems: comparison between gels and solutions, ***J. Phys. France II* 1,** 637-650.
47. Schosseler, F., Moussaid, A., Munch, J.P., and Candau, S.J. (1991) Weakly charged polyelectrolyte gels: temperature and salt effects on the statics and the dynamics, ***J. Phys. II France* 1,** 1197-1219.
48. Schosseler, F., Ilmain, F., and Candau, S.J. (1991) Structure and properties of partially neutralized poly(acrylic acid) gels, ***Macromolecules* 24,** 225-234.
49. Moussaid, A., Candau, S.J., and Joosten, J.G.H. (1994) Structural and dynamic properties of partially charged poly(acrylic acid) gels: nonergodicity and inhomogeneities, ***Macromolecules* 27,** 2102-2110.
50. Schosseler, F., Skouri, R., Munch, J.P., and Candau, S.J. (1994) Swelling and cross-linking density effects on the structure of partially ionized gels, ***J. Phys. II France* 4,** 1221-1239.
51. Shibayama, M., Tanaka, T., and Han, C.C. (1992) Small angle neutron scattering study on weakly charged temperature sensitive polymer gels, ***J. Chem. Phys.* 97,** 6842-6854.
52. Leibler, L. (1980) Theory of microphase separation in block copolymers, ***Macromolecules* 13,** 1602-1617.
53. Erukhimovich, I.Ya. (1982) Fluctuations and the formation of domain structure in heteropolymers, ***Vysokomol. Soedin., Ser.A* 24,** 1942-1949.
54. Lifshitz, I.M. (1968) Some questions of statistical theory of biopolymers, ***Zh. Eksp. Teor. Fiz.* 55,** 2408-2422.
55. Nyrkova, I.A., Doi, M., and Khokhlov, A.R. (1993) Microdomain structures in polyelectrolyte mixtures, ***Polym. Prepr. (Am. Chem. Soc. Div. Polym. Chem.)* 34,** 926-927.
56. Dormidontova, E.E., Erukhimovich, I.Ya., and Khokhlov, A.R. (1994) Microphase separation in poor solvent polyelectrolyte solution: phase diagram, ***Makromol. Chem., Theory Simul.* 3,** 661-675.
57. Ryabina, V.R., Starodubtsev, S.G., and Khokhlov, A.R. (1990) Interaction of polyelectrolyte networks with oppositely charged micelle-forming surfactants, ***Vysokomol. Soedin., Ser.A* 32,** 969-974.
58. Starodoubtzev, S.G. (1990) Influence of topological structure of polyelectrolyte networks on their interaction with oppositely charged micelle-forming surfactants, ***Vysokomol. Soedin., Ser.B* 31,** 925-930.
59. Vasilevskaya, V.V., Kramarenko, E.Yu., and Khokhlov, A.R. (1991) Theory of collapse of polyelectrolyte networks in solutions of ionic surfactants, ***Vysokomol. Soedin., Ser.A* 33,** 1062-1069.
60. Khokhlov, A.R., Kramarenko, E.Yu., Makhaeva, E.E., and Starodubtzev, S.G. (1992) Collapse of polyelectrolyte networks induced by their interaction with oppositely charged surfactants. Theory, ***Makromol. Chem., Theory Simul.* 1,** 105-118.
61. Khokhlov, A.R., Kramarenko, E.Yu., Makhaeva, E.E., and Starodubtsev, S.G. (1992) Collapse of polyelectrolyte networks induced by their interaction with oppositely charged surfactants, ***Macromolecules* 25,** 4779-4783.
62. Bisenbaev, A.K., Makhaeva, E.E., Saletskii, A.M., and Starodubtsev, S.G. (1992) Complexes between sodium polyacrylate networks and cetyltrimethylammonium

bromide. A fluorescent probe investigation, ***Polym. Sci. (USSR)*** **34,** 1059-1062.
63. Philippova, O.E. and Starodoutzev, S.G. (1993) Interaction of slightly cross-linked gels of poly(diallyldimethylammonium bromide) with sodium dodecyl sulfate. Diffusion of surfactant ions in gel, ***J. Polym. Sci., Part B: Polymer Physics*** **31,** 1471-1476.
64. Makhaeva, E.E. and Starodubtzev, S.G. (1993) Swelling of a polyelectrolyte network of sodium methacrylate/acrylamide copolymer in water-2-propanol mixtures in the presence of cetylpyridinium bromide, ***Makromol. Chem., Rapid Commun.*** **14,** 105-107.
65. Machaeva, E.E. and Starodubtzev, S.G. (1993) Reentrant conformational transition of polyelectrolyte network in water alcohol mixtures in the presence of oppositely charged surfactant, ***Polym. Bull.*** **30,** 327-331.
66. Karib'yants, N.S., Starodubtsev, S.G., and Filippova, O.E. (1993) Interaction of the low-cross-linked gel of poly(methacrylic acid) with poly(ethylene glycol) and cetylpyridinium chloride, ***Polym. Sci. (USSR),*** **35,** 471-475.
67. Le Minh Thanh, Makhaeva, E.E., and Starodubtzev, S.G. (1993) Interaction of sodium dodecyl sulfate and salts of $C_{11} - C_{17}$ carboxylic acids with poly(diallyldimethylammonium bromide) gel, ***Vysokomol. Soedin., Ser.A*** **35,** 408-412.
68. Khandurina, Yu.V., Rogacheva, V.B., Zezin, A.B., and Kabanov, V.A. (1994) Interaction of cross-linked polyelectrolytes with oppositely charged surfactants, ***Polym. Sci. (USSR)*** **36,** 184-188.
69. Okuzaki, H. and Osada, Y. (1994) Effects of hydrophobic interaction on the cooperative binding of a surfactant to a polymer network, ***Macromolecules*** **27,** 502-506.
70. Sasaki, S., Fujimoto, D., and Maeda, H. (1995) Effects of salt concentration and degree of ionization on the hydrophobic counterion binding to ionic gel and the contraction of the gel volume, ***Polymer Gels and Networks*** **3,** 145-158.
71. Starodubtzev, S.G., Le Thi Minh Thanh, Makhaeva, E.E., Philippova, O.E., and Pieper, T.G. (in press) Interaction of slightly crosslinked gel of poly(diallyldimethylammonium bromide) with sodium dodecyl sulfate and cetylpyridinium bromide, ***Makromol. Chem.***
72. Chu, B., Yeh, F., Sokolov, E.L., Starodoubtsev, S.G., and Khokhlov, A.R. (in press) Interaction of poly(diallyldimethylammonium chloride) gels with sodium dodecyl sulfate. Phase transitions and structure of polymer-surfactant complexes, ***Macromolecules.***
73. Khandurina, Yu.V., Dembo, A.T., Rogacheva, V.B., Zezin, A.B., and Kabanov, V.A. (1994) Structure of polycomplexes composed of cross-linked sodium polyacrylate and cationic micelle-forming surfactants, ***Polym. Sci. (USSR)*** **36,** 189-194.

EQUILIBRIUM STRUCTURE OF IONIZABLE POLYMER BRUSHES

EKATERINA ZHULINA
Institute of Macromolecular Compounds
Russian Academy of Sciences
St.Petersburg 199004, Russia

1. Introduction

Polymer brushes, i.e. partially organized polymer systems in which macromolecules are fixed (grafted) at one end onto impermeable surfaces of various geometries, continue to attract considerable attention due to both scientific and technological aspects. Typical examples of such systems are the supermolecular structures (mesophases) that are formed in melts and concentrated solutions of block copolymers with incompatible blocks. Depending on the composition and the concentration and selectivity of the solvent, various geometries of mesophases are found. In the strong segregation limit when the mesophases are well-defined and the interfacial regions are small, the blocks of different components form polymer brushes with common interfaces. Thus, polymer brushes of various geometries are found as "elementary units" of the mesophases in the strong segregation limit.

Polymer micelles forming as a result of self-assembling in the dilute solutions of diblock copolymers in selective solvents are also an example of brush-like structures. The core of such micelles is built of the component for which the solvent is poor, whereas the corona formed by the soluble component, prevents the precipitation of the micelles. For sufficiently long polymers the interfacial region between the core and the corona is narrow with respect to the total size of the micelle, and the blocks of both components can be envisioned as grafted onto the common interface. Thus, both the corona and the core of such micelles can be considered as polymer brushes: the core is a dense concave brush with no solvent, whereas the corona is a convex brush swollen with the solvent.

Brush-like structures are found in many other polymer systems: adsorbed monolayers formed by block copolymers with selectively adsorbing blocks, dilute and semi-dilute solutions of regularly branched (star- and

S.E. Webber et al. (eds.), Solvents and Self-Organization of Polymers, 227–258.

comb-like) polymers, thin polymer films at the air-liquid interfaces, etc. Among the main technological applications of polymer brushes are steric stabilization of colloid dispersions, surface modification for adhesion and lubrication and some applications in chromatography. From a scientific point of view, polymer brushes are an interesting example of polymer systems capable of undergoing conformational and phase transitions.

The equilibrium behavior of neutral polymer brushes (i.e. carrying no charge) is rather well understood at present. As shown by Alexander [1] and de Gennes [2], in neutral polymer brushes, chains are stretched in the direction normal to the grafting surface due to the short-range inter-unit interactions. This stretching takes place under the conditions of both good and theta solvents[3] and remains even if the brush is immersed in a poor solvent[4]. This stretching is due to intermolecular repulsion that tends to diminish the concentration of polymer units in the brush. At relatively sparse grafting, where the neighbouring chains do not overlap and intermolecular repulsion is negligible, the chains retain the conformations of individual non-deformed coils (the "mushroom" regime). Correspondingly, two different regimes of behavior are found for neutral polymer brushes: the mushroom regime at sparse grafting and the brush regime at relatively dense grafting, which provides considerable overlap of neighbouring chains. Scaling and self-consistent field (SCF) theories of neutral brushes provide a detailed picture of the brush structure. The most important parameters of the brush structure are the brush thickness and the distribution (profile) of polymer units. Few experimental techniques allow one to measure these characteristics. For neutral brushes, the thickness always decreases with increasing grafting density; the particular law is determined by the solvent quality. The theoretically predicted dependences have been confirmed by low-angle neutron scattering experiments [5-7] and neutron reflectivity measurements [8]. For a more detailed discussion of neutral brushes see, for example, the review of Halperin et al. [9] and references therein.

Charged brushes, where the short-range non-electrostatic interactions are combined with the long-range electrostatic forces, exhibit distinctively novel features with respect to the neutral brushes. Although the theoretical analysis of such systems is far from being complete, a certain understanding of the charged brush behavior has been reached [10-15]. As shown by Pincus [12] and Borisov et al. [14], the chains in polyelectrolyte brushes are also stretched in the direction normal to the grafting surface. However, in contrast to neutral systems, this stretching is determined primarily by electrostatic interactions in the brush, rather than by the short-range repulsion between uncharged units. Moreover, due to the long-range nature of the electrostatic interactions, the grafted polyelectrolyte chains become stretched in the normal direction at grafting densities far below the overlap threshold and exhibit orientational effects absent in neutral brushes.

Other novel features of the polyelectrolyte brushes are induced by the annealing charge effects. One can distinguish two types of grafted polyelectrolytes: with a quenched (fixed) and an annealed (equilibrated) charge distributions. In the quenched case, the number and positions of charges on a chain are fixed and do not change with a variation of external conditions (pH and ionic strength, grafting density, etc). One can envision such a situation for a copolymer (random or regular) of a neutral component and a strong electrolyte. An average distance between neighbouring charges and the corresponding degree of chain ionization are determined by the composition of such a copolymer. A typical example of such a polyelectrolyte is sulphonated polystyrene, where uncharged styrene units are separated by the charged sulphonic groups. In an annealed case, both the number and positions of charges can adjust to variations in external conditions. A typical example of such a polyelectrolyte is a weak polyacid or polybase homopolymer. Variations in pH can easily affect the degree of ionization of such chains. Correspondingly, a brush formed by such polyelectrolytes is expected to be much more sensitive to the variation of external conditions than a quenched brush with a fixed charge. Moreover, the annealing effects can lead to "abnornal" behavior in polyelectrolyte brushes, that is, an increase in the brush thickness with decreasing grafting density. This type of behavior is quite opposite to that of neutral systems.

The goal of this lecture is to compare the equilibrium behavior of quenched and annealed polyelectrolyte brushes. We start with a scaling description of a quenched polyelectrolyte brush and summarize the main laws regulating its behavior. To obtain more insight into the internal brush structure and analyze spacial distributions of charges and polymer units, we use the analytical SCF approach developed earlier [16-21] for neutral systems and the complementary numerical SCF model by Scheutjens and Fleer [22]. We then consider an annealed brush using both scaling and SCF approaches and discuss the peculiarities in the behavior of annealed brushes.

2. Model

Let us consider a brush formed by long, weakly charged polyelectrolyte chains grafted at one end onto an impermeable planar surface and immersed in a dielectric solvent, as shown in Figure 1. The solvent is assumed to be a theta solvent with respect to short-range interactions between the uncharged monomers. Let $N >> 1$ be the number of monomer units in each chain, s be the area per chain ($1/s$ is the grafting density), and m be the average number of uncharged monomer units between two neighbouring (elementary) charges. The total charge on each polyion is $Q = N/m$, and the degree of chain ionization is $\alpha = Q/N = 1/m$. The backbone of the

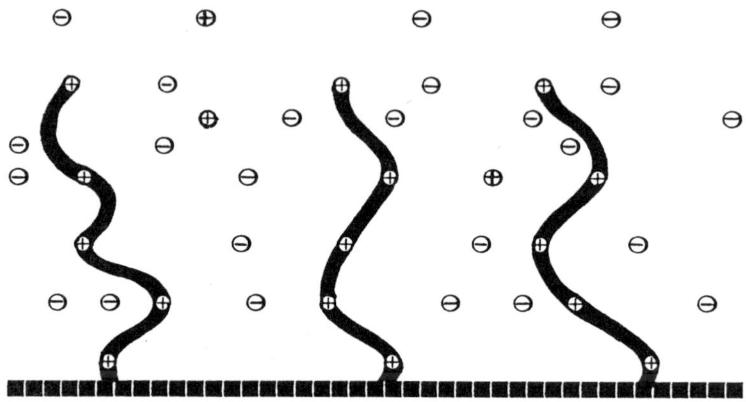

Figure 1. A polyelectrolyte brush.

polyions is assumed to be flexible, so that the Kuhn segment length, A, is equal to the monomer unit length, a, which, in turn, is equal to the chain thickness. The condition of weak charging of the chains

$$m^{1/2} >> u = l_B/a \tag{1}$$

implies that the interactions between neighbouring charges do not lead to the electrostatic stiffening of the chains [23] and that locally they are flexible. Here $l_B = e^2/\epsilon T$ is the Bjerrum length (e is an elementary charge, ϵ is the dielectric constant of the solvent, and T is the temperature in energy units). In water at normal conditions, $u \simeq 1$ for typical flexible polymers, and we omit u in further considerations.

The condition of total electroneutrality of the system results in the presence of $Q = N/m$ small mobile counterions in the solution per grafted polyion (monovalent as well as ionizable groups on the polyions). The solution can also contain certain amounts of added small monovalent co- and counterions, which determine the $pH = -logC_H$ and the ionic strength $I = \frac{1}{2}\sum_i C_i = \frac{1}{2}(C_+ + C_-)$ of solution, where different $i's$ correspond to chemically different ions and C_+ and C_- are the total concentrations of positively and negatively charged ions.

For quenched polyions, the total charge on a chain is $Q = N/m$, and is independent of external conditions, i.e. $Q = const$ for a wide range of I, pH and s values. For annealed polyions, the total charge $Q = \alpha N$ depends on the concentration of counterions in the brush. Let us consider a brush formed by weak polyacids (the generalization for polybases is straightforward). The degree of ionization α of a polyacid chain is determined by the dissociation constant K_D and the local concentration of protons c_H according to the mass action law

$$\frac{\alpha}{1-\alpha} = \frac{K_D}{c_H} \tag{2}$$

We ignore for simplicity the possible dependence of K_D on α (polyelectrolyte effect) and suppose that K_D is constant.

In order to carry out the scaling analysis of the system, we use a simple box model, according to which all the chains in the brush are stretched homogeneously and their free ends are localized at the outer boundary of the brush of thickness H (Alexander - de Gennes model). This model provides the power law dependences for the brush thickness H (and the degree of ionisation α for an annealed system). Later, using the SCF approach, we relax the constraint on the positions of the free ends of the chains and obtain more detailed information about the internal brush structure.

3. Quenched Polyelectrolyte Brush

We start our consideration of a quenched polyelectrolyte brush in a salt-free solution.

3.1. INDIVIDUAL POLYION

As is known [24], an individual polyion in a salt-free solution always loses its counterions, and the interaction of charged groups on the chain has the character of an unscreened Coulombic repulsion. If $Q = N/m$ is the total charge of the polyion and H is its end-to-end distance, the stretching electrostatic force is given by

$$f_{el}/T \simeq aQ^2/H^2 \tag{3}$$

The opposing elastic force related to conformational entropy losses in the stretched chain is given by the Gaussian law

$$f_{conf}/T \simeq H/a^2N \tag{4}$$

The condition of force balance

$$f_{el} = f_{conf} \tag{5}$$

provides an equilibrium value of the end-to-end distance for a polyion chain in a theta solvent [24,25],

$$H_0 \simeq aNm^{-2/3} \tag{6}$$

Here and below, the index "0" refers to the parameters of an isolated polyelectrolyte chain. A comparison of H with the unperturbed dimensions of a Gaussian chain, $R_0 \simeq aN^{1/2}$, shows that polyelectrolyte stretching, $H_0 >> R_0$, takes place at

$$m << N^{3/4} \tag{7}$$

Introducing the concept of a "stretching blob" [26] (the part of the chain that is actually not perturbed by an electrostatic interaction and retains the conformation of an individual coil), we can envision a polyion as a sequence of N_b stretching blobs of size ξ,

$$H \simeq \xi N_b \tag{8}$$

$$\xi \simeq T/f_{el} \tag{9}$$

Taking into account equations (3),(6) and (9), we get

$$\xi_0 \simeq am^{2/3} \tag{10}$$

The chain sequence inside the blob is Gaussian under the conditions of a theta solvent and each blob contains

$$n_b = N/N_b \simeq \xi^2/a^2 \tag{11}$$

polymer units. The dimensions D of the polyions in the direction perpendicular to the end-to-end vector are given by

$$D \simeq \xi N_b^{1/2} \simeq aN^{1/2} \tag{12}$$

3.2. COULOMBIC INTERACTIONS BETWEEN GRAFTED POLYIONS

A sparse grafting of such polyions onto a planar surface does not change the scaling dependences for the polyion dimensions. However, in addition to intramolecular Coulombic repulsion, each grafted polyion experiences a mean "external" electrostatic field created by all the other grafted polyions and their counterions. In contrast to an isolated polyion, the planar grafted polyelectrolyte layer retains its counterions just as an infinite charged plane. The characteristic thickness of the counterion cloud λ near the charged surface is given by [27]

$$\lambda \simeq l_B^{-1}\rho^{-1} \simeq a^{-1}s/Q \tag{13}$$

where ρ is the density of the surface charge. Thus, the distribution of the counterion density and the value of the electrostatic field in the brush are expected to depend on the ratio λ/H. If λ far exceeds the characteristic brush thickness H, almost all the counterions leave the brush. This regime takes place at sufficiently low charge densities, i.e. at low grafting density or low charge of polyions. Under these conditions, grafted polyions and their mobile counterions form an asymmetrical double electrical layer (planar "capacitor") near the grafting surface. The charge related to the polyions is localized in a narrow region of thickness H, whereas the counterion charge is smeared above it over a wide region of characteristic thickness $\lambda >> H$. The fraction of counterions penetrating the brush and partially compensating its charge is small ($\simeq H/\lambda << 1$), and the surface charge per unit area in the double electrical layer is approximately equal to Q/s. The corresponding mean electrostatic force applied to each polyion and acting normally to the surface is given by

$$f_{el}/T \simeq aQ^2/s \tag{14}$$

At sparse grafting densities, $s >> H_0^2$, this mean force is weak in comparison to intramolecular electrostatic repulsion (equation(3)) and causes only an orientation of the stretched polyions (charged "sticks" of blobs) in the normal direction. The mean angle θ between the end-to-end vector of the polyion (the axis of the "stick" of blobs) and the normal to the surface can be estimated from the condition

$$\Delta W_{orient} \simeq f_{el}H_0(1 - \cos\theta) \simeq T \tag{15}$$

where ΔW_{orient} is an increase in the electrostatic energy of the polyion due to a deviation from the most favourable direction $\theta = 0$. As follows from equation(15), the effect of polyion orientation becomes significant ($\theta << 1$) at $f_{el}/T >> H_0^{-1}$, and the mean value of θ in this case is given by

$$\theta \simeq N_b^{-1/2}(s/H_0^2)^{1/2} \tag{16}$$

Chain orientation in the normal direction results also in the corresponding decrease in the lateral dimensions of the polyions. Thus, due to the long-range character of electrostatic interactions, orientational ordering of grafted polyelectrolyte chains starts above the overlap threshold, i.e. at $s >> H_0^2$. Moreover, there is an additional source of orientation for the grafted polyions. The above estimates do not take into account the possible difference in the polarization of the environment above and below the surface, i.e. they are valid at $\epsilon = \epsilon_s$, where ϵ_s is the dielectric constant of the surface. For many cases, however, $\epsilon_s < \epsilon$, and the polarization of the environment, induced by the polyion, is different above and below the surface. This is equivalent to the appearence of an image charge $Q' = Q(\epsilon - \epsilon_s)/(\epsilon + \epsilon_s)$ situated below the surface and interacting with the polyion with a repulsive force

$$f'_{el}/T \simeq \frac{Q'Q}{H_0^2} \tag{17}$$

At $\epsilon >> \epsilon_s$ (the case of water solution), the image charge effects induce strong polyion orientation at arbitrary (sparse) grafting of polyions and increase the orientational ordering in the vicinity of the overlap theshold.

3.3. PINCUS BRUSH

In the vicinity of the overlap threshold, $s \simeq H_0^2$, the mean electrostatic force f_{el} given by equation(14), becomes comparable to the intramolecular force, given by equation(3). At higher grafting densities, $s << H_0^2$, the cooperative (intermolecular) electrostatic interactions in the grafted polyelectrolyte layer predominate over the intramolecular electrostatic interactions and determine the equilibrium structure of the brush.

If the grafting density is not too high so that the condition $\lambda << H$ is still fulfilled, the mean electrostatic force applied to each polyion is given by equation(14). Balance of this stretching force and the chain elastic force, equation(4), leads to the following expression for the brush thickness

$$H \simeq aN\frac{Q^2a^2}{s} \simeq a^3\frac{\alpha^2N^3}{s} \tag{18}$$

As follows from the "capacitor" model, in this regime the chains are stretched by a constant (independent of H) "external" force $f_{el} \sim N^2$, equation(14). This leads to an unusual dependence of the brush thickness on the chain length, $H \sim N^3$. The dependence (18) was obtained earlier by Pincus [12] on the basis of osmotic balance arguments.

Using the concept of stretching blobs, every polyion in the brush can be envisioned as a sequence of blobs of size

$$\xi \simeq T/f_{el} \simeq s/aQ^2 \tag{19}$$

completely stretched in the normal direction, $H \simeq \xi N_b$. The lateral dimensions of the polyions are Gaussian,

$$D \simeq \xi N_b^{1/2} \simeq aN^{1/2} \tag{20}$$

An increase in the surface charge density (increase in Q or decrease in s) results in a simultaneous increase in the brush thickness H, equation(18), and a decrease of the counterion cloud thickness λ, equation(13). When λ becomes comparable to the brush thickness H, more and more counterions localize inside the brush and compensate the charge of polyions.

3.4. OSMOTIC BRUSH

When λ becomes smaller than the brush thickness H, almost all counterions are localized inside the brush so that the brush becomes electroneutral as a whole. Moreover, the condition of the total elecroneutrality of the brush results in its local electroneutrality: at $\lambda << H$ the Debye screening length related to the counterion density in the brush, $r_D \simeq (l_B Q/sH)^{-1/2}$, is much smaller than the brush thickness, $\lambda << r_D << H$.

The decay of the counterion density from a mean value Q/sH inside the brush to zero value occurs in a narrow region $\sim r_D << H$ near the upper boundary of the brush. The fraction of counterions leaving the brush is approximately $r_D/H << 1$. As almost all the counterions are retained inside the brush, the stretching electrostatic force in this case is determined by the osmotic pressure of the gas of counterions to give

$$f_{el}/T \simeq Q/H \tag{21}$$

The above expression for the stretching electrostatic force also follows directly from the solution of the Poisson-Boltzmann equation [28]. Balance of the stretching electrostatic force, equation(21), and the elastic force, equation(4), determines the brush thickness in the "osmotic" regime,

$$H \simeq aNm^{-1/2} \tag{22}$$

In the framework of the blob picture, each chain in the brush can again be represented as a completely stretched sequence of blobs of size $\xi \simeq T/f_{el}$, where f_{el} is now given by equation(21). As follows from equations(21) and (22), in the osmotic regime each blob contains m monomer units and, hence, one charge. The number of blobs N_b coinsides with the number $Q = N/m$ of the charged groups on a chain, and blob size ξ is given by

$$\xi \simeq am^{1/2} \tag{23}$$

The lateral dimensions of polyions remain Gaussian, equation(12).

3.5. QUASINEUTRAL BRUSH

Up to now the effect of non-electrostatic interactions between uncharged units of polyions has been totaly neglected. However, a decrease in s or Q can make the contributions of electrostatic and volume interactions comparable. In the framework of the blob picture, this corresponds to dense packing of stretching electrostatic blobs in the brush so that the boundary of the "quasineutral" regime is determined by the condition

$$\xi^2 \simeq s \tag{24}$$

where the size ξ of the stretching blob is given by equation (19) or equation(23). Since the stretching blob ξ in the quasineutral regime is also determined by volume interactions (ternary contacts in a theta solvent), it remains equal to $s^{1/2}$. Correspondingly, in the quasineutral regime the brush can be always represented as a system of densely packed blobs with each chain of blobs stretched in the normal direction and randomly tangled in the lateral directions. Using the relationship between ξ and N_b, equation(11), we recover the known [3] result for the thickness of a neutral brush in a theta solvent,

$$H \simeq N_b\xi \simeq aN(s/a^2)^{-1/2} \tag{25}$$

3.6. DIAGRAM OF STATES

The above results can be summarized in the scaling type "diagram of states" for a polyelectrolyte brush in the s, m co-ordinates, Figure 2. The diagram contains regions for weakly charged, sparsely grafted Gaussian coils unperturbed by intra- or inte-rchain electrostatic interactions (mushroom regime NC) and individual polyions (sticks of electrostatic blobs), oriented by inter-chain electrostatic repulsion and/or "image charge" interactions (regime OrS), the regime PB for a polyelectrolyte brush stretched by cooperative electrostatic repulsion under the conditions of strong charge separation ($\lambda >> H$), the regime OsB for a polyelectrolyte brush stretched by the osmotic pressure of counterions ($\lambda << H$) and quasineutral regime NB, corresponding to the predominance of non-electrostatic (volume)

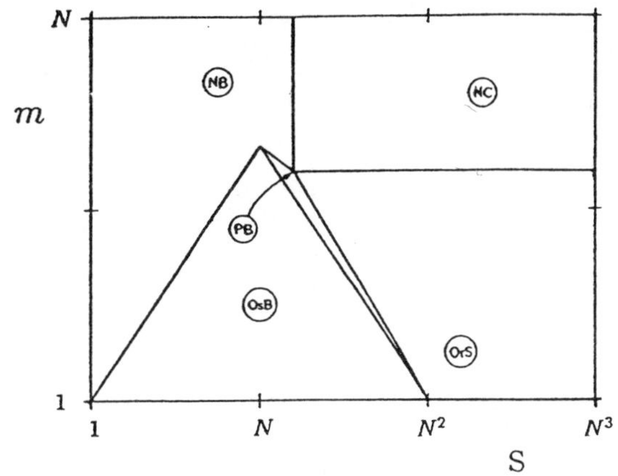

Figure 2. Diagram of states for a charged brush in a salt-free solution.

interactions. The boundaries between any two of the neighbouring regions of the diagram can be obtained from the condition of smooth crossover of corresponding scaling dependences in these regions.

The diagram in Figure 2 reflects rather general features of polyelectrolyte brushes. It comprises the regimes of quasineutral behavior (NB and NC) at weak charging of polyions ($m >> 1$) as well as the polyelectrolyte regimes OsB, PB and OrS, where the long-range electrostatic interactions lead to the novel dependences for the brush thickness. Although this diagram was constructed for a theta solvent conditions, similar regimes are found in solvents of any strength. In athermal solvents, where the assumption of a Gaussian elasticity for stretched chains is not valid and Pincus elasticity [29] is appropriate, we get qualitatively similar diagram [28], though the exponents in the scaling laws are slightly different. In poor solvents the smooth crossover lines between the polyelectrolyte and quasineutral brush regimes are substituted by the lines of phase transitions. As shown by Borisov et al. [14] and Ross and Pincus [13], collapse of a polyelectrolyte brush occurs as abrupt transition with jumpwise decrease in the brush thickness. At sparse grafting such collapse can be also accompanied by brush decomposition, similar to that in neutral collapsed brushes, that lose their lateral homogenuity and form micellar type aggregates [30,31].

As follows from the diagram in Figure 2, below the overlap threshold, $s << H_0^2$, two different polyelectrolyte regimes OsB and PB are found. Whereas the OsB regime is rather wide in the (m, s) space, the PB regime is localized in a narrow gap between the OsB and OrS regimes and is, most probably, hardly detectable. It should be noted that polyelectrolyte brushes are investigated experimentaly very poorly and only few experimenatal results are available in literature [32-34]. Computer simulations could provide another possibility to check the predicted scaling dependences. However, due to complexity of the system incorporating many charged particles with long-range interactions, computer simulations focus at present either on individual polyions or few relatively short charged chains [35,36]. Simulation of a polyelectrolyte brush in various regimes still remains an open problem. However, numerical SCF calculations which are based on direct solution of Poisson-Boltzmann equation, can be helpful in the analysis of the polyelectrolyte brush structure [10,11,37,38].

The detailed comparison of the SCF numerical calcultions with the scaling predictions was carried out by Israels et al. [38] on the basis of Scheutjens and Fleer model. This model was developed initially to describe the adsorption of uncharged polymers [22]. Later it was extended for charged systems [39]. A simple cubic lattice is used to model a polyelectrolyte brush in contact with solution which contains certain amount of salt ions. Electrostatic interactions are handled using the multilayer Stern model, which is a generalization of the Poisson-Boltzmann equation to the case of a lattice containing an arbitrary mixture of molecules. The SCF method permits the exact calculation (within the mean-field approximation) of the statistical weights of all possible conformations of the polymer chains, and the evaluation of the equilibrium distributions of all components present in the system. As a measure of the brush thickness, the weight-averaged thickness, H_{rms}, which is the second moment of the polymer segment distribution, is used. The values of H_{rms} were calculated for sets of the parameters s, m and N, and the apparent values of exponents

$$\beta_s = \frac{\delta \log H_{rms}}{\delta \log s} \qquad \beta_m = \frac{\delta \log H_{rms}}{\delta \log m} \qquad \beta_N = \frac{\delta \log H_{rms}}{\delta \log N} \tag{26}$$

were estimated. Inspection of these exponents reveals that in the brush regimes NB and OsB H_{rms} scales as $H_{rms} \sim N$ ($\beta_N = 1$) in agreement with the scaling predictions. The regime PB with the scaling dependence $H \sim N^3$ was not detected. For OsB regime the values of apparent exponents β_s and β_m are close to the scaling predictions. Figure 3 demonstrates the dependences of H_{rms} on the inverse degree of ionization $m = \alpha^{-1}$ for a polyelectrolyte brush immersed in a nearly salt-free solution. As is seen

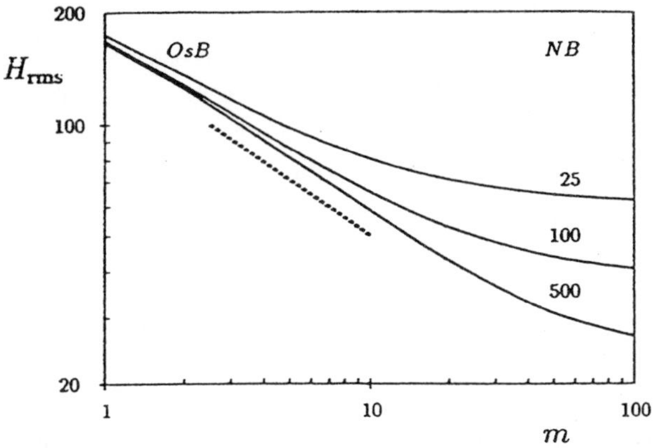

Figure 3. Brush thickness H_{rms} as a function of the inverse charge density m for different values of the area s/a^2 (indicated), $N = 500$, $C_s = 10^{-5}$[38].

from Figure 3, in the range of small m, corresponding to the conditions of the OsB regime, H_{rms} depends only slightly on the value of s (equation(22) predicts $H \sim s^0$), and the slope of the dependences is close to the predicted value $\beta_m = -1/2$ (dashed line in Figure 3). At higher values of m corresponding to the conditions of the NB regime, the value of H_{rms} is determined by the value of the grafting area s and diminishes with s increasing. Inspection of the apparent values of exponents reveals also that the boundary regions between different scaling regimes are rather wide. This may provide an explanation why the PB regime is not detected by the SCF numerical calculations. However, though the predicted exponents ($\beta_N = 3, \beta_s = -1, \beta_m = -2$, equation(18)) were not recovered, closer investigation, using relatively short chains, indicated a regime reminiscent of the PB regime. Moreover, with increasing chain length calculated exponents changed in the direction of the predicted values.

3.7. SALT EFFECT

Up to now we considered the situation when the amounts of added salt ions in the solution were negligible with respect to the concentration of own counterions. Let us consider now the effect of added salt on the brush structure. In the absence of salt, let the brush be swollen in the OsB regime so that all the counterions are contained inside the layer. Their concentration $\sim Q/sH$ determines the Debye screening length in the brush, $r_D \simeq (Ql_B/sH)^{-1/2}$. The effect of added salt on chain conformation becomes considerable when the salt concentration in the bulk solution becomes comparable to the concentration of own counterions inside the brush. When the salt concentration far exceeds this value, a mean electrostatic force applied to grafted chains can be represented as a difference in the osmotic pressure of ions inside and outside the brush. Under the conditions of salt dominance, this mean electrostatic force is given by

$$f_{el}/T \simeq C_s^{-1}Q^2/sH^2 \simeq sa(\frac{Qr_{Ds}}{sH})^2 \tag{27}$$

where $r_{Ds} \simeq (l_B C_s)^{-1/2}$ is the Debye screening length in the salt solution of concentration C_s. The above expression for a mean electrostatic force in the salt dominance regime can be obtained directly from the Poisson-Boltzmann equation[28], but it can be also clarified with the following arguments. In the salt dominance regime the characteristic thickness of the double electrical layer at the upper boundary of the brush is of the order of r_{Ds}. The corresponding separated charge per unit area is approximately equal to Qr_{Ds}/sH so that the stretching force per chain is given by equation(27). Note that in the salt-free, OsB regime the value of r_D is determined by the concentration of own counterions in the brush, leading to equation(21) for the stretching electrostatic force. The equilibrium thickness of the brush is determined by the balance of the electrostatic force, equation(27), and the elastic force, equation(4), that gives

$$H \simeq aNm^{-2/3}(\frac{r_{Ds}}{a})^{2/3}(\frac{s}{a^2})^{-1/3} \tag{28}$$

Equation (28) describes the thickness of polyelectrolyte brush in the so-called salted brush (SB) regime. In this regime the concentrations of the added ions in the brush are close to their bulk values. Screening of the electrostatic repulsion in this regime can be described by an effective second virial coefficient of unit interactions,

$$v_{eff} \simeq \frac{ar_{Ds}^2}{m^2} \sim I^{-1} \tag{29}$$

Recent experimental observations [34] are in reasonable agreement with the predicted dependence $H \simeq C_s^{-1/3}$. The SB regime appears in the diagram of states between the OsB and PB regimes when the salt concentration C_s exceeds the threshold concentration $C_s^* \simeq N^{-2}$. Below C_s^* the salt controlled Debye screening length exceeds the brush thickness at any values of α and there is no salt effect on the brush structure.

Figure 4 demonstrates the results of the SCF numerical calculations [37] for a polyelectrolyte brush immersed in a salt added solution: the dependences of H_{rms} on the combination of parameters (sv_{eff}). At relatively small values of (sv_{eff}), corresponding to the conditions of the salt dominance regime SB, the results collapse on a master curve with the slope 1/3 in accordance with the predicted scaling dependence (28). At high values

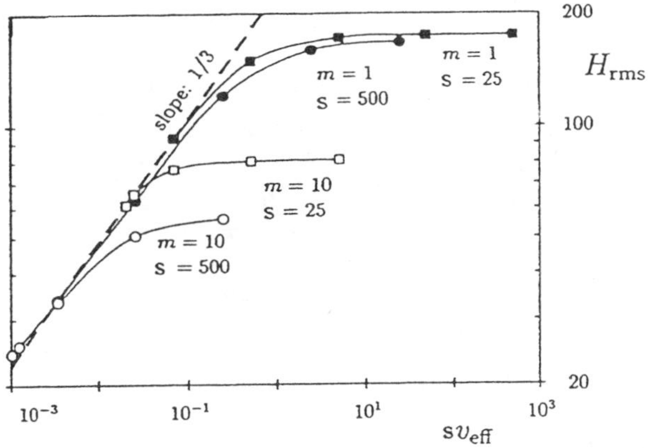

Figure 4. Brush thickness H_{rms} as a function of the salt concentration C_s for different combinations of s and m (indicated) and $N = 500$ [38].

of (sv_{eff}), corresponding to the conditions of the OsB regime, H_{rms} tends to a constant value independent of both s and C_s. This is particularly noticeable for strongly charged chains ($m = 1$ in Figure 4). The calculated values of apparent exponents are close to the scaling predictions given by equation(28), $\beta_N = 1, \beta_m = -2/3, \beta_s = -1/3$.

3.8. PROFILES OF POLYMER UNITS AND MOBILE IONS

The equilibrium distributions of all the components in the system (profiles of units and all mobile ions) can be obtained from a more refined SCF analysis which implies relaxation of the constraint on the positions of the free ends (i.e. their spreading throughout the brush). Two basic assumptions of the analytical SCF theory of polyelectrolyte brushes are: 1) the condition of strong chain stretching, $H >> R_0$, and 2) the local electroneutrality approximation (LEA).

The assumption of strong stretching allows one to ignore the chains fluctuations around the most probable path and reduces a set of various chain conformations to a single "trajectory" specifying the position of each chain monomer above the surface, $x(n)$. As shown by Semenov[16], conformational entropy of the stretched chains can be then represented as

$$\Delta F_{conf}/T = \frac{3}{2a^2}\int_0^H g(x')dx' \int_0^{x'} E(x,x')dx \tag{30}$$

where $g(x')$ is the distribution of the free ends in the brush, whereas $E(x,x') = dx/dn$ is the tension in a chain at the height x above the surface provided that the chain free end is located at the height $x' > x$. The concentration $c(x)$ of polymer units is related to the functions $g(x')$ and $E(x,x')$ as

$$c(x) = \frac{a^3}{s}\int_x^H \frac{g(x')dx'}{E(x,x')} \tag{31}$$

$$\frac{s}{a^3}\int_0^H c(x)dx = N \tag{32}$$

For neutral brushes the interactions between polymer units and other components (if present in the system) are described by the functional

$$\Delta F_{int}/T = s\int_0^H \mathcal{F}[c(x), c_k(x)]dx \tag{33}$$

where $\mathcal{F}$ is the density of the interaction free energy in a system with given distributions $c(x)$ and $c_k(x)$ of the components ($k = 1, 2, ...$). Minimization of the total free energy functional

$$\Delta F/T = \Delta F_{conf}/T + \Delta F_{int}/T \tag{34}$$

under the constraints (31), (32) and the equilibrium coexistence condition of the brush and the bulk solution leads to the following set of equations (see [16,18,19] for more details),

$$\frac{\delta \mathcal{F}}{\delta c_k} = \frac{\delta \mathcal{F}_{bulk}}{\delta C_k} \tag{35}$$

$$\frac{\delta \mathcal{F}}{\delta c} = \Lambda - b^2x^2 \tag{36}$$

Here $\mathcal{F}_{bulk}$ is the density of the interaction free energy in the bulk solution with given concentrations C_k of the components; $b^2 = 3\pi^2/8a^2N^2$ and Λ is an indefinite constant which can be determined from the normalization condition (32). Solution of this set of equations provides the equilibrium distributions of the components in neutral brushes.

For charged brushes, interactions between various components are non-local, and the expression (33) for ΔF_{int} is not valid in general case. However, in the two main brush regimes OsB (osmotic brush) and SB (salted brush) one can use the local electroneutrality approximation (LEA), which implies that spacial distributions of positive $\rho_+ = c_+$ and negative $\rho_- = \alpha c + c_-$ charge, created by the grafted polyions and all the mobile ions, are very close to each other throughout the brush, i.e.$\Delta\rho(x) = \rho_+(x) - \rho_-(x) << \rho_+(x), \rho_-(x)$. Under these conditions the interaction free energy in the brush can be still approximated by equation(33), where $c_k(x)$ is the spacial distribution of ions of type k. In the framework of LEA, the density of the interaction free energy in the brush is determined by the mixing entropy of all the components,

$$\mathcal{F} = \sum_k c_k(\log c_k + \mu_k^*) \tag{37}$$

where μ_k^* is the standard chemical potential of the component of type k. The concentrations of components are subjected to the constraint of local electroneutrality,

$$\alpha c(x) + c_-(x) = c_+(x) \tag{38}$$

where polyions are assumed to be negatively charged and c_- and c_+ denote the total concentrations of negatively and positively charged ions in the brush. Additional constraint follows from the incompressibility of the system. One can also take into account the water dissociation equilibrium, which interconnects the concentrations c_H, c_{OH} and c_{H_2O} in the brush (and bulk solution).

Substituting expression (37) for $\mathcal{F}$ into equations(35) and (36), one gets with the account of the above constraints the following relationships between the profiles for positively and negatively charged mobile ions

$$\frac{c_j^-(x)}{C_j^-} = \frac{C_i^+}{c_i^+(x)} \tag{39}$$

(Donnan rule), and

$$c_j^-(x) = C_j^- \exp\left(-\frac{H^2 - x^2}{h^2}\right) \qquad c_i^+ = C_i^+ \exp\left(\frac{H^2 - x^2}{h^2}\right) \tag{40}$$

where

$$h = (\frac{8a^2N^2\alpha}{3\pi^2})^{1/2} \tag{41}$$

The final expression for the polymer profile, $c(x)$, is given by

$$\alpha c(x) = C_s[\exp(\frac{H^2 - x^2}{h^2}) - \exp(-\frac{H^2 - x^2}{h^2})] \tag{42}$$

where $C_s = C_- = C_+$ is the total concentration of positively or negatively charged ions (ionic strength of the solution). The upper boundary of the brush, H, can be calculated from the normalization condition (32) to give the following equation

$$\gamma = \exp(z^2) \int_0^z \exp(-t^2)dt - \exp(-z^2) \int_0^z \exp(t^2)dt \tag{43}$$

where $z = H/h$ and $\gamma = \alpha N/shC_s$. Two asymptotical solutions of equation(43) corresponding to the cases $\gamma >> 1$ and $\gamma << 1$, determine the value of H in the OsB and SB regimes, respectively. In the first case (OsB regime) we have

$$z \approx [\log(\frac{2\gamma}{\pi^{1/2}})]^{1/2} \qquad \gamma >> 1 \tag{44}$$

and

$$c(x) = \frac{2Na^3}{\pi^{1/2}sh} \exp -(x^2/h^2) \tag{45}$$

Thus, at low ionic strengths $C_s \to 0$ the upper boundary of the brush $H \to \infty$ and decay of the polymer profile is determined by the value of h, equation(41).

In the second case (SB regime) we have

$$z \approx (\frac{3\gamma}{4})^{1/3} \qquad \gamma << 1 \tag{46}$$

and

$$c(x) \approx \frac{3Na^3}{2sH}(1 - x^2/H^2) \tag{47}$$

where

$$H \approx aN(8v_{eff}/\pi^2 as)^{1/3} \tag{48}$$

Earlier the dependences (47) and (48) for the profile and the brush thickness were obtained for the case when added positively charged ions were chemically identical to the own counterions of the brush [15].

As follows from the equations(44) and (45), at low ionic strength of the bulk solution the profile of polymer units is of a Gaussian shape and protrudes far into solution ($H \to \infty$). However, the average brush thickness $H_{rms} \approx h/2^{1/2}$ is finite and scales as predicted by equation(22).

At high ionic strength (SB regime) the profile of polymer units is parabolic, and the average brush thickness $H_{rms} \approx H/5^{1/2}$, where H is given by equation (48). This is not surprising since in the SB regime the electrostatic interactions are described by the positive effective second virial coefficient v_{eff}, and the shape of the profile coincides, thus, with that for a neutral brush immersed in a good solvent [17,20].

As was already mentioned, the above expressions for the profiles are valid under the conditions of the LEA. The numerical SCF calculations [38]

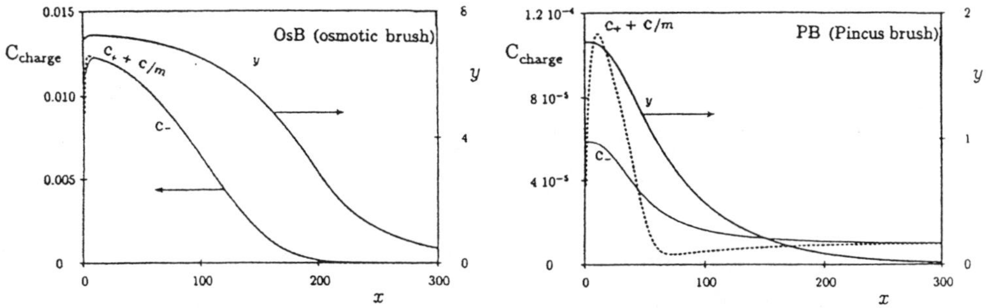

Figure 5. Density profiles of negative and positive charge and the dimensionless electrostatic potential $y = e\Psi/kT$ for an osmotic brush and Pincus brush [38].

based on the direct solution of Poisson-Boltzmann equation, allow one to check the validity of LEA in various brush regimes. Figure 5 demonstrates the calculated profiles of positive and negative charge in the brush and the corresponding normalized electrostatic potential $y(x)$ in the OsB and PB regimes. As follows from Figure 5, under the conditions of the PB regime noticeable decompensation of charge in the brush is found, and the LEA is not applicable. However, for the OsB regime, the profiles of negative and positive charge virtually coincide and slight decompensation is found only near the surface and at the outer edge of the brush. Thus, numerical SCF calculations confirm the validity of LEA in the OsB regime. Evidently, LEA is applicable for the SB regime as well, since in this regime $r_{Ds} < r_D$. The detailed comparison of the distributions of all the components in the brush indicates good agreement between the analytical predictions and the numerical results [38].

3.9. EFFECT OF IONS CHARGE

The above dependences for the brush thickness were obtained for the case of monovalent ions. At high ionic strength, the equation (28) is expected to be valid for multivalent ions as well: in the SB regime the brush structure is determined by the value of the overall ionic strength I of the bulk solution. However, at low and intermediate ionic strength of the solution, the polyelectrolyte brush structure is expected to be sensitive to the sign and the valency of added ions [40]. Let us consider a polyelectrolyte brush immersed in a pure water. The brush is assumed to be in the OsB regime in the absence of added ions, i.e. be swollen by the gas of own counterions (protons). Addition of strongly charged counterions with the charge $q_+ > 1$ results in substitution of protons by the added counterions. Since the number $\alpha N/q_+$ of mobile counterions that is necessary to compensate the polyion charge αN, is smaller by the factor of q_+ than that of protons, the osmotic pressure of counterions gas diminishes, and the brush thickness decreases as

$$H_{rms} \simeq \frac{aN\alpha^{1/2}}{q_+^{1/2}} \tag{49}$$

As follows from equation(49), the effect is noticeable even in the case of bivalent counterions ($q_+ = 2$). In this case the brush thickness decreases by the factor of $2^{1/2}$. Contraction of the brush is expected to be even more strong due to possible complex formation between negatively charged polyions and the added counterions. (We do not consider this effect here). It should be emphasized that the exchange of protons and added counterions takes place at very low concentrations, $C_s << C_H^*$, where $C_H^* = 10^{-7}$ is the concentartion of protons in pure water. Thus, even extremely small addition of multivalent counterions in the bulk solution affects strongly the brush structure in the OsB regime. It should be also noted, that substitution of protons by the added counterions takes place for monovalent counterions (Na^+) as well. For a polyanion brush in pure water this occurs at $C_s \approx C_H^*$. However, due to the same charge of H^+ and Na^+, the concentration of counterions in the brush (osmotic pressure of the counterion gas) remains unaffected by the counterion exchange and the brush thickness is still given by equation(22)(or equation(49) at $q_+ = 1$).

Addition of multivalent co-ions (plus protons as compensating counterions) in the bulk solution does not affect noticeably the brush structure. At low ionic strength the brush is swollen by the gas of protons (OsB regime), whereas at high ionic strength, when the concentration of co-ions is close to that in the bulk solution (SB regime), the brush thickness is given by equation(28). Only for highly charged (polymeric) co-ions, the new regime localized between the OsB and SB regimes, appears [40]. Here, due to high negative charge of co-ions , their concentration in the brush is still much smaller than that in the bulk solution, whereas the concentrations of protons inside and outside the brush are already close. This results in brush contraction,

$$H_{rms} \simeq \frac{N}{msq_- C_s} \tag{50}$$

where $q_- >> 1$ is the charge of a co-ion. However, for low molecular weight co-ions this regime collapses into the boundary region between the OsB and SB regimes and is hardly detectable. The numerical SCF calcultions [40] are in qualitative agreement with this picture.

4. Annealed Polyelectrolyte Brush

Let us now pass to consideration of a polyelectrolyte brush with an annealed distribution of charges on the grafted chains. Evidently, annealing of charge distribution on the polyions can not affect the general physics of brush behavior. One can distinguish the OrS, PB and OsB regimes in a similar way as it was done above for quenched brushes. In particular, depending on the ratio of the screening length λ and the brush thickness H, an annealed brush is found either in the OsB regime ($\lambda/H << 1$) or in the PB and OrS regimes ($\lambda/H >> 1$). Correspondingly, the dependences (22),(18) and (6) for the brush thickness remain valid for annealed brushes as well. However, due to the fact that the degree of chain ionization α is not constant any more, but is determined by the local pH (concentration of protons c_H) according to equation(2), the value of α becomes dependent of the parameters of the system (s, N, C_s, etc). Since the distributions of mobile ions (and protons, in particular) depend on the ratio λ/H, one expects different scaling dependences for α at $\lambda/H << 1$ (OsB regime) and $\lambda/H >> 1$ (PB and OrS regimes). In order to find these dependences we use the box model of polyelectrolyte brushes.

4.1. INDIVIDUAL POLYIONS AND PINCUS BRUSH

As was already discussed above, at sparse grafting of polyions providing $\lambda/H >> 1$, the counterion cloud is spread above the surface and forms together with the grafted polyions a double electrical layer (planar "capacitor") of the thickness $\lambda \simeq s/\alpha N l_B$. The layer is assumed to be electroneutral as a whole. In the absence of added counterions $c_H \simeq \alpha N/s\lambda \simeq \alpha^2 N^2 l_B/s^2$. However, if other monovalent ions are added in the bulk solution and c_+ and c_- are the total concentrations of the added ions in the double layer,

$$c_+ \approx \frac{\alpha N}{s\lambda} + c_- \approx \frac{\alpha N}{s\lambda} \tag{51}$$

The concentration of protons in the double layer is given by the Donnan rule,

$$c_H = c_+ \frac{C_H}{C_s + C_H} \simeq \frac{\alpha^2 N^2 l_B}{s^2} \frac{C_H}{C_H + C_s} \tag{52}$$

where C_H is the concentration of protons in the bulk solution (for pure water $C_H = C_H^* = 10^{-7}$) and C_s is the concentration of the added ions. Using equations (2) and (52), we obtain the equation for α in the PB and OrS regimes [41],

$$\frac{\alpha}{1-\alpha} \simeq \frac{K_D s^2}{\alpha^2 N^2 l_B} \left(\frac{C_H + C_s}{C_H}\right) \tag{53}$$

For weakly charged polyions ($\alpha << 1$) equation(53) gives

$$\alpha \simeq \left[\frac{K_D s^2}{N^2 l_B} \left(\frac{C_H + C_s}{C_H}\right)\right]^{1/3} \tag{54}$$

Substituting expression (54) into equations (18) and (6), we obtain for an annealed brush thickness in the OrS regime

$$H \simeq a N^{5/9} \left(\frac{K_D s^2}{l_B}\right)^{2/9} \left(\frac{C_H + C_s}{C_H}\right)^{2/9} \tag{55}$$

whereas in the PB regime H scales as

$$H \simeq a N^{5/3} (l_B s)^{1/3} K_D^{2/3} \left(\frac{C_H + C_s}{C_H}\right)^{2/3} \tag{56}$$

The most spectacular effect of charge annealing manifests itself in the change of the molecular weight dependence of the brush thickness and the appearence of novel dependences of H on s and C_s. All this is due to the dependence of α on s, N and C_s according to equation(54).

An increase in α with increasing salt concentration C_s in the bulk solution can be understood from the ion exchange arguments. Addition of chemically different ions in the bulk solution leads to substitution of protons by the added counterions. Correspondingly, the concentration c_H of protons decreases and α increases. Similar effect is caused by an increase in s: for more sparse brushes the concentration of neutralizing counterions is smaller and α is, thus, higher. As for the molecular weight dependence of α, an increase in N leads to diminishing of the thickness λ of the double electrical layer and a corresponding increase in the concentration of protons in the brush (a decrease in α). This results in more weak dependences $H = H(N)$ for the annealed polyions: $H \sim N^{5/3}$ (instead of $H \sim N^3$ for a quenched system in the PB regime) and $H \sim N^{5/9}$ (instead of $H \sim N$ for an individual quenched polyion in the OrS regime).

4.2. OSMOTIC BRUSH

Similarly to the case of quenched brushes, at $\lambda/H << 1$ an annealed brush is found in the "osmotic" regime, where the neutralizing counterions are localized mainly inside the brush. Under the conditions of the OsB regime, $H \simeq aN\alpha^{1/2}$ (equation(22)) and

$$c_+ \approx \frac{\alpha N}{Hs} \simeq \frac{\alpha^{1/2}}{as} \tag{57}$$

The Donnan rule gives

$$c_H = c_+ \frac{C_H}{C_H + C_s} \approx \frac{\alpha^{1/2}}{as} \frac{C_H}{C_H + C_s} \tag{58}$$

and, using equations(2)and (58), we get the equation for α in the OsB regime,

$$\frac{\alpha}{1-\alpha} \simeq \frac{asK_D}{\alpha^{1/2}} (\frac{C_H + C_s}{C_H}) \tag{59}$$

For weakly charged polyions ($\alpha << 1$) equation (59) gives

$$\alpha \simeq [asK_D(\frac{C_H + C_s}{C_H})]^{2/3} \tag{60}$$

and the brush thickness scales as

$$H \simeq aN(asK_D)^{1/3}(\frac{C_H + C_s}{C_H})^{1/3} \tag{61}$$

As follows from equations(60) and (61), annealing of charge on the polyions in the OsB regime results in novel scaling dependences of H on s and C_s.

Whereas for quenched brushes H is independent of both s and C_s (equation(22)), in annealed brushes H increases with s and C_s increasing. The origin of this effect is the same as discussed earlier: an increase in s or C_s leads to diminishing of the concentration of protons in the brush and a corresponding increase in α and H. Such type of behavior is quite opposite to the conventional behavior of neutral brushes, where H always decreases with increasing s.

4.3. SALTED BRUSH

With increasing salt concentration $C_s >> \alpha^{1/2}/as$, the salt dominance regime SB takes place. In the SB regime the concentrations of all the ions in the brush are nearly the same as in the bulk solution, and $c_H \approx C_H$. Thus, the degree of ionization of the grafted polyions α is close to that in the bulk solution,

$$\frac{\alpha}{1-\alpha} \approx \frac{K_D}{C_H} \tag{62}$$

and at $\alpha << 1$ the brush thickness H scales in this regime as

$$H \simeq aN(\frac{K_D}{C_H})^{2/3}(asC_s)^{-1/3} \tag{63}$$

As follows from equation(63), at high ionic strength of the solution we recover the conventional dependence $H \sim (C_s s)^{-1/3}$ found for quenched brushes[14].

The annealing effects in polyelectrolyte brushes lead, thus, to an unusual, nonmonotonic dependence for the brush thickness on the ionic strength I of the solution. At low I the degree of polyion ionization is determined by the local $pH' = -\log c_H$ in the brush, which is much lower than the $pH = -\log C_H$ of the bulk solution. This local pH' can be shifted by the addition of small amounts of chemically different ions in the bulk solution. For example, the addition of salt molecules $NaCl$, dissociating into Na^+ and Cl^- ions, leads to substitution of H^+ by Na^+ in the brush, and the local pH' increases, equations(52) and (58). As shown by Birshtein et al. [42], addition of base molecules $NaOH$, dissociating into Na^+ and OH^- ions, leads to a more rapid increase in α (and H) due to an additional increase in the pH of the bulk solution. At high ionic strength, when α is not sensitive to the value of I and is determined only by the pH of the bulk solution, the effect of added ions reduces to conventional screening of the electrostatic interactions in the brush and H diminishes, equation(63).

The maximum on the dependence $H = H(C_s)$ is expected to be located near the boundaries of the OsB - SB or the PB - SB regimes. In the first case, when the brush passes into the SB regime from the OsB regime (relatively dense brushes), the maximum is well pronounced. In the second case , when the transition to the SB regime occures from the PB regime (relatively loose brushes), the maximum transforms into a plato [41]. This plato is expected to exist in the range of C_s providing the value of the Debye screening length $H << r_{Ds} << \lambda$. The existence of a maximum on the $H(I)$ dependences has been also predicted by recent SCF calculations [37,43].

Evidently, at very high values of I , when the non-electrostatic, volume interactions dominate over the screened electrostatic interactions, the brush passes to the quasineutral regimes NB or NC.

4.4. PROFILES OF POLYMER UNITS AND CHARGES

The equilibrium distributions of all the components in the system can be obtained using the analytical SCF approach described above. The effect of charge annealing can be taken into account by introducing additional terms into the interaction part of the free energy functional,

$$\mathcal{F} = \sum_k c_k(\log c_k + \mu_k^*) + \mathcal{F}_\alpha \tag{64}$$

where

$$\mathcal{F}_\alpha = c[\alpha \log \alpha + (1-\alpha)\log(1-\alpha) + \alpha\mu_A^* + (1-\alpha)\mu_{AH}^*] \tag{65}$$

Here $\alpha(x)$ is an unknown profile of the immobilized charge on the polyions and μ_A^* and μ_{AH}^* are the standard chemical potentials of the dissociated and nondissociated AH groups on the grafted chains, which determine the dissociation constant K_D as

$$a^3 K_D = \exp(\mu_{AH}^* - \mu_H^* - \mu_A^*) \tag{66}$$

For quenched brushes

$$s\int_0^H \mathcal{F}_\alpha dx = const \tag{67}$$

and, thus, incorporation of $\mathcal{F}_\alpha$ is not important. For an annealed brush, $\mathcal{F}_\alpha$ takes into account the migration of charges on the grafted chains, and it is necessary to incorporate it into the free energy functional.

Variation of the functional of the free energy (equation(34)), incorporating a new unknown function $\alpha(x)$, leads to equations (35)-(36) for distributions of the mobile ions and polymer units and an additional equation

$$\delta \mathcal{F}/\delta\alpha = 0 \tag{68}$$

which determines the profile of ionization $\alpha(x)$.

For sparse brushes, where volume interactions are negligible, the profile of ionization is given by[44]

$$\alpha(x) = 1 - \frac{C_H}{K_D + C_H} \exp[b^2(H^2 - x^2)] \tag{69}$$

whereas the profile of polymer units can be represented as

$$c(x) = C_+[\frac{K_D}{C_H}\frac{1-\alpha(x)}{\alpha^2(x)} - \frac{C_H}{K_D}\frac{1}{1-\alpha(x)}] \tag{70}$$

Here, as before, H is the outer boundary of the brush, which can be determined from the normalization condition (32). The profiles of the mobile ions are obtained from the Donnan rule (equation(39)) and the condition of local electroneutrality (equation(38)), where $\alpha(x)$ is now given by equation(69).

Figure 6 demonstrates the ionization profiles $\alpha(x)$ at various concentrations of salt in the bulk solution calculated according to equations (69)

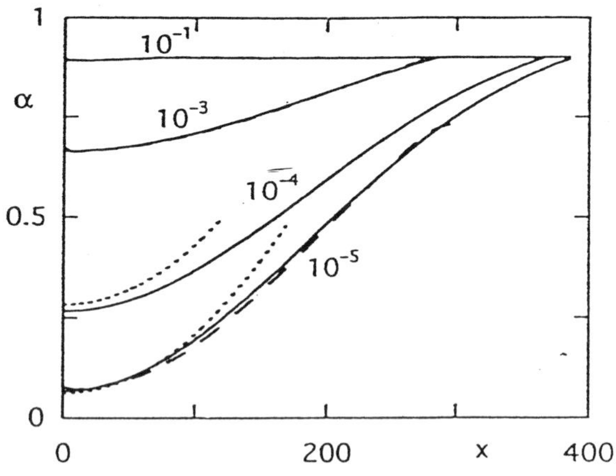

Figure 6. Degree of ionization α as a function of the distance x at different salt concentrations C_s (indicated), $N = 500$, $s/a^2 = 500$ and $K_D = 10^{-7}$ [44].

and (32). Dashed lines indicate the results of the numerical SCF model. Dotted lines show the asymptotical behavior of $\alpha(x)$

$$\alpha(x) \approx \alpha(0) + b^2 x^2 \tag{71}$$

that follows from equation(69) at $x << H$ [44].

Figure 7 shows the profiles of polymer units calculated according to equation(70) for the same values of parameters as in Figure 6. Dashed lines indicate the results of the numerical SCF model. Note that at high

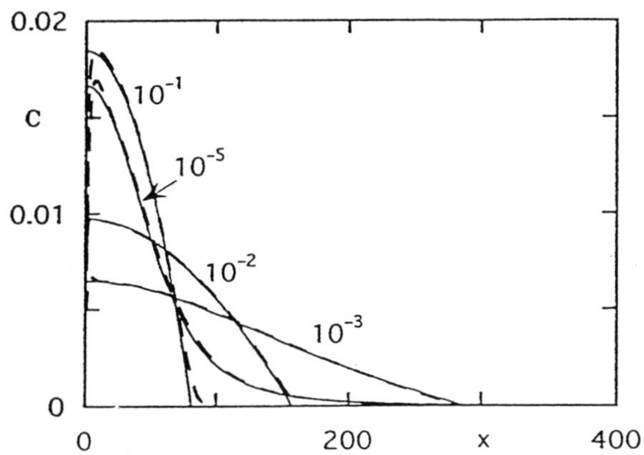

Figure 7. Density profiles of a polyacid brush at different salt concentrations C_s (indicated), parameters as in Figure 6 [44].

salt concentrations when the electrostatic repulsion between monomers is strongly screened, the non-electrostatic (volume) interactions become significant in calculating the polymer profile (see [44] for more details).

As demonstrated by Lyatskaya et al. [44], the predicted scaling asymtotical dependences (61) and (63) for the average brush thickness in the OsB and SB regimes are confirmed by the analytical SCF theory. In the OsB regime

$$H_{rms} \approx aN\left[\frac{8(2+\pi)}{9\pi^4}\frac{asK_D C_+}{C_H}\right]^{1/3} \tag{72}$$

whereas in the SB regime

$$H_{rms} \approx aN\left(\frac{2}{\pi^2}\frac{K_D^2}{asC_+C_H^2}\right)^{1/3} \tag{73}$$

An interesting observation following from the analytical SCF theory [44], concerns the behavior of the overall thickness H and the average brush thickness H_{rms}. Whereas H_{rms} (or any other moment of the polymer profile) exhibits nonmonotonic dependence on C_s, the overall brush size H (i.e. the entire width over which the end points of the grafted chains are found) decreases monotonically with C_s increasing, Figure 8. Such difference between the average and the limiting brush size can not be indicated by the

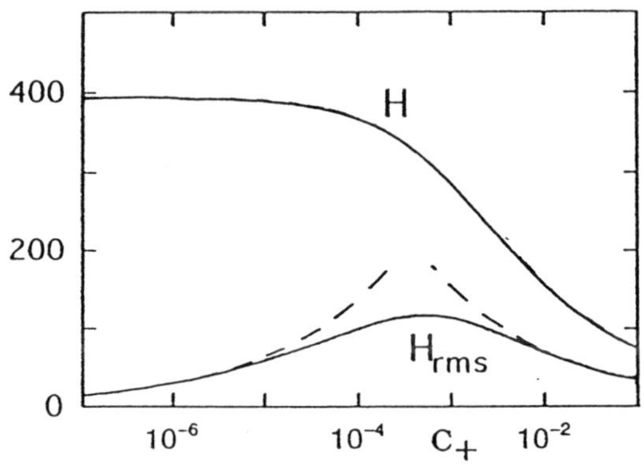

Figure 8. Brush thicknesses H and H_{rms} as functions of salt concentration $C_s \approx C_+$, parameters as in Figure 6. Dashed lines indicate the asymptotical dependences (72) and (73) [44].

simple box model, which provides scaling asymptotics for the averaged brush parameters. It is interesting to note, that for quenched polyelectrolyte brushes as well as for neutral brushes, both thicknesses H and H_{rms} always vary in a similar manner. An annealed brush is a first example, where the qualitative difference in the behavior of H and H_{rms} is predicted.

5. Concluding Remarks

Using the scaling and SCF analysis of the charged brushes, we demonstrated that the annealing (equilibration) of charge on the grafted polyions can noticeably affect the brush structure and properties. At low ionic strength I of the solution, annealing effects can lead to "abnormal" behavior: an increase in the brush thickness H with increasing area per chain s and ionic strength I. The reason for this is an increase in the local brush

pH' and a corresponding increase in the degree of chain ionization α.

The considered annealing effects are of rather general nature. They are possible in various polymer systems, such as polyelectrolyte gels, charged micelles, biological cells, etc. One can formulate the general requirements for the systems where the described annealing effects can be expected:

- The system can be subdivided into two subsystems separated by a "membrane", which is permeable for the mobile ions and the solvent, but is impermeable for the immobilized charge. The two subsystems are in thermodynamic equilibrium with each other;
- Presence of the immobilized charge within one of the subsystems. The value of the immobilized charge is not fixed, but is determined by the equilibrium requirements for both subsystems;
- The "membrane" is flexible and permits variation of the subsystem dimensions. The geometry and the dimensions of the subsystem with immobilized charge provide its electroneutrality.

Though particular scaling laws will be different for different systems, the sign of the predicted annealing effects is expected to be invariant to the physical nature of the system.

Acknowledgement

The material of this lecture comprises the results obtained in collaboration with many people. I am deeply grateful to all coauthors of the joint publications: Dr. O.V. Borisov, Prof. T.M. Birshtein, Dr. Yu.V. Lyatskaya, Dr. R. Israels, Dr. F.A.M. Leemakers and Prof. G.J. Fleer. I am also grateful to Prof. A.C. Balazs at the University of Pittsburgh (USA) for her hospitality and critical reading of the manuscript. The financial support from the Materials Research Center, funded by AFOSR, through grant number F49620-95-1-0167 is acknowledged.

References

1. Alexander, S.J. (1977) Adsorption of chain molecules with a polar head. A scaling description, *J.Physique* **38**, 983-987.
2. de Gennes, P.G. (1980) Conformations of polymers attached to an interface, *Macromolecules* **13**, 1069-1075.
3. Birshtein, T.M., and Zhulina, E.B. (1983) Conformations of polymer chains grafted to impermeable planar surface, *Polymer Science USSR* **25**, 2165-2174.
4. Borisov, O.V., Zhulina, E.B., and Birshtein, T.M. (1988) Diagram of state and collapse of grafted chains layers, *Polymer Science USSR* **30**, 772-779.
5. Auroy, P., Auvray, L., and Leger, L. (1991) Structure of end-grafted polymer layers: a small-angle neutron scattering study, *Macromolecules* **24**, 2523-2528.
6. Auroy, P., Auvray, L., and Leger, L. (1991) Characterization of the brush regime for grafted polymer layers at the solid-liquid interface, *Phys.Pev.Lett.* **66**, 719-722.
7. Auroy, P., Auvray, L., and Leger, L. (1990) The study of grafted polymer layers by neutron scattering, *J.Phys.:Condens.Matter* **2**, SA 317-321.
8. Karim, A., Satija, S.K., Dauglas, J.F., Ankner, J.F., and Fetters, L.J. (1994) Neutron reflectivity study of the density profile of a model end-grafted polymer brush: influence of solvent quality, *Phys.Rev.Lett.* **73**, 3407-3410.
9. Halperin, A., Tirrell, M., and Lodge, T.P. (1992) Tethered chains in polymer microstructures, *Advances in Polymer Science* **100**, 31-68.
10. Micklavic, S.J., and Marcelia, S.J. (1988) Interaction of surfaces carrying grafted polyelectrolytes, *J.Phys.Chem.* **92**, 6718-6722.
11. Misra, S., Varanasi, S., and Varanasi, P. (1989) A polyelectrolyte brush theory, *Macromolecules* **22**, 4173-4179.
12. Pincus, P. (1991) Colloid stabilization with grafted polyelectrolytes, *Macromolecules* **24**, 2912-2919.
13. Ross, R., and Pincus, P. (1992) The polyelectrolyte brush: poor solvent, *Macromolecules* **25**, 2177-2183.
14. Borisov, O.V., Birshtein, T.M., and Zhulina, E.B. (1991) Collapse of grafted polyelectrolyte layer, *J.Physique II* **1**, 521-526.
15. Zhulina, E.B., Borisov, O.V., and Birshtein, T.M. (1992) Structure of grafted polyelectrolyte layer, *J.Physique II* **2**, 63-74.
16. Semenov, A.N. (1985) Contribution to the theory of microphase layering in block copolymer melts, *Sov.Phys. JETP* **61**, 733-742.
17. Skvortsov, A.M., Gorbunov, A.A., Pavlushkov, I.V., Zhulina, E.B., Priamitsyn, V.A., and Borisov, O.V. (1988) Structure of dense-grafted polymer monolayers, *Polymer Science USSR* **30**, 1706-1715.
18. Zhulina, E.B., Priamitsyn, V.A., and Borisov, O.V. (1988) Structure and conformational transitions in grafted chain layers: new theory, *Polymer Science USSR* **31**, 205-215.
19. Zhulina, E.B., Borisov, O.V., and Brombacher, L. (1991) Theory of planar grafted chain layer immersed in solution of mobile polymer,

Macromolecules **24**, 4679-4690.
20. Milner, S.M., Witten, T.A., and Cates, M.E. (1988) Theory of the grafted polymer brush, *Macromolecules* **21**, 2610-2619.
21. Milner, S.M. (1991) Polymer brushes, *Science* **251**, 905-914.
22. Fleer, G.J., Cohen Stuart, M.A., Scheutjens, J.M.H.M., Cosgrove, T., and Vincent, B. (1993) *Polymers at Interfaces*, Chapman and Hall, London.
23. Odijk, T. (1978) Electrostatic persistence length and its relation to a unified theory of polyelectrolytes in solution, *Polymer* **19**, 989-990.
24. de Gennes, P.G., Pincus, P., Velasko, R.M., and Brochard, F.J. (1976) Remarks on polyelectrolyte conformation, *J.Physique* **37**, 1461-1498.
25. Khokhlov, A.R., and Khachaturian, K.A. (1882) On the theory of weakly charged polyelectrolytes, *Polymer* **23**, 1742-1750.
26. de Gennes, P.G. (1985) *Scaling Concepts in Polymer Physics*, Kornell University Press, Ithaca NY.
27. Israelachvili, J.N. (1985) *Intermolecular and Surface Forces*, Academic Press, London.
28. Borisov, O.V., Zhulina, E.B., and Birshtein, T.M. (1994) Diagram of the states of a grafted polyelectrolyte layer, *Macromolecules* **27**, 4795-4803.
29. Pincus, P. (1976) Excluded volume effects and stretched polymer chains,*Macromolecules* **9**, 386-388.
30. Klushin, L.I. (1992) Lateral segregation in anchored polymeric monolayers, Preprint.
31. Williams, D.R.M. (1993) Grafted polymers in bad solvents: octopus surface micelles, *J.Physique II* **3**, 1313-1318.
32. Auroy, P., and Auvray, L. (1993) Private communication.
33. Watanabe, H., Patel, S.S., Argiller, J.F., Patsonage, E.E., Mays, J., Dan-Bradon, N., and Tirrell, M. (1992) , Manipulating solid-surface properties with polymer agents, *Mat.Res.Soc.Proc.* **249**, 255-265.
34. Guenoun, P., Schalchli, A., Sentenac, D., Mays, J.W., and Benattar, J.J. (1995) Free standing black films of polymers: a model of charged brushes in interaction, *Phys.Rev.Lett.* **74**,3628-3631.
35. Srivastava, D., and Mathukumar, M. (1994) Interpenetration of interacting polyelectrolytes, *Macromolecules* **27**, 1461-1465.
36. Stevens, M.J., and Kremer, K. (1995) The nature of flexible linear polyelectrolytes in salt free solution: a molecular dynamics study, *J.Chem.Phys.* **103**, 1669 - 1690.
37. Misra, S., and Varanasi, S. (1991) A model for the permeation characteristics of porous membranes with grafted polyelectrolyte brushes, *J.Colloid Interface Sci.* **146**, 251-275.
38. Israels, R., Leermakers, F.A.M., Fleer, G.J., and Zhulina, E.B. (1994) Charged polymer brushes: structure and scaling relationships, *Macromolecules* **27**, 3249-3261.
39. Bohmer, M.R., Evers, O.A., and Scheutjens, J.M.H.M. (1990)

Weak polyelectrolytes between two surfaces: adsorption and stabilization, *Macromolecules* **23**, 2288-2301.
40. Zhulina, E.B., Israels, R., and Fleer, G.J. (1994) Theory of planar polyelectrolyte brush immersed in solution of asymmetric salt, *Colloids Surf.A: Physicochem.Eng.Asp.* **86**, 11-24.
41. Zhulina, E.B., Birshtein, T.M., and Borisov, O.V. (1995) Theory of ionizable polymer brushes, *Macromolecules* **28**, 1491-1499.
42. Birshtein, T.M., Zhulina, E.B., and Borisov, O.V. (1995) Polyelectrolyte brushes: fixed charge distribution and ionizable brushes, *Polymer Science*, in press.
43. Israels, R., Leermakers, F.A.M., and Fleer, G.J. (1995) On the theory of grafted weak polyacids, *Macromolecules* **27**, 3087-3093.
44. Lyatskaya, Yu.V., Leermakers, F.A.M., Fleer, G.J., Zhulina, E.B., and Birshtein, T.M. (1995) Analytical self-consistent-field model of weak polyacid brushes, *Macromolecules* **28**, 3562-3569.

STRUCTURE AND INTERACTIONS IN TETHERED CHAINS

Polymeric Micelles

ALICE P. GAST
Department of Chemical Engineering
Stanford University
Stanford, CA 94305

1. Introduction

Over the past fifteen years there has been considerable interest in polymeric stabilization of colloidal suspensions [1–3]. This process, termed steric stabilization, requires the attachment of a layer of strongly bound polymer to the colloidal surface to prevent aggregation due to van der Waals attractions. The desire to produce a thick, essentially irreversibly bound polymer layer has led to the investigation of polymers attached to surfaces, either physically or chemically, by one end only. It is these chains anchored by one end that have come to be known as "tethered chains".

Block copolymers in a selective solvent have a tendency to self-assemble at surfaces and into micelles [4–7]. This interfacial activity makes them attractive as colloidal dispersants and surface modifiers prompting many studies of block copolymer adsorption on solid surfaces, [8–13] forces between tethered layers [14–17] and the self-assembly of block copolymers into micelles [4, 18–20]

Theoretical efforts in self-consistent field models [21–26], scaling concepts [27–29] and simulations [30, 31] have contributed to our understanding of tethered chains. This large body of work has given us a reasonable view of the structure and interactions of polymers tethered to surfaces by one end only. The structure of chains tethered at high grafting densities differs from that of free polymer chains because they assume highly stretched configurations. In the course of the considerable work on adsorbed block copolymers it has become clear that the influence of the topology of the adsorbing substrate could have a substantial influence on the structure and resulting interactions in the adsorbed layer. Block copolymers have been adsorbed onto smooth mica surfaces [9, 15], onto small colloidal grains [3], inside random porous media [12, 13] and inside rhombohedral pores [32, 33].

S.E. Webber et al. (eds.), Solvents and Self-Organization of Polymers, 259–280.

When chains are tethered to large curved surfaces such that the radius of curvature is larger than the tethered chain length, they resemble planar polymer brushes. The structure of planar polymer brushes, described by self-consistent mean field (SCF) theory, adopts a parabolic profile [34]. This result has been supported by scaling analysis [27,29], numerical SCF calculations [22–26], molecular dynamics simulations [35], and Monte Carlo simulations [30,31]. The parabolic profile is also consistent with the layer structure found from neutron reflectivity [9,10], optical reflectometry [8] and small angle neutron scattering (SANS) [11–13] experiments.

Chains tethered to a small spherical core have a concentration profile that differs qualitatively from that of planar brushes. In the limit of a star-polymer configuration where chains are tethered to a single point, scaling analysis [36] provides a density profile that decays from the center as $\rho(r) \approx (r/b)^{1/\nu-3}$ where ρ is the segment density, ν is the Flory exponent, and b is the polymer segment length. This power-law chain profile has been found in molecular dynamics simulation [37] and self-consistent field lattice calculations [38–40] and matches small angle x-ray scattering on polymeric micelles with very small cores [19,20].

The important influence of the curvature of the adsorbing surface was recognized and first investigated for cylindrical surfaces by self-consistent field models and simulations [41,42]. The effect of attachment to spherical surfaces has more recently been illustrated by self-consistent field calculations [43–46]. These calculations show that the chain profiles evolve from the planar parabolic "brush" profile to the power-law "star" profile and have been supported with Monte Carlo simulations [47]. Thus far, there have been few experimental tests of these calculations.

We use the self-assembling nature of diblock copolymers in selective solvents to produce monodisperse spherical micelles. These micelles provide a valuable model system to study chains tethered to curved interfaces. They provide a means to alter the topology of the anchoring interface as well as the tethered chain length through variations in the diblock copolymer degree of polymerization. A polymer chain layer that is thin compared to the core radius has more highly stretched chains than those in larger layers. Intuitively, one also expect a change in the nature of the interactions between these micellar structures. The micelles form very monodisperse suspensions that are readily studied with scattering experiments. Our work on the ordering of polymeric micelles into cubic arrays demonstrates the profound influence of the micellar structure on their interactions [48].

In this chapter we review the theoretical and experimental investigations of tethered chain structure and interactions. We focus on the influence of surface curvature on tethered chain behavior and we illustrate the two limits of planar, brush structures and highly curved, star polymer geometries. We then present a study of a model system of tethered chains in the form of polymeric micelles.

2. Theoretical Developments

2.1. SCALING MODELS

Some of the most useful models of complex polymeric structures arise from the application of scaling concepts to the description of the polymer concentration profiles and free energies [49]. Through these models we develop an intuitive picture of the structure of tethered chains [29, 50, 51]. Largely restricted to extremely long polymer chains in very good solvents, scaling models are complemented by more detailed mean-field calculations and molecular simulations.

2.1.1. *Planar Layers*

The pioneering view of the structure of adsorbed block copolymers began with the scaling concepts for chains tethered to planar surfaces developed by Alexander [27] and deGennes [29]. In this model the chains are anchored to a surface at a dimensionless surface density σ such that the distance between anchoring points is $D = b/\sqrt{\sigma}$. All chain ends are assumed to exist at the outer edge of the polymer layer and the concentration profile takes on a step-function form. The layer thickness $L \approx Nb\sigma^{1/3}$ varies linearly with the degree of polymerization of the tethered chain, N and depends on the cube root of the surface density. This scaling result has been tested in numerous experiments [7] and generally holds for systems where the surface density remains constant as the block length, N changes [6, 7, 52].

The interaction between two such layers is simply calculated from the osmotic repulsion due to concentration changes as two layers approach within a separation z

$$\frac{u(z)}{kT} = Nw\sigma b^{2/3}(\frac{z}{2L} - \frac{2L^2}{z^2}) \tag{1}$$

where $w = b^3(1 - 2\chi)$ is the excluded volume parameter defined in terms of the Flory interaction parameter, χ.

2.1.2. *Curved Interfaces*

Early analyses of chain profiles on highly curved spherical surfaces came from the study of polymeric micelles. Initially, a uniform core-corona model provided a useful approximation to the micellar structure for further thermodynamic analysis [53–57]. Recent experimental studies on micelles and star-shaped polymers [6,19,20,58–60] confirm that such a step-concentration

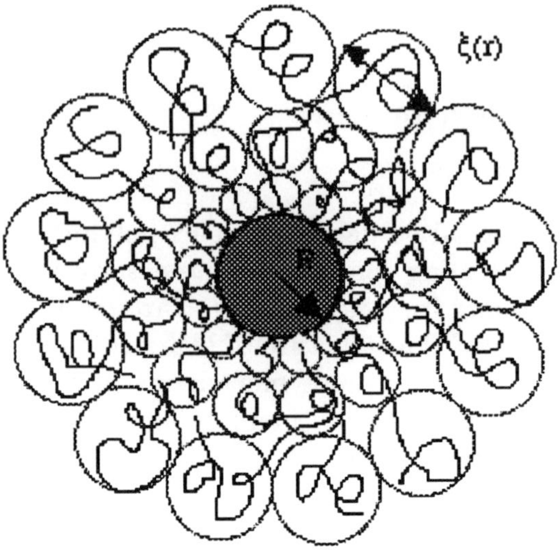

Figure 1. Star polymer model for chains tethered to highly curved surfaces. Chains are represented as a string of close packed blobs of size $\xi(r)$ extending radially from the core of the star.

profile is not accurate and theoretical models [38,40,61,62] have evolved to include the details of the tethered chain profile.

Daoud and Cotton provided the scaling analysis of chains tethered to a small spherical core [36]. Consideration of the tethered arms as confined to a string of close-packed blobs provides a star-like polymer density profile that decays from the center as $\rho(r) \approx (r/b)^{1/\nu-3}$ as illustrated in Figure 1. This power-law profile is supported by molecular dynamics simulation [37], Monte Carlo simulations [47], and lattice calculations [38,39] and describes polymeric micelles having small insoluble core blocks [20,19].

Leibler and Pincus [63] used Flory-de Gennes arguments to predict the repulsive intermicellar interaction potentials for two overlapping uniform coronas. Witten and Pincus [7,62] proposed an interaction potential between spherical particles of radius R carrying tethered polymer chains. Based upon the Daoud-Cotton description of a star polymer, this interaction potential scales with the free energy of a tethered chain layer

$$\frac{u(r)}{kT} \sim 4\pi R^3 \sigma^{3/2} \ln(r/R_{part}) \tag{2}$$

and has a logarithmic dependence of the potential on the separation scaled on the overall radius. The SCF calculations discussed below show that the interactions rise too steeply for the particles to attain the small separation distances implied in the scaling model. The scaling model does demonstrate that more densely tethered short chains will create a steeper repulsion than fewer longer chains.

2.2. SELF-CONSISTENT FIELD THEORY

The scaling models presented above are lacking in detail and are unable to include finite chain effects and polymer-solvent interactions. The development of the self-consistent mean field formalism by Edwards [64] provided the means to calculate polymer concentration profiles in a relatively tractable form. Each polymer is described as a freely jointed chain with N statistical segments of length b. In self-consistent mean field theories, the configurational statistics of the chain are modeled as a random walk within a mean field potential. The potential in turn depends upon the polymer segment distribution, thus requiring a self-consistent solution in analogy to other physical problems such as Hartree-Fock. The mean field is an average potential field due to the configurations of all other segments and ignores correlations between polymer segments and is thus applicable to dilute solutions where correlations are weak or concentrated solutions having short-ranged correlations. The fundamental quantity calculated in mean field studies is the polymer segment probability distribution function, $G(\mathbf{z}, \mathbf{z}'|s)$, which represents the probability of finding segment s at the position, $\mathbf{z}$, given that the segment started at the position, $\mathbf{z}'$. A number of methods are available to solve for G, including lattice methods and continuum approaches.

The polymer chain is modeled as a random walk in a potential field such that the probability distribution function, G, is governed by the forced diffusion equation

$$\frac{\partial G}{\partial s} - \frac{b^2}{6}\nabla^2 G + \omega(\mathbf{z})G = \delta(s)\delta(\mathbf{z} - \mathbf{z}') \tag{3}$$

where ω is the mean field potential due to the interaction of the chain with all other polymer chains.

The volume fraction profile of the tethered chain layer is determined from G as

$$\phi_A(\mathbf{z}) = \frac{\rho_A(\mathbf{z})}{\rho_{oA}} = \frac{C}{\rho_{oA}W}\int_0^N ds\; G(\mathbf{z},R|s)\int_R^\infty d\mathbf{z}'\; G(\mathbf{z},\mathbf{z}'|N-s) \qquad (4)$$

where

$$W = \int_R^\infty d\mathbf{z}'\; G(\mathbf{z}',R|N) \qquad (5)$$

is the configurational partition function representing the configurations of a tethered polymer chain, ρ_{oA} is the bulk number density of the polymer, and $\rho_A(\mathbf{z})$ is the local number density at $\mathbf{z}$. Equation 4 is the normalized product of two probability distribution functions, one describing a polymer segment of length s, that starts at the core surface and ends at the position $\mathbf{z}$ and one describing a polymer segment of length $N-s$ that starts in the bulk and also ends at the position $\mathbf{z}$. The normalization constant C ensures conservation of polymer segments. The solution is assumed to be locally incompressible and the volume fraction of the solvent is $\phi_S(\mathbf{z}) = 1-\phi_A(\mathbf{z})$.

The self-consistency in the equations arises from the direct relationship between the volume fraction profiles and the mean field potential, ω, through

$$\omega(\mathbf{z}) = \frac{\rho_{oS}}{\rho_{oA}}\{-\ln\phi_S(\mathbf{z}) + \chi[\phi_S(\mathbf{z}) - \phi_A(\mathbf{z})]\} \qquad (6)$$

where ρ_{oA} is the bulk number density of the chain segments, ρ_{oS} is the bulk number density of solvent molecules and χ is the Flory-Huggins interaction parameter [25, 65].

There have been two primary directions in self-consistent field theory: direct solution of the diffusion equation either numerically or analytically [24, 34, 64–67] and a lattice representation of the field problem developed by Scheutjens and Fleer [68–71]. Both approaches have been fruitful in the study of polymers tethered to flat and curved interfaces.

2.2.1. *Planar Layers*

The first mean-field descriptions of chains tethered to planar surfaces came from solution of the self-consistent field equations [23, 24] for end-anchored polymers with and without excluded volume interactions. From these studies it is clear that a step-function was not a completely accurate picture of the segemental conentration profile. More recently, Milner and coworkers and Zhulina and coworkers [34, 72] developed an analytical approach to the problem of highly stretched end-anchored chains having excluded volume. This solution provides a parabolic profile

$$\rho(z) = \frac{\pi^2}{8wN^2}(L^2 - z^2) \qquad (7)$$

whose thickness,

$$L = (\frac{2w}{\pi^2})^{1/2} N b \sigma^{1/3} \tag{8}$$

retains the same scaling as before. There is a marked contrast between these two profiles; however, the general scaling behavior remains unchanged.

This concentration profile results in a different repulsive force law for two tethered chain layers

$$\frac{u(z)}{kT} = \frac{5Nw\sigma b^{2/3}}{9}(\frac{z}{2L} + \frac{4L^2}{z^2} - \frac{32L^5}{5z^5}) \tag{9}$$

an expression that compares well with experimental measurements of force profiles [16].

One feature of the SCF calculations has been the elucidation of the distribution of chain ends in the layer. The SCF formalism allows the ends to exist anywhere in the brush, and in contrast to the restriction in the scaling model, there is a broad distribution of end segments throughout the brush.

More precise self-consistent field calculations of chains tethered to planar interfaces have been made with the lattice calculation approach [22,71]. In these detailed analyses it is clear that while the parabolic profile holds qualitatively, there are important differences at the inner and outer edges of the layer.

2.2.2. *Curved Interfaces*

In many systems of interest such as colloidal particles or polymeric micelles polymers are tethered to curved surfaces. Tethered chains with curved interfaces have largely been studied in the limiting case of star polymers where polymer chains are tethered to a small spherical core. Less work has concentrated on probing the curvature regime in between the limits of polymer brushes on planar interfaces and star polymers. Dan and Tirrell [43] have performed SCF calculations elucidating the transition between the parabolic polymer brush profile and the star polymer density profile. These results are supported by recent Monte Carlo simulation results [47]. Wijmans and Zhulina [44] compared lattice SCF calculations to analytic expressions for the density profiles with different curvatures, solvent conditions, and chain lengths. Li and Witten [73] have used a variational approach to solve the mean field equations analytically.

We have recently presented a theoretical investigation of the structure and interactions between tethered chain layers on curved interfaces [45,46]. We extended the continuum SCF treatment [64,65] of curved layers by Dan and Tirrell [43] to determine the pair interaction potential between curved layers.

We illustrate the results of these calculations with volume fraction profiles for $N = 200$ and $\sigma = 0.1$ under theta ($\chi = 0.5$) conditions with different core curvatures in Figure 2. These profiles demonstrate the large influence of core curvature on the structure of the tethered chain layer. The shape of the volume fraction profiles smoothly varies from the power law decay of a star polymer when $R/b = 4$ to the parabolic profile characteristic of polymer brushes at a flat interface when $R = \infty$, in good agreement with the profiles determined by Dan and Tirrell [43] and Wijmans and Zhulina [44]. The tail in the segment density distribution lacking in the analytic model is clearly illustrated in the calculations shown in Figure 2

The distribution of chain ends is also of interest in SCF calculations. As the curvature of the anchoring surface increases, the chain ends are excluded from a region near the interface [41,43]. The end-distribution also becomes narrower as illustrated in Figure 2.

Since the tethered layer density profiles qualitatively change with increasing curvature from a parabolic form to a power law decay, one expects the resulting interactions between these structures to change as well. The interactions between polymer layers at curved interfaces has been studied in a number of different contexts. Recently, Genz et al. [74] used the parabolic analytic SCF solution in combination with the Derjaguin approximation to model the structure and thermodynamics of sterically stabilized colloids. Detailed SCF lattice calculations performed by Wijmans, Leermakers, and Fleer [75] extended the lattice SCF formalism to two dimensions to calculate the interaction potential between small colloidal particles carrying polymer chains of 50 segments.

We used the concentration profile calculations described above to predict the pair interaction energy between two approaching curved layers. Given the volume fraction profiles from equation 4, the Helmholtz free energy, F, for the tethered chain layer [25] is

$$\frac{F}{kT} = -f \ln W - \rho_{oS} \int_R^\infty [-\ln \phi_S(\mathbf{z}') - \chi \phi_A^2(\mathbf{z}')] d\mathbf{z}' \tag{10}$$

where k is the Boltzmann constant, T is the temperature, and $f = 4\pi R^2 \sigma$ is the number of chains tethered to the surface. The first term results from the single chain partition function and the second term corrects for an overcounting of the segment-segment interactions arising in mean field theories [76]. Unlike the one dimensional isolated layer, the case of interacting spherical particles becomes a two dimensional problem. To reduce the computational load, we approximate the particle pair interaction potential, $u(r)$ from the free energy. We improve upon the traditional method for calculating the interaction energy between curved surfaces through the Derjaguin approximation [77] where the free energy per unit area from a

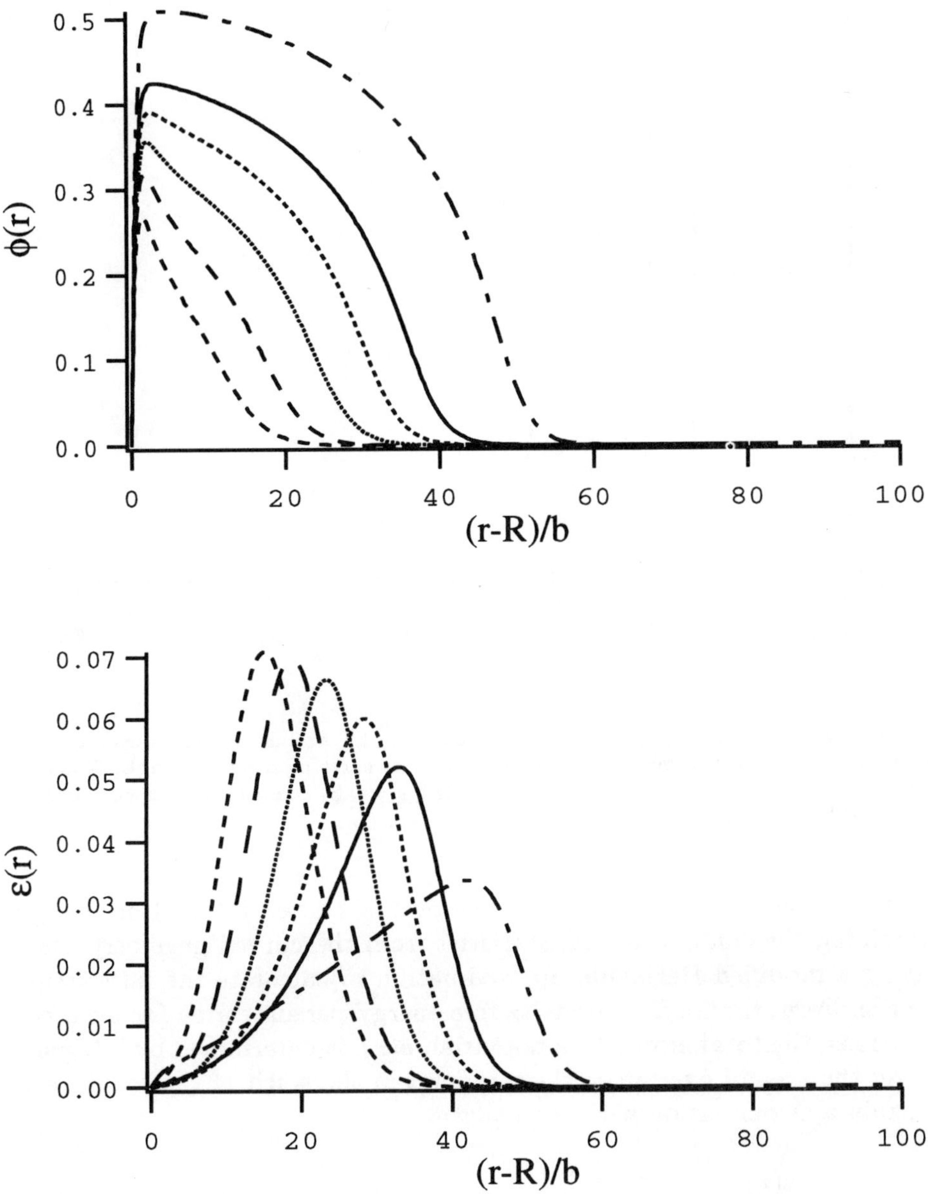

Figure 2. **Top: Volume fraction profiles for chains of length $N = 200$ tethered at a surface density $\sigma = 0.1$ under theta ($\chi = 0.5$) conditions. The core curvatures are $R/b = 4, 8, 16, 32, 64, \infty$ from left to right. Bottom: Segment end distributions for the same conditions.**

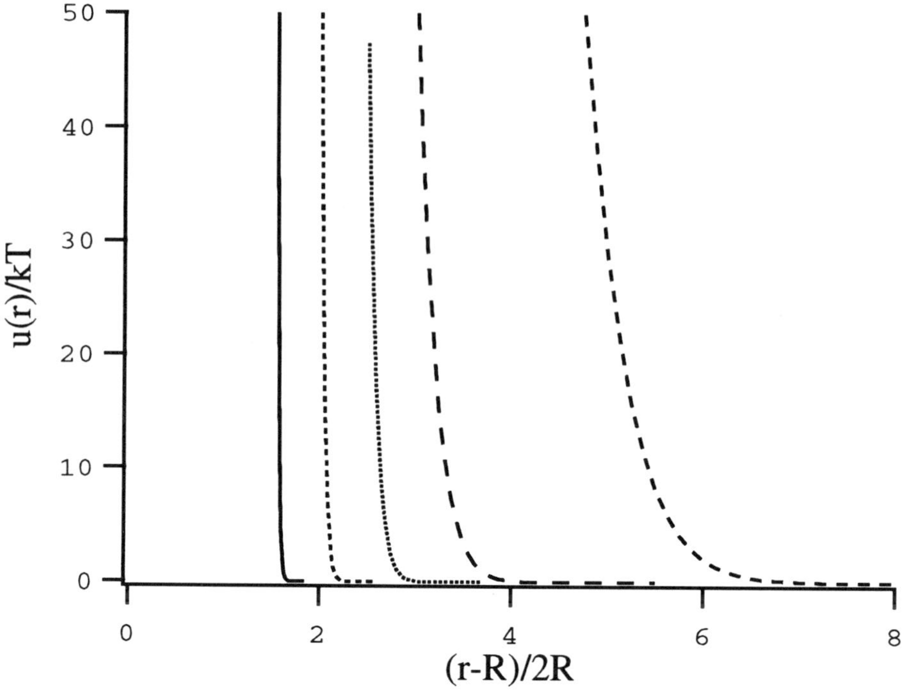

Figure 3. Pair interaction potential energies scaled on kT against center-to-center separation scaled on particle diameter for spherical particles carrying chains of length $N = 200$ tethered at a surface density $\sigma = 0.1$ under theta ($\chi = 0.5$) conditions. The core curvatures are $R/b = 4, 8, 16, 32, 64$ from left to right.

flat interface geometry is related to the overall curved interaction energy by retaining the configurational statistics from the curved geometry, then applying a modified Derjaguin approximation to calculate the interaction potential. From the local Helmholtz free energy per unit area for different separations, the total interaction potential, $u(r)$, is determined by integrating over the area of overlap at the midpoint in the spirit of the traditional Derjaguin approximation with the equation

$$\frac{u(r)}{kT} = 2\pi \int_{H^o}^{R+H^o} [R - H' + H^o] \frac{\Delta F_{mid}(H')}{kT} dH' \tag{11}$$

where $2H^o$ is the core surface to surface separation along the centerline and $2H'$ is the parallel surface separation off the centerline. The range of the interaction potential is then scaled on the particle core diameter, $2R$. We show the pair interaction potentials in Figure 3 for the curved layers depicted in Figure 2.

3. Experimental Studies: Chains Tethered to Curved Surfaces

3.1. STRUCTURE

The development of scaling models for star polymers [36] and the concurrent synthesis of star branched polymers [78] expanded our view of polymeric structure in highly curved, constrained geometries. Both the scaling analysis and simulations [37] pointed toward a rapidly decaying radial concentration profile. Micelles formed from diblock copolymers having small insoluble core blocks are good models for testing these predictions. One can create, by changing the solution chemistry or the block copolymer architecture, a variety of star-like structures for study via small angle scattering.

The primary measurements of diblock copolymer micelle structure are the radius of gyration from small angle light, x-ray and neutron scattering and the hydrodynamic radius determined from dynamic light scattering experiments. In addition, the decay of the intensity in the intermediate scattering regime can provide a picture of the decay of the concentration profile in a polymeric structure. The latter information clearly illustrates chain conformation in polymeric stars and micelles [19].

The radius of gyration can be calculated for a given scattering contrast from the star concentration profile

$$R_g^2 = \frac{\int_0^\infty r^4 \rho(r) dr}{\int_0^\infty r^2 \rho(r) dr} \tag{12}$$

where ρ is the scattering density for the core when r is less than R and follows the scaling prediction through the corona [6, 20].

Studying a solution of polystyrene-poly(ethylene oxide) in cyclopentane at the theta point for the polystyrene, we were able to observe the star-like structure in polymeric micelles. This system is highly sensitive to the presence of trace amounts of water [20, 79] allowing us to manipulate the micellar aggregation number by adding microgram amounts of water to our solutions. We found that the polystyrene chains forming the corona of the micelle followed the star scaling provided the chain length was long relative to the core radius. The radius of gyration for micelles from one of these polymers is presented in Table 1 for three aggregation numbers. We find good agreement with the star scaling for this system. We also use the ratio of the radius of gyration to the hydrodynamic radius as a measure of the structure of the micelle. Gaussian coils have a ratio R_g/R_h of 1.27 while stars at theta conditions are more compact having R_g/R_h=0.71. Addition of the solid core, $R_g/R_h = 0.775$, perturbs this value slightly resulting in ratios generally around 0.7-0.8 depending on the scattering contrast. In a system having relatively short polystyrene blocks, the chains were substantially stretched from the expected random walk configuration in the star geometry. This deviation from the star behavior is not surprising when the core radius becomes comparable to the length of the tethered chain.

In a series of polystyrene-polyisoprene diblock copolymers suspended in decane, a good solvent for polyisoprene, we further probe the influence of core curvature on polymer layer structure. These diblocks form monodisperse micelles having a dense core of polystyrene [48, 80]. Measuring the radius of gyration and micelle molecular weight of these micelles via small angle x-ray and neutron scattering, we find reasonable agreement with that predicted from a starlike micelle model as shown in Table 1. The radius of gyration for the largest micelles is underpredicted by the star model due to the transition toward a brush profile for these coronal chains.

TABLE 1. Comparison between radius of gyration for polymeric micelles measured via small angle x-ray and neutron scattering and that determined from star-like micelle model

System B/A	N_B	N_A	f	R_g nm	$R_{g,star}$ nm	R_g/R_h
PEO/PS dry	170	1730	17	27.5	24.0	0.79
PEO/PS wet	170	1730	77	30.8	35.4	0.70
PEO/PS wet	170	1730	103	37.4	37.8	0.75
PS/PI x-ray	96	281	5-10	10.8	12.2	0.71
PS/PI neutron	96	281	5-10	12.0	13.3	0.78
PS/PI x-ray	96	187	40	11.9	13.3	0.57
PS/PI neutron	96	187	40	15.0	14.9	0.70
PS/PI x-ray	288	187	150	16.5	16.8	0.58
PS/PI neutron	288	187	150	19.9	19.6	0.70
PS/PI x-ray	442	187	300	21.1	19.6	0.64

3.2. INTERACTIONS

In order to assess the influence of curvature on the interactions between layers of tethered chains, we studied the liquid structure produced in concentrated solutions of micelles. We create spherical micelles from a series of monodisperse deuterated polystyrene-polyisoprene (dPS/PI) diblock copolymers [45, 81] dissolved in n-decane a preferential solvent for polyisoprene. The molecular weights and the polydispersity indices for these polymers are listed in Table 2. We prepared solutions in mixtures of deuterated (Cambridge Isotopes, 99% deuteration) and protonated n-decane (Fisher Scientific, 99.8% purity) to match the scattering length density of either the d-polystyrene core or the polyisoprene corona. We measured the small

TABLE 2. Parameters for polystyrene-polyisoprene polymeric micelles studied via small angle neutron scattering. Core radii from the form factor are compared with space-filling spheres and the corona radius of gyration is compared with and that determined from self-consistent field calculations

N dPS/PI	f	R(Å) P(q)	R(Å) solid	R/b	σ $fb^2/4\pi R^2$	R_g^{corona} SANS(Å)	R_g^{corona} SCF(Å)
134/140	80	74	75	8.9	0.08	158.6	143.1
170/75	270	116	121	14.0	0.11	155.2	164.5
180/94	235	117	119	14.1	0.09	182.9	171.7
295/206	335	150	160	18.1	0.08	-	246.6
320/337	245	142	145	17.1	0.07	-	276.9
393/206	420	204	186	24.6	0.06	-	295.7
357/375	90	106	108	12.8	0.04	234.7	231.2
402/422	130	137	127	16.5	0.04	-	276.6

angle neutron scattering on beamline NG7 at the National Institute of Standards and Technology (NIST) [82]. Measurements were taken at an incident wavelength of λ=7.00 Å ($\Delta\lambda/\lambda = 0.11$ at full width, half maximum).

The resulting scattered intensities are a product of intra- and intermicellar interference

$$I(q) = nP(q)S(q) \tag{13}$$

where $P(q)$ is the form factor accounting for intramicellar interference, n is related to the number of scatterers, and $S(q)$ is the static structure factor. The static structure factor details the interference that arises from short range correlations in the micellar suspension and depends explicitly on the radial distribution function

$$S(q) = 1 + \rho \int e^{-i\mathbf{q}\cdot\mathbf{r}}[g(r) - 1]d\mathbf{r} = 1 + \rho\hat{H}(q). \tag{14}$$

We measure $P(q)$ from dilute micellar solutions where interactions between micelles are negligible and $S(q) = 1$ and then examine the static structure factor $S(q)$ at higher concentrations where interactions between micelles become important.

Like classical or simple monatomic liquids, polymeric micelles display short range correlations indicative of their interactions. The local density of micelles at a radial distance from a reference micelle is described by the radial distribution function, $g(r)$. We use liquid state theory to calculate $g(r)$ from the interaction potential $u(r)$. Several integral equations provide the radial distribution function from $u(r)$. In the integral equations, the radial distribution function, $g(r)$, is decomposed into the direct correlation function $c(r)$ describing the direct influence of any one particle on another and the indirect correlation function $\gamma(r)$ describing the influence of one particle on another through a third particle. All the correlation functions decay to zero at large separations. Ornstein and Zernike [83] related the direct and indirect correlation functions

$$\gamma(r_{12}) = h(r_{12}) - c(r_{12}) = \rho \int c(r_{13})h(r_{23})dr_3 \tag{15}$$

where the total correlation $h(r_{23}) = g(r_{23}) - 1$ can be written as a series expansion in powers of density. A major advantage of this formalism is that its Fourier transform now has the simple form

$$\hat{H}(q) = \hat{C}(q) + \rho\hat{C}(q)\hat{H}(q) \tag{16}$$

where the static structure factor, $S(q) = 1 + \rho\hat{H}(q)$, offers a direct link to scattering experiments. The Ornstein-Zernike equation can be solved for $g(r)$ given $u(r)$; however, the integral equation is an infinite series. Closures relating $u(r)$ to the correlation functions are required to solve the integral equation.

Successful closures requiring thermodynamic consistency include the Rogers-Young closure to the Ornstein-Zernike equation [84]. The Rogers-Young closure is a hybrid approach mixing the Percus-Yevick and Hyper-Netted Chain closure relations

$$g(r) = \exp[-\beta u(r)](1 + \frac{\exp[\gamma(r)f(r)] - 1}{f(r)}) \tag{17}$$

where $0 \leq f(r) \leq 1$, with $f(0) = 0$ and $f(\infty) = 1$. At $r = 0$, $f(0) = 0$ produces the PY equation. As r increases, $f(r)$ approaches unity, reducing equation 17 to the HNC approximation. The mixing equation is

$$f(r) = 1 - \exp(-\alpha r) \tag{18}$$

where α is adjusted to achieve thermodynamic consistency in the osmotic compressibility.

Since we combine a discrete model of the interaction potential from our SCF calculation and the Rogers-Young closure, we numerically solve for the radial distribution function via Gillan's method [85]. Liquid state theory thus allows us to take the SCF $u(r)$, use integral equations and closure relations to calculate $g(r)$ and $S(q)$ for comparison to the scattering experiments.

We apply our theoretical model to the series of dPS/PI diblock copolymer micelles described in Table 2. The core radii, R, determined by fitting the core-contrast dilute solution (0.5 wt% polymer) scattering profiles to polydisperse solid sphere form factors with a Schulz distribution are presented in Table 2. The reasonable agreement between the polydisperse solid sphere form factor and experimental data suggest that the micelles have dense spherical cores with relatively sharp interfaces. The polydispersity in the hard sphere form factor (standard deviation = 0.1 for both systems) is partly due to the wavelength dispersity ($\Delta\lambda/\lambda$=0.11). The aggregation number, f, calculated from the absolute intensity at zero angle through Zimm analysis [86] provides the radius of a space-filling core of polystyrene chains (Table 2). Agreement between these radii suggests that the solvent swelling of the core is minimal, less than 10% by volume. The surface density $\sigma = fb^2/4\pi R^2$ of tethered polyisoprene is based upon the core radius determined from $P(q)$ and the aggregation number from Zimm analysis. These diblock copolymer systems have a range of core radii and chain lengths forming tethered chain structures that fall between the limiting cases of a star-like structure and the planar parabolic structure.

The profiles calculated with the above parameters can be compared to the measured radii of gyration, R_g, of the micellar coronas from SANS on dilute suspensions in corona contrast decane as shown in Table 2. The radii of gyration calculated from the chain profiles compare favorably with the available experimental values; all are within the 10% uncertainty in the experimental parameters used in the SCF model. Scattering experiments in corona contrast n-decane offer a reasonable test of the SCF calculations because the distribution function, $P(q)$ for the corona, depends only on the tethered chain conformation. Our calculated results suggest that we have the correct length scale for the interaction potential; however, the R_g measurements provide little detail about the chain profile.

The potentials are repulsive with a soft-sphere character; short-range potentials generally correspond to micelles with brush-like coronal layers smaller than the core diameter and softer potentials correspond with more star polymer-like micelles. These potential energies provide the theoretical input for our characterization of the liquid structure in strongly interacting suspensions. We compare two diblock copolymers, dPS/PI 393/206 and dPS/PI 402/422 to investigate the interactions in parabolic and star-like systems, respectively.

The structure factors for both micellar systems are shown in Figures 4 and 5. The oscillatory form of $S(q)$ clearly reflects a liquid-like structure. As expected, we see that $S(q)$ approaches unity as q increases, indicating that our division of the intensity by $P(q)$ is reasonable. At higher concentrations, we see increases in the peaks of the structure factor, indicating increasing correlations in the suspension. Increasing the concentration eventually leads to $S(q)$ for the ordered micellar structure. These structure factors are a great improvement over the results from a similar study on concentrated

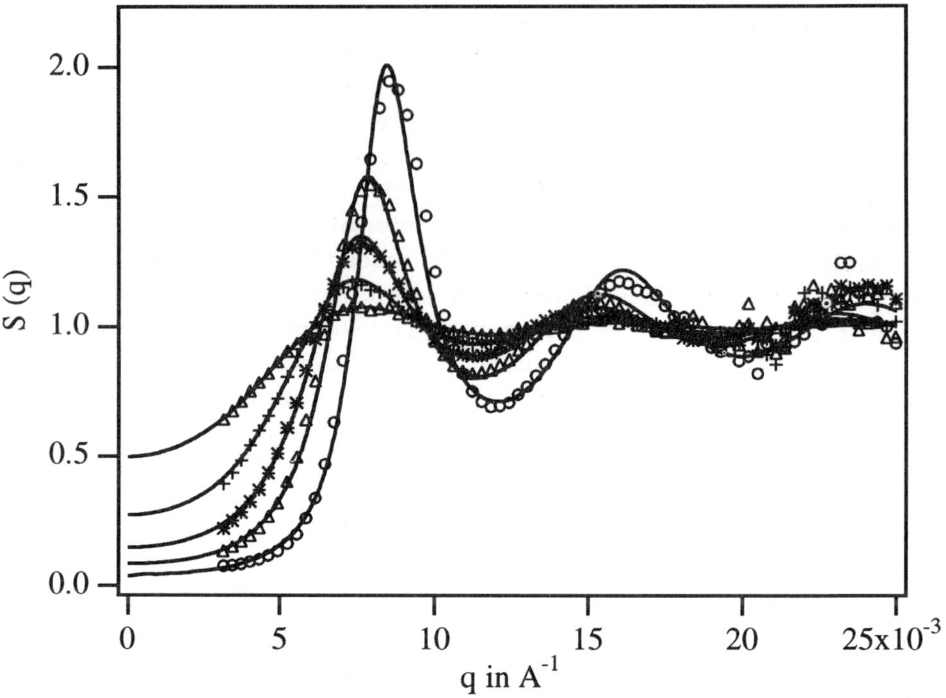

Figure 4. Structure factors for dPS/PI micelles of 393/206 repeat units at core volume fractions of 0.012 (△), 0.02 (+), 0.03 (∗), 0.04 (△) and 0.05 (o). The lines are the theoretical fits from the SCF interaction potentials and the Rogers Young closure to the OZ equation.

star polymer solutions [87] where no liquid-like structure could be obtained even at high concentrations. Unlike the star polymer solutions, polymeric micelles have a dense spherical core providing a sharp contrast with the background solution. We also obtain $S(q)$ for a series of concentrations unavailable in earlier studies on micellar suspensions [18,88].

We compare the experimental $S(q)$ with structure factors calculated with the SCF potentials using Gillan's method, and the Rogers-Young closure. The liquid state theory provides the radial distribution function $g(r)$ for the SCF interaction potentials at a proposed volume fraction. Taking the Fourier transform of $(g(r) - 1)$, we generate a structure factor with a scattering vector scaled on the particle diameter.

We first look at the micellar system formed from 393/206 PS/PI diblocks where the core is relatively large compared to the corona at polymer concentrations from 3 wt% to 12.5 wt%. The resulting interactions are of relatively short range and should be well-characterized with the SCF theory and the Derjaguin approximation. A direct comparison of theoretical

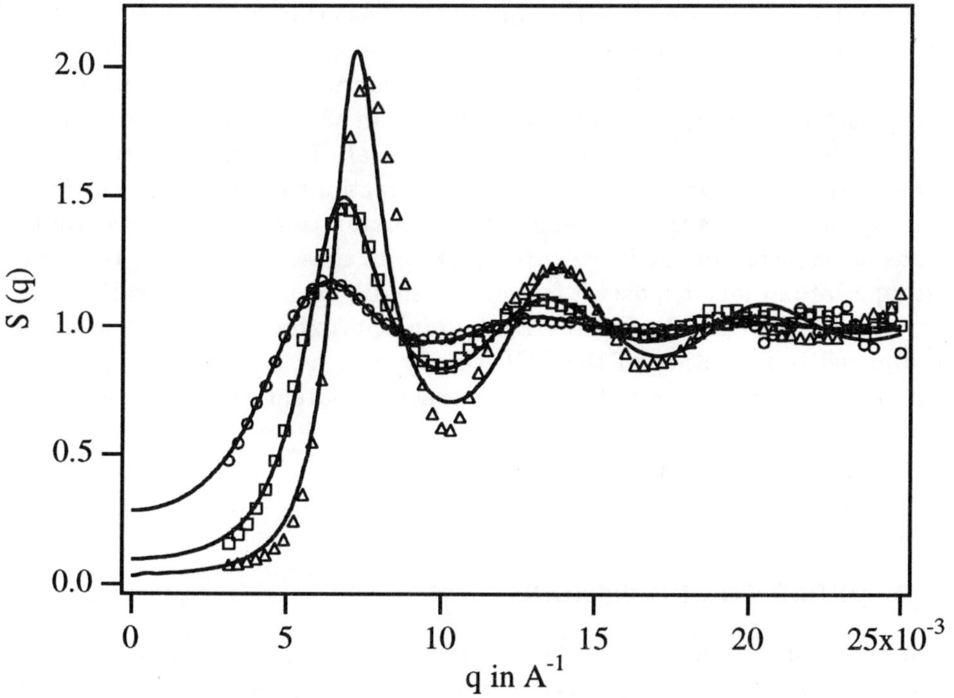

Figure 5. Structure factor for dPS/PI micelles of 402/422 repeat units at core volume fractions of 0.006(○), 0.013 (□), 0.019 (△) The lines are the theoretical fits from the SCF interaction potentials and the Rogers Young closure to the OZ equation.

and experimental structure factors for core volume fractions from 0.012 to 0.050 are shown in Figure 4. We find excellent agreement between our experimental and theoretical structure factors. The excellent agreement, even for a highly correlated liquid suspension observed at a core volume fraction of .050, suggests that the calculated pair-interaction potential is a good approximation. The importance of the decaying character of the potential is seen when $S(q)$ is fit with an effective hard sphere potential. This provides a much poorer fit to the data and requires a systematic decrease in the hard sphere radius as the concentration increases.

The second system, dPS/PI 402/422, has softer interactions due to its long polyisoprene chains and relatively small core. We have fewer experimental structure factors for this system. The strong correlations in $S(q)$ occur at smaller core volume fractions because of the long range of the interactions. Again, as Figure 5 demonstrates, we find reasonable agreement with our liquid state theory fit with the SCF potential and experiment. A comparison of the two structure factors at the highest concentration shows a mismatch in the phase of oscillation. In this case the theoretical structure factor has its first maximum slightly before the experimental while the third maximum is slightly after the experimental structure factor. By testing this system with a strongly correlated liquid, we can see that some error exists in the estimate of the pair-interaction potential. This may be attributed to the Derjaguin approximation for the interactions in this highly curved geometry or to errors in the mean field approximation for the more dilute regions at the edge of the large corona. Despite these limitations, they are inviting approximations, useful for moderate, liquid-like concentrations. A more detailed model relaxing the Derjaguin approximation or using a full two-dimensional solution of the SCF theory may be necessary to provide a truly accurate estimate of the pair-interaction potential applicable at all liquid concentrations.

4. The Influence of Tethered Chain Structure on Micellar Ordering

4.1. DISORDER TO ORDER TRANSITION

A more profound influence of the structure in tethered chain particles appears when the system reaches a concentration where it orders. The disorder to order transition occurs in many systems having purely repulsive interactions. Known as the Kirkwood-Alder transition, this crystallization was discovered through computer simulations of hard spheres by Alder et al. [89]. Hard spheres order into face-centered-cubic (FCC) arrays; several model colloidal systems provide experimental support for the simulations [90]. In the twenty seven years since the discovery of hard-sphere ordering, interaction potentials having finite range have been studied both theoretically and experimentally. Generally, repulsions acting over long range can produce body centered cubic (BCC) arrays while short range repulsions always favor the FCC. Particles interacting via a power law repulsion, $u(r) = (1/r)^n$, order into FCC arrays for powers $n > 7$, but form BCC arrays for $n \leq 7$ [91]. Electrostatically stabilized colloidal particles interact via a screened Coulombic repulsion well-described by a Yukawa potential [92]. Experimental and computational studies of Yukawa systems show ordering into FCC structures when the repulsions are of short-range and into BCC structures when the range of the repulsion greatly exceeds the particle size [90, 92–97]. It is reasonable to expect, then that spherical particles interacting through tethered chains would order into cubic arrays.

As we increase the concentration of solutions of dPS/PI micelles, they order into cubic lattice structures. The soft-sphere star-like micelles form BCC arrays while the harder brush micelles crystallize into an FCC structure. Thus, the two systems examined above form FCC (dPS/PI 393/206) or BCC (dPS/PI 402/422) arrays reflecting the range and steepness of their repulsions [48]. An interesting feature of ordering in star polymers is the finding that at least 16 arms, or tethered chains per star, are required to produce an ordered array. Stars comprising fewer chains produce only liquid structures [60]. In addition, 16-arm stars undergo a disorder to order transition upon increasing concentration; however, they then disorder again when the overall polymer concentration exceeds the average concentration in the star. At elevated concentrations star polymers do not produce sufficent repulsion to order; interpenetration or collapse of the arms occurs instead.

Prediction of the disorder to order transition from the interaction potential energies is difficult; the spatial correlations in a liquid near the disorder-order transition are very sensitive to the details of the interaction potential. We are currently applying density functional models to this system [98–101]; these models have been successful in predicting the hard-sphere and Lennard Jones ordering transitions.

5. Acknowledgments.

This work was carried out by an enthusiastic group of graduate students and visitors including Kathleen Cogan-Farinas, Lena Vagberg, Frans Leermakers, Eric Lin and Glen McConnell. Additional help with this work came from Stephen Nilsen and David Marr. We have benefitted from collaborations with Steven D. Smith, John S. Huang, Min Y. Lin and Malcolm Capel. This work was partially supported by the NSF-MRL Program through the Stanford Center for Materials Research. Additional financial support from the Exxon Educational Foundation is gratefully acknowledged. We acknowledge the support of the National Institute of Standards and Technology, U.S. Department of Commerce, and the Department of Energy for providing the facilities used in these studies.

References

1. D. H. Napper, *Polymeric Stabilization of Colloidal Dispersions*, Academic Press, 1983.
2. T. C. R. Buscall and J. Stageman, *Polymer Colloids*, Elsevier, 1985.
3. J. Clarke and B. Vincent, *J. Colloid Interface Sci.* 1981 **82**, 208.
4. Z. Tuzar and P. Kratochvil, *Adv. In Colloid and Interface Science* 1976 **6**, 201.
5. P. B. G. Riess and G. Hurtrez, *Encyclopedia of Polymer Science and Engineering*, volume 2, chapter Block Copolymers, pages 324–434, Wiley, 1985.
6. A. P. Gast, *Scientific Methods for the Study of Polymer Colloids and Their Applications*, chapter Block Copolymers at Interfaces, pages 311–328, Kluwer, 1990.
7. A. Halperin, M. Tirrell, and T. P. Lodge, *Adv. Poly. Sci.* 1992 **100**, 31.
8. F. Leermakers and A. Gast, *Macromolecules* 1991 **24**, 718.
9. T. Cosgrove, T. G. Heath, J. S. Phipps, and R. M. Richardson, *Macromolecules* 1991 **24**, 94.

10. J. B. Field et al., *Macromolecules* 1992 **25**, 434.
11. T. Cosgrove and K. Ryan, *Langmuir* 1990 **6**, 136.
12. P. Auroy, L. Auvray, and L. Leger, *Macromolecules* 1991 **24**, 2523.
13. P. Auroy, Y. Mir, and L. Auvray, *Phys. Rev. Lett.* 1992 **69**, 93.
14. G. Hadziioannou, S. Patel, S. Granick, and M. V. Tirrell.
15. M. Tirrell, S. Patel, and G. Hadziioannou, *Proc. Natl. Acad. Sci. U.S.A.* 1987 **84**, 4725.
16. H. J. Taunton, C. Toprakcioglu, L. J. Fetters, and J. Klein, *Nature* 1988 **332**, 712.
17. H. J. Taunton, C. Toprakcioglu, and J. Klein, *Macromolecules* 1988 **21**, 3333.
18. J. S. Higgins, J. V. Dawkins, G. G. Maghami, and S. A. Shakir, *Polymer* 1986 **27**, 931.
19. K. Cogan, M. Capel, and A. Gast, *Macromolecules* 1991 **24**, 6512.
20. L. Vagberg, K. Cogan, and A. Gast, *Macromolecules* 1991 **24**, 1670.
21. S. T. Milner, *Science* 1991 **251**, 905.
22. T. Cosgrove, T. Heath, B. van Lent, F. A. M. Leermarkers, and J. M. H. M. Scheutjens, *Macromolecules* 1987 **20**, 1692.
23. A. K. Dolan and S. F. Edwards, *Proc. R. Soc. Lond. A.* 1974 **337**, 509.
24. A. K. Dolan and S. F. Edwards, *Proc. R. Soc. Lond. A.* 1975 **343**, 427.
25. M. D. Whitmore and J. Noolandi, *Macromolecules* 1990 **23**, 3321.
26. C. Wijmans, J. M. H. M. Scheutjens, and E. B. Zhulina, *Macromolecules* 1992 **25**, 2657.
27. S. Alexander, *J. Phys. (Paris)* 1977 **38**, 983.
28. S. Alexander, *J. Phys. (Paris)* 1977 **38**, 977.
29. P. G. de Gennes, *Macromolecules* 1980 **13**, 1069.
30. A. Chakrabarti and R. Toral, *Macromolecules* 1990 **23**, 2016.
31. P. Y. Lai and K. Binder, *J. Chem. Phys.* 1991 **95**, 9288.
32. R. M. Webber and J. L. Anderson, *Langmuir* 1994 **10**, 3156.
33. P. F. McKenzie and J. L. Anderson, *Colloids and Surfaces A* 1994 **86**, 263.
34. S. T. Milner, T. A. Witten, and M. E. Cates, *Macromolecules* 1988 **21**, 2610.
35. M. Murat and G. Grest, *Phys. Rev. Lett.* 1989 **63**, 1074.
36. M. Daoud and J. P. Cotton, *J. Phys. (Paris)* 1982 **43**, 531.
37. G. Grest, K. Kremer, and T. Witten, *Macromolecules* 1987 **20**, 1376.
38. K. Cogan, F. Leermakers, and A. Gast, *Langmuir* 1992 **8**, 429.
39. B. van Lent and J. M. H. M. Scheutjens, *Macromolecules* 1989 **22**, 1931.
40. F. Leermakers, C. M. Wijmans, and G. J. Fleer, *Macromolecules* 1995 **28**, 3434.
41. R. C. Ball, J. F. Marko, S. T. Milner, and T. A. Witten, *Macromolecules* 1991 **24**, 693.
42. M. Murat and G. Grest, *Macromolecules* 1991 **24**, 704.
43. N. Dan and M. Tirrell, *Macromolecules* 1992 **25**, 2891.
44. C. M. Wijmans and E. B. Zhulina, *Macromolecules* 1993 **26**, 7214.
45. G. A. McConnell et al., *Faraday Discussions of The Royal Society of Chemistry* 1995 **98**, 121.
46. E. K. Lin and A. P. Gast, *Macromolecules* 1995 **in press**.
47. R. Toral and A. Chakrabarti, *Phys. Rev. E* 1993 **47**, 4240.
48. G. A. McConnell, A. P. Gast, J. S. Huang, and S. D. Smith, *Phys. Rev. Lett.* 1993 **71**, 2102.
49. P. G. de Gennes, *Scaling Concepts in Polymer Physics*, Cornell University Press, Ithaca, New York, 1979.
50. P. G. de Gennes, *C.R. Acad. Sci. (Paris)* 1985 **301**, 1399.
51. P. G. de Gennes, *Adv. Colloid and Interface Sci.* 1987 **27**, 189.
52. A. P. Gast and M. R. Munch, *Polymer Communications* 1988 **30**, 324.
53. J. Plestil and J. Baldrian, *Makromol. Chem.* 1975 **176**, 1009.
54. J. Noolandi and K. M. Hong, *Macromolecules* 1983 **16**, 1443.
55. L. Leibler, H. Orland, and J. C. Wheeler, *J. Chemical Physics* 1983 **79**, 3550.
56. M. R. Munch and A. P. Gast, *Macromolecules* 1988 **21**, 1360.

57. R. Nagarajan and K. Ganesh, *J. Chemical Physics* 1989 **90**, 5843.
58. M. Adam, L. J. Fetters, W. W. Grassley, and T. A. Witten, *Macromolecules* 1991 **24**, 2434.
59. B. J. Bauer, L. J. Fetters, W. Graessley, N. Hadjichristidis, and G. Quack, *Macromolecules* 1989 **22**, 2337.
60. D. Richter et al., *Journal de Physique IV - C8* 1993 **3**, 3.
61. A. Halperin, *Macromolecules* 1987 **20**, 2943.
62. T. A. Witten and P. A. Pincus, *Macromolecules* 1986 **19**, 2509.
63. L. Leibler and P. A. Pincus, *Macromolecules* 1984 **17**, 2922.
64. S. F. Edwards, *Proc. Phys. Soc.* 1965 **85**, 613.
65. K. M. Hong and J. Noolandi, *Macromolecules* 1981 **14**, 727.
66. P. G. de Gennes, *Rep. Prog. Phys.* 1969 **32**, 187.
67. H. J. Ploehn and W. B. Russel, *Advances in Chemical Engineering* 1990 **15**, 137.
68. J. M. H. M. Scheutjens and G. J. Fleer, *J. Phys. Chem.* 1979 **83**, 1619.
69. J. M. H. M. Scheutjens and G. J. Fleer, *J. Phys. Chem.* 1980 **84**, 178.
70. J. M. H. M. Scheutjens and G. J. Fleer, *Macromolecules* 1985 **18**, 1882.
71. G. J. Fleer, J. M. H. M. Scheutjens, M. A. C. Stuart, T. Cosgrove, and B. Vincent, *Polymers at Interfaces*, Elsevier, 1993.
72. E. B. Zhulina, O. V. Borisov, and V. A. Pryamitsin, *J. Colloid and Interface Sci.* 1990 **137**, 495.
73. H. Li and T. Witten, *Macromolecules* 1993 **27**, 449.
74. U. Genz, B. D'Aguanno, J. Mewis, and R. Klein, *Langmuir* 1994 **10**, 2206.
75. C. M. Wijmans, F. A. M. Leermakers, and G. J. Fleer, *Langmuir* 1994 .
76. W. B. Russel, D. A. Saville, and W. R. Schowalter, *Colloidal Dispersions*, Cambridge University Press, Cambridge, Great Britain, 1989.
77. B. V. Derjaguin, *Kolloid Z.* 1934 **69**, 155.
78. N. Khasat, R. W. Pennisi, N. Hadjichristidis, and L. J. Fetters, *Macromolecules* 1988 **21**, 1100.
79. K. A. Cogan and A. P. Gast, *Macromolecules* 1990 **23**, 745.
80. C. Harkless, M. Singh, S. Nagler, G. Stephenson, and J. Jordan-Sweet, *Physical Review Letters* 1990 **64**, 2285.
81. S. D. Smith, in *Polymer Solutions, Blends, and Interfaces*, edited by I. Noda and D. N. Rubingh, pages 43–64, Elsevier, 1992.
82. B. Hammouda, S. Krueger, and C. J. Glinka, *J. Res. Natl. Inst. Stand. Technol.* 1993 **98**, 31.
83. D. A. McQuarrie, *Statistical Mechanics*, Harper & Row, New York, New York, 1976.
84. F. J. Rogers and D. A. Young, *Phys. Rev. A* 1984 **30**, 999.
85. M. J. Gillan, *Mol. Phys.* 1979 **38**, 1781.
86. O. Glatter and O. Kratky, *Small Angle X-Ray Scattering*, Academic Press, New York, New York, 1982.
87. W. D. Dozier, J. S. Huang, and L. J. Fetters, *Macromolecules* 1991 **24**, 2810.
88. J. S. Higgins et al., *Polymer* 1988 **29**, 1968.
89. B. Alder, W. Hoover, and D. Young, *J. Chem. Phys.* 1968 **49**, 3688.
90. P. Pusey and W. van Megen, *Nature* 1986 **320**, 340.
91. W. G. Hoover, D. A. Young, and R. Grover, *J. Chem. Phys.* 1972 **56**, 2207.
92. Y. Monovoukas and A. P. Gast, *J. Coll. Int. Sci.* 1989 **128**, 533.
93. M. O. Robbins, K. Kremer, and G. S. Grest, *J. Chem. Phys.* 1988 **88**, 3286.
94. Y. Monovoukas and A. P. Gast, *Langmuir* 1991 **7**, 460.
95. Y. Monovoukas, G. G. Fuller, and A. P. Gast, *Journal of Chemical Physics* 1991 **93**, 8294.
96. R. Carlson and S. Asher, *Applied Spectroscopy* 1984 **38**, 297.
97. H. Lindsay and P. Chaikin, *Journal of Chemical Physics* 1982 **76**, 3774.
98. W. Curtin, *Physical Review B* 1989 **39**, 6775.
99. W. Curtin and N. Ashcroft, *Physical Review A* 1985 **32**, 2909.

100. W. Curtin and N. Ashcroft, *Physical Review Letters* 1986 **56**, 2775.
101. D. Marr and A. P. Gast, *Physical Review E* 1993 **47**, 1212.

BLOCK COPOLYMER SELF-ASSEMBLY AT SURFACES: STRUCTURE AND PROPERTIES

MATTHEW TIRRELL
Department of Chemical Engineering and Materials Science
University of Minnesota, Minneapolis, Minnesota 55455

1. Introduction

The adsorption behavior of copolymers is naturally richer than that of polymers containing only a single kind of segment since more than one kind of affinity can be built into different parts of the same macromolecule. The interfacial activity of such polymers can be enhanced and manipulated by incorporating segments tailored to interact in some specific way with each of the contacting phases. Architecture of the copolymer plays an important role since it influences the configuration adopted by the molecule at the interface.

Among many interesting and useful possibilities, selective adsorption of block copolymers from selective solvents merits study since it is one means of binding a polymer chain to a surface on which it would not itself adsorb. This can be used to modify the surface of a solid and thereby to manipulate the interactions that this surface experiences. In the case of diblock copolymers, this mode of adsorption is one means of creating a polymer "brush", where polymer chains are tethered by their ends onto an interface.[1,2] The adsorbing block serves as the "anchor" and the tethered chain as the "buoy".[3] A brush results when the density of adsorption is sufficiently high that the buoy chains begin to overlap substantially and stretch normal to the surface to alleviate the build up of osmotic pressure in the dense brush. Alexander[4] and de Gennes[5] first analyzed this situation and showed that the osmotic stretching of the brush would be balanced by the build-up of stretching energy in the chain. Subsequent analyses have developed these ideas more fully.[6-9]

Block copolymers are more than means toward polymer brushes, however; they are versatile, useful polymeric amphiphiles. Amphiphilicity and surface activity are often synonymous molecular characteristics. Such is not the case for macromolecular systems. Homopolymers are often strongly adsorbing and, in this sense, surface active. This is because the interaction energy between the polymer and the surface grows with the polymerization degree, N, while the translational entropy is independent of N. Consequently, even polymers comprising weakly adsorbing monomers, characterized by small monomeric adsorption energy $\varepsilon \ll kT$, may adsorb strongly, provided N is large

S.E. Webber et al. (eds.), Solvents and Self-Organization of Polymers, 281–308.

enough. However, many surface active polymers are not amphiphilic, exhibiting no selective affinities and no tendency to aggregate and form mesophases. Two molecular designs yield proper polymeric amphiphiles. In one, short chain surfactants are incorporated into the backbone of the polymer. The second category includes block copolymers and end functionalized chains. Among the polymers belonging to the second group diblock copolymers are of special interest since they are structurally analogous to low molecular weight surfactants. Most research in surface science is concerned with short chain, "monomeric," surfactants. Much less attention has been given to macromolecular amphiphiles in general, and to block copolymers in particular. Yet, the study of such macromolecules is of interest since, apart from practical considerations, polymeric model systems afford important benefits in advancing our understanding of self-assembly processes. This article aims to bring out the characteristics that distinguish block polymer surfactants and to develop the concepts that generalize their behavior as amphiphiles.

Polymeric surfactants are of interest because of both their similarity and their dissimilarity to short chain amphiphiles. The similarity is important since it provides grounds for comparison. Both species exhibit comparable phenomenology, that is self-assembly into micelles, lamellae, etc. The overall phase behavior determining the coexistence of these aggregates is also hardly distinguishable. Altogether, the similarities concern large scale behavior. On the other hand, the fine structure of the aggregates formed by the two surfactant species is markedly different. The study of macromolecular amphiphiles is worthwhile because of their special, polymeric, characteristics. Two such features are of special importance. First, the theoretical analysis of polymeric surfactants is much simpler. Because the chains are long enough, it is possible to base the analysis on their long chain character. The resulting, scaling type, description is simpler since it can be constructed in universal terms and molecular details are avoided. The second feature is the option to vary the number of constituting monomers over a wide range and study the ensuing trends. This provides a simple test for theoretical predictions and a method for the fine tuning of surfactant properties. Molecular weight becomes a thermodynamic parameter that both determines the number of interactions per molecule and sets the size scale of the structures. While a similar approach exists in principle for monomeric surfactants, the accessible range is orders of magnitude smaller.

The macromolecular nature of polymeric surfactants also affects the nature of their amphiphilicity. Two polymeric characteristics are involved: The polymeric propensity to phase separation and their enhanced surface activity. Both effects are traceable to the polymerization degree, N. While the translational entropy is independent of N, the number of interactions increases with N. Thus, hydrogenated and deuterated polymers, of large enough N, can phase separate and a deuterated-hydrogenated diblock copolymer will act as the corresponding amphiphile. No monomeric surfactant is capable of such "fine tuned" amphiphilicity.

Phase behavior of polymeric amphiphiles is also distinctive. Monomeric amphiphiles form lyotropic liquid crystalline phases in the presence of selective solvents. The corresponding thermotropic liquid crystals are typically formed by neat mesogens in the absence of solvents. Block copolymers are distinct in their ability to form liquid crystalline phases in melt as well as in solution. Consequently, it is possible to study the phase behavior and the structure of aggregates over a much wider range of compositions. Furthermore, for certain polymeric systems there is an exact correspondence between the thermotropic and lyotropic liquid crystalline phases. In such cases one may study the system while the "thermotropic" phase is modified into the corresponding "lyotropic" one by change of solvent content.

The richness of their molecular architecture is, potentially, the most exciting feature of polymeric surfactants. While all types of monomeric amphiphiles have macromolecular analogs, the converse statement is untrue. Diblock copolymers and end functionalized homopolymers correspond to single chain monomeric surfactants while linear triblock copolymers and star block copolymers have no monomeric counterparts. In turn, these uniquely polymeric surfactants afford new options for materials design.

Finally, there are fundamental differences in the theoretical description of monomeric and polymeric aggregates. The concept of optimal packing of the head groups is of crucial importance in the description of monomeric aggregates. It accounts for the close packing of headgroups of definite geometries. This concept is of limited applicability to polymeric surfactants. For example, the head groups in aggregates of flexible diblock copolymers, the immiscible blocks, are deformable and interpenetrating. It is thus impossible to assign definite geometry to such moieties. In other cases, involving, for example, monolayers of end functionalized homopolymers, the surface density is often determined by the repulsion between the flexible tail chains. Under such conditions the geometry of the head groups is irrelevant. In this article, we focus on aggregation at surfaces. The phenomena of micellization are distinct but closely related.

2. Block copolymer adsorption: Theory and experiment

2.1. NEUTRAL COPOLYMERS: MODELLING

Several models have been proposed for block copolymer adsorption from selective solvents.[10] As we shall demonstrate, the model proposed by Marques et al.[3] makes some analytical predictions that can be tested with the data we have obtained. Furthermore, by rearrangement of the model, we have uncovered a new prediction for the effect of block size asymmetry on adsorbed amount arising from the physics of their theory. We therefore focus our introductory discussion on this theory.

The model of Marques et al. (MJL) imagines adsorption to be as depicted in Figure 1. The block copolymer is dissolved in a solvent that is selective for one of the

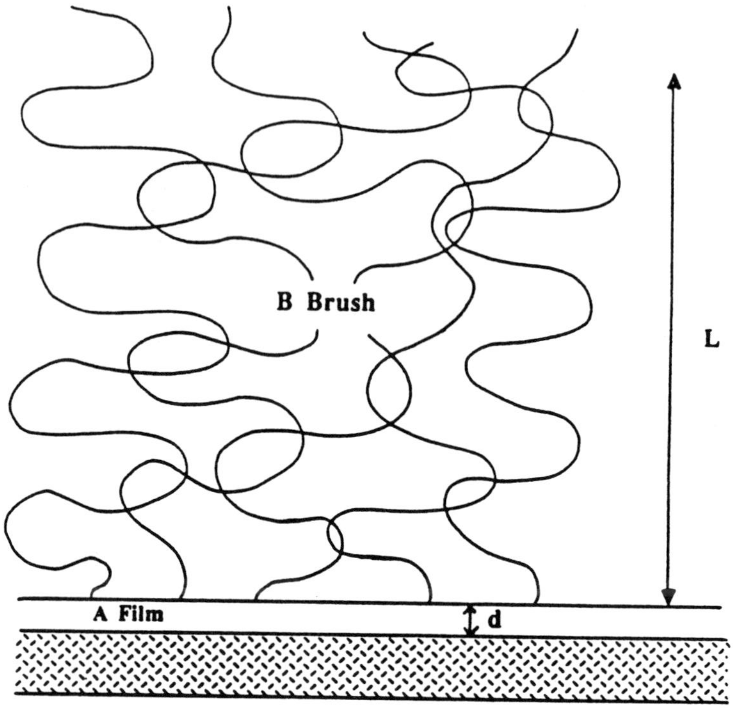

Figure 1: Adsorption of AB block copolymers from a selective sol-vent. The poorly solvated A block forms a thin adherent wetting film on the substrate while the well-solvated B blocks form a terminally attached brush.

blocks. That is, for single chains in solution, the well-solvated block (B) is assumed to take on the swollen coil configuration ($R_B = aN_B^{3/5}$) while the poorly solvated block (A) is in solution in the form of a collapsed globule, which may be plasticized by the solvent to a polymer A volume fraction c ($R_A = a(3/4\pi N_A/c)^{1/3}$) . (We use α here to denote the segment size for both blocks.) Upon adsorption, the A blocks form a thin concentrated film, which completely wets the substrate surface. The thickness of the film d is related to the polymer volume fraction c and surface density σ in chains per unit area by:

$$N_A a\sigma = cd \tag{1}$$

where σ is related to the average distance, δ, between adsorbed chains by $\sigma = (a/\delta)^2$. The adsorbed layer is then assumed to be in equilibrium with solution having polymer chemical potential μ_{ex}.

The free energy of the well-solvated B block is assumed to be that of polymer brush chains in a good solvent. Initial theoretical treatment of grafted brushes assumed a uniform density of monomers within the brush.[4,5] This situation is anticipated at surface densities sufficiently high to produce overlap among the swollen, grafted coils $(\delta < R_B \text{ or } \sigma > a^2 / R_B^2)$. Scaling arguments based on the blob picture[5] predict an equilibrium average brush height $L_B = a\,\sigma^{1/3} N_B$. The corresponding free energy per unit area for the equilibrated brush is given by:

$$F_B = \frac{T}{a^2} N_B \, \sigma^{11/6} \tag{2}$$

More recently, Milner et al.[6] has solved the self-consistent mean-field problem for the segment density profile. In the limit of strong stretching, an analytical solution is obtained, which shows the density profile to be parabolic rather than a step function.[6] However, the scaling of brush height and free energy with molecular weight and surface density is unchanged from that predicted by the global energy balance analysis.[5]

Analyses of grafted brushes treat surface density as a parameter,[4-6] but for block copolymer adsorption, the equilibrium surface density is also dependent on the free energy of the adsorbed anchoring block F_A. In the MJL analysis, energetic contributions to the adsorption of the A block are the dominant factors. Two contributions to the energy of adsorption of the A block are considered. The first contribution is the spreading coefficient S defined as:

$$S = \gamma_{WS} - \gamma_{WA} - \gamma_{AS} \tag{3}$$

where γ_{WS}, γ_{WA}, and γ_{AS} are the surface tensions for the wall and solution, wall and concentrated A film, and A film and the solution, respectively. In essence, S is the free energy gain (S > 0) per unit area when a wall/solution interface is replaced by the wall/A film and A film/solution interfaces.

The second component of the A film free energy is the long-range van der Waals attraction of the substrate for the A film. A similar problem is the interaction between two semi-infinite slabs separated in vacuo by a distance *d*. The energy of interaction is given by

$$P(d) = -\frac{A_H}{12\pi d^2} \tag{4}$$

where A_H is the Hamaker constant. Lifshitz[13] developed the theory for the interaction between two bodies partitioned by a thin liquid film. For thin films, the energy of interaction has the same form as eq 4. However, with a medium in between, the relative interaction between the two bodies can be either attractive ($A_H > 0$) or repulsive ($A_H < 0$) depending on the nature of the intermediate fluid. "Repulsive" in this context means that the film of anchor blocks in Figure 1 tends to thicken. The free energy for a thin wetting A film of thickness *d* per unit area is thus given by

$$F_A = -S + \frac{A}{12\pi d^2}\ ;\ A_H = -A \tag{5}$$

where the negative Hamaker constant reflects the relative attraction of the A film to the substrate.

The grand canonical free energy (*G*) per unit area contains the solution chemical potential μ_{ex}, brush free energy F_B, and a film free energy F_A

$$G = -\mu_{ex}\,\sigma\, a^{-2} - S + \frac{A}{12\pi d^2} + \frac{T}{a^2} N_B\, \sigma^{11/6} \tag{6}$$

where the film thickness *d* is related to the surface density σ through eq 1. Minimization with respect to σ then gives an expression for the equilibrium surface density:

$$\mu_{ex} = -\frac{A}{6\pi} c^2 \frac{1}{N_A^2} \frac{1}{\sigma^3} + \frac{11}{6} T\, N_B\, \sigma^{5/6} \tag{7}$$

This is essentially the result of MJL for one of the regimes they studied that is most applicable to the present work. It expresses a testable relationship between adsorbed amount and the molecular weights of the two blocks.

To compare experimental measurements with the theoretical analysis of Marques et al., we cast eq 7 in terms of a normalized surface density σ^*, scaled to σ_{o1}, the density necessary for the buoy chains to begin to overlap sufficiently to form a brush $\approx a^2 / R_g^2 \approx 1 / \pi N^{6/5}$:

$$\sigma^* = \frac{\sigma}{\alpha_{o1}} = \pi\sigma N_{PS}^{6/5} \tag{8}$$

The resulting rearrangement of eq 7 gives:

$$\mu_{ex} = -\frac{A\pi^2}{6} c^2\, \beta^3 \frac{1}{\sigma^{*3}} + \frac{11}{6\pi^{5/6}} \tag{9}$$

where $\beta = N_{PS}^{6/5} / N_{PVP}^{2/3}$ is an asymmetry ratio for the copolymer in solution that arises naturally from this analysis. This ratio measures the solvent-induced asymmetry of the radii of gyration of the solvated block copolymer. According to eq 9, the molecular weight dependence of σ^* is principally in β. That is, at fixed μ_{ex} and Hamaker constant A, the theory predicts σ^* to be a unique function of β.

The effect of solution chemical potential on the adsorption properties can be evaluated by measuring adsorption isotherms. If it is observed that, over a range of dilute concentrations, σ is fairly insensitive to the concentration of the incubation solution, as has been seen by Parsonage, et al.,[14] this suggests that a balance between the van der Waals energy of the PVP film and the free energy of the PS brush (the first and second terms in the right-hand side of eq 9 mainly determines σ and that μ_{ex} can be neglected. This is the behavior expected for the van der Waals-buoy regime discussed by MJL[3] For such cases, eq 9 predicts

$$\sigma^* = K \left(\frac{A}{T} \right)^{\frac{6}{23}} c^{\frac{12}{23}} \beta^{\frac{18}{23}} \tag{10}$$

where K is a numerical constant of order unity.[14] The conclusion of this MJL analysis, as modified by Parsonage, et al.,[14] is that the degree of crowding of block copolymers on the surface, measured by σ^*, is a unique function of the radius of gyration asymmetry, measured by β.

It is essential to note that the foregoing analysis applies only to one of the conceivable regimes of block copolymer adsorption, namely, adsorption from a very selective solvent share $\beta >> 1$. MJL studied several others, including the cases where $\beta \approx 1$ and crowding of the buoys does not occur; and where the solvent is <u>non</u>selective but the A blocks still adsorb on the surface. The latter case has received some attention experimentally,[15,16] but less than the strongly selective solvent case strongly selective solvent case to be discussed in detail in the experimental section.

2.2. COPOLYMERS WITH CHARGED BUOY BLOCKS: MODELLING

Colloidal particles are often suspended in aqueous solutions rather than in nonpolar solvents. For such systems, water-soluble chains, such as charged polymers, are required as stabilizers or flocculants. The chain configurations and, therefore, aggregation characteristics of polyelectrolytes differ from those of neutral polymers in good solvents due to the effect of electrostatic interactions. At high charge densities along the backbone and low ionic strength (corresponding to low salt concentration) in solution, electrostatic correlations between segments dominate chain configurations, leading to an increase in both the effective persistence length associated with the chain and the excluded volume in solution.[17-20] In the limit of no added salt, it is generally accepted that a highly charged polyelectrolyte at high dilution is nearly fully stretched. As the salt concentration is increased, the range of electrostatic interactions begins to be reduced by screening effects, and the chain starts to recover its flexible nature. The interactions are completely screened at very high salt concentrations and the chain conformation resembles that of a neutral polymer. For example, poly(styrene sulfonate) has a θ point (at which a neutral polymer chain exhibits a Gaussian configuration in solution) at an aqueous sodium chloride concentration of 4.2 *M* at 25°C,[21] and behaves like a neutral polymer in a good solvent at 0.15 *M* at 25°C.[20]

The conformation of the polyelectrolyte chain can be reasonably expected to affect its adsorption behavior. This means that two additional factors, charge density along the chain and salt concentration, must be considered in the adsorption of diblock copolymers with charged buoys at interfaces. For example, for adsorption of homopolymer poly(styrene sulfonate) on silica from aqueous solution, there is no detectable adsorption in the absence of salt, a consequence of the strong repulsion between charged groups on the polyelectrolyte, and adsorbed amounts are found to increase with increasing ionic strength (salt concentration), a consequence of the screening effect of salt on electrostatic interactions.[22] The range of electrostatic interactions is also expected to affect the adsorption characteristics of charged diblock copolymers on surfaces from aqueous solutions.

The dependence of the equilibrium adsorbed amount on the size of the polyelectrolyte block and the salt concentration can be predicted by constructing an expression for the grand canonical free energy, *G*, for the polymer brush in contact with a reservoir of solution. Similar to the analysis of MJL for adsorption of neutral diblock copolymers. The polyelectrolyte chains in this analysis are envisioned as being grafted on the solid surface via the adsorbed hydrophobic block. The total free energy per unit area is given in units of kT by:

$$G = \frac{\sigma}{a^2}\left(-\Delta + \frac{N_B \sigma^{5/6}}{\phi_s^{2/3}} - \mu_{ex}\right) \tag{11}$$

where σ ~ number density of chains on the surface, i.e. the number of chains per unit surface area, - Δ is the energy gain per chain due to attachment of the hydrophobic block on the surface, and μ_{ex} is the chemical potential of the solution.$N_B\sigma^{5/6}/\phi_s^{2/3}$ is the free energy per grafted polyelectrolyte block,[23, 24] N_B segments in size, in a swollen brush. In order to compare this equation with eq 6, we note that Δ in eq 11 plays the role of the second and the third terms on the right hand side of eq 6. In polyelectrolyte buoy block copolymer adsorption, we are seldom dealing with values of b sufficiently small, that is, anchor blocks so large, that a continuous wetting film of anchor block covers the adsorption surface. Rather, the anchors adsorb individually and discontinuously with energy Δ. (Contrast with Figure 1.) The factor of $\phi_s^{2/3}$ in eq 11 gauges how adding salt, thereby screening electrostatic interactions an decreasing the stretching energy, can lower the free energy of adsorption.[23] The brush is in contact with an aqueous solution at a salt concentration given by ϕ_s. The inclusion of the salt concentration in the free energy expression indicates that, in the case of polyelectrolytes in contact with ionic solvents, electrostatic interactions between charged segments contribute to the stretching energy of the grafted chain. An assumption in this analysis is that the energy of interaction of the hydrophobic block with the surface can be approximated by a constant and does not depend on the size of the block.

We again assume[25] that the chemical potential in the solution, μ_{ex}, which should change with c_0, is small and can be neglected in the overall energy balance. Thus, the equilibrium surface coverage is covered by the balance between the energy of adsorption of a diblock and the free energy of the polyelectrolyte brush. Minimizing G with respect to σ gives:

$$\sigma \sim \phi_s^{4/5} \left(\frac{\Delta}{N_B}\right)^{6/5} \tag{12}$$

The height of the equilibrium brush, L_B, is then given by:

$$L_B = N_B \left(\frac{\sigma^{1/3}}{\phi_s^{2/3}}\right) \tag{13}$$

or

$$L_B = N_B^{3/5} \left(\frac{\Delta}{\phi_s}\right)^{2/5} \tag{14}$$

The adsorbed amount, A, in weight of polymer adsorbed per unit area, is given by:

$$A \sim \sigma N_B \tag{15}$$

or

$$A \sim \phi_s^{4/5} \left(\frac{\Delta^{6/5}}{N_B^{1/5}}\right). \tag{16}$$

2.3. EXPERIMENTAL STUDIES OF BLOCK COPOLYMER ADSORPTION

Experimental work on the adsorption of block copolymers includes the measurement of adsorbed amounts includes a variety of systems in the strongly selective solvent limits. The kinetics of adsorption of PVP-PS block copolymers from toluene solutions onto silver substrates has been studied by Tassin et al.[26] using surface plasmon oscillations. Their results indicate the kinetics of assembly to be very slow and dependent on the microstructure in solution. Stouffer and McCarthy[27] have measured adsorbed amounts for the adsorption of PS-poly(propylene sulfide) (PPS) block copolymers onto gold substrates from THF solutions. Adsorbed amounts were measured by scintillation counting from adsorbed layers of tritium-labeled polymers. Their results indicate a decrease in the surface density with increasing molecular weight of the anchoring PPS block. Furthermore, the measured surface densities become less sensitive to PPS molecular weight for polymers having a relatively low PPS content.

Other experimental work on the adsorption of block copolymers has included the measurement of forces between adsorbed layers[9,28-31] using the surface force apparatus.[32] The measurement of forces between adsorbed layers has been reviewed.[33] The surface force measurements are relevant to the present work in that they provide support for the form of layer thickness and stretching energy embodied in eq 2.

Watanabe and Tirrell[30] measured adsorbed amounts for several PVP-PS and PVP-PI block copolymers. The measured surface densities were sufficiently large such that the PS and PI blocks are overlapping in the adsorbed layers. Adsorption of PVP-poly(*tert*-butylstyrene)[34] from toluene solutions onto mica substrates have also been studied by using surface forces. Webber et al.[35] have studied the hydrodynamic layer thickness of a PVP-PS adsorbed layer and obtained values for the extension of the nonadsorbing block normal to the surface that are in fair agreement with those obtained from surface force measurement.

The issue remains of the proper interpretation of how the sizes of the blocks in a copolymer each affect the adsorption process. This understanding is essential to comprehend all other aspects of block copolymer adsorption, such as surface forces or hydrodynamic thickness.

Table I gives a summary of mass coverages, s (mg/m^2), for both tritium-labeled and nonlabeled diblock copolymers of PVP-PS as measured by scintillation counting and XPS. The code for individual samples represents the molecular weight of the constituent PVP-PS blocks in units of thousands. Also shown in column 3 is the adsorbed amount represented in terms of surface density σ ($\sigma = sN_{Av}/(M_{PVP} + M_{PS})$). The final column gives the surface density σ_{ol} above which the PS blocks are overlapping and form a semidilute solution in the adsorbed layer.

$$R_{PS} = aN_{PS}^{\nu};\ a = 1.86\text{Å},\ \nu = 0.595\ \sigma_{ol} = 1/\pi R_{PS}^2 \tag{17}$$

where RPS represents the radius of gyration of the PS blocks in dilute toluene solutions. The data for RPS in toluene has been reported by Higo et al.[36]

$$R_{PS} = aN_{PS}^{\nu};\ a = 1.86\text{Å},\ \nu = 0.595 \tag{18}$$

where a and NPS represent the segment length and degree of polymerization of PS, respectively.

The adsorbed amounts are in good agreement with several values reported by Watanabe et al.[30] In those experiments, the thickness of dried adsorbed layers of PVP-PS deposited from toluene onto mica was measured with the surface force apparatus. Furthermore, with only three exceptions where the PVP is much larger than the PS block (PVP-PS 124-31, t124-60, and 62-31), the measured surface densities (σ) are larger

TABLE I. Summary of adsorbed amounts of PVP-PS diblock copolymers

sample	mass coverage	surface density	$1/\left(\pi R_{PS}^2\right)^a$
PVP-PS	s, mg/m^2	σ, m^{-2} x 10^{-16}	σ_{ol}, m^{-2} x 10^{-16}
3-36	2.16	3.32	0.84
9-36	2.12	2.84	0.84
16-31	1.79	2.30	1.00
18-36	1.83	2.05	0.84
15-92	2.02	1.13	0.27
15-152	1.97	0.71	0.15
31-31	1.66	1.61	1.00
36-36	1.49	1.25	0.84
31-92	1.82	0.89	0.27
30-152	1.89	0.63	0.15
62-31	1.36	0.88	1.00
61-92	1.65	0.65	0.28
61-152	2.05	0.58	0.15
72-36	1.67	0.93	0.84
124-31	1.57	0.61	1.00
t52-63	1.61	0.84	0.43
t9-32	1.95	2.90	0.98
t124-60	1.13	0.37	0.45

[a] The surface density ($\sigma_{ol,}$) above which the PS buoys will overlap in the adsorbed layer.

than that required for the PS blocks to overlap in the adsorbed layer (σ_{ol}). On comparison of these results with the adsorption conditions in the surface force measurements of Patel et al.[9] and Watanabe et al.,[30] it appears that the osmotic interactions in the crowded brush layer at the measured surface density are sufficient to stretch the PS blocks and to be in the range of validity of eq 2.

Overall, with the range of molecular weights available to us, there is only a modest variation in the mass coverage. The fundamental information to be gleaned from the body of data in Table I is the effect of block molecular weight on the surface density. Figure 2 shows a plot of the experimentally measured surface densities of Table I as a function of both the PS and PVP degrees of polymerization. A reasonable expectation is that the surface density is some function of the relative sizes of the two blocks. One generalization that can be drawn from Figure 2 is that s generally decreases with the molecular weights of each block.

Figure 3 shows the collapse in the experimental data of Table I and Figure 2 predicted from eq 10. As suggested by eq 10, Figure 4 shows a plot of σ^* against $\beta^{18/23}$. The data are in good agreement with the linear predictions of eq 10, although deviations from the theoretical predictions start to appear for the five largest values of β ($\beta > 100$). This corresponds to the break in the curve of Figure 3. If these five values are held out of consideration, a linear least-squares fit to the power law region of a logarithmic plot of σ^* vs β gives a power law of 0.80, within experimental error of 18/23 (=0.78).

Watanabe et al.[30] have published adsorbed amounts for the adsorption of PVP - poly(isoprene) (PI) block copolymers from toluene solutions onto smooth mica substrates. This system is similar to the adsorption of PVP-PS blocks in that toluene is a preferential solvent for the PI blocks. PVP-PI adsorbed amounts were determined by measuring the thickness of the dried adsorbed layers using the surface force apparatus. Shown in Figure 3 is a plot of σ^* versus β for the adsorption of both PVP-PS and PVP-PI block copolymers as given in Table I and Reference 31. Good agreement is found between the adsorption behavior of these two systems. This is probably in part due to

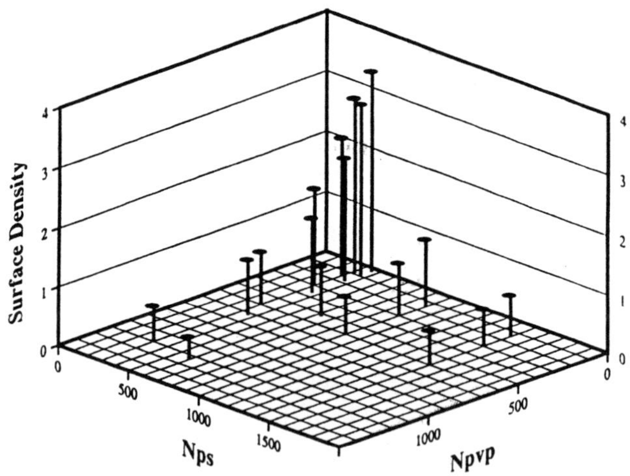

Figure 2: Measured surface densities σ ($m^{-2}x10^{-16}$) of Table I as a function of both the PS (N_{PS}) and PVP (N_{PVP}) degrees of polymerization. Vertical lines are used to indicate heights of data points above their coordinates in the N_{PS}-N_{PVP} plane.

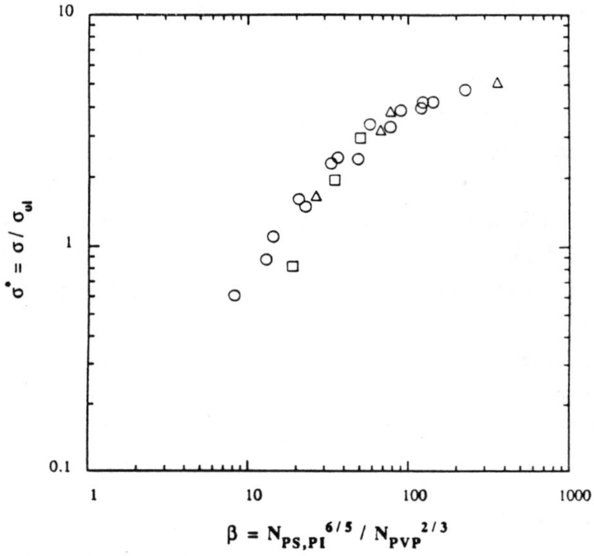

Figure 3: Normalized surface density σ* versus copolymer asymmetry ratio β for the adsorption of PVP-PS block copolymers as measured by scintillation counting (squares) and XPS (circles) and for adsorption of PVP-PI block copolymers as measured by Watanabe et al.[30] using the surface force apparatus (triangles).

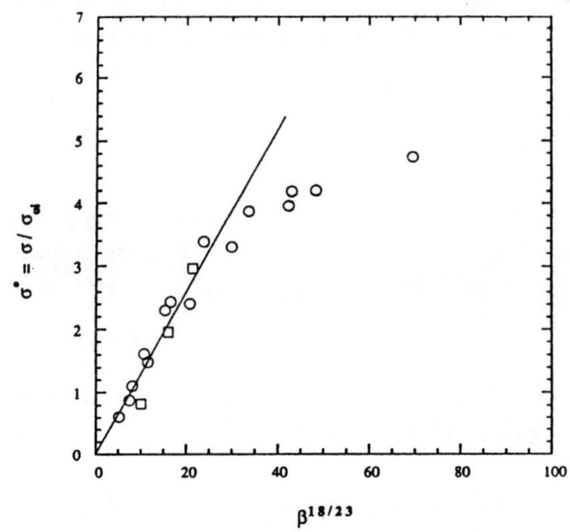

Figure 4: Normalized surface density σ* versus $\beta^{18/23}$ as measured by scintillation counting from tritium-labeled adsorbed layers (squares) and XPS (circles). The solid line represents a least-squares fit to the linear portion of the data (excluding five data points at highest β values) as predicted from eq 10.

the anchoring PVP block for these two copolymers and the good solvent characteristics of toluene for both PS and PI. Surface force measurements for PVP-PI in solution show similar behavior to the PVP-PS systems.[30] We have also found that the data of Stouffer and McCarthy[27] fit on the correlation of Figure 3; however, we are not sure that the interaction of the adsorbing PPS block in that case is similar to that of PVP in our work.

The effect of block molecular weight on the adsorbed amount is summarized by the experimental correlation of the normalized surface density σ^* with the asymmetry ratio β. Within this correlation we see two regimes of behavior. For polymers of moderate asymmetry ($10 < \beta < 100$), σ^* varies with β in such a way that the actual surface density σ is determined primarily by the molecular weight of the anchoring block. These results in this range of β are in good agreement with the theory of Marques et al.[3] where the surface density is determined by a balance between the van der Waals energy of the anchoring PVP film and the free energy of the PS brush. The correlation of σ^* with β is also in good agreement with adsorbed amounts for PVP-poly(isoprene)block copolymers deposited from toluene solutions as measured by Watanabe et al.[30] using the surface forces apparatus. The agreement between the data may be partly attributed to the identity of the anchoring block and the fact that toluene is a very good solvent for both PS and PI.

For samples of large asymmetry ($\beta > 100$), we see behavior that is somewhat the converse of that described above. That is, for samples of very large PS blocks (relative to PVP), the surface density σ is determined primarily by the molecular weight of the well-solvated PS block; being somewhat insensitive to the molecular weight of the PVP block. An explanation of this regime will require future work, although it may be due to the effect of solution chemical potential or possibly breakup of the wetting PVP film as a result of the large asymmetry of these samples.

The dominance of the nonadsorbing block in determining the adsorbed amounts in this high size asymmetry situation is consistent with the observations made by Munch[10] and by Taunton et al.[37] in explaining the adsorption of copolymers with small anchors.

2.4. EXPERIMENTAL STUDIES OF ADSORPTION OF BLOCK COPOLYMERS WITH POLYELECTROLYTE BUOYS.

Recently, some studies of the adsorbed amounts of block copolymers with charged buoy blocks have been published.[25] Diblock copolymers of poly(*tert*-butylstyrene)-sodium poly(styrene sulfonate) (PtBS-NaPSS) were synthesized. To ensure the water solubility of the copolymers, these diblocks are highly asymmetric with a small hydrophobic PtBS block relative to the hydrophilic PSS block. In a water solution, the hydrophobic block will form a collapsed globule while the hydrophilic block will be solvated with a characteristic coil dimension that will depend on the size of the block and the salt concentration in solution.[20] Disassociation of the sodium counterions on the polyelectrolyte blocks will occur in solution. A NaPSS homopolymer of molecular weight 10^5 g/mol was studied for comparison. The structure of these diblock copolymers is shown in Figure 5 and the molecular characteristics of the polymers used in this study are summarized in Table II.

Salt solutions were prepared from pure water obtained by distillation, deionization and reverse osmosis. Polymer solutions at a concentration of 1 mg/mL were prepared by dissolution in distilled water and equilibration for at least one week before any adsorption experiment. The polymer solutions were then diluted by the addition of water or 1M salt solution to the required concentration and used for the adsorption experiments on silica surfaces. Adsorption experiments were performed using a SOPRA® ES4G spectroscopic ellipsometer with a rotating polarizer and fixed analyzer

The ellipsometric angles Δ and Ψ correspond to the change, on reflection, of the relative phase and amplitude of *p* (parallel to the plane of incidence) and *s* (perpendicular to the plane of incidence) components of the light wave. Δ and Ψ are determined

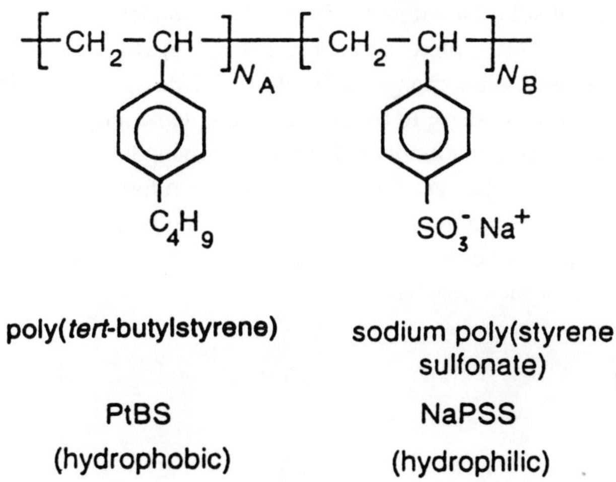

Figure 5: Molecular structure of neutral-charged block copolymer used in this study.

experimentally by the analyzer setting *A* with respect to the plane of incidence and the variation of the light intensity as the polarizer is rotated.

Adsorption measurements were first made in water without added salt. The following procedure was used: ellipsometry measurements were taken on a silica substrate immersed in pure water. The adsorption cell was then drained, the cell was refilled with a block copolymer solution of concentration 60 ppm in water, and measurements were taken as a function of time.

A similar experiment was conducted in a 1M aqueous *NaCl* solution. In this case, initial measurements were taken on a silica substrate immersed in salt solution with no added polymer. The adsorption cell was then drained, the cell was refilled with a 1M aqueous NaCl solution containing block copolymer at a concentration of 60 ppm, and measurements were taken as a function of time.

Representative data sets are shown in this section. Figure 6 shows the variation with time of the ellipsometric parameters $\tan\Psi$ and $\cos\Delta$ for the MT3/water and MT3/aqueous salt solution systems. Values were recorded every 11 s for the first 1.5 hours and then less frequently, as appropriate. Data for the first 1.5 h are shown in this plot. The zero on the *x* axes marks the point in time when polymer solution was added to the cell (t_0). Values of $\tan\Psi$ and $\cos\Delta$ shown at t_0 are for initial measurements in water or aqueous salt solution without polymer. We can clearly deduce from this figure that, in the absence of salt, there is insufficient aggregation of polymer at the silica surface to cause an observable change in the ellipsometric parameters. However, at a salt concentration of 1M, a decrease in $\cos\Delta$ and a complementary increase in $\tan\Psi$ occur after addition of the polymer solution, indicating a growing layer.

Figure 7 compares the variation of $\tan\Psi$ and $\cos\Delta$ for MT2 and PSS as a function of time, under similar solution conditions. Clearly, the homopolymer polyelectrolyte, whose molecular size is comparable to that of the charged block in MT2, shows a smaller extent of adsorption than MT2.

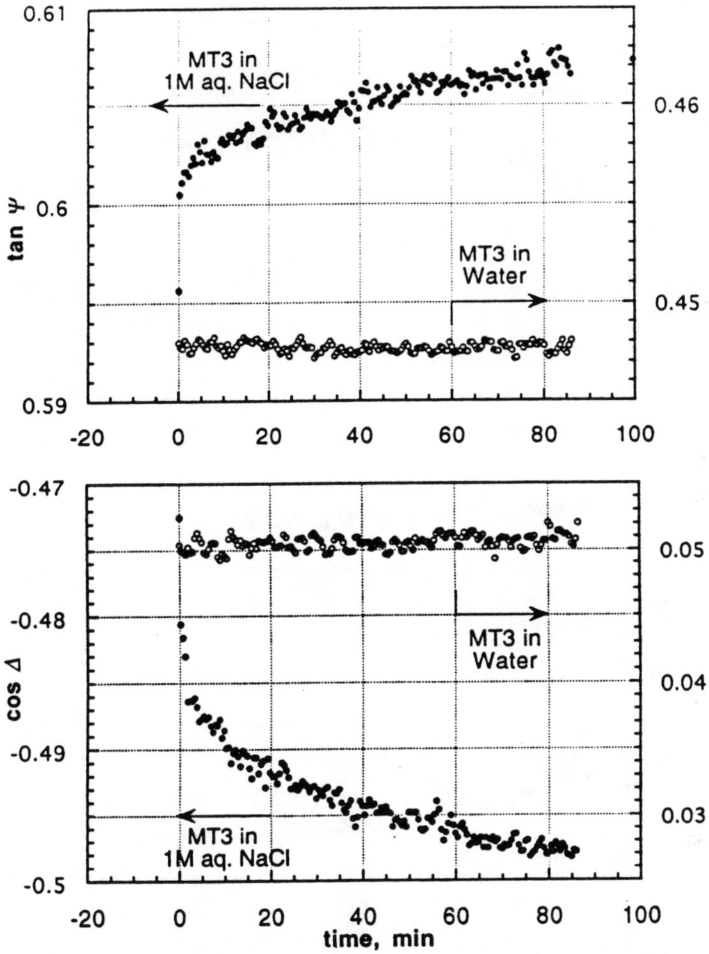

Figure 6: Variation of ellipsometric parameters as a function of time for adsorption of MT3 (neutral-charged diblock) on a silica surface from aqueous solutions of polymer. The polymer concentration in solution was 60 ppm (weight of polymer/weight of solution).

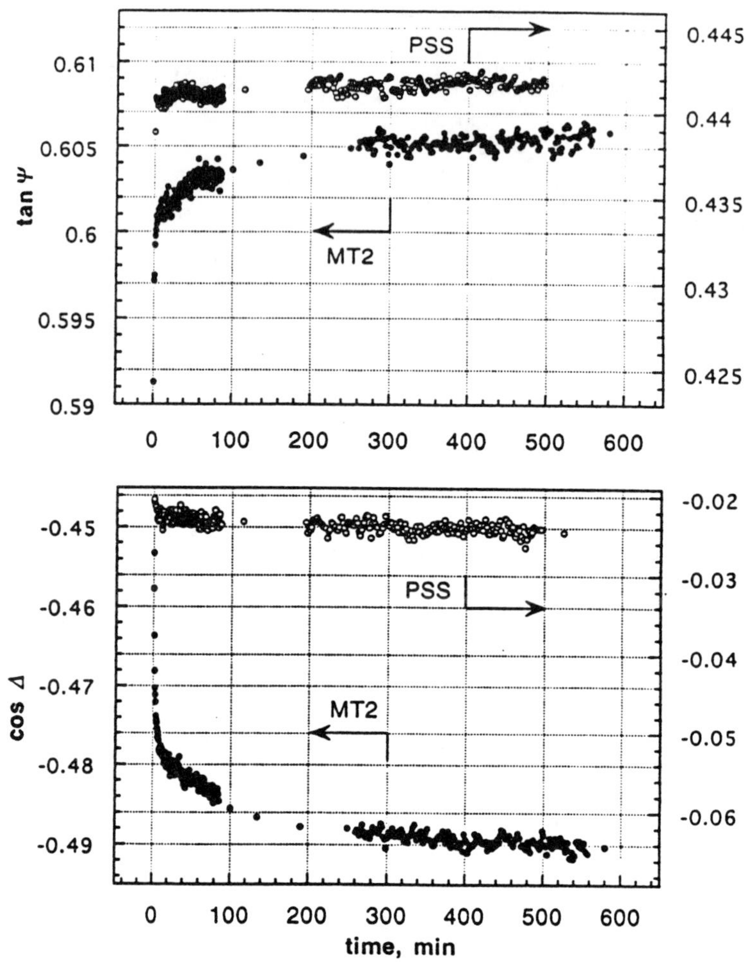

Figure 7: Variation of ellipsometric parameters as a function of time for adsorption of MT2 diblock and PSS homopolymer of comparable size from aqueous solutions of polymer. The polymer concentration in solution was 60 ppm in both cases, and NaCl was present at a concentration of 1 M.

Table II shows the adsorbed amount calculated from the experiments described in the previous section for each of the polymer systems studied. Only *final* adsorbed amounts calculated from data taken at long times under different adsorption conditions are listed here.

TABLE II. Polymer molecular characteristics

Sample	$10^3 M_w$, g/mol	M_w/M_n	% of phenyl groups sulfonated on PSS block	N_{NaPSS}	N_{PtBS}
MT2	87	1.03	89	404	26
MT3	160	1.04	87	757	27
NaPSS	100	1.1	90	485	-

The following observations can be made from these results:

a) No adsorbed amount could be measured for polyelectrolyte homopolymer or neutral-polyelectrolyte block copolymer in pure water without added salt. It seems logical to conclude that repulsion between charged groups on the polyion chain or block in the absence of electrostatic screening inhibits polymer accumulation at the solid-solution interface.

b) At a salt concentration of 1 M, electrostatic interactions are screened sufficiently to allow aggregation of polymer at the solid surface; adsorbed amounts can now be measured from ellipsometric data. Semiquantitatively, this can be understood as follows: the range of repulsive interactions in the presence of dissolved salt is given by the Debye length κ^{-1} where $\kappa^2 = 8\pi I\, l_B$; I is the ionic strength of the solution, and the Bjerrum length, l_B, is given by $l_B = e^2/\varepsilon T$, where e is the electronic charge, ε is the dielectric constant of the solvent and T is the temperature in energy units. For distances larger than κ^{-1}, electrostatic interactions in solution are considered to be screened. The Debye length should be compared to the effective distance between repulsive charges on the polyelectrolyte chain. This would be the monomer length (~3 Å) for the charge densities considered, except that counterion condensation[38] reduces the effective polymer charge for distances smaller than l_B (~8 Å). The Bjerrum length is, therefore, the appropriate length scale for electrostatic repulsion along the chain. At 25°C, κ^{-1} changes from around 30 Å at a sodium chloride concentration of 0.01 *M* to 10 Å at 0.1 *M*, and 3 Å at 1 *M*. At 1 *M*, then, the Debye screening length is smaller than l_B, and the conformational and aggregation properties of the polyelectrolyte chain or block should resemble those of a neutral chain in moderate-to-good solvent conditions.

c) Adsorbed amounts are 4-5 times higher for MT2 than for PSS homopolymer of molecular size comparable to that of the charged block in MT2, for adsorption from a

60 ppm solution in 1M salt. This indicates that the short, hydrophobic, PtBS block in MT2 plays a role in anchoring the block copolymer to the wafer surface: such a configuration is logical since it would minimize the (unfavorable) contact between the hydrophobic block and the aqueous solution. However, a more detailed description of the structure of the adsorbed layer is difficult to deduce from the adsorbed amount alone.

TABLE III. Adsorbed amounts for polymers

Polymer	Adsorption medium	Conc., ppm	Adsorbed amounts,* A, mg/m^2	
			Before rinsing with water	After rinsing with water
NaPSS	water	60	-#	
	1M aq. salt soln.	60	0.45	0.45
MT2	water	60	-#	-
	1M aq. salt soln.	60	2.1	1.9
MT3	water	60	-#	-
	1M aq. salt soln.	60	1.8	1.9
	1M aq. salt soln.	300	1.8	1.8

* Calculated adsorbed amounts have an average associated error of 10%. # No adsorption could be measured.

Experimental results from this study are in qualitative agreement with predictions: According to Equation 15, mass of polymer adsorbed per unit area (A) should be *ca.* 13% higher for MT2 than for MT3 on account of the difference in PSS block length between these systems, all solution conditions being similar. Experimentally, a range of 5-15% is calculated from different experiments. It needs to be noted that values of A are measured with an average error of 10%, which is comparable to the expected difference in A between the two systems from scaling arguments. However, experimental values appear to be systematically lower for MT3, as can be seen in Figure 11. More importantly, Equation 10 predicts that the ratio of the *number of chains* per unit area (σ) should be around 2:1 for MT2 versus MT3. This is the result that is seen experimentally, with statistical significance well outside the bounds of experimental error.

The effect of salt concentration is expected to become significant when local flexibility in the polyelectrolyte brush starts to be affected by electrostatic interactions. Quantitatively, one may compare the intrinsic segment length of PSS, L_P, which is around 12 Å,[20] with the electrostatic persistence length, L_E, which varies inversely with the salt concentration. For PSS, L_E is given by:[20]

$$L_E \approx 3.5\times10^{-2}/\kappa^2, \tag{19}$$

where κ^{-1} is the Debye screening length. L_E is only about 0.3 Å for PSS at a salt concentration of 1 *M*, rising to 3 Å at 0.1 *M*, and 30 Å at 0.01 *M*. For $L_E > L_P$, therefore, adsorbed amounts at equilibrium are expected to increase with increasing salt concentration along with a decrease in brush height for a given surface coverage, as predicted by eq 12 and 13. At 1 M salt, charges on the polyelectrolyte are sufficiently screened so that the aggregation properties of the neutral-charged diblocks resemble those of an uncharged diblock in a non-polar solvent.

2.5. PROPERTIES OF ADSORBED BLOCK COPOLYMER LAYERS: SURFACE FORCE-DISTANCE PROFILES.

TABLE IV. Characteristics of block copolymer layers used for force measurements

copolymer		$2L_{contact}$ (Å)[a]	$10^{-16}\sigma$ (m^{-2})	$10^{-16}\sigma^*$ (m^{-2})[b]
PVP-PI	26-50	33	1.25	0.39
	30-217	30	0.34	0.067
	38-69	37	1.00	0.26
	69-39	31	0.87	0.52
PVP-PS	60-60[c]	31	0.82	0.48
	60-90[d]	43	0.88	0.28

[a] $2L_{contact} = D$ measured at molecular contact of dry layers.
[b] $\sigma^* = 1/\pi\langle s^2\rangle$.
[c] $10^{-3}\ M_{PVP} = 60$, $10^{-3}\ M_{PS} = 60$.
[d] $10^{-3}\ M_{PVP} = 60$, $10^{-3}\ M_{PS} = 95$.

Figure 8 shows the dependence of the forces on distance between the mica substrates for the six polymers for which the molecular and adsorption characteristics are given in Table IV. All of the block copolymers, in good solvent conditions for the nonadsorbing blocks, exhibit monotonic repulsion, as has been reported previously.[28] Detectable repulsion is seen for all of the polymers at a range which exceeds (by more than a factor of 10 in some cases) the dimension in free solution of the nonadsorbing blocks of comparable molecular weights. Table IV documents that all of these copolymers adsorb on the surface at an average distance between chains which is less that the average dimension of the same molecule in solution. The surface number density of chains exceeds (by a factor of 1.5-5) the density required for mutual overlap, creating a semidilute surface layer or polymer brush.[3]

In order to understand the relationship among the different sets of data for different molecular weight and surface densities in Figure 8 in more detail, we can compare the data with models of the force-distance profiles of tethered brushes.[6,8,9]

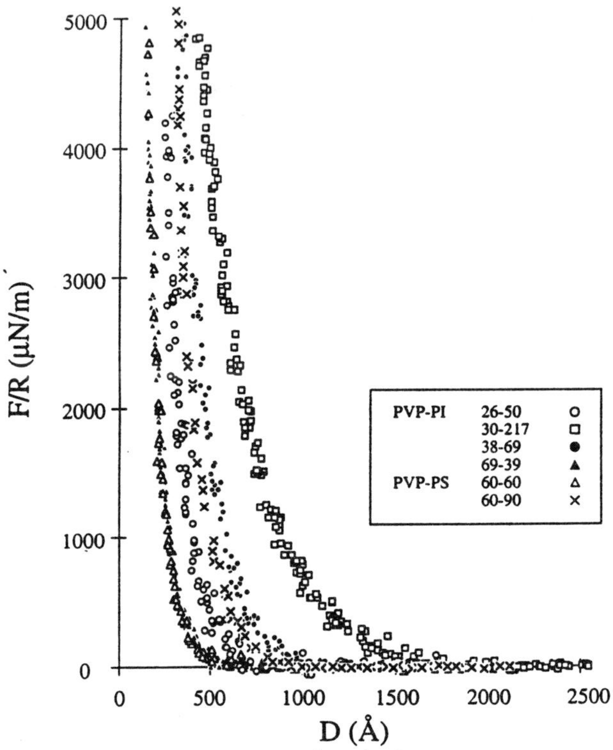

Figure 8: Force-distance profiles obtained for the PVP-PI and PVP-PS block copolymer layers in toluene at 32°C.

The first model inviting comparison with these data is that of Patel et al.[9] where excluded volume effects are treated but the density profile of segments within the brush is assumed to be constant (that is, a step-function profile). In this respect, it is the direct extension of the Alexander-de Gennes model of single brushes to two interacting brushes. It is also related to a model of interacting brushes proposed by de Gennes.[8] We shall term the model of Patel et al.[9] the PTH model and use it here in a form most suitable for comparison with the data.

The PTH model uses the essential Alexander-de Gennes idea that the equilibrium interaction between two proximal brushes is determined by the balance between the osmotic and elastic energies of the brushes. The basis on which we will write the model is first in terms of the energy of one tethered layer per unit area. Then, with the Derjaguin approximation for relating force to energy per unit area,[39] the model for the force distance profiles is conveniently expressed in terms of dimensionless, reduced force, **f**

$$\mathbf{f} = (F / R)\ [kT\sigma^{(2v+1)/2v} a^{1/v} N_B] \tag{20}$$

and reduced to distance, ∂:

$$\partial = [D - 2L_{PV} P^{\circ}] / [2\sigma^{(1-v)/2v} a^{1/v} N_B] \tag{21}$$

The reduced force is (4π times) the measured interaction energy per unit area divided by the energy per unity area (energy per blob times the number of blobs per unit area) in the uncompressed, infinite separation brushes. The reduced distance is the separation between the surfaces to which the nonadsorbing blocks are attached.

Figure 9 compares the data shown in Figure 8 with the variables in reduced form. We note that the data for the block copolymer layers having different N, σ, and chemical

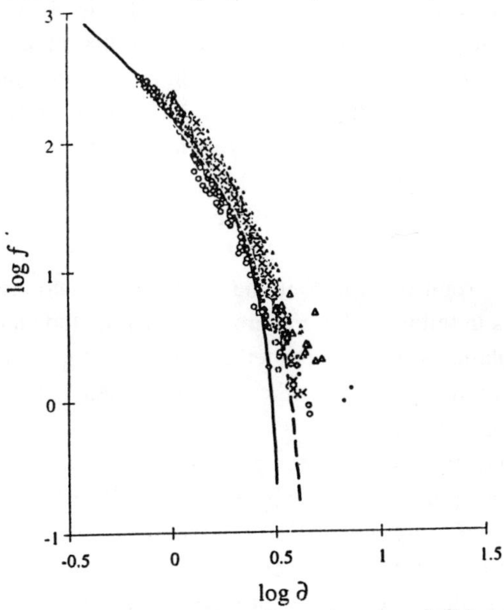

Figure 9: Comparison of the surface force data (symbols) and the prediction of the model of Patel et al.[9] (thick solid line) on double logarithmic scales. The sense of the symbols is the same as in Figure 8. The thick dashed line indicates the prediction of the model of Milner et al.[6]

composition (PI and PS) are well-collapsed in a single curve with the reduced variables **f** and ∂. While there remains a certain dispersion in the plot, it is not systematic with respect to tethered chain length, surface density, or chain type. Some of this small dispersion may be due to our lack of accounting for the selling of the PVP anchor blocks or the fact that the covalent bond joining the anchor and the nonadsorbing block is not rigidly fixed at a precise elevation above the mica surface. The point of onset of detectable forces is particularly well-scaled through the reduced variable ∂.

Milner et al.[6] have developed a model (MWC) that treats the segment density profile in the tethered brush more realistically. Osmotic and elastic contributions to the free energy are still the ingredients of the model. In the MWC model, however, the balance between osmotic repulsion and elastic resistance is done locally, at each point along the chain, and self-consistently, in contrast to the PTH model which assumes that every chain is equally stretched and balances effects over the chain as a whole. The MWC model is based on a less restrictive assumption than uniform stretching; namely, it assumes that every region of the chain is strongly stretched locally to the point that fluctuations around the most probable chain trajectory are unimportant. This model produces a segment density distribution that varies with distance from the tethering surface.

In Figure 9, the heavier, dashed line indicates the free energy calculated from the MWC semidilute model using parameters equivalent to the PTH model. These two models converge in the high compression limit since, in this region of compression, the more detailed treatment of stretching by MWC, and the resultant differences in the segment density profiles between the two models, have been rendered irrelevant by the imposition of confinement by squeezing the layers and the consequent uniform segment density in the gap. On the other hand, in the larger ∂ limit, corresponding to **f** less than about 10, the MWC agrees better with the data plotted in this universal format. The nearly parabolic profile predicted by the MWSC model is necessary for an accurate prediction of the onset of repulsive forces in the weak compression regime.

Future directions in tailoring these interactions will include the study of controlled distribution in the molecular weights of the tethered chains (e.g., bimodal brushes[40]) and interactions among the tethered chains other than excluded volume (e.g., electrostatic interactions in tethered polyelectrolytes[41]).

Recently, we have shown how force profiles between adsorbed block copolymer layers can be tailored for specific purposes by mixing lengths of polymers.[42] We created a bimodal distribution of polymer chains on the solid surface by adsorbing from solutions containing either a binary distribution of diblock copolymers[43] or a triblock copolymer in which a middle block is adsorbing and the two end blocks, differing in chain length, are nonadsorbing. We found that the bimodal layer force profiles[42] did not conform to the universal curve profile of the monodisperse diblock copolymers[28,29] except in the high compression, osmotic pressure limit, where the undeformed configurations of the tethered chains are no longer relevant. This deviation was attributed to the presence of two characteristic length scales in the bimodal layer, instead of the single characteristic length determined by chain length for the monodisperse case.

2.6. SURFACE FORCE MEASUREMENT BETWEEN LAYERS OF HYDROPHOBICALLY ANCHORED POLYELECTROLYTE CHAINS.

The range of interactions between surfaces carrying the ionic buoys strongly depend on the concentration of salt in the solution, as shown in Figure 10 in qualitative agreement with the predictions for such chains. As concentration of the salt, NaCl, decreases from 1 M to $6*10^{-5}$ the range of interactions, approximately equal to 2L, increases from 400Å to 1600Å for the 4-83 PtBS-PSS diblock. The 4-160 diblock

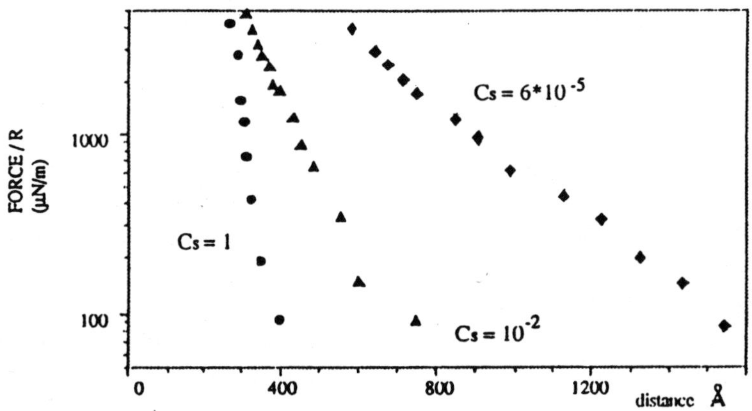

Figure 10: Influence of the salt concentration on the range of the forces between PtBS-PSS diblock copolymers (MT2).

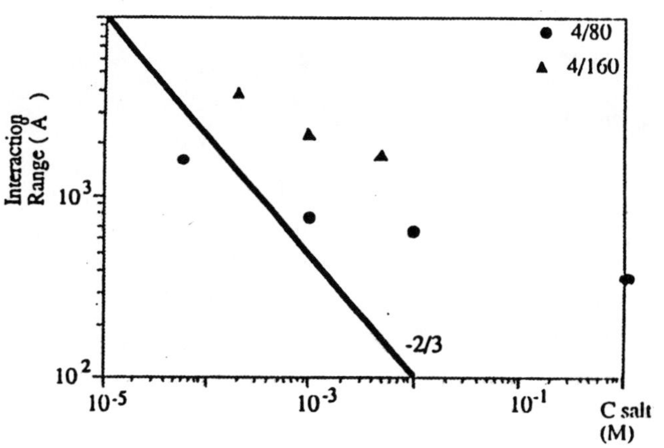

Figure 11: The effect of salt concentration on interaction range for PtBS-PSS layers.

shows a similar trend. However, plotting the interaction range as a function of salt concentration does not show the expected -2/3 power law (Figure 11), but a much weaker dependence of the interaction range on the salt concentration.

3. Conclusions and Directions for Future Work

Block copolymer adsorbed layers afford a simple and powerful means of manipulating the characteristics of surfaces and interfaces. Work of the last decade, reviewed here, has developed great insight into the structure of these layers. Several tools, beyond those discussed already, such as neutron reflectometry,[44-48] have provided accurate data on the complete segment density profile in the brush layer. Furthermore, the universality embodied in eqs 10, 20 and 21 mean that, with *a priori* knowledge of just a few parameters: N_A,N_B, segment size and solvent quality for the B block; we can **predict** with very good accuracy, the adsorbed amount, the segment profile and the force-distance profile, for neutral diblock copolymer layers.

This means that the study of the structures of these layers is now a mature field. Effort should now be invested in other directions such as: kinetics of self-assembly at surfaces, where a few pioneering studies have appeared,[10] tailoring the structure of layers by molecular architecture and multicomponent assembly; investigations of the **properties** of layers and their relation to structure, rather than exclusive concentration on structure (The work of Klein and coworkers[41] on shearing self-assembled polymer brushes is a good example of this.); experimental investigations of layers of charged, tethered chains.

In a broader sense, the assembly of block copolymer layers at surfaces is a geometrically simple model of a wide class of aggregation phenomena and resulting properties of self-organizing polymers.[1]

4. Acknowledgments

This chapter summarizes work, much of it from our own group at Minnesota, that has been done in collaboration with many coworkers whom I would like to acknowledge with thanks: Georges Hadziioannou, Hiroshi Watanabe, Jimmy Mays, Sanjay Patel, Ed Parsonage, Nily Dan, Avi Halperin, Jean-François Argillier, Patrick Guenoun, Cathy Amiel, Mohan Sikka, Sunil Dhoot, Larry Chen, Guang-Zhao Mao, Jin Schneider and Yuan-Yuan Zhang.

5. References

1. Halperin, A., Tirrell, M., Lodge, T. (1992)*Adv. Polymer Sci.* **100**, 31.

2. Milner, S.T. (1991) *Science* **251**, 905.
3. Marques, C.M., Joanny, J.-F., Leibler, L. (1988) *Macromolecules* **21**, 1051.
4. Alexander, S. (1977) *J. Phys. (Paris)* **38,** 977 & 983.
5. de Gennes, P.-G. (1976) *J. Phys. (Paris)* **37**, 1443; (1980) *Macromolecules* **13**, 1069.
6. Milner, S.T., Witten, T.A., Cates, M.E. (1988) *Macromolecules* **21**, 2610; (1988) *Europhys. Lett.* **5,** 413; (1989) *Macromolecules* **22**, 853; Milner, S.T., Wang, Z.G., Witten, T.A. (1989) *Macromolecules* **22**, 489; Hirz, S.J., (1986) M.S. Thesis, University of Minnesota; Skvortsov, A.M., Pavlushkov, I.V., Gorbunov, A.A., Zhulina, E.B., Borisov, O.V., Pryamitsin, V.A. (1988) *Polym. Sci. USSR (Engl. Transl.)* **30**, 1706.
7. Shim, D.F.K., Cates, M.E. (1989) *J. Phys. (Paris)* **50**, 3535.
8. de Gennes, P.-G. *C.R.* (1985) *Acad. Sci., Ser. 2* **300**, 839.
9. Patel, S., Tirrell, M., Hadziioannou, G. (1988) *Colloids Surf.* **31**, 157.
10. Munch, M.R., Gast, A.P. (1988) *Macromolecules* **21**, 1366; van Lent, B., Scheutjens, J.M.H.M. (1989) *Macromolecules* **22**, 1931; Evers, O.A., (1989) Wageningen Agricultural University.
11. Balazs, A.C., Lewandowski, S. (1990) *Macromolecules* **23**, 839.
12. Hamaker, H.C. (1937) *Physica* **4**, 1058.
13. Lifshitz, E.M. (1960) *Sov. Phys. JETP* **37**, 161.
14. Parsonage, E., Tirrell, M., Watanabe, H., Nuzzo, R.G. (1991) *Macromolecules* **24**, 1987.
15. Leermakers, F.A., Gast, A.P. (1991) *Macromolecules* **24**, 718.
16. Guzonas, D., Boils, D., Hair, M.L. (1991) *Macromolecules* **24**, 3383.
17. de Gennes, P.G., Pincus, P., Velasco, R.M., Brochard, F. *J. Phys. Fr.* **1976**, *37*, 1461.
18. Odijk, T., Houwaart, A.C. (1978) *J. Polym. Sci., Polym. Phys. Ed.* **16**, 627.
19. Fixman, M., Skolnick, J. (1978) *Macromolecules* **11**, 863.
20. Wang, L., Yu, H. (1988) *Macromolecules* **21**, 3498.
21. Takahashi, A., Kato, T., Nasagawa, M. (1967) *J. Phys. Chem.* **71**, 2001.
22. Marra, J., van der Schee, H.A., Fleer, G.J., Lyklema, J. (1983) in *Adsorption From Solution*, Ottewill, R.H., Rochester, C.H., Smith, A.L., Eds., Academic Press: New York.
23. Dan, N., Tirrell, M. (1993) *Macromolecules* **26**, 4310.
24. Pincus, P. (1991) *Macromolecules* **24**, 2912.
25. Amiel, C., Sikka, M., Schneider, J.W., Tsao, Y-H., Tirrell, M., Mays, J.W. (1995) *Macromolecules* **28**, 3125.
26. Tassin, J., Siemens, R., Tang, W., Hadziioannou, G., Swalen, J., Smith, B. (1989) *J. Phys. Chem.* **23**, 2106.
27. Stouffer, J., McCarthy, T. (1988) *Macromolecules* **21**, 1204.
28. Hadziioannou, G., Patel, S., Granick, S., Tirrell, M. (1986) *J. Am. Chem. Soc.* **108**, 2869.
29. Tirrell, M., Patel, S., Hadziioannou, G. (1987) *Proc. Natl. Acad. Sci. U.S.A.* **84**, 4725.
30. Watanabe, H., Tirrell, M. (1993) *Macromolecules* **26**, 6455.
31. Marra, J. Hair, M. (1989) *Colloids Surf.* **34**, 215.

32. Israelachvili, J. (1973) *J. Colloid Interface Sci.* **44**, 259.
33. Patel, S.S., Tirrell, M. (1989) *Annu. Rev. Phys. Chem.* **40**, 597.
34. Ansarifar, M.A., Luckham, P.F. (1988) *Polymer* **29**, 329.
35. Webber, R.M., Anderson, J.L., John, M.S. (1990) *Macromolecules* **23**, 1026.
36. Higo, Y., Ueno, N., Noda, I. (1983) *Polym. J.* **15**, 367.
37. Taunton, H.J., Toprakcioglu, C., Fetters, L.J., Klein, J. (1988) *Nature* **332**, 712; (1990) *Macromolecules* **23**, 571.
38. Manning, G.S. (1969) *J. Chem. Phys.* **51**, 924.
39. Derjaguin, B.V. (1934) *Kolloid. Z.* **69**, 155.
40. Parsonage, E.E., Watanabe, H., Dhoot, S., Tirrell, M. (1991) *Polym. J. (Tokyo)* **23**, 641.
41. Watanabe, H., Patel, S.S., Argillier, J.-F., Parsonage, E.E., Mays, J.M., Dan, N., Tirrell, M. (1992) *Mater. Res. Soc. Symp. Proc.* **249**, 255.
42. Dhoot, S., Watanabe, H., Tirrell, M. (1994) *Colloids Surf.* **86**, 47.
43. Dhoot, S. and Tirrell, M. (1995) *Macromolecules* **28**, 3692.
44. Cosgrove, T., Phipps, J.S., Richardson, R.M., Hair, M.L. and Guzonas, D.A. (1993) *Macromolecules* **28**, 4363
45. Mansfield, T.L., Iyengar, D.R., Beaucage, G., McCarthy, T.J., Stein, R.S. and Composto, R.J. (1995) *Macromolecules* **28**, 492
46. Perahia, D., Wielser, D.B., Satija, S.K., Fetters, L.J., Sinha, S.K. and Milner, S.T. (1994) *Phys. Rev. Lett.* **72**, 100.
47. Karim, A., Satija, S.K., Douglas, J.F., Ankner, J.F. and Fetters, L.J. (1994) *Phys. Rev. Lett.* **73**, 3407.
48. Kent, M.S., Lee, L.T., Factor, B.J., Rondelez, F. and Smith, G. (1993) *J. Phys. IV (France)* **3**, 49.
49. Klein, J. (1992) *Pure Appl. Chem.* **64**, 1577.

COPOLYMER MICELLES IN AQUEOUS MEDIA

Z. TUZAR
Institute of Macromolecular Chemistry
Academy of Sciences of the Czech Republic
162 06 Prague 6, Czech Republic

1. Introduction

Until a few years ago, most of the studies on copolymer micelles were done using organic selective solvents [1]. In recent years, more attention has been paid to micelles of hydrophobic/hydrophilic block copolymers in water, in aqueous buffers, and in water-rich mixtures with organic solvents. Reasons for this delayed interest could have been, e.g., a belief that block copolymer micelles in aqueous media can add nothing substantially new to the classical surfactants, relative difficulties in the preparation of well-defined hydrophobic/hydrophilic block copolymers, or the experience that many hydrophobic/hydrophilic block copolymers failed to form micelles upon direct dissolution in water.

In last few years, we are witnessing of an avalanche of papers dealing with block copolymer micelles in aqueous media, most probably due to their practical application as emulsifiers, dispersing and foaming agents, thickeners, solubilizers and, mainly, as vehicles for a targetted drug delivery (e.g., [2-7]).

Since the structure and various properties of these systems are subjects of several Chapters in this book, I am giving here a sketchy overview only, emphasizing experimental contribution of my co-workers to the topic.

2. Micelle Formation

First systematic study of the micellizing of amphiphilic copolymers concerned block and graft copolymers, polystyrene-b-poly(4-vinyl-N-ethylpyridinium bromide) and polystyrene-g-poly(4-vinyl-N-ethylpyridinium bromide), respectively, in mixtures methanol/water with a small amount of LiBr [8-11]. It was found that block copolymers formed polymolecular

S.E. Webber et al. (eds.), Solvents and Self-Organization of Polymers, 309–318.

micelles above a certain content of water in the mixture [8-10], whereas graft copolymers were molecularly dissolved regardless of the solvent composition [11]. As solvent mixtures with a higher amount of water are better solvents for grafts and stronger precipitants for polystyrene backbones, the observed decrease in the intrinsic viscosity values without any change in molar mass could have been interpreted as follows [11]: In methanol-rich mixtures, graft copolymer molecules assume a loose, comb-like conformation, (Fig. 1a) and in water-rich mixtures a star-like conformation (Fig. 1b) with the backbone collapsed.

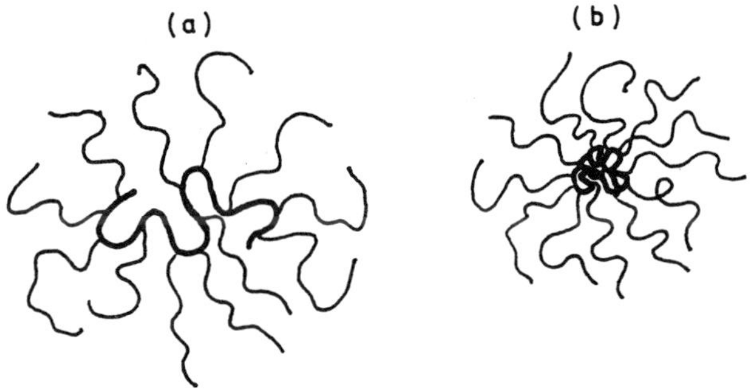

Figure 1. Schematic representation of comb-like (a) and star-like (b) conformations of a graft copolymer molecule.

Most of the amphiphilic copolymers studied in recent years can be devided into two groups: poly(ethylene oxide) -b-poly(propylene oxide)-b-poly(ethylene oxide) (further PEO-PPO-PEO) copolymers, known under commercial names Pluronics or Poloxamers, and laboratory prepared di- and triblock copolymers AB and ABA, where A is a hydrophilic block as poly(methacrylic acid) or poly(ethylene oxide), and B polystyrene.

2.1. MICELLES OF PEO-PPO-PEO COPOLYMERS

PEO-PPO-PEO copolymers represent a bridge between classical surfactants and block copolymers. Due to relatively low molar masses (a few thousands) and temperature-dependent phase transitions, these copolymers can form, depending on temperature and concentration, true solutions, micelles of different shapes or physical gels of various structures. PEO-PPO-PEO are presently the most studied associating copolymers.

Experimental and theoretical results are summarized in several review papers (e.g., [2,7,12]).

2.1.1. *Micellization of PEO-PPO-PEO*

Early results on PEO-PPO-PEO micellization were often controversial. The so called anomalous micellization, i.e., a presence of extremely large particles, orders of magnitude larger then regular micelles (cloudy appearence of the solution) has been observed at the onset of micellization (in the vicinity of the critical micelle concentration, cmc, or the critical micelle temperature, cmt). It has been demonstrated by Zhou and Chu [13,14] that a strong heterogeneity in chemical composition is responsible for this annoying effect. When a fraction rich in PPO was removed, an anomaly disappeared.

2.1.2. *cmc and cmt*

One of the similarities with micellar systems of soaps and surfactants is a relatively high value of cmc (often several per cent), which in case of PEO-PPO-PEO strongly depends on temperature. Since the solubility of PPO decreases with increasing temperature, typical copolymers like Pluronic L-64, PEO_{13}-PPO_{30}-PEO_{13} (subscripts denote numbers of monomer units), are molecularly dissolved at room temperature and start forming micelles at higher temperatures. Generally, cmc decreases with increasing temperature [15,16].

It is evident that the same (by label) Pluronic or Poloxamer copolymers, being commercial products, may slightly differ from batch to batch. Removal of the PPO-rich fraction (mainly by filterring-off the turbid portion) does not lead to identical samples, either. This is the reason why the cmc values, reported by different laboratories, even those determined by the same experimental technique, differ [17].

2.1.3. *Micellar structure*

Particles resulting from a spontaneous association have been studied mainly by static and dynamic light scattering, small-angle X-ray and neutron scattering, NMR, rheological methods, etc. (e.g., [7,12]. Unlike high-molar-mass block copolymers in selective organic solvents, PEO-PPO-PEO copolymers in water form not only spherical but, depending on the block lengths, temperature, and concentration, also rod-like micelles [7,18], the so-called secondary large micelles [12], gels, and various liquid crystalline structures [7].

The spherical core/shell micelles are the most frequent ones. While the radii of typical surfactant micelles are 2–3 nm and those of regular block copolymer micelles several tens of nm [1], radii of PEO-PPO-PEO micelles are about 10 nm [7]. It has also been shown that the core/shell interface is rather sharp in case of regular block copolymer micelles [1], but in case of PEO-PPO-PEO micelles is very diffuse [7]. In some cases, spherical micelles may form at a certain concentration and temperature a quasi-crystalline body-centered cubic lattice [19]. Similar arrangement has been observed also with regular block copolymer micelles in organic solvents [20,21] but at much lower concentrations.

2.2. MICELLES OF HIGH-MOLAR-MASS BLOCK COPOLYMERS

Here, micelles of AB and ABA di- and triblock copolymers, where A is poly(ethylene oxide) (PEO) or poly(methacrylic acid) (PMAc) and B polystyrene (PS), are briefly discussed, mainly their micelle formation and electrophoretic mobility. Other properties, namely micelle structure and dynamics on segmental level, probed by fluorescence techniques, will be dealt with in detail in other Chapters.

2.2.1. *Micelle Formation*

All types of copolymers mentioned above form micelles with PS cores and hydrophilic shells by direct dissolution in water only when their molar mass is relatively small (below ca. 2×10^4 g mol^{-1}) and the mass fraction of PS is below ca. 0.2 [22,23]. Otherwise, due to the strong hydrophobic character of polystyrene, copolymer samples are undispersable in water. Micelles must first be prepared in a mixture of water with some organic co-solvent, and then they can be transferred into water. In case of PS-*b*-PEO copolymers, a mixture with THF was used and then THF was removed by azeotropic distillation [24].

In case of PS-*b*-PMAc, it was possible to prepare aqueous micellar solutions with copolymers having molar mass up to 70×10^3 g mol^{-1} and mass fraction of polystyrene up to 0.60 [22,25]. First, a given copolymer had to be dissolved in a mixture of water and dioxane where copolymer molecules spontaneously associated to micelles with swollen polystyrene cores. These micelles, which were in a dynamic equilibrium with unimers, could be transferred to organic-solvent-free aqueous solutions by a stepwise dialysis. Naturally, during this process, the dynamic equilibrium became frozen and micelles with glassy polystyrene cores behaved like stable particles. During the dialysis, changes in molar mass and size of micelles, as well as the deswelling of the core, were monitored by SLS, DLS, and fluorescence techniques [25-27], the disappearance of the micellar equilibrium by sedimentation velocity [28].

2.2.2. *Electrophoretic Mobility*

Block copolymer micelles with polyelectrolyte shells represent particles, size and various properties of which are strongly influenced by many factors, e.g., by the degree of dissociation, pH of the solvent, polar interactions, etc. These micelles provide a unique material, in which polyelectrolyte properties can be studied at a very high segment concentration. Although there are numerous theoretical studies in this field, e.g. [29-31], experimental studies are rare (see Webber's Chapter).

In our recent study [32], we have studied micelles (association number ca. 200) with PS cores and PMAc shells (unimer molar mass 6.7×10^4, mass fraction of PS 0.54) in various buffers (pH5 – pH10) by DLS and isotachophoresis.

The results can be summarized as follows: In spite of a relatively large size (ca. 20 nm, Fig. 2), micelles migrate as fast as, e.g., IO_4^-

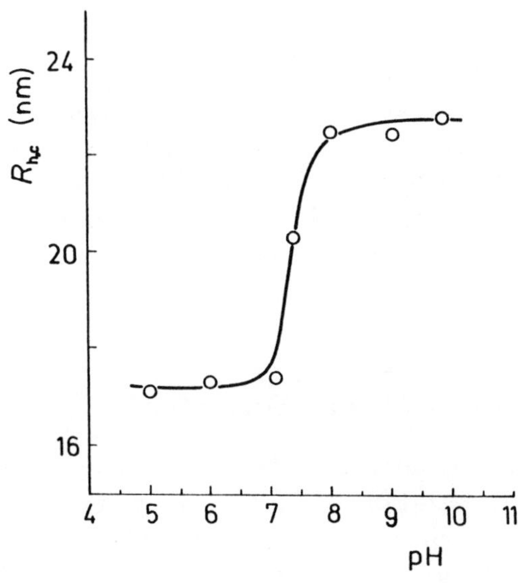

Figure 2. Hydrodynamic radius of a micelle as a function of pH.

anions. It means that the effective charge of a micelle outweights its friction. The typical S-shaped curve of the electrophoretic mobility vs. pH (Fig. 3) with a steep rise close to pH7 indicates a considerable increase in the dissociation of carboxylic groups in this region. The position of the inflection point, shifted a few pH units from a linear poly(methacrylic acid) homopolymer (below pH5), may indicate that a part of the shell experiences

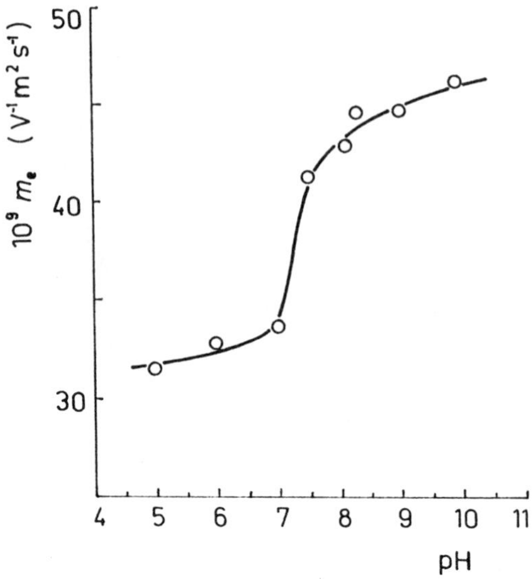

Figure 3. Electrophoretic mobility of micelles as a function of pH.

lower "effective pH" than has a given buffer. This assumption has been corroborated by a fluorescence study [33]: According to the steady-state fluorescence and anisotropy decays of dansyls attached to the ends of PMAc blocks, a certain fraction of dansyls (probably those on the blocks that were folded back to the core/shell interface) monitored a lower polarity of the environment, and their mobility was substantially restricted. It can be concluded that both degree of dissociation and an "effective pH" decrease from the shell perifery to the core/shell interface.

3. Thermodynamics of Micelle Formation

It has been shown in my first Chapter (Z. Tuzar: Overview of polymer micelles) that the micellization of block copolymers in organic selective solvents is enthalpy driven process. Micellization of PEO-PPO-PEO and PPO-PEO-PPO copolymers in water has been proved to be an entropy driven process [34,35], like in case of soaps and surfactants.

Conclusions in [34] are the following: A positive standard enthalpy of micellization shows that the transfer of a unimer into a micelle is an enthalpically unfavorable endothermic process. The negative Gibbs energy confirms the spontaneous nature of micelle formation. Thus, a negative entropy contribution must be the driving force for micellization. The presence of hydrophobic blocks in water causes a significant decrease in the water entropy, suggesting that it induces an increase in the degree of structuring of water molecules, owing to cavity formation. When the hydrophobic blocks aggregate forming micellar cores, the structure of water is restored and water entropy increases; this overcomes the loss of entropy due to the localization of the hydrophobic blocks in the micelles.

PPO-PEO-PPO forms micelles in water only et higher concentrations and at a narrow region of temperatures [35]. Although the experimental data showed that the association process was endothermic and thus entropy driven, the Gibbs energy was less negative than in case of the micellization of PEO-PPO-PEO copolymer. The effect has been interpreted as follows: In micelles of PPO-PEO-PPO, the cores are formed by the outer PPO blocks and the shells are formed by PEO which assume a loop geometry. These loops cause an entropy loss that is responsible for the reduced driving force for micellization.

4. Solubilization

Solubilization means an enhanced solubility of some compounds (solubilizates) brought about by the presence of other compounds (solubilizers). The phenomenon of solubilization was studied mostly in aqueous solutions of soaps and surfactants. Solutions of these compounds above cmc dissolve a certain amount of hydrocarbons that are only sparingly soluble in water. Molecules of the solubilizate enter into micelles, increasing somewhat their association number. Further addition of the solubilizate above an amount called the saturation limit leads to the formation of an additional phase, usually in the form of a macroemulsion. A macroemulsion is also formed by the solubilizate (hydrocarbon) in the presence of a solubilizer with a concentration below cmc. Such a case can be documented by the behavior of aqueous solutions of PEO-PPO-PEO with a small amount of hexane [36] . At 25°C, when the copolymer was dissolved molecularly, solutions were cloudy due to the presence of a macroemulsion, i.e., unsolubilized hexane droplets. At 35°C micellization of the copolymer took place, and hexane solubilized into micelles in an amount up to the saturation limit. Unlike micelles with a solubilizate below the saturation limit that are thermodynamically stable at given conditions, the macroemulsions are unstable.

In a comprehensive study dealing with the solubilization by block copolymer micelles, PEO-PPO and poly(N-vinylpyrrolidone)-*b*-polystyrene were employed as solubilizers and a series of aliphatic and aromatic hydrocarbons as solubilizates [37]. A comparison of the solubilization by surfactants and both above copolymers leads to an important conclusion: Unlike surfactants, block copolymers show a high specificity towards solubilizates. The amount of a solubilized compound is controlled by the Flory-Huggins parameter characterizing the interaction between the solubilizate and the core-forming copolymer blocks. So, aromatic hydrocarbons were solubilized into micelles with polystyrene cores to a much greater extent than the aliphatic ones. When a mixture of benzene and heptane was solubilized, the block copolymer micelles selectively solubilized benzene. Thus, block copolymer micelles may become an effective material for separation of various low-molar-mass mixtures.

Polymer micelles have lately been considered as a potential vehicle for controlled release of organic substances into aqueous media. Prior to release, it is desirable to solubilize the maximum possible amount of a given substance into micellar cores. In a recent study [38], the system of micelles with polystyrene cores and poly(methacrylic acid) shells, and a model solubilizate, pyrene, was explored. It has been shown that the amount of solubilized pyrene depended on the method of loading. When the process was performed in an aqueous solution, the maximum amount solubilized was ca. 250 pyrene molecules per one polystyrene core containing 23000 styrene units. When pyrene was loaded into the same micelles (having the same association number) but with swollen cores in a water/dioxane mixture, the amount of solubilized pyrene was several times higher. After dioxane had been removed by stepwise dialysis, still triple amount of pyrene, i.e., about 800 molecules, was found in a polystyrene core. In the first method, 250 pyrene molecules per one core can be considered an equilibrium amount. In the second method, the enhanced "solubilization" is caused by trapping pyrene molecules during dialysis in the glassy cores of the non-equilibrium, frozen micelles. The process of release of pyrene into water was studied by fluorescence methods. The release was found to proceed on the time scale of thousands of seconds.

Solubilization (which has been also treated theoretically [39,40]) and a controlled release of organic substances into and from block copolymer micelles, are subject of a special Chapter (S. Webber: Kinetics of hydrophobe absorption and controled release).

Acknowledgements. This research was supported by the Collaborative Research Grant 920166 from the Scientific and Environmental Division of the NATO and by the U.S.-Czech Science and Technology Joint Fund No. 95010.

References

1. Tuzar, Z. and Kratochvíl, P. (1993) Micelles of block and graft copolymers in solutions, *Surface and Colloid Science* **15**, 1–83.

2. Schmolka, I.R. (1991) Poloxamers in the pharmaceutical industry, in P.J. Tarcha (ed.), *Polymers in controlled drug delivery*, CRC Press, Boca Raton, 189–214.
3. Rolland, A., O'Mullane, J., Goddard, P., Brookman, and Petrak, K. (1992) New macromolecular carriers for drugs. I. Preparation and characterization of poly(oxyethylene-b-isoprene-b-oxyethylene) block copolymer aggregates, *J. Appl. Polym. Sci.* **44**, 1195–1203.
4. Hurter, N. and Hatton, T.A. (1992) Solubilization of polycyclic aromatic hydrocarbons by poly(ethylene oxide – propylene oxide) block copolymer micelles: Effect of polymer structure, *Langmuir* **8**, 1921–1299.
5. Kataoka, K. (1994) Design of nanoscopic vehicles for drug targeting based on micellization of amphiphilic block copolymers, *J. Macromol. Sci., Pure Appl. Chem.* **A31**(11), 1759–1769.
6. Chu, B. and Wu,G. (1994) Supramolecular formation of triblock coplymer in polar/nonpolar solvents, *Makromol. Chem., Macromol. Symp.* **87**, 55–67.
7. Almgren, M., Brown, W., and Hvidt, S. (1995) Selfaggregation and phase behavior of poly(ethylene oxide) - poly(propylene oxide) – poly(ethylene oxide) block copolymers in aqueous solution, *Colloid Polym. Sci.* **273**, 2–15.
8. Selb, J. and Gallot, Y. (1980) Micellisation de copolymères sèquences polystyrène-polyvinylpyridinium, 1. Méthodes d'études. Influence de la nature du solvent, *Makromol. Chem.* **181**, 1605–2624.
9. Selb, J. and Gallot, Y. (1980) Micellisation de copolymères sèquences polystyrène-polyvinylpyridinium, 2. Influence des caractéristique moléculaire des copolymères, *Makromol. Chem.* **182**, 1491–1511.
10. Selb, J. and Gallot, Y. (1980) Micellisation de copolymères sèquences polystyrène-polyvinylpyridinium, 3. Influence de la concentration en sel et de la température, *Makromol. Chem.* **182**, 1513–1524.
11. Selb, J. and Gallot, Y. (1980) Comportement de copolymères greffés en milieu solvent sélective, des greffons, *Makromol. Chem.* **182,** 1775–1786.
12. Chu, B. (1995) Structure and dynamics of block copolymer colloids, *Langmuir* **11**, 414–421.
13. Zhou, Z. and Chu, B. (1987) Anomalous association behavior of an ethylene oxide/propylene oxide ABA block copolymers in water, *Macromolecules* **20**, 3089–3091.
14. Zhou, Z. and Chu, B. (1988) Anomalous micellization and composition heterogeneity of a triblock ABA copolymer of (A) ethylene oxide and (B) propylene oxide in aqueous solution *Macromolecules* **21**, 2548–2554.
15. Linse, P. and Malmsten, M. (1992) Temperature dependent micellization in aqueous block copolymer solutions, *Macromolecules* **25**, 5434–5439.
16. Pospíšil, H., Pleštil, J., and Tuzar, Z. (1993) Small-angle neutron scattering study of poly(oxyethylene)-*block*-poly(oxypropylene)-*block*-poly(oxyethylene) in aqueous solutions, *Collect. Czech. Chem. Commun.* **58**, 2428–2436.
17. Kabanov, A.V., Nazarova, I.R., Astafieva, I.V., Batrakova, E.V., Alakhov, V.Yu., Yaroslavov, A.A., and Kabanov, V.A. (1995) Micelle formation and solubilization of fluorescent probes in poly(oxyethylene-*b*-oxypropylene-*b*-oxyethylene) solutions *Macromolecules* **28**, 2303–2314.
18. Schillén, K, Brown, W., and Johnsen, R.M. (1994) Micellar sphere-to-rod transition in an aqueous triblock copolymer system. A dynamic light scattering study of translational and rotational diffusion, *Macromolecules* **27**, 4825–4832.
19. Mortensen, K., Brown, W., and Nordén, B. (1992) Inverse melting transition and evidence of three-dimensional cubatic structure in a block-copolymer micellar system, *Phys. Rev. Letters* **68**, 2340–2343.
20. Watanabe, H. and Kotaka, T. (1984) Rheology and structure of styrene-butadiene diblock copolymers dissolved in selective solvents, *Polym. Eng. Rev.* **4**, 73–122.
21. Pleštil, J., Hlavatá, D. Hrouz, J., and Tuzar, Z. (1990) Dilute and semidilute solutions of ABA block copolymer in solvents for A or B blocks: 1. Small-angle

X-ray scattering study, *Polymer* **31**, 2112–2117.
22. Tuzar, Z., Kratochvíl, P., Procházka, K., and Munk, P. (1993) Block copolymer micelles in aqueous media (1993) *Coll. Czech. Chem. Commun.* **58**, 2362–2369.
23. Xu, R., Winnik, M.A., Riess, G., Chu, B., and Croucher, M.D. (1992) Micellization of polystyrene-poly(eyhylene oxide) block copolymers in water, *Macromolecules* **25**, 644–652.
24. Xu, R., Winnik, M.A., Hallett, R.F., Riess, G., and Croucher, M.D. (1991) Light scattering study of the association behavior of styrene-ethylene oxide block copolymers in aqueous solutions, *Macromolecules* **24**, 87–93.
25. Tuzar, Z., Webber, S.E., Ramireddy, Ch., and Munk, P. (1991) Association of polystyrene-poly(methacrylic acid) block copolymers in aqueous media, *Polymer Preprints* **32**, 525–526.
26. Procházka, K., Kiserow, D., Ramireddy, Ch., Tuzar, Z., Munk, P., and Webber, S.E. (1992) Time-resolved fluorescence studies of the chain dynamics of naphtalene-labelled polystyrene-*block*-poly(methacrylic acid) micelles in aqueous media, *Macromolecules* **25**, 454–460.
27. Kiserow, D., Procházka, K., Ramireddy, Ch., Tuzar, Z., Munk, P., and Webber, S.E. (1992) Fluorimetric and QELS study of the solubilization of nonpolar low molar mass compounds into water-soluble block-copolymer micelles, *Macromolecules* **25**, 461–469.
28. Tian, M., Qin, A., Ramireddy, C., Webber, S.E., Munk, P., Tuzar, Z., and Procházka, K.(1993) Hybridization of block copolymer micelles, *Langmuir* **9**, 1741–1748.
29. Rabin, Y. and Marko, J.F. (1991) Microphase separation in charged diblock copolymers: The weak segregation limit, *Macromoloecules* **24**, 2134–2136.
30. Marko, J.F. and Rabin, M., Microphase separation of charged diblock copolymers: Melts and solutions, *Macromolecules* **25**, 1503–1509.
31. Dan, N. and Tirrell M. (1993) Self-assembly of block copolymers with a strongly charged and a hydrophobic block in a selective, polar solvent. Micelles and adsorbed layers, *Macromolecules* **26**, 4310–4315.
32. Tuzar, Z., Procházka, K., Zusková, I., and Munk, P. (1993) Some properties of polyelectrolyte micelles, *Polym. Prepr.* **34**(1), 1038–1039.
33. Procházka, K, Kiserow, D., Ramireddy, Ch., Webber, S.E., Munk, P., and Tuzar, Z. (1992) Time-resolved fluorescence and light scattering studies of water-soluble block-copolymer, *Makromol. Chem., Macromol. Symp.* **58**, 201–207.
34. Alexandridis, P., Holtzwarth, J.F., and Hatton, T.A. (1994) Micellization of poly(ethylene oxide)-poly(propylene oxide)-poly(ethylene oxide) triblock copolymers in aqueous solutions: Thermodynamics of copolymer association, *Macromolecules* **27**, 2414–2425.
35. Zhou, Z. and Chu, B. (1994) Phase behavior and association properties of poly(oxypropylene)-poly(oxyethylene)-poly(oxypropylene) triblock copolymer in aqueous solution, *Macromolecules* **27**, 2025–2033.
36. Al-Saden, A.A., Whateley, T.L., and Florence, A.T., (1982) Poloxamer association in aqueous solutions *J. Colloid Interface Sci.* **90**, 303–309.
37. Nagarajan, R. Barry, M., and Ruckenstein, E. (1986) Unusual selectivity in solubilization by block copolymer micelles, *Langmuir* **2**, 210–215.
38. Cao, T., Munk, P., Ramireddy, Ch., Tuzar, Z., and Webber, S.E. (1991) Fluorescence studies of amphiphilic poly (methacrylic acid)-b-polystyrene-b-poly(methacrylic acid) micelles, *Macromolecules* **24**, 6300–6305.
39. Nagarajan, R. and Ganesh, K. (1989) Block copolymer self assembly in selective solvents. Theory of solubilization in spherical micelles, *Macromolecules* **22**, 4312–4325.
40. Hurter, P.N., Scheutjens, J.M.H.M., and Hatton, A. (1993) Molecular modelling of micelle formation and solubilization in block copolymer micelles. 1. A self-consistent mean-field theory, *Macromolecules* **26**, 5592–5601.

THE STRUCTURE OF TELECHELIC ASSOCIATING POLYMERS IN WATER

AHMAD YEKTA, BAI XU AND MITCHELL A. WINNIK*
Department of Chemistry and Erindale College, University of Toronto, Toronto, Ontario, M5S 3H6, Canada

Abstract

Hydrophobically modified urethane-ethoxylate (HEUR) thickeners (AT) containing alkyl end groups ($C_{16}H_{33}$) are purified and used to study the association behavior of the AT chains in aqueous solutions. Fluorescence data demonstrate that micelles are formed even at very dilute concentrations (about 10 ppm) with around 20 end-groups per micelle. This end-group aggregation number is found to be independent of polymer concentration. The flow behavior of the AT aqueous solutions is also determined. The analysis of the rheological data suggests that large structures are formed through secondary aggregation when polymer concentration is beyond 1 wt%. By combining fluorescence and rheology, we propose a two-step association model to account for our experimental observations and calculate some key parameters to get a picture of the topology in AT solutions.

1. Introduction

Associative thickeners [AT's] are water-soluble polymers containing hydrophobic end-groups with unique rheological properties. [1-5] At modest concentrations (1 to 2 wt%), they impart high viscosity to aqueous solutions, and these solutions undergo shear thinning without storing elastic energy at high shear rates. Because AT polymers may act as rheology modifiers in paint additives, oil recovery, steric stabilizers in ink and pigment vehicle and drug delivery system, [1, 6-8] their study is of great importance both from the fundamental and the industrial applications point of view. Much effort has been put into trying to understand the behavior of these systems. [2, 9-10] There is a wide-spread belief that these materials promote viscosity through end-group association to form micelles. The polymer chains bridge the micelles, creating a network which resists macroscopic deformation of the fluid. At high enough concentration, the network spans the solution, and the system gels.

The micelles, or the aggregates of hydrophobic groups, that are present in AT solutions have been characterized by static and dynamic light scattering, flow birefringence, nuclear magnetic resonance (NMR), rheology, fluorescence and neutron-scattering experiments. [10-16] Rheological measurements indicate that these solutions have, in certain concentration regions, similar dynamic properties as entangled polymer systems. [1-3] The data are consistent with the existence of a supermolecular transient network which can be formed through end-group interactions. To have a full picture of

S.E. Webber et al. (eds.), Solvents and Self-Organization of Polymers, 319–330.

the topology of AT aqueous solution, we need to determine several parameters. Among the parameters we need to determine are the critical micelle concentration (CMC), describing the onset of association, the micelle size, and the end-group aggregation number N_R (i.e. the number of end groups per micelle), all as a function of polymer concentration. Another parameter which is crucial to understand the flow behavior of AT solution is the functionality of the transient network. Functionality is defined as the number of mechanically effective chains connected to each crosslink point. The determination of these parameters is essential for understanding the mechanism of rheology modification of this class of material, and for understanding their role in paints and coatings technologies.

In this paper, we will review our spectroscopic study on telechelic associative thickeners. [4, 5] We describe the application of fluorescence quenching experiment as a means to obtain N_R values, [4, 18] and point out why incorrect assumptions in previous work led to values of N_R that are too small. We discuss rheology measurements and show that micelles and not free polymer molecules are the building block of the network. Finally, we calculate the fraction of bridging chains as a function of polymer concentration. We believe that this parameter is important in understanding the topology of AT solutions, and it helps to explain their macroscopic properties from a microscopic point of view.

2. Experimental

The polymeric associative thickeners are from Union Carbide, sample numbers 46RCHX22-2 and 46RCHX22-3. [18] They are urethane-coupled poly(ethylene oxide) polymers end-capped with $C_{16}H_{33}O$- groups as shown below. We refer to them as **AT22-2** and **AT22-3**, respectively.

$C_{16}H_{33}O$-IPDU-[PEO-IPDU]$_y$-$OC_{16}H_{33}$

AT 22-2 (M_n = 34,000, y = 4, 1.3 C_{16}/polymer)
AT 22-3 (M_n = 51,000, y = 6, 1.7 C_{16}/polymer)

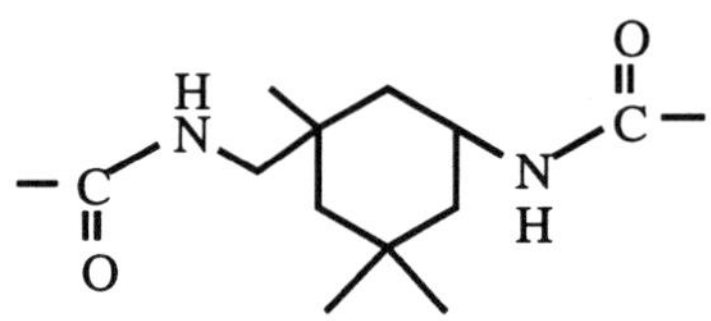

IPDU

The polymers described here are purified by recrystallization of the samples supplied by Union Carbide. **AT22-2** and **AT22-3** have number-averaged molecular weights of 34,000 and 51,000 respectively determined by GPC.

Distilled water purified through a Millipore Milli Q^{TM} purification system was used to prepare polymer solutions. Pyrene (Aldrich) was recrystallized three times from ethanol. In all these experiments, the polymer solutions were mixed with the pyrene 3-5 days before measuring the fluorescence to ensure a complete dissolution of pyrene label in micelles. Concentrations of pyrene solubilized in the micelles are checked by UV absorption, and are expressed as the total macroscopic concentration. All solutions studied were aerated, temperature controlled at 20 ± 0.5 °C during all measurements. UV and fluorescence measurements were carried out as described previously. [4]

Low shear viscosities and oscillatory shear response were measured on a Rheometrics RAA with a geometry of cone and plate (50 mm diameter). Steady state viscosity was obtained by measuring the torque as a function of the shear rate.

Viscoelastic properties were measured in oscillatory shear mode. The frequency range used was 0.01-500 rad/sec, with strains of 20-80%, chosen so that the given data is within the linear viscoelastic region.

3. Results and Discussions

3.1 ONSET OF AGGREGATION

In Figure 1 we present the intensity of pyrene emission as a function of polymer concentration. The increase in pyrene fluorescence intensity with polymer concentration indicates that pyrene partitions into a hydrophobic phase (i.e. a micellar phase). At low polymer concentrations, or in pure water, pyrene fluorescence decays exponentially with $\tau° = 150$ ns. At polymer concentrations above 1 g/L and low pyrene concentrations, where essentially all of the pyrene has partitioned into the micelles, the fluorescence decay profiles are also exponential, but with a longer lifetime, because the nonradiative decay rate in the micellar phase is suppressed. The 300 ns lifetime found at high polymer concentration is typical of that found for pyrene in, for example, sodium dodecyl sulfate micelles. [19] The data in Figure 1 indicate that polymer association begins at concentrations about 0.1 g/L. Beyond this concentration, the polymer forms micelle in its aqueous solution.

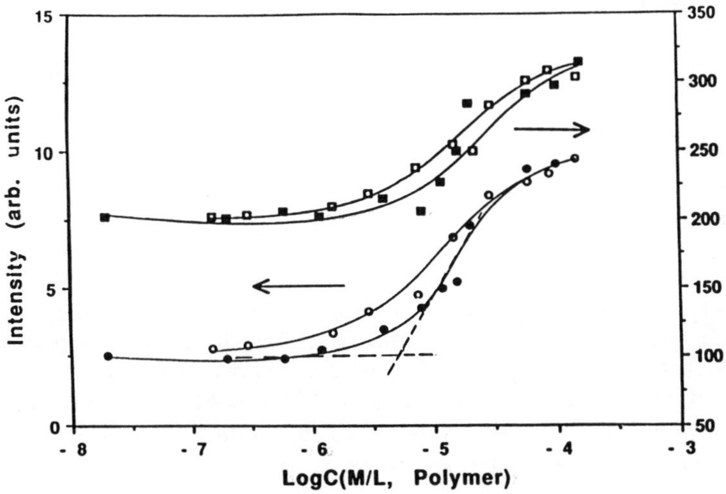

Figure 1. Determination of the onset of AT association. Plots of fluorescence intensity (lower two curves) and mean fluorescence decay time) (upper two curves) as a function of polymer concentration in moles per liter based upon M_n for AT22-2 (open symbols)and AT 22-3 (closed symbols), taken from ref. 5.

3.2 EVIDENCE FOR "ROSETTES," I.E. LOOPED STAR-LIKE MICELLES

Dynamic light scattering (DLS) and pulsed-gradient spin-echo (PGSE) NMR experiments in the enhanced viscosity polymer concentration regime indicate complex dynamic motion of the system. [10, 13] At lower concentrations, rather simple behavior is observed. Both DLS and PGSE NMR study show that there is a narrow distribution of diffusing species in solution, with a diameter 40 nm in **AT22-2** solution, when the polymer concentration is slightly above its cmc (i.e. 0.1 g/L). [4, 13] A reasonable explanation for this observation is that the aqueous AT polymer contains micelles formed by the association of individual polymer chains. These micelles have a diameter of 40-50 nm depending on the polymer chain length. [4]

3.3 DETERMINATION OF THE AGGREGATION NUMBER OF THE MICELLES

Hydrophobic dyes in aqueous micellar systems normally exhibit Poisson quenching kinetics. [17] Both the fluorophore and quencher dispersed among the micellar environment follow a Poisson distribution. Under such circumstances, the fluorophore decay rate I(t) is given by:

$$I(t) = I(0)\, exp\{ - \frac{t}{\tau^{o}} - n\,(1 - exp\,(-k_1 t))\, \} \qquad (1)$$

Here,

$$n = \frac{[Q]}{[micelle]} \qquad (2)$$

In this expression, [Q] and [micelle] refer respectively to the molar concentration of quenchers and micelles in the solution, and n is the mean number of quenchers per micelle. When pyrene (Py) is used as the fluorescence probe, ground-state Py molecules act as the quenchers in a reaction that leads to excimer formation. In this case, [Q] = [Py]. The t/τ^{o} term describes the exponential decay of fluorophore in micelles containing no quencher. The second term describes the contribution to the total observable fluorescence intensity from micelles containing i quenchers (i = 2, 3, ...), and k_1 is the corresponding pseudo-first-order rate coefficient for the intramicellar reaction with one quencher. When k_1 is very large, so that intramicellar quenching is very efficient, integration of eq. (1) yields eq. (3), the classic static quenching model used to analyze micellar quenching by steady-state fluorescence measurements.

$$ln\frac{I^{o}}{I} = \frac{[Q]}{[micelle]} = \frac{[Q]\,N_R}{[surfactant] - cmc} \qquad (3)$$

However, when the intramicellar quenching reaction is not efficient enough, the aggregation number calculated from eq. (3) will have significant error. A careful analysis of the quenching kinetics in these AT systems reveals that quenching is in fact rather inefficient in terms of the relatively large amount of quencher per micelle required. The rate constant k_1 that we obtained by fitting experimental data to eq. (1) is typically around 3 μs^{-1}, independent of polymer concentration as shown in Figure 2. This is nearly an order of magnitude slower than in traditional surfactant micelles. Figure 3 shows the plots of *n* vs. pyrene concentration [Py] for three **AT22-2** polymer concentrations. The slope gives the inverse of the micelle concentration. From the slope, after correction for end-group content, [18] we can calculate the end-group aggregation number. In this way we find that the end-group aggregation number N_R is 18 for **AT22-2** and 22 for **AT22-3**. This aggregation number is found to be the same at all three polymer concentrations.

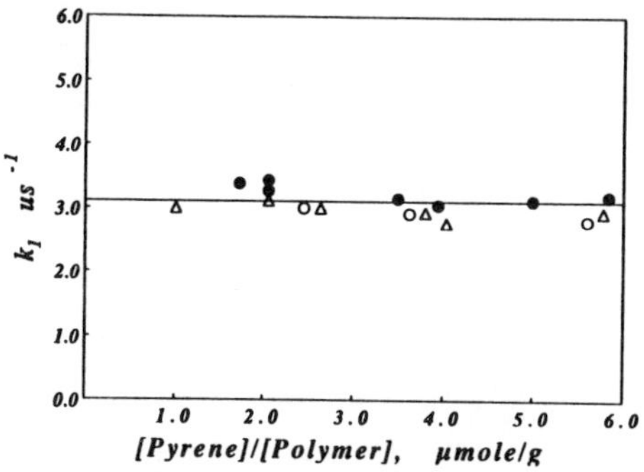

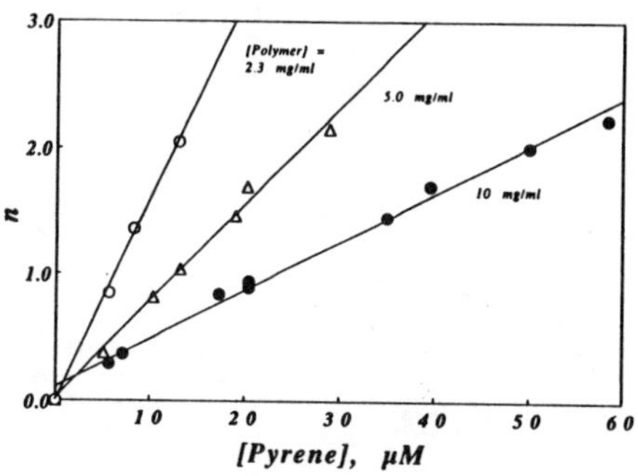

Figure 2. Plot of rate constant obtained by fitting the fluorescence decay data of **AT22-2** to equation (1) for different pyrene local concentration $[Py]/c_{pol}$. Taken from ref. 18.

Figure 3. Plots of n vs pyrene concentration in **AT22-2** solution for different polymer concentrations. c_{pol} = 10 g/L, 5 g/L, 2.3 g/L. Taken from ref. 18.

3.4 THE RIGIDITY OF THE CORE OF MICELLES

The local "viscosity" of microheterogenous domains can often be probed with molecules that form excimers intramolecularly. In Figure 4 we present fluorescence spectra of dipyme dissolved in aqueous SDS micelles and also solubilized by a 0.5 wt% solution of **AT22-2**. One notes immediately that the excimer intensity is much lower in the AT polymer solution, indicating a much more viscous environment than that found in the SDS micelles. By comparison with other experiments, the microviscosity of the AT micelles is much closer to that of phospholipid bilayers than that of typical surfactant micelles. [20]

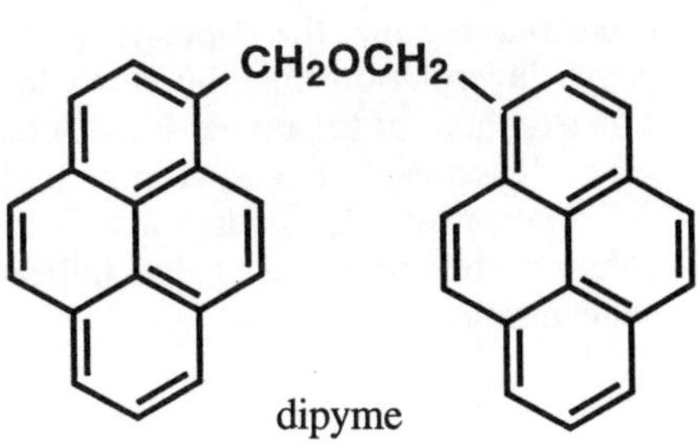

dipyme

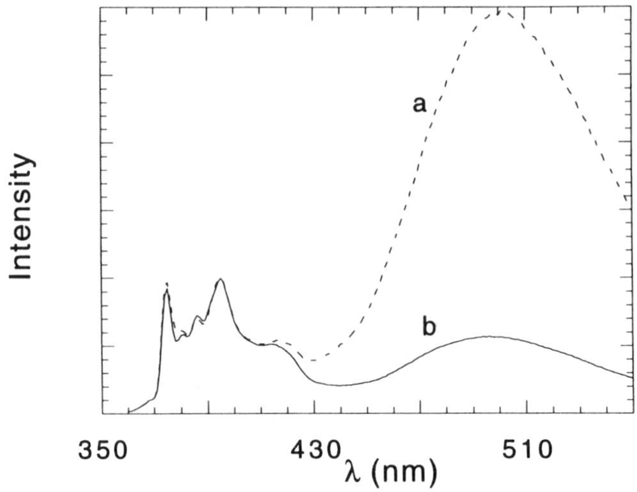

Figure 4. Fluorescence spectra of 1 x 10^{-6} M dipyme molecularly solubilized in aqueous solutions of 5.0 g/L **AT22-2** (solid line) and 0.05 M SDS (dashed line). Excitation wavelength is at 346 nm.

3.5 CONCENTRATION AND SHEAR DEPENDENCE OF THE ZERO-SHEAR VISCOSITY OF AT SOLUTIONS

It is well documented that the low shear viscosity of **AT22-3** and similar polymers grows rapidly with increasing concentration above ca. 1 wt% as shown in Figure 5. [1, 3] These solutions exhibit Newtonian behavior over a range of low shear rates, and then exhibit pronounced shear thinning at high shear rates. An example of this behavior is shown in Figure 6. For increased polymer concentrations, the low shear viscosity is enhanced. One sees that for certain intermediate concentrations in the vicinity of 1 wt%, the solutions exhibit noticeable shear thickening prior to shear thinning. In the high shear rate regime, the dependency of the viscosity on shear rate can be described by a power law relationship, similar to the behavior found in concentrated polymer solutions. The exponent of the power law is close to -1.

The onset of shear-thinning shifts smoothly to lower shear rate with increasing AT concentration. If this onset is considered as the breaking rate of AT junctions in solution, then this observation indicates a decrease of breaking rate with increasing AT concentration.

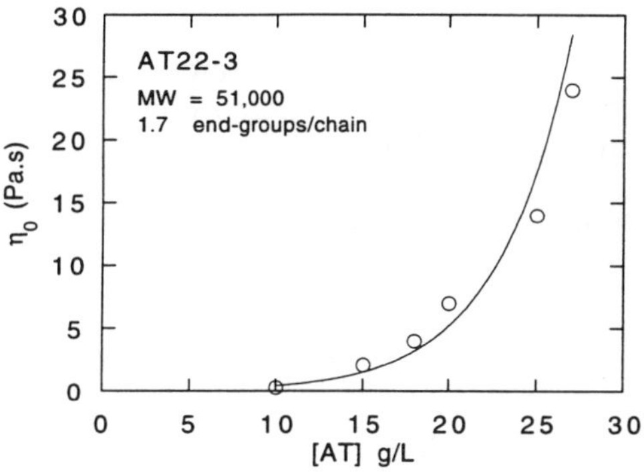

Figure 5. Plot of zero shear viscosity of **AT22-3** vs c_{pol} for a series of polymer concentrations.

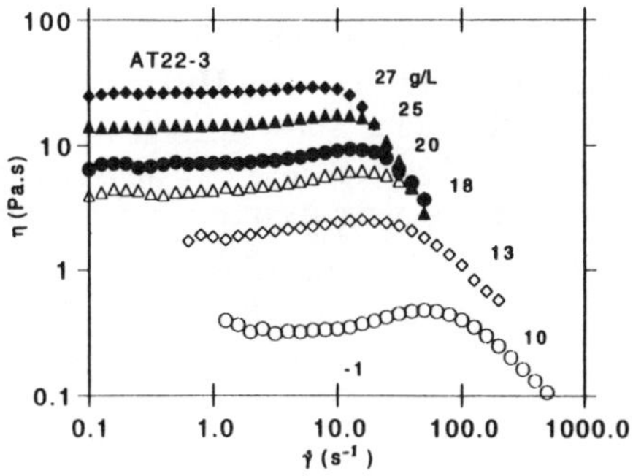

Figure 6. The steady shear viscosity (h) of **AT22-3** at different concentrations as a function of shear rate.

3.6 RELATING MACROSCOPIC VISCOSITY TO MICROSTRUCTURE OF AT SOLUTIONS

In Figure 7, we attempt to relate the viscosity profile with microstructures that may form in solution using ideas from our micelle-microgel model. [4,18] The graph presents a viscosity profile (η vs. shear rate $\dot{\gamma}$) for a 1 wt% solution of **AT22-3**. At this concentration, a pronounced shear thickening occurs in the range of $\dot{\gamma} \approx 10$ Hz. At higher shear rates, shear thinning becomes dominant. According to Annable, micelle aggregation forms localized networks at lower concentrations. [3] These networks, which we have referred to as "microgels", are interconnected by bridging chains. Dilution of the solution with water results in the breakup of these structures to smaller microgels, accompanied by a transition of some bridging chains to looped configurations (bridge-to-loop transition). Increasing the polymer concentration leads to larger microgels interconnected by bridging chains (loop-to-bridge transition). This transformation is entropically driven and gives rise to the strong initial concentration dependence of the viscosity and the viscoelastic modulus. Annable's picture is quite similar to the microgel model that we have proposed. [4]

In Figure 7 we superpose over the viscosity profile illustrations of our view of the structure of the solution. At low shear rates, the system is comprised of aggregates of micelles, held together by bridging chains. The size of these aggregates depends upon the polymer concentration. Here we draw an aggregate made up of eight micelles. In the shear thickening region, the structure is stretched by the stretching of the bridging chains. Increasing the shear rate increases the stress on such structures, and once a critical force is exceeded, fragmentation occurs. Fragmentation involves pull-out of one end of a bridging chain (bridge-to-loop transition). One end of the bridging chain exits from its micelle and loops back to join its other end in an adjacent micelle.

This model receives strong support from micellar core-based fluorescence experiments carried out on HEUR systems under shear [21] or extensional flow. [22] These experiments show no changes of signal under conditions where both shear thickening and strong shear or extensional thinning are observed. The results require that the average size and structure of the micellar core be conserved throughout the transition region.

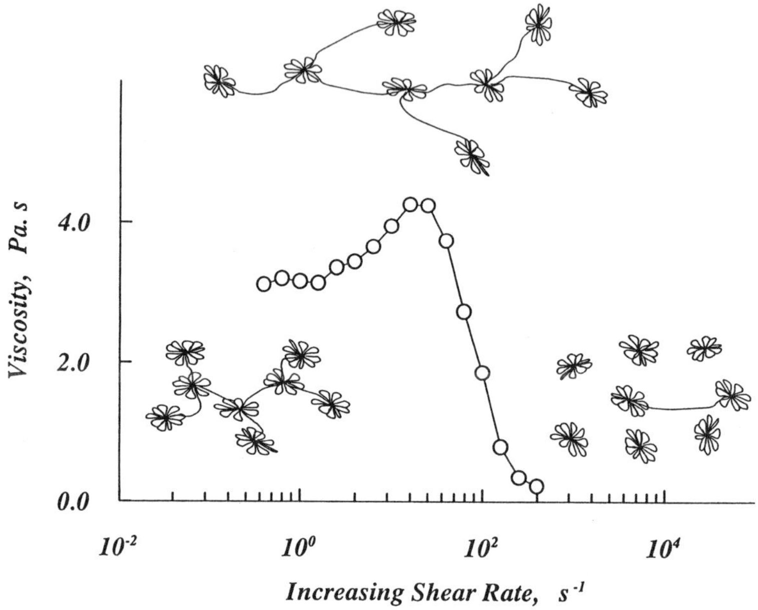

Figure 7. A plot of steady shear viscosity vs shear rate for a 10 g/L **AT22-3** solution measured at 20 °C. Shown on the figure is a illustration of the solution topology that might form at this concentration through association of 8 micelles. At higher shear rates, this structure is stretched, and shear thinning occurs when the structure fragments and the polymers rearrange. A key feature of the model is that the mean aggregation number of chain ends in the micellar core remains at 22 throughout the range of shear rates. Taken from ref. 18.

It is useful at this point to recall the model for the association structure suggested by Maechling-Strasser et. al. [23] based upon light scattering measurements to determine the apparent M_W of the system as a function of concentration. They proposed an *open association* model in which the fundamental building blocks were unassociated polymer chains. In our model, individual polymer chains associate to form micelles of a well-defined structure (i.e. *closed association*), but these micelles undergo secondary association to form clusters. Since the size of the structure formed increases with polymer concentration, cluster formation follows an *open association* process. Thus one might find similarities between our model and that of Maechling-Strasser et. al., with the key difference that in our model, the fundamental building blocks of the associated structure are the rosette-like micelles.

3.7 PLATEAU MODULUS OF AT SOLUTIONS UNDER OSCILLATORY SHEAR

Aqueous solutions of **AT 22-3** show linear viscoelastic behavior below strains of 200% . Within this linear viscoelastic region, we carried out oscillatory shear experiments. The storage modulus as a function of the angular frequency ω, is plotted in Figure 8 for different AT concentrations. At high frequencies, the storage modulus (G') approaches a constant value and bears strong resemblance to the classical rubber plateau modulus (G_N^o). This type of behavior implies that at high frequencies the solution behaves as an elastic body. At low frequencies, G' varies approximately as ω^2. In this region, the solutions behave as viscoelastic fluids.

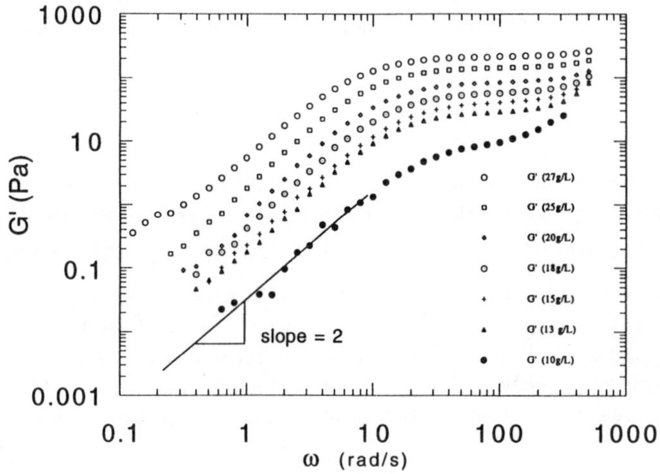

Figure 8. The storage modulus (G') of **AT22-3** as a function of shear rate. Polymer concentrations are 27, 25, 20, 18, 15, 13, 10 g/L.

One of the curious features of these systems is that in spite of their complexity, the moduli of these AT solutions can be well fitted to a model involving a single Maxwell component. [3, 24] From the parameters of the fit, one obtains the plateau modulus and the relaxation time of the system. The magnitude of plateau modulus G_N^o depends upon the average chain length between entanglements, i.e. on the mesh size. As the mesh size decreases, G_N^o increases.

To calculate the functionality of the network and the fraction of bridging chains, we assume that elasticity in the system is a consequence of bridging of chains between micelles, which are the mechanically active components of the system. The crosslink points of the network are micelles formed by aggregation of end groups of about 10 polymer chains. This assumption is supported by our pyrene fluorescence experiments which monitor the Poisson distribution of pyrene among the hydrophobic end groups of the system. If the linking unit allowing two or more chains in the sequence to build a bridge between micelles involved small-unit association of the end groups, this would perturb the end-group distribution and affect the fluorescence experiment. [4, 18]

By analogy with the classical theory of rubber elasticity, the pseudo-equilibrium modulus for an ideal network is given by:

$$G_N^o = g\nu RT \qquad (4)$$

where ν is the number of moles of elastically effective chains per unit volume, R is the gas constant, and T is the absolute temperature. According to Jenkins, the numerical factor g is approximately equal to unity for transient networks of associative polymers. [1]

We can utilize equation (4) in conjunction with our fluorescence-based knowledge of the micellar structure to calculate the functionality of the network. The parameter ν in equation (4) measures the molar density of elastic chains in the transient network formed by association. If we take the micelle as the crosslink point, the number of elastically effective chains connected to each micelle is given by the expression:

$$f_n = 2\nu/[micelle] = 2\nu N_R / q_R c_{pol} \qquad (5)$$

where f_n is the functionality of the system, i.e., the number of elastically effective chains attached to one cross-link point. N_R is the aggregation number of the end-groups per micelle; q_R is the end-group content of the polymer in mol/g; and c_{pol} is the polymer concentration in g/L. The molar value of q_R which equals to 1.7 for **AT22-3** is determined by ^{1}H NMR and expresses the mean number of hydrophobic end groups per polymer chain. [18] Typical values of f_n are found to be below 4 for **AT22-3** solutions below 3 wt%. These f_n values represent an average over the entire system, and increase linearly with increasing polymer concentration. [24] To explain how the system can act as a gel with f_n values less than 3, we recognize that at low c_{pol}, not all micelles in the system make elastic contributions to the network. The micellar phase consists of fragments of microgels. The existence of a plateau modulus implies the percolation of the micellar phase across the macroscopic volume separating the cone and plate in our rheometer. As the polymer concentration is increased, the micellar phase fills a greater fraction of the total volume, and above c*, the overlap concentration, the micelle cores move closer together. This is accompanied by an increase in the fraction of bridging chains, which leads to an increase in f_n.

We can use the calculated values of functionality to estimate the fraction of chains involved in bridging. There are 22 chain ends per micelle for **AT22-3**, and this number is independent of polymer concentration. Because the polymers contain 1.7 end groups per chain, the fraction of bridging chains is given by

$$f_{bridge} = \frac{f_n/2}{22/1.7} \qquad (6)$$

As an example, we estimate that a 10 g/L solution has 3% of the chains acting as bridges, whereas a 20 g/L solution has 10% of the chains involved in bridging.

What these discussions imply is the idea that the effect of dilution is to create more void space in the system. Thus the bridging percentage is reduced while the basic micelle structure is retained. In our view, high shear causes some of the bridging chains to rearrange to looped chains, without affecting the individual micelle structure. From this point of view, it is the star-like micelle which is the fundamental building block of the network structure that is formed at higher concentrations.

4. Conclusions

Our fluorescence data from $C_{16}H_{33}$ telechelic HEUR associating polymers indicate that these polymers form micelles at very dilute concentrations (about 10 ppm) with ca. 20 end-groups per micelle. This end-group aggregation number is independent of polymer concentration. We interpret the rheological behavior of HEUR associating polymer solutions in terms of a model for the association structure. This model invokes a closed association process for the formation of looped star-like micelles, followed at higher concentration by a secondary association process of the micelles. The secondary association is an "open association" process, in which the size of the secondary aggregates microgels increases with polymer concentration.

From this point of view, it is the star-like micelle which is the fundamental building block of the network structure that is formed at higher concentrations. We use the simple theory of rubber elasticity to calculate the number of elastically effective chains per unit volume from the measured plateau modulus. By comparing this value to the density of micelles in the system, we obtain the effective functionality of the network and the fraction of bridging chains in the system.

5. Acknowledgments

The authors thank NSERC Canada, Union Carbide, and Aqualon for their support of this work. We also thank Dr. D. Bassett at Union Carbide for the sample of **AT22-2** and **AT 22-3** examined here.

6. References

1. Jenkins, R. D. (1990) *The Fundamental Thickening Mechanism of Associative Polymers in Latex Systems: A Rheological Study*, Lehigh University.
2. Huldén, M. (1994) Hydrophobically Modified Urethane-ethoxylate (HEUR) Associative Thickeners 1. Rheology of Aqueous Solutions and Interactions with Surfactants, *Colloids and Surfaces A: Physicochemical and Engineering Aspects* **82**, 263-277.
3. Annable, T.; Buscall, R.; Ettelaie, R.; Whittlestone, D. (1993) The Rheology of Solution of Associating Polymers: Comparison of Experimental Behavior with Transient Network Theory, *J. Rheol.* **37**, 695-726.
4. Yekta, A.; Duhamel, J.; Brochard, P.; Adiwidjaja, H.; Winnik, M. A. (1993) A Fluorescent Probe Study of Micelle-like Cluster Formation in Aqueous Solutions of Hydrophobically Modified Poly(ethylene oxide), *Macromolecules* **26**, 1829-1836.
5. Wang, Y.; Winnik, M. A. (1990) Onset of Aggregation for Water-Soluble Polymeric Associative Thickeners: A Fluorescence Study, *Langmuir* **6**, 1437-1439.
6. Chu, D. Y. and Thomas J. K. (1991) Characterization of Polymer By Excited State Techniques, *Photochemistry and Photophysics* **3**, 49-102.
7. Von Bünau, G.; Wolff, T. (1988) In *Advances in Photochemistry*; D. H. Volman, G. S. Hammond and K. Gollnick, Eds.; John Wiley & Sons: New York.
8. Zana, R. (1986) In *Surfactant Solutions: New Methods of Investigation*; R. Zana, Ed.; Marcel Dekker: New York.

9. Jenkins, R. D.; Silebi, C. A.; El-Aasser, M. S. (1989) *Proceeding of American Chemical Society, Division of Polymer Materials and Science and Engineering* **61**, 629-633.
10. Persson, K.; Abrahmsen, S.; Stilbs, P.; Hansen, F. K.; Walderhaug, H. (1992) The Association of Urethan-Polyethyleneoxide (HEUR) Thickeners, as Studied by NMR Self-Diffusion Measurements *Colloid and Polymer Science* **270**, 465-469.
11. Fonnum, G.; Bakke, J.; Hansen, F. K. (1993) Associative Thickeners. Part I: Synthesis, Rheology and Aggregation Behavior, *Colloid Polymer Science* **271**, 380-389.
12. Nyström, B.; Walderhaug, H.; Hansen, F. K. (1993) Dynamic Crossover Effects Observed in Solutions of a Hydrophobically Associating Water-Soluble Polymer, *J. Phys. Chem.* **97**, 7743-7752.
13. Rao, B. H.; Uemura, Y.; Dyke, L.; Macdonald , P. M. (1995) Self-Diffusion Coefficients of Hydrophobic Ethoxylated Urethane Associating Polymers Using Pulsed-Gradient Spin Echo Nuclear Magnetic Resonance, *Macromolecules* **28**, 531-538.
14. Glass, J. E. Editor (1986) *Water Soluble Polymers, ACS Adv. Chem. Ser.* **Vol. 213**.
15. Lundberg, D. J.; Ma, Z.; Alahapperuna, K.; Glass, J. E. (1991) Surfactant Influences on Hydrophobically Modified Thickener Rheology, *ACS Symp. Ser.* , **No 462**, 234-253.
16. Ottewill, personal communications.
17. Yekta, A.; Aikawa, M.; Turro, N. J. (1979) Photoluminescence Methods for Evaluation of Solubilization Parameters and Dynamics of Micellar Aggregates. Limiting Cases Which Allow Estimation of Partition Coefficients, Aggregation Numbers, Entrance and Exit Rates, *Chem. Phys. Lett.* **63**, 543-548.
18. Yekta, A.; Xu, B.; Duhamel, J.; Adiwidjaja, H.; Winnik, M. A. (1995) Fluorescence Studies of Associating Polymers in Water: Determination of The Chain End Aggregation Number and A Model for the Association Process *Macromolecules* **28**, 956-966.
19. Lianos, P.; Zana, R. (1981) Fluorescence Probe Studies of the Effect of Concentration on the State of Aggregation of Surfactants in Aqueous Solution, *J. Colloid Interface Sci.* **84**, 100-107.
20. Turro, N. J.; Aikawa, M; Yekta, A. (1979) A Comparison of Intermolecular and Intramolecular Excimer Formation in Detergent Solutions. Temperature Effect and Microviscosity Measurements, *J. Am. Chem. Soc.* **101**, 772-774.
21. Richey, B.; Kirk, A. B.; Eisenhart, E. K.; Fitzwater, S.; Hook, J. (1991) Interactions of Associative Thickeners with Paint Components as Studied by the Use of a Fluorescently Labeled Model Thickener, *J. Coatings Technology* **63**, 31-40.
22. Vlahiotis, D. (1992) Extensional Flow Behaviour of Associative Polymer Solutions, *Ph. D. Thesis*, University of Toronto: Toronto, Canada.
23. (a) Maechling-Strasser, C.; François J.; Clouet, F.; .Tripette, C. (1992) Hydrophobically End-capped Poly(ethylene Oxide) Urethanes: 1. Characterization and Experimental Study of Their Association in Aqueous Solution, *Polymer* **33**, 627636. (b) Maechling-Strasser, C.; Clouet, F.; François, J. (1992) Hydrophobically End-capped Poly(ethylene Oxide) Urethanes: 2. Modeling Their Association in Water, *Polymer* **33**, 1021-1025.
24. Xu, B.; Yekta, A.; Li L; Masoumi Z. and Winnik, M. A., (1995) The Functionality of Associative Polymer Networks: The Association Behavior of Hydrophobically Modified Urethane-Ethoxylate (HEUR) Associative Polymers in Aqueous Solution (submitted to *Advances in Colloid and Interface Science*)
25. Semenov, A. N.; Joanny, J.-F.; Khokhlov, A. R. (1995) Associating Polymers: Equilibrium and Linear Viscosity, *Macromolecules* **28**, 1066-1075.

UNIMOLECULAR MICELLES OF HYDROPHOBICALLY MODIFIED POLYELECTROLYTES

YOTARO MORISHIMA
Department of Macromolecular Science, Faculty of Science, Osaka University, Toyonaka, Osaka 560, Japan

1. Introduction

Water-soluble hydrophobically associating polymers have been the subject of extensive studies over the past decade because of fundamental interests in their molecular self-organization phenomena in relevance to biopolymer systems and also in their nanoscopic molecular architectures in connection to materials science. This class of polymers is also the focus of industrial research because of its utility in a variety of applications such as paints, coatings, drugs, cosmetics, personal care goods, flocculation, colloid stabilization, oil field formulation, and rheology modification.[1,2]

Hydrophobically modified polyelectrolytes were first synthesized by Strauss and Jackson[3] in 1951 by quaternization of poly(2-vinylpyridine) with *n*-dodecyl bromide. They showed that the long alkyl side chains attached to partially quaternized poly(2-vinylpyridine) associated in aqueous solution to form a compact conformation. This compact conformation showed an analogy to surfactant micelles in that it solubilized hydrophobic small molecules in dilute aqueous solution, and they termed this "polysoap".[3] Later, Dubin and Strauss[4,5] synthesized a series of alternating copolymers of maleic acid and alkyl vinyl ethers, and they showed that the copolymer with *n*-octyl vinyl ether adopted a "hypercoiled" structure even at high degrees of ionization.

In spite of the pioneering work of Strauss and Jackson[3] over forty years ago, very few reports on amphiphilic polyelectrolytes have followed. It was not many years ago when the volume of published work on this subject rapidly increased.

In amphiphilic polyelectrolytes, the hydrophobic interaction competes with electrostatic repulsion. Therefore, the balance of the contents of the hydrophobic and charged units in a polymer chain is a critical factor to determine whether or not the polymer collapses as a result of the cooperative hydrophobic associations. For example, amphiphilic polycarboxylic acids, which adopt an extended chain conformation at high pH, would collapse to form a compact conformation upon decreasing pH.[6-9] This transition from an extended to a compact structure, a typical of cooperative processes that

S.E. Webber et al. (eds.), Solvents and Self-Organization of Polymers, 331–358.

can be viewed as a two-state transition, takes place within a narrow pH range. On the other hand, amphiphilic polysulfonic acids, strong polyelectrolytes, show no such pH dependence of their conformation because they are fully ionized even at very low pH.[10,11]

In general, intrapolymer hydrophobic association may be dominant in dilute polymer solutions, whereas interpolymer association may occur in a semi-dilute or concentrated regime. In the case of polymers with a strong tendency for intrapolymer association, unimolecular micelles (unimers) would be formed regardless of the polymer concentration. In contrast, polymers with a strong propensity for interpolymer association give rise to multipolymer aggregates that may or may not accompany phase separation.

This chapter deals with associating behavior of polysulfonates modified with bulky hydrophobes and their unimers with an emphasis put upon experimental techniques that have been used to investigate the self-organization phenomena and to characterize the unimers. In the last section of this chapter, the functionalization of the unimers with some photoactive dyes will briefly discussed.

2. Self-Organization of Hydrophobically Modified Polyelectrolytes

Water molecules around hydrophobes in amphiphilic polymers in aqueous solution have an ice-like structure.[12] The solvent structuring occurs because the water molecules surrounding the hydrophobes have a tendency to form the largest possible number of hydrogen bonds.

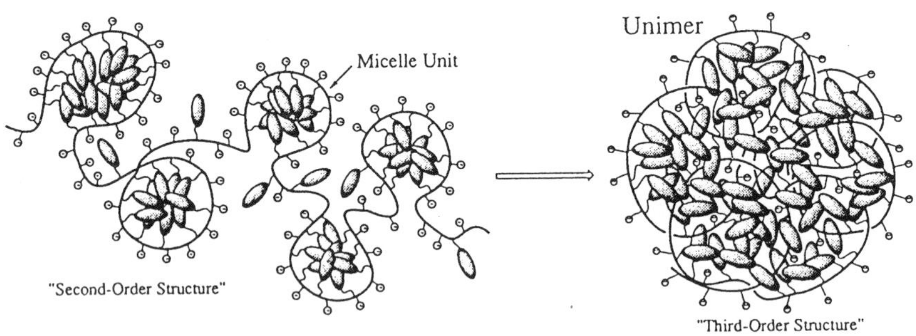

Figure 1. A schematic model of unimers.

If hydrophobes in an amphiphilic polymer come close to each other, they would associate such that they can have the smallest possible contact with water by releasing the structured water molecules into free water in the bulk phase. The principal driving force for this hydrophobic association is a large increase in entropy to overcome an enthalpy loss due to the breakup of the hydrogen bonds.[13,14] As schematically illustrated in Figure 1, if the content of the hydrophobe in an amphiphilic polyelectrolyte is above a certain critical content, the polymer would undergo self-organization in aqueous solution. The first step of the self-organization may be intrapolymer hydrophobic association to form micelle units along a polymer chain. However, the conformation thus formed, which may be viewed as a "second-order structure", may not be stable to stay as such because a significant portion of the surface of the hydrophobic cluster is exposed to water. Consequently, the micelle units may congregate to form a higher-order structure and thus the polymer may end up with a micelle made up with a single macromolecule (unimer).

In general, unimers must be preferentially formed in highly dilute aqueous solutions. However, as the concentration is increased, the hydrophobic association may not necessarily occur only in an intrapolymer mode. If interpolymer hydrophobic association occurs, then multipolymer aggregates would be formed instead of the unimers. Whether the intrapolymer association predominates over the interpolymer association depends primarily on the chemical structure(the first-order structure) of the amphiphilic polyelectrolyte. Thus, it is important to establish structural requirements for polymers that would form either unimers or multimolecular aggregates.

Chart 1

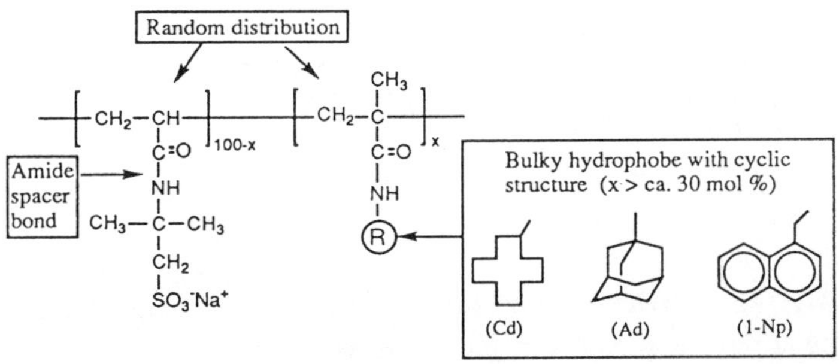

Morishima *et al.*[15] have recently shown that, in the type of polymers shown in Chart 1, the hydrophobic association occurs completely in an intrapolymer mode and thereby the polymer forms a unimer regardless of the concentration of the polymer. The sequence distribution of charged and hydrophobic units along the polymer chain is a very important structural factor to determine the mode of the hydrophobic association. Block sequences have a strong tendency for interpolymer association, whereas random and alternating sequences tend to associate in an intrapolymer mode.[16-20] Chang and McCormick[21] have recently shown that even a subtle difference in the sequence distribution in a random copolymer has a significant effect on the mode they associate; if the sequence distribution is blocky in nature, there is a tendency for intermolecular association. Another important structural factors for unimer-forming polymers is that the hydrophobes should be bulky and should have cyclic structures, such as those illustrated in Chart 1, and their contents should be higher than ca. 30 mol %. Furthermore, amide spacer bonds may be important because they form hydrogen bond networks that may contribute to retaining the compact unimer structure.[15]

These unimers are very different from the classical surfactant micelles in that (a) all charged and hydrophobic groups are covalently linked to the polymer backbone, (b) the unimers are "static" in nature as oppose to the "dynamic" nature of the surfactant micelles which exist in equilibrium between association and dissociation, and (c) the unimer structure is retained as it is even at very low and high concentrations.

3. Synthesis of Unimer-Forming Amphiphilic Polyelectrolytes

Unimer-forming amphiphilic polyelectrolytes can be synthesized by free-radical copolymerization of 2-acrylamido-2-methylpropanesulfonate (AMPS) and methacrylamides substituted with bulky hydrophobes of cyclic structures such as *N*-cyclododecylmethacrylamide (CdMAm), *N*-(1-adamantyl)methacrylamide (AdMAm), and *N*-(1-naphthylmethyl)methacrylamide (1NpMAm).[22] The composition of the unimer-forming copolymers ranges 40-70 mol % in the AMPS unit and 30-60 mol % in the hydrophobic monomer unit.

For spectroscopic characterization, these polymers can be labeled with a small amount of pyrene moieties by terpolymerization using *N*-(1-pyrenylmethyl)-methacrylamide (PyMAm).[22]

These co- and terpolymerizations can be carried out in a vacuum-sealed glass ampule containing respective monomers and 2,2'-azobis(isobutyronitrile) (AIBN) (0.5 mol % based on the total monomers) in *N*, *N*-dimethylformamide (DMF) solution at 60 °C for 12 h.[22] After the polymerization, the polymers are reprecipitated from a methanol solution into an excess of ether and then dissolved in a dilute NaOH aqueous solution to neutralize the sulfonic acid groups. The alkaline solution is dialyzed against pure water for several days. The compositions of the co- and terpolymers can be determined by N/C and S/C ratios and also by absorption spectroscopy for the labeled polymers.

These polymerizations are characterized as the "ideal copolymerization" where the monomer reactivity ratios are practically unity, resulting in copolymer compositions equal to monomer feed compositions and the completely random distribution of the monomer units along the polymer chain.[22,23] Therefore, one can determine the terpolymer compositions directly by the molar ratios of the monomers in feed, which allows one to prepare co- and terpolymers with well defined compositions and completely random distributions of the monomeric units along the polymer chain, which is a prerequisite for unimer-forming polymers. The terpolymers with up to ca. 60 mol% hydrophobic monomer units are completely soluble in water to give optically clear solutions.

In the text, the terpolymers are denoted as poly(A/R/P), where A, R, and P represent the AMPS, hydrophobe, and label units, respectively (Chart 2).

Chart 2

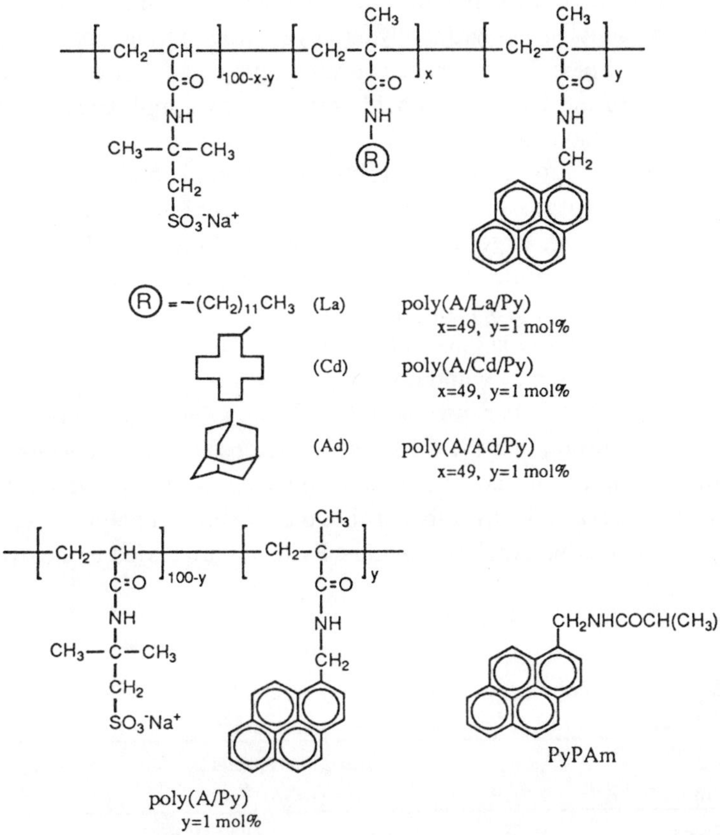

4. Characterization of Unimers

Characterization of unimers and multipolymer aggregates of hydrophobically modified polyelectrolytes is a challenging task. Spectroscopic methods, such as fluorescence, NMR, and IR spectroscopies, provide microscopic information on the structure and dynamic behavior in a short dimensional range, while scattering methods, such as static light scattering (SLS), dynamic light scattering (DLS), and small-angle X-ray scattering (SAXS) techniques, provide information based on a medium or long range structure. Therefore, use of a variety of techniques in combination is of importance for successful characterization.

4.1. FLUORESCENCE METHODS

4.1.1. *Fluorescence Spectra and Lifetime of Pyrene Labels*

For the fluorescence studies, the polymers may be labeled with a small amount of 1-pyrenyl (Py) or 2-naphthyl (2-Np) groups by terpolymerization techniques as described in the previous section. Since Py species have a strong propensity for excimer formation, the loading amount of the Py labels should be as low as 1 mol % so that the labels show only monomeric fluorescence.

It is known that the ratio of the third to the first vibronic bands (I_3/I_1) in pyrene fluorescence spectra depends on the polarity in microenvironments where pyrene exists,[24] I_3/I_1 values being larger in less polar media. In a parallel experiment, this empirical rule has been confirmed to apply to the model compound, PyPAm (Chart 2), which possesses a substituent group similar to the Py label in the polymers (Chart 2).[25]

The I_3/I_1 ratio for PyPAm is 0.59, reflecting the polarity of water. The reference copolymer poly(A/Py) (Chart 2) without hydrophobes shows the same value as that of PyPAm in water.[26] This is because the reference copolymer assumes an extended conformation in pure water, and the Py labels are exposed to the aqueous phase. In contrast, the I_3/I_1 ratios of the terpolymers poly(A/La/Py), poly(A/Cd/Py), and poly(A/Ad/Py) (Chart 2) are all larger than that of the reference copolymer poly(A/Py) as compared in TABLE 1, which indicates that the Py labels are buried in the hydrophobic clusters.[22]

TABLE 1. Steady-state fluorescence and decay parameters for the fluorescence of the Py-labeled co- and terpolymers in aqueous solution[22]

polymer	$I_3/I_1{}^a$	fitting parameters τ_i(ns)/$\alpha_i{}^b$		$<\tau>^c$ (ns)
poly(A/Py)	0.59	29/0.08	148/0.916	136
poly(A/La/Py)	0.80	55/0.258	207/0.742	168
poly(A/Cd/Py)	0.83	57/0.284	238/0.716	187
poly(A/Ad/Py)	0.76	41/0.336	219/0.664	159

a. Intensity ratio of peak 3 to peak 1 in the fluorescence spectra.
b. Fitting function; $I(t)=\Sigma\alpha_i\exp(-t/\tau_i)$.
c. Average lifetime defined by $<\tau>=\Sigma\alpha_i\tau_i$.

The fluorescence decays for the Py labels in the terpolymers are significantly slower than that for the reference copolymer, reflecting the hydrophobic microenvironments where the Py labels reside.[22] The decay curves for all these polymers are best-fitted with double-exponential functions: the values of the relative weight of the pre-exponential factors α_i and the lifetimes τ_i are listed in TABLE 1. The presence of the shorter life component suggests that the microenvironments about the Py labels are not uniform. The long lifetimes for the longer life component, which is the main component, indicate that the Py labels are encapsulated in the hydrophobic clusters in the unimers.

Among the hydrophobes in the terpolymers, there are significant differences in the micropolarity in the hydrophobic clusters: the steady-state fluorescence and fluorescence decay data indicate that the micropolarity decreases in the order of Ad > La > Cd clusters. A reason for the lowest micropolarity of the Cd cluster may be that the Cd groups can intimately be packed together with the Py labels in the hydrophobic cluster, leading to an effective protection of the Py labels from the aqueous phase.

Chart 3

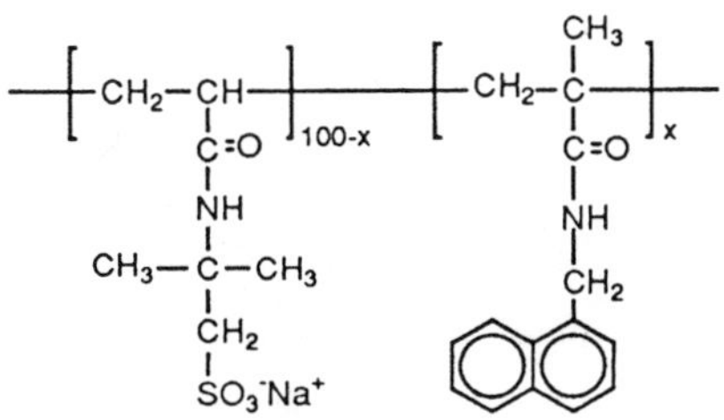

x=13 - 58 mol %

poly(A/1-Np(x))

If naphthalene moieties are employed as the hydrophobes in place of the bulky aliphatic hydrocarbons, the naphthalene cluster in the unimer shows characteristic photophysical behavior. Figure 2 shows fluorescence spectra of the copolymers of AMPS and 1NpMAm, poly(A/1-Np(x)) (Chart 3), with varying 1-Np contents in aqueous solution at room temperature.[23]

The fluorescence spectra depend markedly on the copolymer composition. As the 1-Np content increases from 13 to 28 mol %, fluorescence due to the monomeric 1-Np chromophore peaking at 337 nm decreases, and excimer fluorescence peaking at 390 nm increases systematically. The copolymers with > 41 mol % 1-Np content emit strong excimer fluorescence which dominates over the monomer fluorescence. Figure 3 shows a plot of the ratio of the excimer to monomer fluorescence intensities (I_E/I_M) as a function of the 1-Np content in the copolymer. The I_E/I_M ratio sharply increases at a 1-Np content between 28 and 41 mol %, indicating that the critical content of the 1-Np group for the unimer formation lies between 28 and 41 mol %.

Since the fluorescence spectra shown in Figure 2 were measured at the same molar concentration of the 1-Np residue (20 μM), the relative fluorescence intensity can approximately be related to the relative fluorescence quantum yield. The total fluorescence quantum yield increases as the excimer component increases with increasing

1-Np content at room temperature; e.g., the copolymers with 13 and 58 mol % 1-Np units show quantum yields of 0.09 and 0.16 at room temperature, respectively.[23] This is rather unusual as compared with the tendency shown by polymer-bound aromatic chromophores studied previously.[27,28]

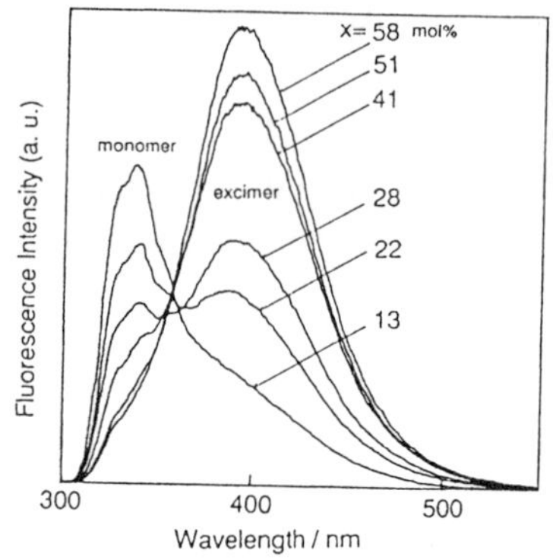

Figure 2. Fluorescence Spectra of poly-(A/1-Np(x)) with varying 1-Np contents in aqueous solution. Excitation wavelength, 290 nm.[23]

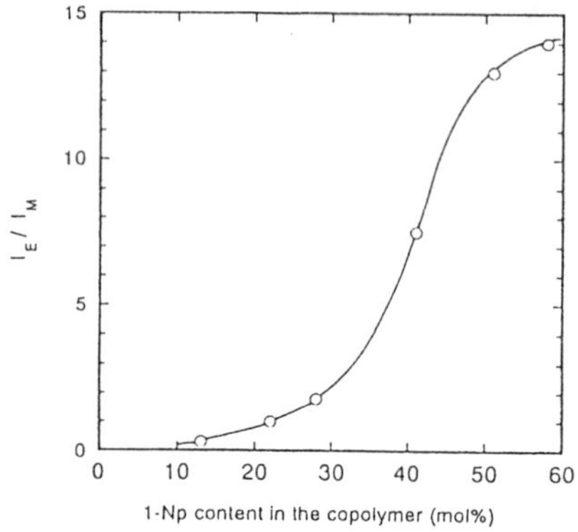

Figure 3. Intensity ratio of Excimer to monomer fluorescence as a function of the 1-Np content in poly(A/1-Np(x)) in aqueous solution.

Unlike conventional random copolymers with comparable 1-Np contents, there exist little or no self-quenching sites in poly(A/1-Np(x)), despite the fact that the 1-Np residues are tightly packed in the cluster in aqueous solution. From studies of fluorescence decay, it has become evident that extremely rapid migration of singlet-excited energy occurs throughout the cluster of the 1-Np chromophores in the unimer, and that the migrating singlet energy is thoroughly trapped by the preformed excimer sites which exist in a very small amount within the cluster.[23]

4.1.2. *Interpolymer Nonradiative Energy Transfer*

Nonradiative energy transfer (NRET) between polymers labeled with an energy donor and an energy acceptor on separate polymer chains provides a useful tool to see if interpolymer association occurs.[29-31] Naphthalene and pyrene labels have been successfully used as a singlet energy donor and acceptor, respectively, because they have a large spectral overlap, and the naphthalene can be almost selectively excited at a wavelength near 290 nm.

The NRET technique has been successfully applied to the La-, Cd-, and Ad-containing polysulfonates to examine whether the hydrophobic association is an intra- or interpolymer event.[15] For this particular experiment the La-, Cd-, and Ad-containing polymers labeled with the 2-Np groups (Chart 4) have been employed together with the corresponding Py-labeled polymers (Chart 2). If the self-association of the hydrophobes is a completely intrapolymer event to form a unimer, the fluorescence labels are confined to the hydrophobic clusters of the individual polymers. Therefore, in an aqueous solution of a mixture of the 2-Np-labeled and Py-labeled polymers possessing the same aliphatic hydrophobes, the possibility of NRET from the singlet excited 2-Np label in one unimer to the Py label in another unimer should be precluded. On the other hand, if the hydrophobic association is an interpolymer event, it should be possible for the 2-Np and Py labels to come close to each other within the Förster radius (R_0=2.86 nm for NRET from 2-methylnaphthalene to pyrene[32]), which allows NRET to occur. If the interpolymer NRET takes places, fluorescence from the Py label should be observed when the 2-Np label is selectively excited.

Chart 4

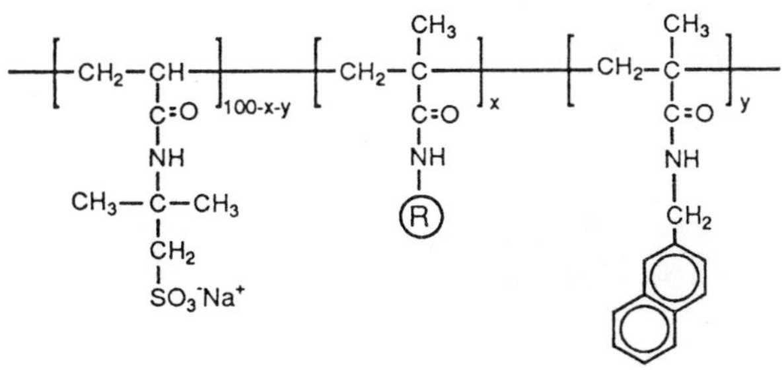

Ⓡ = $-(CH_2)_{11}CH_3$ (La) poly(A/La/2-Np) x=32, y=4 mol%

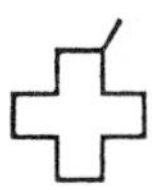

(Cd) poly(A/Cd/2-Np) x=27, y=3 mol%

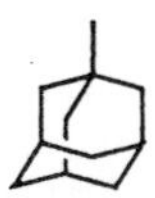

(Ad) poly(A/Ad/2-Np) x=32, y=4 mol%

Aqueous solutions of the mixture of the 2-Np labeled polymer and the Py-labeled polymer, each possessing the same aliphatic hydrophobes, were irradiated at 290 nm, and fluorescences from the 2-Np and Py labels were monitored at 340 and 395 nm, respectively.[15] In Figure 4 are plotted the ratios of the intensities of the fluorescence emitted by the Py and 2-Np labels as a function of the total polymer concentration for the La-, Cd-, and Ad-containing terpolymer systems.[15] The 2-Np labels can be predominantly excited at 290 nm, but a slight contribution of direct excitation of the Py labels cannot be eliminated. The polymer concentrations were varied by adding the nonlabeled copolymers having the same aliphatic hydrophobes, keeping the concentrations of the labeled polymers constant. In the case of the La-containing polymer system, the intensity of the Py fluorescence increases significantly with increasing the total polymer concentration following onset of the Py fluorescence at ca.

0.2 wt %. In contrast, the Cd- and Ad-containing polymer systems show no such increase in the Py fluorescence until the concentration is increased to ca. 7 wt %. These observations indicate that the Cd and Ad residues have much stronger tendency for intrapolymer association than does the La residue. In the La-containing polymer, the intrapolymer hydrophobic association occurs only at concentrations below ca. 0.2 wt %. In contrast, in the Cd- and Ad-containing polymers, the intrapolymer association is dominant and the polymers exist as the unimers over a concentration range < ca. 7 wt %. These experiments lead to an important conclusion that the mode of the hydrophobic association is very different depending on whether the structure of the hydrophobes is cyclic (Cd or Ad) or linear (La), although the numbers of the carbon atoms in these hydrophobes are more or less the same.

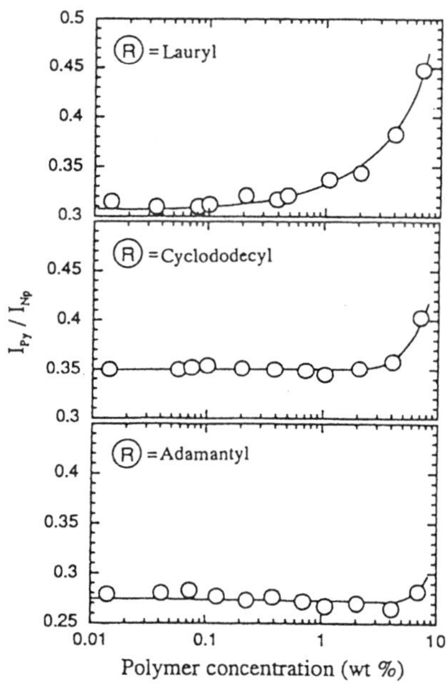

Figure 4. Intensity ratio of the fluorescence of the Py and 2-Np labels as a function of the concentration of the total polymers. Hydrophobes in the polymers are indicated in the figures.[15]

In methanol solutions of the Cd- and Ad-containing polymers, the interpolymer NRET between the 2-Np and Py labels begins to occur at much lower concentrations than in aqueous solution.[15] These polymers adopt a random coil or slightly extended conformation in methanol because the hydrophobic interaction is absent and electrostatic interaction is much less than that in water. Therefore, it is likely that the polymer chains interpenetrate in methanol, which facilitates interpolymer NRET. A comparison of the interpolymer NRET in water and in methanol makes it clearer that the Cd- and Ad-containing polymers remain as the unimers at much higher concentrations than does the La-containing polymer.

4.1.3. *Fluorescence Quenching by Thallium Nitrate*

The encapsulation of the Py labels in the hydrophobic clusters and their protection from the aqueous phase are clearly indicated by a sharp suppression of fluorescence quenching by Tl^+ ions as shown in Figure 5. The fluorescence of pyrene is known to be quenched by Tl^+ due to an external heavy atom effect that requires a short range interaction.[33] In the reference copolymer poly(A/Py) which is in an open chain conformation, Tl^+ ions are electrostatically concentrated in the vicinity of the anionic polymer chain and can come into contact with the Py labels. Thus, the Py fluorescence in the reference copolymer is efficiently quenched by Tl^+. In contrast, because the Py labels in the terpolymers are buried inside the hydrophobic clusters, Tl^+ ions cannot reach the Py sites, giving rise to a sharp suppression of the fluorescence quenching.

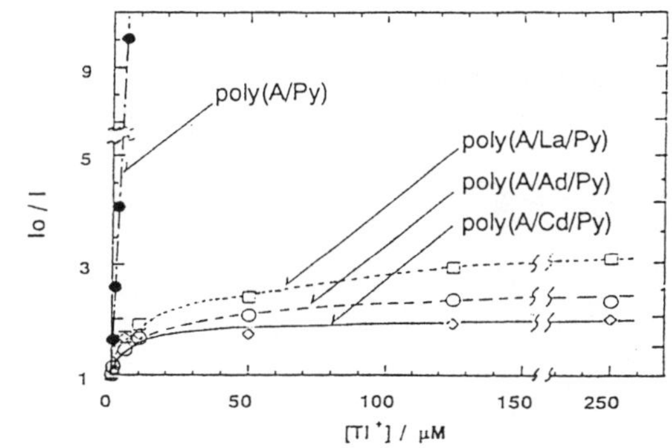

Figure 5. Stern-Volmer plots for fluorescence quenching by Tl^+ ions for the Py labels in the terpolymers and the reference copolymer in aqueous solution.[15]

In the case where there are two chromophore sites, one is accessible to quenchers and the other is not, the Stern-Volmer equation can be modified as[34]

$$I_0 / (I_0 - I) = 1 / f K \, [Tl^+] + (1 / f) \qquad (1)$$

where I and I_0 are the fluorescence intensity in the presence and absence of the quencher, respectively, K is the Stern-Volmer constant for the accessible chromophores, and f is the fraction of the accessible chromophores. Figure 6 shows the quenching data plotted according to eq 1 for the terpolymers. From the intercepts, f values were estimated to be 0.76, 0.45, and 0.58 for poly(A/La/Py), poly(A/Cd/Py), and poly(A/Ad/Py), respectively. These values imply that 55 and 42 % of the Py labels are protected from the access of Tl^+ in the Cd- and Ad-containing terpolymers, respectively, while only 24 % of the Py labels are protected in the La-containing terpolymer. The protection is less effective in the La-containing terpolymer than in the Cd- and Ad-containing terpolymers, the Cd-containing terpolymer resulting in the most effective protection. The polymer concentration for these quenching experiments is low enough (0.01 wt %) even for the terpolymer poly(A/La/Py) to exist as the unimer.

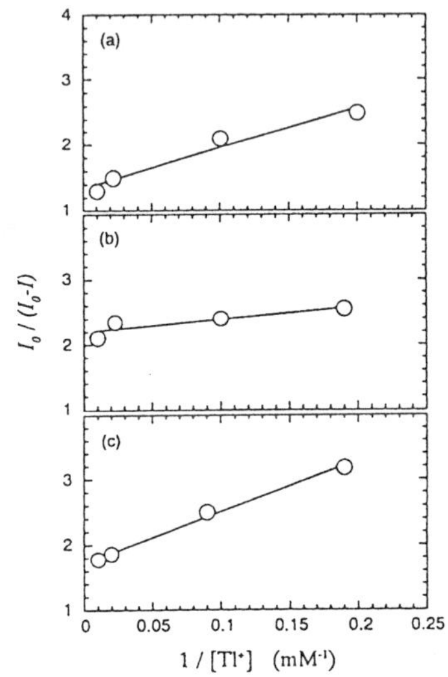

Figure 6. Modified Stern-Volmer plots for fluorescence quenching by Tl^+ ions for the Py labels in poly(A/La/Py) (a), poly(A/Cd/Py) (b), and poly(A/Ad/Py) (c) in aqueous solution.[15]

Although a dominant part of the fluorescence quenching by Tl^+ is due to the heavy atom interaction, quenching by electron transfer may be involved to some extent. Since electron transfer can occur at a longer distance than does the heavy atom interaction, fluorescence from the Py labels buried near the surface in the hydrophobic cluster may be quenched by Tl^+ via the electron transfer, though the Py sites are too far from Tl^+ for the heavy atom interaction to occur. If there is a contribution of such electron transfer quenching, f values must be overestimated.

4.2. NMR METHODS

4.2.1. *Resonance Line Broadening*

^{1}H-NMR spectroscopy provides useful information on the self-organization of amphiphilic polyelectrolytes. In D_2O, resonance bands for the protons on the hydrophobes become extremely broad when they from a hydrophobic cluster because the mobility of the hydrophobes is highly restricted.

In Figure 7 are compared ^{1}H-NMR spectra for the 1-Np protons in poly(A/1-Np(x)) with varying 1-Np contents.[23] These spectra were measured in DMSO-d_6 and in D_2O at a constant polymer concentration (7 wt %) under identical instrumental conditions.[23] In DMSO-d_6, the peak intensity increases systematically with increasing 1-Np content in the copolymer. In D_2O, however, the peak intensity sharply decreases with increasing 1-Np content at room temperature when the 1-Np content exceeds 28 mol %. This is due to the motional line broadening of the 1-Np proton resonance, indicating that the motion of the naphthalene ring becomes highly restricted when the copolymer forms a unimer in aqueous solution.

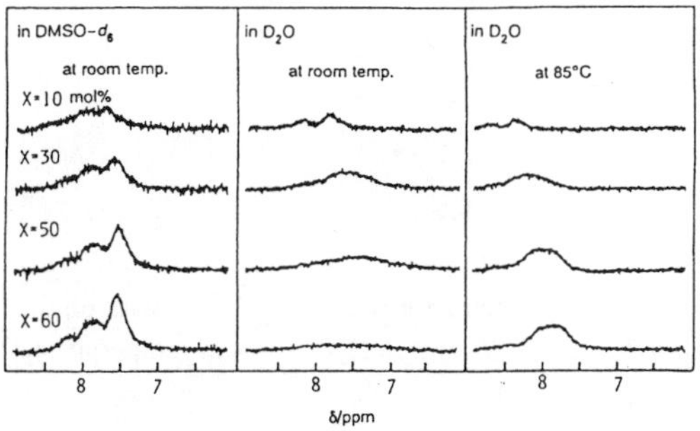

Figure 7. ^{1}H-NMR spectra of poly(A/1-Np(x)) with varying 1-Np contents in DMSO-d_6 and in D_2O.[23]

A marked line broadening occurs at a 1-Np content between 28 and 41 mol % in the copolymer, indicating that the critical content of the 1-Np group for the unimer formation exists between 28 and 41 mol %. This is consistent with the tendency observed in the intensity ratio of the excimer to monomer fluorescence described in the previous section. When the temperature is raised to 85 °C, the NMR peak in D_2O becomes apparent as shown in Figure 7. At this temperature the structured water diminishes, and thereby the hydrophobic interaction is destabilized. Thus, the packing of the 1-Np groups in the cluster is loosened, and the motional restriction of the hydrophobes is alleviated.

4.2.2. *Two-dimensional Nuclear Overhauser Enhancement Spectroscopy*

Two-dimensional nuclear Overhauser enhancement spectroscopy (2D-NOESY) gives qualitative information about the spatial proximities of protons in a molecule. Figure 8 compares 2D-NOESY spectra of poly(A/1-Np(x)) with 60 mol % 1-Np content recorded in DMF-d_7 and in D_2O at room temperature.[26] The resonance lines in D_2O are much broader than those in DMF-d_7. A broad peak at 1.7 ppm observed in D_2O is attributed to the overlap of the methyl (in AMPS and at the α-position in the methacrylamide units) and methylene (in main chain) protons, while a broader peak at 7.3 ppm is assigned to the naphthalene protons. In D_2O, cross peaks between the aromatic and aliphatic protons appear owing to dipolar interactions, whereas cross peaks are absent in DMF-d_7. These observations are indicative of a highly compact unimer conformation and of highly restricted mobility of the 1-Np residues in the hydrophobic cluster in aqueous solution.

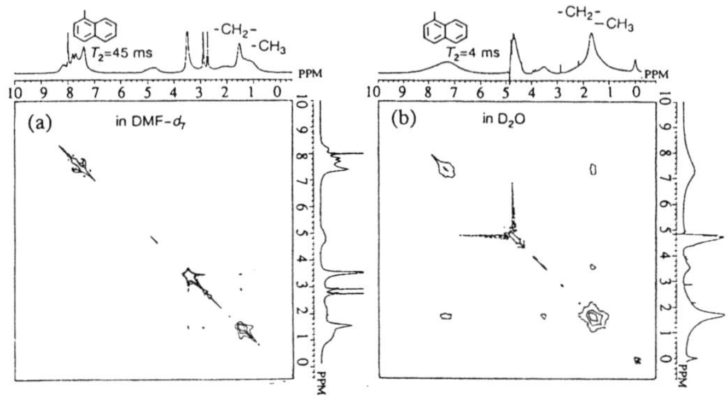

Figure 8. 2D-NOESY spectra for poly(A/1-Np(x)) with 60 mol % 1-Np content recorded in DMF-d_7 and in D_2O at room temperature.[26]

4.2.3. *Spin-lattice and Spin-spin Relaxation Times*

The NMR relaxation techniques are a useful tool to look into local segment motions in polymers.[35-37]

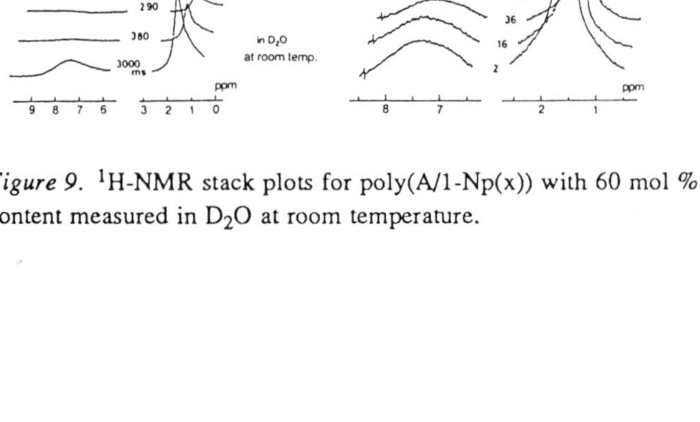

Figure 9. ^{1}H-NMR stack plots for poly(A/1-Np(x)) with 60 mol % 1-Np content measured in D_2O at room temperature.

Figure 9 shows stack plots of the ^{1}H-NMR spectra for poly(A/1-Np(x)) with 60 mol % 1-Np content in D_2O at 25 °C measured by the inversion-recovery technique (with a 180°-τ-180° pulse sequence) for the determination of the spin-lattice relaxation time (T_1) and by the CPMG method[38] for the determination of the spin-spin relaxation time (T_2). T_2 for the naphthalene protons at 7.5 ppm and aliphatic protons at 1.7 ppm are calculated to be 4 and 9 ms, respectively. These values are 1 order of magnitude smaller than those observed in DMF-d_7. The values of T_1 for the naphthalene and aliphatic protons in D_2O are estimated to be 726 and 304 ms, respectively.

The spin-lattice relaxation occurs most efficiently through molecular motion whose frequency is comparable to the NMR frequency.[39] Therefore, T_1 decreases concurrently with T_2 as a molecular motion decreases. Reaching a minimum value, T_1 then increases with a further decrease in the molecular motion, while T_2 remains as a minimum value. The large T_1 value, along with the small T_2 value, for the naphthalene protons is indicative of highly restricted motions of the naphthalene rings in the unimer.

4.3. FTIR STUDIES

In the amphiphilic terpolymers and reference copolymer shown in Chart 2, the hydrophobic pendant groups are linked to the main chain via amide spacer bonds.

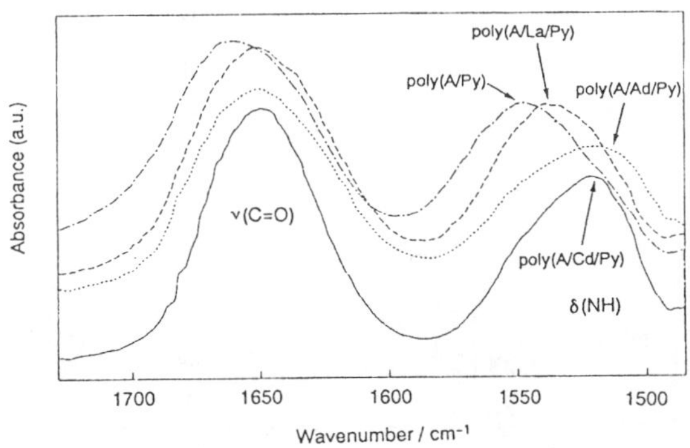

Figure 10. FTIR spectra for the terpolymers and the reference copolymer measured as KBr pellets.[15]

Figure 10 shows FTIR spectra measured as KBr pellets prepared with the polymers recovered from their aqueous solutions by a freeze-drying technique.[15] All the polymers show characteristic IR absorption bands due to ν(C=O) and δ(NH) of the amide bond. These bands for the terpolymers are shifted toward lower wavenumber as compared to those for the reference copolymer, more significant shifts observed in the δ(NH) band than in the ν(C=O) band. These lower-wavenumber shifts indicate the presence of the hydrogen bonds between the amide spacer bonds in the terpolymers. Furthermore, there are significant differences in the extent of the lower-wavenumber shift, particularly in the δ(NH) band, among the terpolymers. The La-containing terpolymer exhibits a lower-wavenumber shift by 8 cm^{-1} as compared with the δ(NH) band of the reference copolymer, while the Cd- and Ad-containing terpolymers show much larger shifts of ca. 30 cm^{-1}. These observations suggest that hydrogen bonds are formed between the spacer amide bonds more strongly in the Cd- and Ad-containing terpolymers than are in the La-containing terpolymer.

The polymers recovered from ca. 5 wt % aqueous solutions by a freeze drying technique were employed to prepare the KBr pellets. In general, conformations of polymers in solution may not always be retained in freeze-dried solid polymers. In the case of the amphiphilic polysulfonates, however, the microphase-separated structures, either unimers or multimolecular aggregates, may be retained, if not perfectly, because the self-organized structures have essentially no dynamic nature and may be viewed as solid particles sustained in water by negatively charged surface.

4.4. SCATTERING METHODS

4.4.1. *Light Scattering*

A characteristic feature of the unimer of the amphiphilic polysulfonates in aqueous solution is that the hydrodynamic volume is extremely small for their high molecular weights.

TABLE 2. Light scattering and SAXS data for the La-, Cd-, and Ad-containing terpolymers in aqueous solution[15]

polymer	M_w[a]	R_S[b] (nm)	d[c] (nm)
poly(A/La/Py)	1.2×10^5	7.0	...
poly(A/Cd/Py)	5.1×10^5	5.5	11
poly(A/Ad/Py)	3.5×10^4	6.2	7.3
poly(A/1-Np/Py)	1.3×10^5	...	...

a. Weight average molecular weight determined by SLS.
b. Average Stokes radius determined by DLS.
c. Spacing calculated from the scattering angle in SAXS.

In TABLE 2 are listed SLS and DLS data for poly(A/La/Py), poly(A/Cd/Py), poly(A/Ad/Py), and poly(A/1-Np/Py) in 0.1 M NaCl solutions.[15] Weight average molecular weights (M_w) of poly(A/La/Py), poly(A/Cd/Py), and poly(A/1-Np/Py) determined by SLS are on the order of 10^5, and that of poly(A/Ad/Py) is on the order of 10^4. In contrast, the Stokes radii (R_S) determined by DLS are extremely small for their molecular weights. This is an experimental manifestation of the highly compact conformation of the unimers in aqueous solution. The unimer of poly(A/Cd/Py) is particularly compact, as indicated by the highest ratio of mass to dimension (M_w=5.1x10^5 versus R_s=5.5 nm).

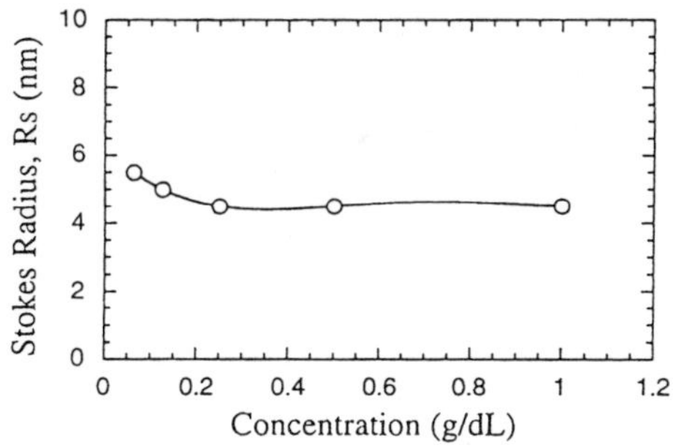

Figure 11. Stokes radius for poly(A/Cd/Py) in 0.1 M NaCl aqueous solution as a function of the polymer concentration at room temperature.

Important evidence for the unimer is that the Stokes radius is independent of the concentration. Figure 11 shows a plot of the Stokes radius as a function of the concentration. The stokes radius stays fairly constant over the concentration range up to 1 g/dL. If interpolymer association occurs, the size increases significantly as the polymer concentration is increased.

4.4.2. *Small-angle X-ray Scattering*

Concentrated aqueous solutions of the Cd-containing and Ad-containing terpolymers show X-ray scattering peaks in the small angle region, while dilute and semi-dilute solutions show no such scattering.

Figure 12b shows SAXS data for poly(A/Cd/Py) and poly(A/Ad/Py) in a concentrated (ca. 10 wt %) aqueous solution. Scattering peaks were observed at 2θ=0.8 and 1.2° for the Cd- and Ad-containing terpolymers, respectively. From these peak angles the spacings are calculated to be 11 and 7.3 nm for the Cd-containing and Ad-containing terpolymers, respectively.[15] A spacing of 11 nm for the Cd-containing terpolymer coincides with the Stokes diameter determined by DLS (TABLE 2). These scattering peaks can be interpreted to be due to closely packed unimer particles in the concentrated aqueous solutions. The Cd- and Ad-containing terpolymers remain as the unimers even at high concentrations as discussed in the previous section, and the unimers exist independently (without interpenetrating) even at the high concentration. As can be seen in Figure 12a, no SAXS peaks exist in methanol solutions. These terpolymers adopt random coils in methanol, and the random coils interpenetrate each other at such a high concentration (ca. 10 wt %) employed for the SAXS measurements.

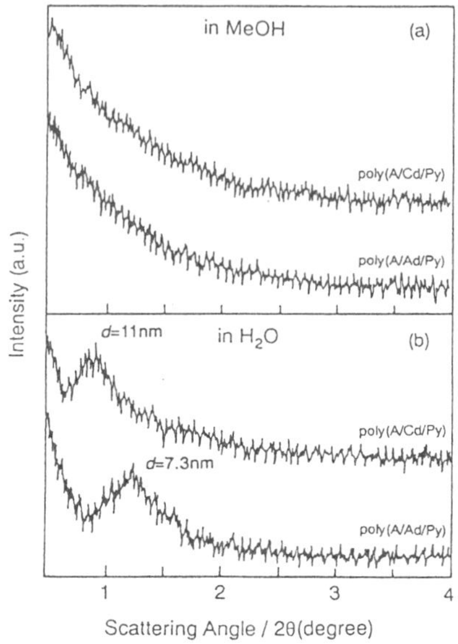

Figure 12. SAXS patterns for poly(A/Ad/Py) and poly(A/Cd/Py) in methanol (a) and in water (b). Concentration of the polymers, ca. 10 wt %.[15]

It should be noted, however, that in the case of poly(A/La/Py), no scattering peaks were recognized in aqueous solution in the small angle region.[15] This is because the La-containing terpolymer can only exist as the unimer at concentrations below ca. 0.2 wt % as described in the previous section, and the interpolymer association occurs at higher concentrations.

5. Compartmentalization of Photoactive Chromophores in Unimers

If a small amount of hydrophobic dyes is covalently incorporated into the unimer of hydrophobically modified polysulfonates, the dyes are tightly encapsulated in the hydrophobic cluster in the unimer (Morishima *et al.*[40-42] termed this "compartmentalization"). The chromophores compartmentalized in the unimers are completely isolated from one another in highly constraining nonpolar microenvironments, and are protected from the aqueous phase.[22] Unlike the conventional molecular assembly systems, the unimer induces a large modification of the photophysical and photochemical behavior of the compartmentalized dyes.[40-48]

Webber *et al.*[49-51] have shown that the overall yield and kinetics of photoinduced electron transfer (ET) for a polyelectrolyte-bound chromophore are modified by steric effects arising from hydrophobic interactions between the polymer and chromophore, and they termed this "hydrophobic protection". More recently, Nowalowska and Guillet[52] have shown that partially sulfonated poly(vinylnaphthalene)s form "hypercoiled" structures in water, and photoexcitation energy migrates through the naphthalene units in the hypercoil. Such "antenna" polyelectrolytes with photochemical reaction centers entrapped inside of the hypercoil exhibited efficient photosensitized reactions due to the "antenna effect", and they termed "photozyme"[52]. The "hydrophobic protection" and "photozyme" are based on the same principles as the "compartmentalization".

5.1. ENERGY MIGRATION AND TRAPPING WITHIN THE UNIMER

Very rapid and efficient energy migration and trapping occur within the unimer consisting of naphthalene clusters and pyrene labels.[43] A terpolymer of 40 mol % AMPS , 59 mol % 1NpMAm, and 1 mol % PyMAm , poly(A/1-Np/Py) (Chart 5), forms a unimer in water, the Py species being compartmentalized in the 1-Np clusters. From the weight average molecular weight (M_w=1.3x10^5) and the compositions of the 1-Np and Py units in the terpolymer, one can roughly estimate that there exist 300 1-Np and 5 Py units in a unimer.

Chart 5

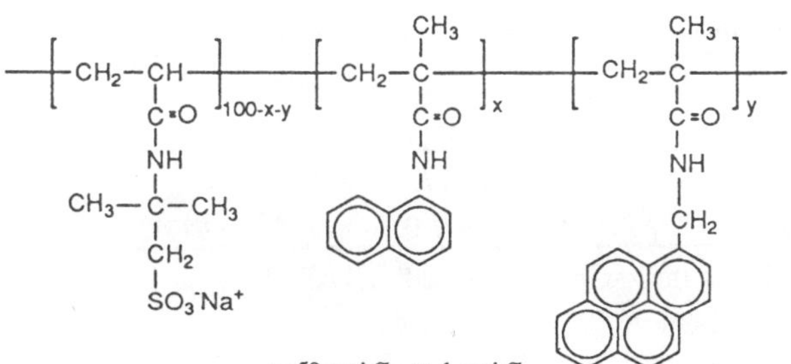

x=59 mol %, y=1 mol %

poly(A/1-Np/Py)

Figure 13 compares fluorescence spectra of poly(A/1-Np/Py) with selective excitation of the 1-Np chromophore (at 290 nm) in methanol and in water[43]. In methanol, in which poly(A/1-Np/Py) adopts a random coil conformation, the terpolymer exhibits both 1-Np fluorescence and Py fluorescence, the former being predominant over the latter. The random coil allows NRET from the singlet excited 1-Np to the Py sites to some extent, as can be seen from the significant intensity of the Py fluorescence (Figure 13a). In contrast, poly(A/1-Np/Py) emits only Py fluorescence in water (Figure 13b), indicating that the singlet excited energy migrates over the 1-Np clusters within the unimer and is thoroughly trapped by the Py sites.

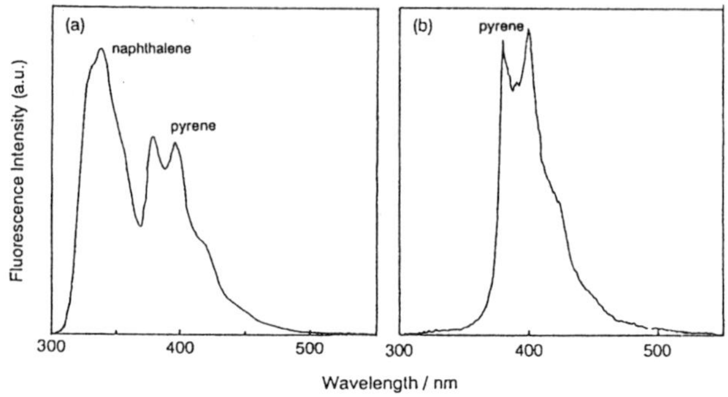

Figure 13. Fluorescence spectra for poly(A/1-Np/Py) in methanol (a) and in water (b) by excitation of the 1-Np moieties at 290 nm.[43]

This observation suggests that, even if there are a number of cluster units in a unimer (Figure 1), all the cluster units are in contact with each other such that the 1-Np singlet energy can migrate from one cluster unit to another.

TABLE 3. Fluorescence decay and rise times on the picosecond time scale for poly(A/1-Np/Py) in aqueous solution at room temperature[43]

λ_{ex} [b] (nm)	λ_{em} [c] (nm)	fitting parameters[a] τ_i(ps)/α_i			χ^2
290	330	19/0.771,	167/0.175,	2550/0.054	1.03
290	400	20[d]/-0.300,	298[d]/-0.160,	25500/1.43	1.04

a. Fitting function is the same as indicated in TABLE 1.
b. Excitation wavelength.
c. Wavelength at which fluorescence was monitored.
d. Rise time.

Fluorescence decay and rise data for poly(A/1-Np/Py) in aqueous solution monitored at 330 nm (1-Np fluorescence) and at 400 nm (Py fluorescence) on a picosecond time scale are fitted with a three-exponential function. The lifetimes τ_i and the relative weight of the pre-exponential factors α_i for the three-exponential function are listed in TABLE 3. The 1-Np fluorescence decays extremely rapidly; a large portion of the fluorescence decays with a lifetime of 19 ps. Importantly, there are rapid rises in the Py fluorescence with rise times of 20 and 298 ps. This indicates that extremely rapid energy migration occurs throughout the 1-Np cluster and the energy is rapidly trapped by the Py sites in the unimer of poly(A/1-Np/Py).

Such extremely fast energy migration and trapping phenomena cannot reasonably be explained by the NRET theories based on dipole-dipole mechanisms. If chromophores are aligned with a sufficiently small spacing and a similar orientation, photoexcitation energy may be delocalized over a large space as an exciton.[53] This may be the case, at least locally, for the chromophore cluster in the unimer of poly(A/1-Np/Py).

5.2. PHOTOINDUCED CHARGE SEPARATION

The primary photoprocess in natural photosynthetic systems involves a fast photoinduced electron transfer (ET) and a subsequent charge separation. An understanding of the molecular mechanism for such critical functions in the biological systems and design of relevant model systems have been a challenge for chemists for decades. For efficient charge separation of photochemically generated ion pairs to be achieved, the recombination of the geminate ion pair should be suppressed. An important factor that governs the charge recombination rate is the tightness of the geminate ion pair; the back ET occurs very rapidly in "tight" ion pairs whereas it becomes slow in "loose" ion pairs.[54]

In the Py labeled amphiphilic polysulfonates (Chart 2), the Py moieties are compartmentalized within the hydrophobic clusters in the unimers in aqueous solution. If methylviologen (MV^{2+}), an electron acceptor for singlet excited Py, is added to the aqueous solutions of these terpolymers, MV^{2+} would be electrostatically bound to the surface of the hydrophobic cluster. When Py labels are excited with light, ET occurs from Py to MV^{2+} across the charged layer of the hydrophobic cluster to generate a geminate ion pair from which either charge recombination or charge escape occurs.[22]

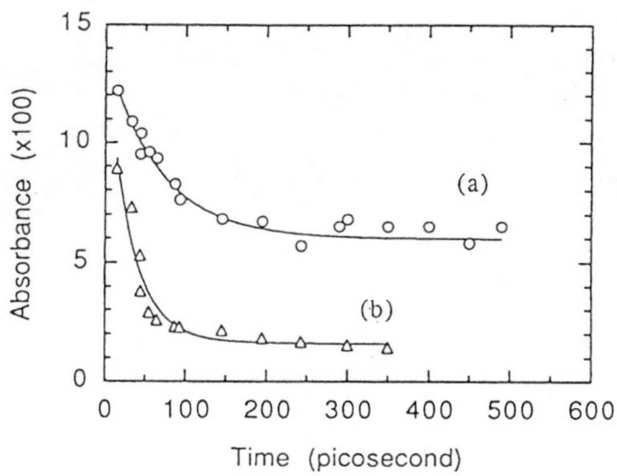

Figure 14. Time profiles of transient absorbances at 600 nm (due to $MV^{+\cdot}$) for (a) poly(A/Cd/Py)-MV^{2+} and (b) poly(A/Py)-MV^{2+} systems.[22]

Surprisingly, the forward ET occurs extremely rapidly despite the fact that the Py chromophores are compartmentalized in the hydrophobic cluster and geometrically separated from MV^{2+}. For example, in a poly(A/Cd/Py) aqueous solution containing MV^{2+}, the build-up of the reduced MV^{2+} ($MV^{+\cdot}$) species is completed within ca. 20 ps. The decay of $MV^{+\cdot}$ monitored at 600 nm is compared with that of the reference copolymer poly(A/Py) in Figure 14. In the poly(A/Cd/Py) system, following a fast decay due to the recombination of the geminate ion pairs over a period of ca. 100 ps after the laser pulse, the decay is considerably slowed. The forward ET occurs with a first-order rate constant on the order of 10^{11} s^{-1}, while the geminate recombination in the time domain of <100 ps is two orders of magnitude slower than the forward ET rate. On the other hand, in the poly(A/Py) system, the decay of $MV^{+\cdot}$ due to the geminate recombination occurs one orders of magnitude faster than the poly(A/Cd/Py) system.[22]

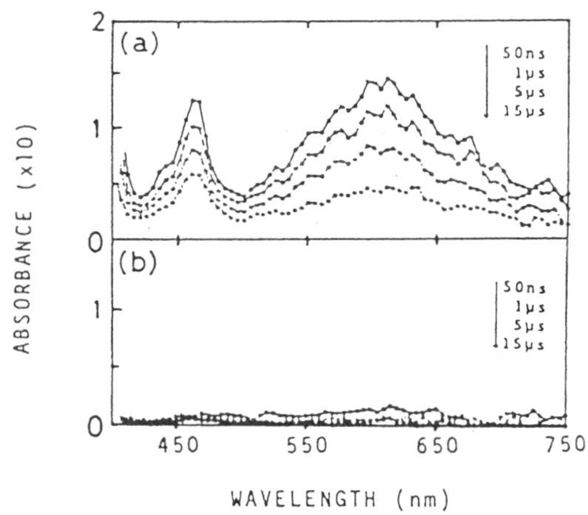

Figure 15. Nano- and microsecond time-resolved transient absorption spectra for (a) poly(A/Cd/Py)-MV^{2+} and (b) poly(A/Py)-MV^{2+} systems. Absorbances of the sample solutions were adjusted to 1.0 at 355 nm (excitation wavelength in the laser photolysis).[22]

Figure 15 shows the time-resolved transient absorption spectra in a microsecond time domain. In the Cd-containing terpolymer system, the yields of $MV^{+\cdot}$ and $Py^{+\cdot}$ in the microsecond time regime is much higher than that in the reference copolymer system.

A reason for the efficient charge separation observed in the terpolymer system may be that loose geminate ion pairs are formed, and rapid charge escape may take place due to fast electron self-exchange between $MV^{+\cdot}$ and MV^{2+} concentrated on the surface of the unimer.

5.3. PHOTOISOMERIZATION

There has been increasing attention, in recent years, focused on photochromic isomerization because of its potential utility in an application for a photochemical switch.[55,56] In this relevance, the photoisomerization of azobenzene (Abz) moieties in various macromolecular microenvironments has been a focus of studies.[57-59]

If Abz species are confined within the hydrophobic cluster in the unimer, the photoisomerization is considerably suppressed because of highly constraining microenvironments.[44,45]

Chart 6

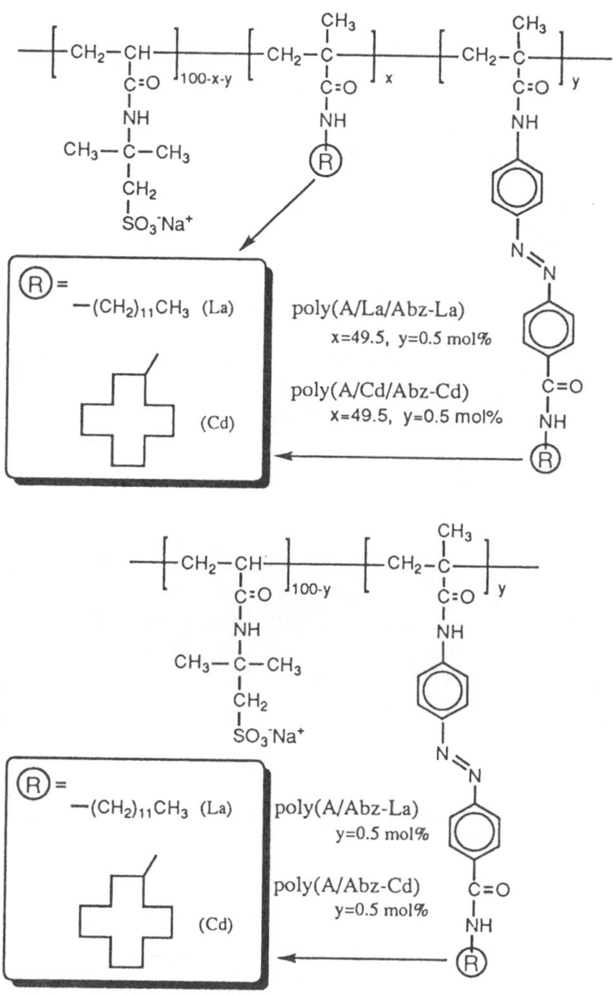

Amphiphilic terpolymers consisting of 50 mol % AMPS, 49.5 mol% La or Cd residues, and 0.5 mol % of a moiety of Abz substituted with a La (Abz-La) or with a Cd group (Abz-Cd) have been synthesized (Chart 6).[45] In these terpolymers, the Abz-La and Abz-Cd residues are compartmentalized in the hydrophobic clusters in respective unimers in aqueous solution. ^{1}H-NMR relaxation times indicate that the local mobility of the Abz species in the hydrophobic cluster is highly restricted. In D_2O, T_2 for the phenyl protons in the Abz-Cd moieties in the terpolymer poly(A/Abz-Cd/Cd) is 5 ms as compared to 133 ms in the reference copolymer poly(A/Abz-Cd) (Chart 6). On the other hand, the corresponding $T2$ values in the La-containing terpolymer poly(A/Abz-La/La) and reference copolymer poly(A/Abz-La) are 11 and 114 ms, respectively.[45] The local motion of the Abz moieties is much more restricted in the Cd cluster than in the La cluster.

Trans azobenzene, the stable conformation, undergoes photoisomerization to the cis isomer which thermally isomerizes back to the trans isomer with an appreciable rate at room temperature. A large change in the molecular shape is involved during the isomerization, which requires a certain free volume around azobenzene.

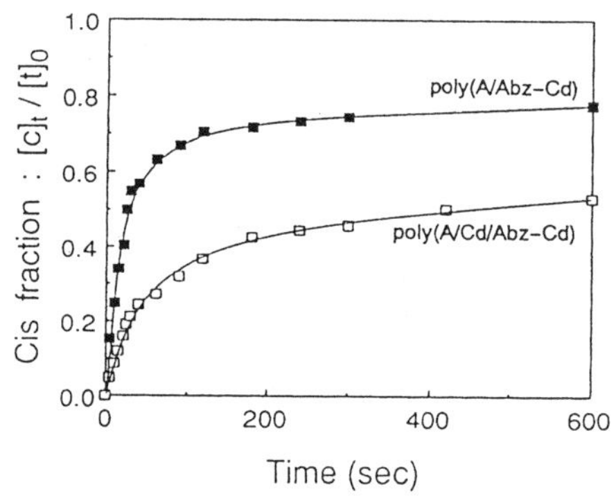

Figure 16. Changes in the cis fractions as a function of irradiation time for trans-to-cis photoisomerization of poly(A/Cd/Abz-Cd) and poly(A/Abz-Cd) in aqueous solution at 298 K.[45]

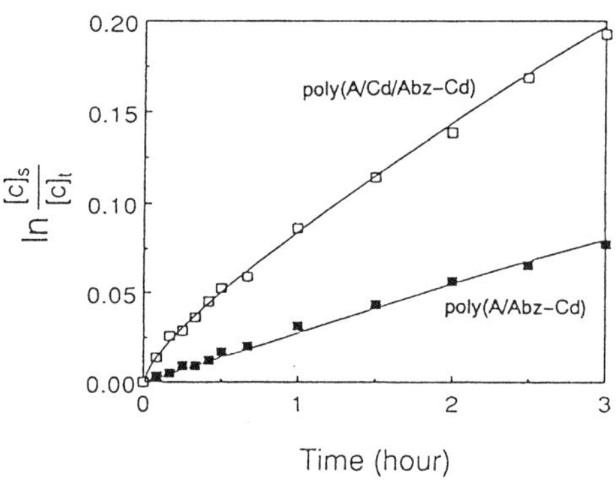

Figure 17. First-order plots for cis-to-trans thermal back-isomerization of poly(A/Cd/Abz-Cd) and poly(A/Abz-Cd) in aqueous solution at 298 K.[45]

In Figure 16 are plotted the fractions of the cis isomer in poly(A/Cd/Abz-Cd) as a function of irradiation time of 355-nm light. It can be seen that the photoisomerization of the Abz-Cd residue in the Cd-containing terpolymer are impeded as compared to that in the reference copolymer poly(A/Abz-Cd).[45] It should be noted here that there is a significant difference in the extent of the impeding effect between the Cd and La groups in the terpolymers: the rate of photoisomerization of the Abz-Cd residue is much slower than that of the Abz-La residues in the terpolymers. This indicates that the Cd cluster imposes more pronounced restriction on the motion of the Abz species than does the La cluster, being consistent with the NMR relaxation data.

In contrast to the photoisomerization, the thermal back isomerization of the cis form of the Abz-Cd species in the Cd-containing terpolymer becomes faster than that in the reference copolymer (Figure 17). Since the microenvironment about the Abz-Cd species confined in the Cd cluster is so rigid that the photogenerated cis isomer may not be accommodated by the matrix in its most stable conformation, and therefore it may isomerize back to the trans isomer with a faster rate.[45]

5.4. ANOMALOUS BEHAVIOR OF TRIPLET EXCITED PORPHYRINS

Chart 7

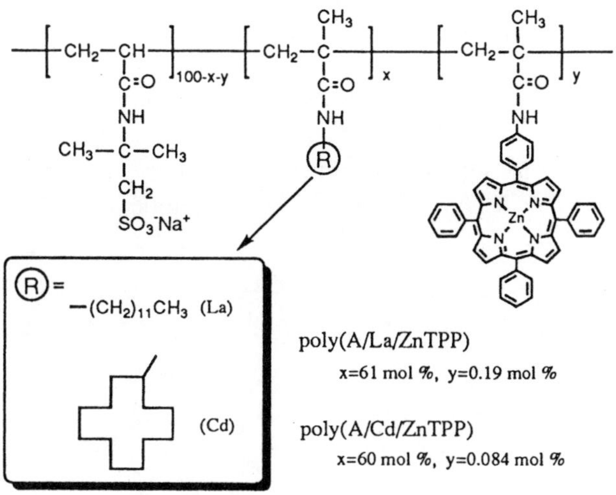

poly(A/La/ZnTPP)
x=61 mol %, y=0.19 mol %

poly(A/Cd/ZnTPP)
x=60 mol %, y=0.084 mol %

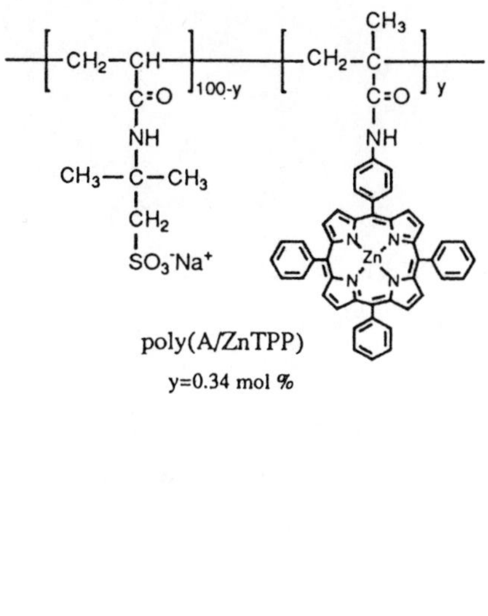

poly(A/ZnTPP)
y=0.34 mol %

Studies of the effects of generic microenvironments on photophysical and photochemical behavior of metalloporphyrin in model systems may provide mechanistic insight into the biological photosynthesis. In addition, such studies may provide a guideline for designing of porphyrin-based artificial photosystems such as photon energy conversion into electronic energy or chemical potential, information processing, and organic photoimaging systems.

If a small amount of zinc(II) tetraphenylporphyrin (ZnTPP) is compartmentalized in the highly constraining microenvironments in the unimer in aqueous solution, the photophysical behavior of the ZnTPP can be greatly modified.[46-48]

Most remarkably, the lifetime of the triplet excited ZnTPP gets extraordinarily long at ordinary or higher temperatures. This is particularly true in the Cd-containing terpolymer (Chart 7). The triplet excited lifetimes at ordinary or higher temperatures were determined from the decays of triplet-triplet absorption in laser photolysis experiments.[48] The decay profiles for the compartmentalized ZnTPP systems were fitted with double-exponential functions, from which average triplet lifetimes were estimated. The triplet lifetime of ZnTPP in the reference copolymer (Chart 7) is more or less 2 ms, which is about the same as the lifetime of small molecular weight ZnTPP in organic solvents. In contrast, the average lifetime in the Cd cluster is 58 ms, whereas it is 17 ms in the La cluster. The reason for the longer triplet lifetime in the Cd cluster may be due to the rigidity of the matrix where ZnTPP is compartmentalized.[48]

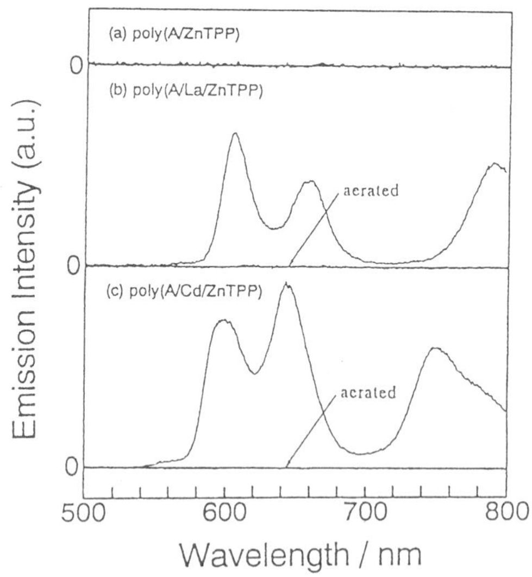

Figure 18. Delayed emission spectra in aqueous solution at 30 °C at a delay time of 5 ms.[46,47]

Because the triplet lifetime of the ZnTPP species in the Cd cluster is extremely long, the terpolymer emits phosphorescence and "E-type" delayed fluorescence in aqueous solution at room temperature (Figure 18).[46,47] This is a very unusual phenomenon for Zn-porphyrin dyes. The reference copolymer does not emit such delayed fluorescence and phosphorescence in aqueous solution at room temperature. The delayed emissions show a characteristic temperature dependence; with increasing temperature, the delayed fluorescence increases whereas the phosphorescence decreases. Analysis of the temperature dependence data indicates that the delayed fluorescence is due to the thermal activation of the triplet excited state back up to the singlet excited state.

ZnTPP encapsulated in the La cluster also shows phosphorescence and delayed fluorescence in aqueous solution at room temperature, although their intensities are much lower than those in the Cd cluster. The phosphorescence and fluorescence spectral profiles for the ZnTPP in the Cd cluster are greatly distorted as compared with those in the La cluster; in the Cd case, phosphorescence peak is blue-shifted by about 40 nm and the intensities of the 0-0 and 0-1 fluorescence bands are reversed as compared with the spectra in the La cluster.[47] This difference in the microenvironment between the Cd and La cluster is also reflected in the absorption spectra; the absorption band profile in the Cd cluster is broader than that in the La cluster.[47] As described above, the Cd cluster is more rigid than the La cluster, and because of the rigid matrix the porphyrin ring may be "pinned down" to a distorted conformation in the Cd cluster which may be responsible for the unusual photophysical behavior.[48]

6. References

1. Schulz, D. N., Bock, J., and Valint Jr, P. L. (1994) Synthesis and characterization of hydrophobically associationg water-soluble polymers, in P. Dubin, J. Bock, R. M. Davies, D. N. Schulz, and C. Thies (eds.), *Macromolecular Complexes in Chemistry and Biology*, Springer-Verlag, Berlin Heidelberg, p. 3-13.
2. Varadaraj, R., Branham, K. D., McCormick, C. L., and Bock, J. (1994) Analysis of hydrophobically associating copolymers utilizing spectroscopic probes and labels, in P. Dubin, J. Bock, R. M. Davies, D. N. Schulz, and C. Thies (eds.), *Macromolecular Complexes in Chemistry and Biology*, Springer-Verlag, Berlin Heidelberg, p. 15-31.
3. Strauss, U. P. and Jackson, E. G. (1951) *J. Polym. Sci.* **6**, 649.
4. Dubin, P. and Strauss, U. P. (1967) *J. Phys. Chem.* **71**, 2757.
5. Dubin, P. and Strauss, U. P. (1970) *J. Phys. Chem.* **71**, 2842.
6. Kotin, L. and Nagasawa, M. (1962) *J. Chem. Phys.* **36**, 873.
7. Nagasawa, M., Murase, T., and Kondo, K. (1965) *J. Phys. Chem.* **69**, 4005.
8. Joyce, D. E. and Kurucse, T. (1981) *Polymer* **22**, 415.
9. Morcellet-Sauvage, J., Morcellet, M., and Loucheux, C. (1981) *Makromol. Chem.* **182**, 949.
10. Morishima, Y., Itoh, Y., and Nozakura, S. (1981) *Makromol. Chem.* **182**, 3135.
11. Guillet, J. E., Wang, J., and Gu, L. (1986) *Macromolecules* **19**, 2793.
12. Nemethy, G. and Scheraga, H. A. (1962) *J. Chem. Phys.* **36**, 3382.
13. Nemethy, G. and Scheraga, H. A. (1962) *J. Phys. Chem.* **66**, 1773.
14. Jencks, W. P. (1969) *Catalysis in Chemistry and Enzymology*; McGraw-Hill, New York, pp 393.
15. Morishima, Y., Nomura, S., Ikeda, T., Seki, M., and Kamachi, M. (1995) *Macromolecules* **28**, 2874.
16. McCormick, C. L. and Chang, Y. (1994) *Macromolecules* **27**, 2151.
17. Kamioka, K., Webber, S. E., and Morishima, Y. (1988) *Macromolecules* **21**, 972.

18. Prochazka, K., Kiserow, D., Ramireddy, C., Tuzar, Z., Munk, P., and Webber, S. E. (1992) *Macromolecules* **25**, 454.
19. Morishima, Y., Lim, H. S., Nozakura, S., and Strutevant, J. L. (1989) *Macromolecules* **22**, 1148.
20. McCormick, C. L. and Salazar, L. C. (1992) *Polymer* **33**, 4617.
21. Chang, Y. and McCormick, C. L. (1993) *Macromolecules* **26**, 6121.
22. Morishima, Y., Tominaga, Y., Kamachi, M., Okada, T., Hirata, Y., and Mataga, N. (1991) *J. Phys. Chem.* **95**, 6027.
23. Morishima, Y., Tominaga, Y., Nomura, S., and Kamachi, M. (1992) *Macromolecules* **25**, 861.
24. Kalyanasundaram, K. and Thomas, J. K. (1977) *J. Am. Chem. Soc.* **99**, 2039.
25. Morishima, Y., Seki, M., Tominaga, Y., and Kamachi, M. (1992) *J. Polym. Sci., Polym. Chem. Ed.* **30**, 2099.
26. Seki, M., Morishima, Y., and Kamachi, M. (1992) *Macromolecules* **24**, 6540.
27. Morishima, Y. (1990) *Prog. Polym. Sci.* **15**, 949.
28. Webber, S. E. (1990) *Chem. Rev.* **90**, 1469.
29. Ringsdorf, H., Simon, J., and Winnik, F. M. (1992) *Macromolecules* **25**, 7306.
30. Ringsdorf, H., Simon, J., and Winnik, F. M. (1992) *Macromolecules* **25**, 5353.
31. Winnik, F. M. (1990) *Polymer* **31**, 2125.
32. Berlman, I. B. (1973) *Energy Transfer Parameters of Aromatic Compounds*; Academic Press, New York.
33. Hashimoto, S. and Thomas, J. K. (1985) *J. Am. Chem. Soc.* **107**, 4655.
34. Lakowicz, J. R. (1983) *Principles of Fluorescence Spectroscopy*; Plenum Press, New York, pp 281.
35. Erdmann, K. and Gutsze, A. (1987) *Colloid Polym. Sci.* **265**, 667.
36. Raby, P., Budd, P. M., Heatley, F., and Price, C. (1991) *J. Polym. Sci., Polym. Phys. Ed.* **29**, 451.
37. Brereton, M. G., Ward, I. M., Boden, N., and Wright, P. (1991) *Macromolecules* **24**, 2068.
38. Meiboom, S. and Gill, D. (1958) *Rev. Sci. Instrum.* **29**, 688.
39. Pake, G. E. (1965) *Solid State Physics*; Seitz, F. and Turnbull, D. (eds.), Academic Press, New York,*Vol. 2*, pp. 1-92.
40. Morishima, Y., Furui, T., Nozakura, S., Okada, T., and Mataga, N. (1989) *J. Phys. Chem.* **93**, 1643.
41. Morishima, Y. (1992) *Adv. Polym. Sci.* **104**, 51.
42. Morishima, Y. (1994) *Trends Polym. Sci.* **2**, 31.
43. Morishima, Y., Tominaga, Y.;,Nomura, S., Kamachi, M., and Okada, T. (1992) *J. Phys. Chem.* **96**, 1990.
44. Morishima, Y., Tsuji, M., Kamachi, M., and Hatada, K. (1992) *Macromolecules* **25**, 4406.
45. Morishima, Y., Tsuji, M., Seki, M., and Kamachi, M. (1993) *Macromolecules* **26**, 3299.
46. Aota, H., Morishima, Y., Kamachi, M. (1993) *Photochem. Photobiol.* **57**, 989.
47. Morishima, Y., Saegusa, K., and Kamachi, M. (1995) *Macromolecules* **28**, 1203.
48. Morishima, Y., Saegusa, K., and Kamachi, M. (1995) *J. Phys. Chem.* **99**, 4512.
49. Delaire, J. A., Rodgers, M. A. J., and Webber, S. E. (1986) *Eur. Polym. J.* **22**, 189.
50. Delaire, J. A., Sanquer-Barrie, M., and Webber, S. E. (1988) *J. Phys. Chem.* **92**, 1252.
51. Stramel, R. D., Webber, S. E., and Rodgers, M. A. J. (1988) *J. Phys. Chem.* **92**, 6625.
52. Nowakowska, M. and Guillet, J. E. (1991) *Macromolecules* **24**, 474.
53. Philpot, M. R. (1975) *J. Chem. Phys.* **63**, 485.
54. Mataga, N., Kanda, Y., and Okada, T. (1986) *J. Phys. Chem.* **90**, 3880.
55. Irie, M. and Tanaka, H. (1983) *Macromolecules* **16**, 210.
56. Irie, M. and Schnabel, W. (1985) *Macromolecules* **18**, 394.
57. Eisenbach, C. D. (1980) *Bunsenges. Phys. Chem.* **5**, 680.
58. Sung, C. S. P., Gould, I. R., and Turro, N. (1984) *Macromolecules* **17**, 1447.
59. Mita, I., Horie, K., and Hirano, K. (1989) *Macromolecules* **22**, 558.

THERMALLY-INDUCED PHASE SEPARATION OF POLY(ETHOXY-ETHYL VINYL ETHER) STUDIED BY THE FLUORESCENCE METHOD

JUNICHI HORINAKA, KEIKO ONO, MASAHIDE YAMAMOTO*,
Division of Polymer Chemistry, Graduate School of Engineering, Kyoto University, Sakyo-ku, Kyoto 606-01, Japan

SADAHITO AOSHIMA, and EIICHI KOBAYASHI
Department of Industrial Chemistry, Faculty of Science and Technology, Science University of Tokyo, Noda, Chiba 278, Japan

1. Introduction

The physical properties of thermally-responsive polymers change sharply at a certain temperature. Such a response is interesting from both basic and applied viewpoints [1-12, 22-25]. For example, methylcellulose [2], poly(vinyl ether) [1], and poly(acrylamide) derivatives [2-8] are thermally-induced phase separation polymers, the solubility of which changes in water with the temperature. Although in general, the polymers become more soluble with the increase in temperature, in polymer/solvent systems having LCST phase separation occurs at a certain temperature and the polymers become insoluble above it. The dynamics of phase separation has been studied extensively[1,3,5,8,25], but the molecular details of the phase separation behavior remain to be solved.

Recently, structure-controlled poly(vinyl ether) derivatives have been synthesized by living cationic polymerization and the effects of MW and MW distribution on the thermal behavior were reported [9,10]. By using a unique terminator, we synthesized poly(ethoxyethyl vinyl ether) (PEVE) labeled with anthracene in the middle of the main chain. The phase separation temperature of PEVE (*Mw* was ca. 20,000) aqueous solution is 20 °C [10]; the PEVE aqueous solution suddenly becomes turbid at this temperature.

We have examined local chain dynamics in dilute solutions mainly by the fluorescence depolarization method [14-21]. By this method, microscopic chain mobility can be estimated through the motion of the fluorescence probe labeled in the middle of the main chain. Application of this technique to the study of phase separation gives us valuable microscopic information in the vicinity of the phase separation temperature. Herein we studied the microscopic behavior of PEVE in an aqueous solution around the phase separation point by the fluorescence method, including the fluorescence depolarization method. We also used the light scattering technique.

S.E. Webber et al. (eds.), Solvents and Self-Organization of Polymers, 359–366.

2. Experimental

2.1. SAMPLE

The poly(ethoxyethyl vinyl ether) (PEVE) sample was synthesized by living cationic polymerization. By coupling the living ends with 9,10-anthracenedithiol [13] we obtained PEVE labeled with anthracene in the middle of the main chain (Figure 1).

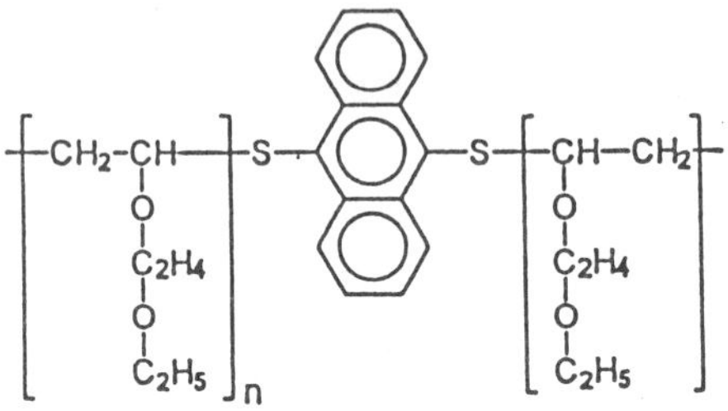

Figure 1. Anthracene-labeled PEVE sample.

The obtained polymer had a *Mw* of ca. 11,000 and *Mw/Mn* of 1.13. From this polymer, we fractionated the high molecular weight part by GPC. This high MW part had a *Mw* of ca. 2×10^4. As mentioned in the introduction, the phase separation temperature of PEVE aqueous solution was determined to be 20 °C by monitoring the transmittance of a 500 nm light beam [10]. Then we carried out the measurement at 5 to 25 °C. The polymer concentration was prepared to be ca. 0.02 wt% and was degassed by freeze-to-pump-thawing. We selected such dilute solutions to minimize the influence of turbidity at the phase separation.

2.2. MEASUREMENT

First, measurement was carried out under continuous illumination using a Hitachi 850 fluorescence spectrophotometer. We examined the temperature dependence of excitation spectrum, emission spectrum, and fluorescence intensity. Excitation spectra were measured at 457 nm and the excitation wavelength for emission spectra was set at 396 nm.

Next we carried out time-resolved fluorescence measurement by the single photon counting technique and estimated the fluorescence lifetime and anisotropy ratio of anthracene-labeled PEVE in aqueous solutions. A diode laser PLP-101 (Hamamatsu Photonics, 411 nm, FWHM; ca. 50 ps) was used as a light source and a microchannel plate-photomultiplier tube as a detector. The excitation light was polarized vertically, and the parallel fluorescence component, $I_{\parallel}(t)$ and the perpendicular fluorescence component, $I_{\perp}(t)$, to the plane of polarization were measured.

Light scattering was measured to examine polymer aggregation. The sample aqueous solution was heated or cooled at the rate of less than 0.1 °C/min. Light intensity scattered at an angle of 30 degree was recorded as a function of temperature by using a light scattering apparatus (Otsuka Electronics DLS-7000).

3. Analysis for Time-Resolved Measurement

The total fluorescence intensity is equivalent to $I(t)$ defined as eq. 1

$$I(t) = I_{||}(t) + 2GI_{\perp}(t) \tag{1}$$

where G is the compensating factor. To the experimental $I(t)$ data we fitted the equation in which eq. 2 was convoluted with the instrumental function by nonlinear-least-squares method.

$$I(t) = x_1 \exp(-t/\tau_1) + x_2 \exp(-t/\tau_2) + (1 - x_1 - x_2) \exp(-t/\tau_3) \tag{2}$$

We calculated the mean fluorescence lifetime $<\tau>$ by eq. 3 with the parameters obtained from the fitting.

$$<\tau> = x_1\tau_1 + x_1\tau_2 + (1-x_1-x_2)\tau_3 \tag{3}$$

The fluorescence anisotropy ratio $r(t)$ is defined by eq. 4.

$$r(t) = [I_{||}(t) - GI_{\perp}(t)] / [I_{||}(t) + 2GI_{\perp}(t)] \tag{4}$$

$r(t)$ was fitted with eq. 5 in a similar way as eq. 1. Empirically eq. 5 consisting of three exponential terms represents the experimental data well.

$$r(t) = r_0 [x_1 \exp(-t/T_1) + x_2 \exp(-t/T_2) + (1-x_1-x_2) \exp(-t/T_3)] \tag{5}$$

The mean relaxation time T_m can be calculated by eq. 6.

$$T_m = x_1 T_1 + x_2 T_2 + (1-x_1-x_2) T_3 \tag{6}$$

4. Results and Discussion

4.1. FLUORESCENCE INTENSITY AND LIFETIME

Figure 2 shows the excitation and emission spectra at 12.2 °C. We checked the spectra at other temperatures, but observed no change. This means that the anthracene probes do not interact with each other at any measurement temperature.

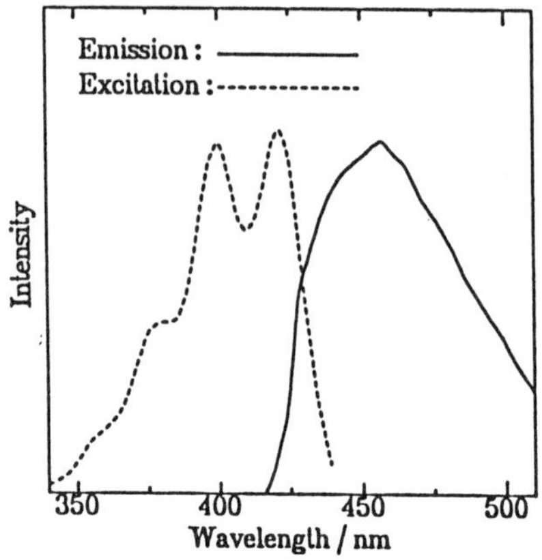

Figure 2. Emission and excitation spectra at 12.2 °C.

Figure 3 shows the plot of fluorescence intensity against temperature. A marked decrease can be seen at 16 °C. This phenomenon will be discussed in the section on the fluorescence lifetime. The fluorescence lifetime also changed at ca. 16 °C as did fluorescence intensity as shown in Figure 3.

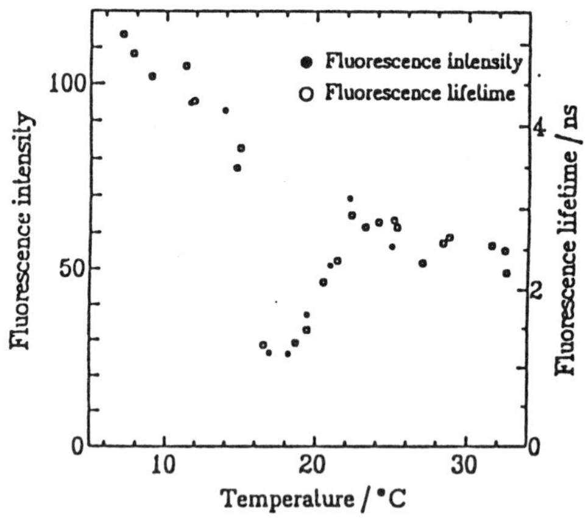

Figure 3. Fluorescence intensity and lifetime against temperature.

Temperature dependence of lifetime was not observed in our previous anthracene-labeled polystyrene or poly(methyl methacrylate) system. The change observed on PEVE indicates that the micro-environment around anthracene in solution changes at this temperature. Furthermore it is noteworthy that this change occurred at a few degrees below the phase separation temperature (20 °C).

Halary et al. showed that poly(vinylmethylether) (PEVE) quenches the fluorescence of anthracene labeled in the middle of polystyrene main chain in the ternary blend film of PVME, polystyrene, and labeled polystyrene [1]. They considered that the mechanism of the fluorescence quenching is static with the lifetime remaining constant. Since the molecular structure of our PEVE is similar to that of PVME, it is reasonable that PEVE quenches fluorescence emission of labeled anthracene. However, contrary to their finding, fluorescence intensity in our study was proportional to fluorescence lifetime as indicated in Figure 3. At this stage we propose the dynamic quenching of the anthracene probe with PEVE segments since strong quenching was observed preceding the macroscopic phase separation.

4.2. CHAIN MOBILITY

Figure 4 shows the plot of relaxation time for local motion of the PEVE chain against temperature. The relaxation time was normalized by solvent viscosity at that temperature to eliminate the influence of viscosity on chain mobility. The relaxation time increased an order of magnitude in the vicinity of 20 °C. This indicates that chain mobility decreases at 20 °C or above.

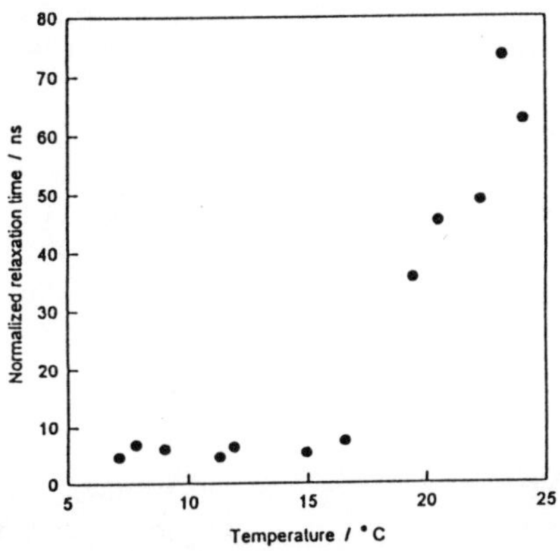

Figure 4. Normalized relaxation time vs. temperature.

4.3. SCATTERING LIGHT INTENSITY

Figure 5 shows the relationship between scattering light intensity and temperature. On heating, the intensity increased sharply above 19 °C. This indicates that the polymer chains start to aggregate at 19 °C and form large aggregates.

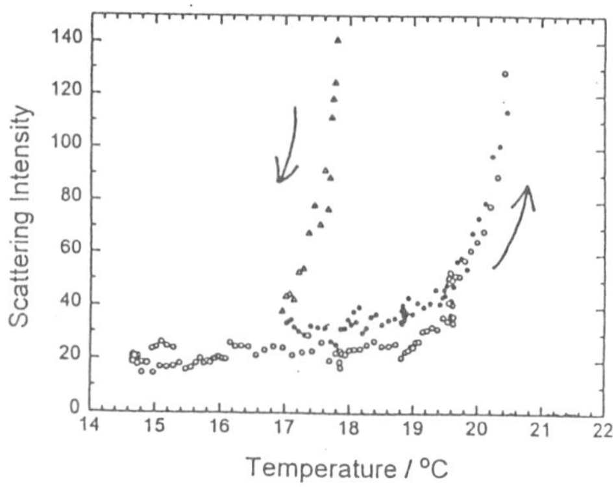

Figure 5. Scattering light intensity on heating (circle) or cooling (triangle).

On cooling, the temperature at which the scattering intensity started to decrease was lower than that on heating, i.e., the phase separation and/or the dissolution of the aggregate cannot follow the continuous temperature variation of 0.1 °C/min.

4.4. FURTHER DISCUSSION

Note two different temperatures, i.e., ca. 16 °C and ca. 20 °C. The fluorescence intensity and the lifetime decreased sharply at ca. 16 °C and increased at ca. 20 °C. On the other hand, the relaxation time and the scattering intensity increased at ca. 20 °C and no change was observed around 16 °C. The process of phase separation in PEVE aqueous solution is speculated as follows.

At 16 °C, the absence of a change in both the relaxation time and the scattering intensity indicates that no aggregation occurs at this temperature. If aggregation occurs and the polymer density increases, the mobility of the polymer chain decreases. We explain the observed apparent fluorescence quenching as follows. At this temperature, which is just below the phase separation temperature, thermal fluctuation of local segment density becomes vigorous preceding the phase separation, and the local density of the PEVE chain becomes momentarily high. The PEVE segment (quencher) comes in contact with the anthracene probe frequently, then fluorescence quenching occurs effectively. Although the segment density fluctuates widely with time, the average segment density may not change in this narrow temperature range.

The increase of light scattering intensity at 19 °C shows the intermolecular aggregation on a large scale. Aggregation of polymer chains suppresses chain motion. The marked increase of relaxation time at (roughly) 19 °C indicates the decrease of chain mobility. On the other hand, the decrease of chain mobility results in a lower frequency of the collision between quencher segments with an anthracene probe, i.e., the fluorescence quenching becomes weak. The observed increase of fluorescence intensity and lifetime at 19 °C is due to the intermolecular aggregation.

Finally, we will compare our microscopic measurements with macroscopic observation. The light scattering measurement revealed the presence of large aggregates in the intermolecular aggregation process. We believe that this aggregation corresponds to macroscopic phase separation. The microscopic intermolecular aggregation temperature detected by the fluorescence measurement is in good agreement with the macroscopic phase separation temperature. The fluorescence method is useful not only for observation of intermolecular aggregation but also the precedent thermal fluctuation. The phenomenon preceding phase separation is difficult to observe by light scattering techniques. We conclude that the fluorescence technique is very useful to examine the phase separation behavior.

5. Conclusion

We examined the phase separation behavior of an aqueous PEVE solution which becomes turbid abruptly at 20 °C. Fluorescence intensity, lifetime, relaxation time (chain mobility) and scattering intensity measurements showed two noticeable temperatures. In the vicinity of 16 °C, thermal fluctuation of segment density is thought to become large, which precedes phase separation. At ca. 19 °C, microscopic intermolecular aggregation increases corresponding to the macroscopic phase separation. The fluorescence technique used in this study is very sensitive and useful to examine phase separation.

6. References

1. Halary, J.L., Ubrich, J.M., Nunzi, J.M., and Monnerie, L. (1984) Phase separation in polystyrene-poly(vinylmethylether) blends: A fluorescence emission study, *Polymer* **25**, 956-962.

2. Fujishige, S., Kubota, K., and Ando, I. (1989) Phase transition of aqueous solutions of poly(N-isopropylacrylamide) and poly(N-isopropylmethacrylamide), *J. Phys. Chem*, **93**, 3311-3313.

3. Hu, Y., Horie, K., Ushiki, H., and Tsunomori, F. (1993) Fluorescence studies of microenvironment and dynamic fluctuation for volume phase transition of polyacrylamide gels with pyrenyl probe at crosslinks, *Eur. Polym. J.* **29**, 1365-1372.

4. Hirokawa, Y. and Tanaka, T. (1984) Volume phase transition in a nonionic gel, *J. Chem. Phys.* **81**, 6379-6380.

5. Tanaka, T. and Fillmore, D.J. (1979) Kinetics of swelling gels, *J. Chem. Phys.* **70**, 1214-1218.

6. Binkert, Th., Oberreich, J., Meewes, M., Nyffenegger, R., and Ricka, J. (1991) Coil-globule transition of poly(N-isopropylacrylamide): A study of segment mobility by fluorescence depolarization, *Macromolecules* **24**, 5806-5810.

7. Meewes, M., Ricka, J., de Silva, M., Nyffenegger, R., and Binkert, Th. (1991) Coil-globule transition of poly(N-isopropylacrylamide): A study of surfactant effects by light scattering, *Macromolecules* **24**, 5811-5816.

8. Winnik, F.M. (1990) Phase transition of aqueous poly(N-isopropylacrylamide) solutions: A study by non-radiative energy transfer, *Polymer* **31**, 2125-2134.

9. Aoshima, S. and Higashimura, T. (1989) Living cationic polymerization of vinyl monomers by organoaluminiumhalides. 3. Living polymerization of isobutyl vinyl ether by EtAlCl2 in the presence of ester additives, *Macromolecules* **22**, 1009-1013.

10. Aoshima, S., Oda, H., and Kobayashi, E. (1992) Synthesis of thermally-induced phase separating polymer with well-defined polymer structure by living cationic polymerization. I., *J. Polym. Sci.: Part A: Polym. Chem.* **30**, 2407-2413.

11. Kobayashi, E. and Sadahito, A. (1993) *Kagaku to Kogyo* **46**, 62-64.

12. Aoshima, S., Oda, H., and Kobayashi, E. (1992) *Kobunshi Ronbunshu* **49**, 937-941.

13. Kobayashi, E., Jiang, J., Ohta, H., and Furukawa, J. (1990) Novel conjugated polymer containing anthracene backbone: Addition polymer of 9,10-diethylanthracene with 9,10-anthracenedithiol and its properties, *J. Polym. Sci.: Part A: Polym. Chem.* **28**, 2641-2650.

14. Sasaki, T., Yamamoto, M., and Nishijima, Y. (1988) Chain dynamics of poly(methyl methacrylate) in dilute solutions studied by the fluorescence depolarization method, *Macromolecules* **21**, 610-616.

15. Yokotsuka, S., Okada, Y., Tojo, Y., Sasaki, T., and Yamamoto, M. (1991) Activation energy of local polymer motions estimated from the fluorescence depolarization measurements, *Polym. J.* **23**, 95-101.

16. Sasaki, T., Arisawa, H., and Yamamoto, M. (1991) Fluorescence depolarization study on local motions of anthracene-labeled poly(alkyl methacrylate)s in dilute solutions and evaluation of their chain stiffness, *Polym. J.* **23**, 103-115.

17. Ono, K., Okada, Y., Yokotsuka, S., Ito, S., and Yamamoto, M. (1994) Chain dynamics of polystyrene in high viscosity solvents studied by the fluorescence depolarization method, *Polym. J.* **26**, 199-205.

18. Ono, K., Okada, Y., Yokotsuka, S., Sasaki, T., and Yamamoto, M. (1994) Chain dynamics of styrene polymers studied by the fluorescence depolarization method, *Macromolecules* **27**, 6482-6486.

19. Ono, K., Ueda, K., and Yamamoto, M. (1994) Local chain dynamics of poly(cis-1,4-isoprene) in dilute solutions studied by the fluorescence depolarization method, *Polym. J.* **26**, 1345-1351.

20. Horinaka, J., Ono, K., and Yamamoto, M. (1995) Local chain dynamics of syndiotactic poly(methyl methacrylate) studied by fluorescence depolarization method, ***Polym. J.* 27**, 429-435.

21. Ono, K., Sasaki, T., Yamamoto, M., Yamasaki, Y., Ute, K., and Hatada, K. (1995) Local chain motion of isotactic and syndiotactic poly(methyl methacrylate)s studied by the fluorescence depolarization method, *Macromolecules* **28**, 5012-5016.

22. Ptitsyn, O.B., Kron, A.K., and Eizner, Y.Y. (1968) The models of the denaturation of globular proteins. I. Theory of globule-coil transitions in macromolecules, *J.Polym. Sci. PartC* **16**, 3509-3517.

23. de Gennes, P.G. (1975) Collapse of a polymer chain in poor solvents, *J. Phys. Lett.* **36**, L-55-57.

24. Norisuye, T., and Nakamura, Y. (1993) Remarks on smoothed-density theories for flexible chains with three-segment interactions, *Polymer* **34**, 1440-1443.

25. Chu, B., Ying, Q., and Grosberg, A.Y. (1995) Two-stage kinetics of single-chain collapse. Polystyrene in cyclohexane, *Macromolecules* **28**, 180-189.

CLASSICAL METHODS FOR THE STUDY OF BLOCK COPOLYMER MICELLES

PETR MUNK
Department of Chemistry and Biochemistry
and Center for Polymer Research,
The University of Texas at Austin, Austin, Texas 78712, U.S.A.

1. Introduction

Block copolymers when dissolved in a selective solvent that is thermodynamically good for one block and poor for the other arrange themselves into a structure in which the insoluble blocks segregate themselves into domains that are rather concentrated with respect to the insoluble blocks. When the overall concentration of the copolymer is small, these domains adopt a spherical shape (a micellar core) that is surrounded by a much less concentrated micellar shell consisting from the soluble blocks.

At equilibrium, the structure of the solution is such as to provide the lowest value of the Gibbs free energy G for the whole system. This thermodynamic criterion usually leads to a solution containing two types of dissolved particles: 1. Individual copolymer molecules called unimers (their insoluble block may exist in a collapsed form) and 2. block copolymer micelles (BCM) that are composed from several tens to several hundreds of copolymer molecules. The thermodynamics predicts that the distribution of aggregation numbers within a given sample is rather narrow.

In this respect the situation is similar to the well-known detergent micelles. However, there are important differences. In detergent solutions, the micelles start forming when the concentration reaches the critical micelle concentration (CMC). At higher overall concentrations, the unimers are present at concentrations more or less equal to CMC. In BCM solutions, the CMC may be undetectably low. More importantly, when the block copolymer is not monodisperse, the concentration of the unattached unimers may change (increase) with the overall concentration of the copolymer even above CMC [1].

S.E. Webber et al. (eds.), Solvents and Self-Organization of Polymers, 367–381.

Other micellar phenomena that are not observed for detergent micelles include the possibility that the micellar solutions are not at full thermodynamic equlibrium. For example, the micelles may be transfered (e. g., by dialysis) into a different solvent in which the equilibration processes are slowed down or arrested completely. We will call this situation a kinetically frozen equilibrium.

Another scenario, not yet analysed theoretically, invokes a possibility that some insoluble blocks are not fully integrated into a single micellar core, but are partially incorporated into two cores forming a tie-line between two micelles. While such an arrangement is not favorable enthalpically, entropy considerations would certainly allow it to some degree. This phenomenon may explain the frequently observed micellar clustering [2]. The situation is similar to the micellization of triblock copolymers in which the outer blocks are insoluble in the selective solvent. At high dilutions, such copolymers form well-behaving spherical micelles, in which both outer blocks reside in the same core and the central soluble blocks form loops within the shell. At higher concentrations, the end blocks may reside in different cores and the central blocks then form the ties that lead to extensive clustering and eventual precipitation of such systems [3].

It is thus obvious that BCM systems may become quite complicated and that we need a battery of experimental techniques in order to obtain a dependable information about their structure. In turn, such an information is needed for understanding of the rules that govern the process of micellization. This article is devoted to the description of such techniques, primarily of those that had been the trusted tools for the study of more usual polymeric systems. The more complex nature of the micellar solutions demands many theoretical and experimental improvements of these techniques as well as a development of new ways of combining experimental data obtained by several techniques.

2. Static Light Scattering

The method of static light scattering (SLS) using the procedure developed by Zimm (measuring the dependence of the intensity of the scattered light on the scattering angle θ and on the concentration of the solute c and extrapolating to vanishing values of θ and c) is one of the most valuable characterization techniques for polymer solutions. It provides the weight average molecular weight of the solute M_W , second virial coefficient A_2 , and the z-average of the radius of gyration R_G. Numerous experimental studies have shown that SLS can be successfully applied to micellar solutions as well. However several precautions must be taken during the evaluation of the experimental data.

2.1. UNIMER-MICELLE EQUILIBRIA

The routine procedure of extrapolating the experimental data toward the vanishing concentration of the solute is based on a tacit assumption that the distribution of the masses of the solute particles is independent of the solute concentration. This assumption is not fulfilled when the solute participates in an equilibrium aggregation process, e. g., in unimer-micellar equilibrium. We may distinguish two situations:

1. The CMC is so low that the unimer concentration is completely negligible with respect to the micellar concentration at all overall concentrations at which the experiments are performed. Then it is legitimate to neglect the presence of the unimers altogether and to treat the solution as a solution of micelles only.

2. When the concentration of the unimers is not negligible then the M_w value depends on the overall concentration c. The experimentally accessible quantity Kc/R_0 (where K is the well-known light scattering constant and R_0 is the Rayleigh ratio extrapolated to vanishing scattering angle) must be then interpreted as

$$Kc/R_0 = 1/M_{w,t} + 2A_2c + ... = 1/(M_{w,m}f_m + M_{w,u}f_u) + 2A_2c + ... \quad (1)$$

where $M_{w,t}$, $M_{w,m}$, and $M_{w,u}$ are the molecular weight averages of the total solute, micelles, and unimer, respectively. Mass fractions of the micelles and unimers are f_m and f_u, respectively. Equation (1) is an equation with three unknowns: $M_{w,m}$, f_m, and A_2 ($M_{w,u}$ is known and $f_u = 1 - f_m$). All these unknowns are possibly concentration dependent. Tuzar *et al.* [4] studied by SLS styrene-butadiene-styrene copolymer in tetrahydrofuran/allyl alcohol mixed solvent system. They evaluated $M_{w,m}$ from equation (1) assuming that A_2 could be taken as zero; they obtained f_m from sedimentation velocity experiments. Their results have shown that $M_{w,m}$ is essentially independent of the overall concentration of the polymer. However, the concentration of the unimer was increasing with the overall concentration of the polymer in defiance of the simplest CMC theory. Apparently, the polydispersity of the copolymer led to this result.

2.2. SECOND VIRIAL COEFFICIENT

When evaluating the Zimm plots, the initial slope of the dependence of Kc/R_0 on concentration is assumed to be equal to twice the second virial coefficient $2A_2$. We have shown recently that this value has to be corrected for the turbidity of the solution [5]. The Rayleigh ratio of a liquid is the ratio of the intensity of the scattered light leaving a small scattering volume of the liquid to the intensity of the primary light entering this element. Experimentally, we do not measure this ratio. We are measuring the ratio of light intensity reaching the detector and intensity of the primary light entering the scattering cell. The primary beam, before reaching the scattering element, travels a distance L_p through the scattering solution and is weakened by the turbidity τ of the solution by a factor $\exp(-L_p\tau)$. Similarly the scattered beam travels a distance L_s before reaching the wall of the cell and is weakened by the factor $\exp(-L_s\tau)$. Thus, the actually measured value $R_\Theta{}^*$ is related to the true Rayleigh ratio at angle Θ , R_Θ as

$$R_\Theta{}^* = R_\Theta \exp[-(L_p + L_s)\,\tau] \tag{2}$$

The turbidity of a nonabsorbing solution containing solute molecules that are smaller (or at least not much larger) than about $\lambda/10$ is related to its Rayleigh ratio as

$$\tau = (16\pi/3)\, R_0 \tag{3}$$

Developing the exponential function in equation (2) into Taylor series and substituting equations (2) and (3) into equation (1) we obtain finally

$$Kc/R_0 = 1/M_w + (2A_2 + A_T)c + \ldots \tag{4}$$

$$A_T = (16\pi/3)\, K\, (L_p + L_s) \tag{5}$$

We have evaluated the correction term A_T for a typical experiment on a polymer solution using green mercury light, λ_0 = 546 nm and a cell with 2 cm diameter (i. e., $L_p = L_s$ = 1 cm). Selecting refractive increment equal to 0.169 (polystyrene in cyclohexane) we obtained $K = 2.15\text{x}10^{-7}$ g^{-2}cm^{2}mol and A_T = $7.2\text{x}10^{-6}$ g^{-2}cm^{3}mol.

Correction term of this magnitude is usually negligible when typical polymers are measured in thermodynamically good solvents. (It may become important for measurements in the vicinity of the θ-temperature.) However, block copolymer micellles behave thermodynamically like hard spheres [2], their virial coefficients are of the order of 10^{-6} to 10^{-7}, and the corrections become significant. In some instances the turbidity term becomes dominant and the true virial coefficient cannot be obtained from SLS experiments at all.

2.3. PREFERENTIAL ADSORPTION OF SOLVENT COMPONENTS

When the selective solvent chosen for the preparation of the micellar solution is a very poor solvent for the insoluble block then the copolymer may not dissolve in it at all. Typical examples are hydrophilic/hydrophobic block copolymers with polystyrene hydrophobic blocks. Such copolymers are usually not soluble in water or aqueous buffers. However, they are soluble in mixed solvents in which water is combined with a good solvent for polystyrene, e. g., with tetrahydrofuran or dioxane.

In such mixed solvents, the polystyrene micellar core is swollen by the good solvent. From the viewpoint of light scattering, such particles should be considered as exhibiting preferential adsorption of a solvent component. It is a well-known fact that it is necessary, when evaluating the light scattering constant K, to use the refractive increment of the solute particles measured at constant chemical potential of the solvent components in order to obtain the correct molecular weight of the (unsolvated) particles [6]. Such a refractive increment is usually measured using solutions that were dialysed against the mixed solvent. When the refractive increment at constant composition of the solvent is also available (it is measured on solutions prepared by dissolution of a dry polymer in the mixed solvent) the extent of the preferential adsorption may be evaluated also.

The above considerations fully apply to micellar solutions. When it is possible to assume that the nonsolvent (water) cannot enter the micellar core (polystyrene) then the preferential adsorption is a direct measure of the swelling of the core. It is not necessary that the components of the mixed solvent are mutually miscible in all proportions. In fact, in water saturated by a sparcely soluble solvent (e. g., a hydrocarbon), the chemical potential of this solvent is virtually equal to its chemical potential in its pure form. Consequently, it can swell the micellar cores to a considerable degree. We have used this phenomenon for measuring the scavenging capacity of micelles formed by polystyrene-*block*-poly(methacrylic acid) in buffer solutions for a number of hydrophobic solvents [7].

In our experiments, we prepared a micellar solution in a buffer and measured the intensity of the scattered light I_0 . Then we equilibrated (by overnight shaking) the micellar solution with the hydrophobic solvent and measured again the intensity of the scattering I. We have assumed that the identity of the micelles is not disturbed by this procedure, i. e., that the number concentration of the micelles in the solution does not change as well as the distribution of the aggregation numbers. Under these circumstances it is legitimate to express the scattered intensities I_0 and I as

$$I_0 = \alpha\ [m_{mic}(dn/dc)_{mic}]^2 \tag{6}$$

$$I = \alpha\ [m_{mic}(dn/dc)_{mic} + m_{sol}(dn/dc)_{sol}]^2 \tag{7}$$

where α is a constant, m_{mic} is the mass of a single micelle, m_{sol} is the total mass of organic molecules solubilized by this micelle, $(dn/dc)_{mic}$ is the refractive increment of the micelles, and $(dn/dc)_{sol}$ is the refractive increment of the solubilizate that can be evaluated as

$$(dn/dc)_{sol} = (n_{sol} - n)\ \rho_{sol} \tag{8}$$

Here n is the refractive index of the main solvent; n_{sol} and ρ_{sol} are the refractive index and density of the solubilizate, respectively.

Consequently, the extent of solubilization by the micelles δ can be evaluated as

$$\delta \equiv m_{sol}/m_{mic} = [(dn/dc)_{mic}/(dn/dc)_{sol}]\ [(I/I_0)^{1/2} - 1] \tag{9}$$

2.4. RADIUS OF GYRATION

The initial slope of the scattering envelope at vanishing concentration of the scattering particles is routinely interpreted in terms of the radius of gyration of the scattering particles R_G . However, this interpretation is correct only when the scattering particle can be considered as composed of scattering entities that all have the same ratio of scattering power and mass. For homopolymers, monomeric units are such entities. When the particle contains two or more types of scattering entities then the value calculated from the slope is an apparent radius of gyration $R_{G,app}$. In light scattering from solutions, the ratio of the scattering power and mass is proportional to the refractive increment of the scattering entities. The theory of light scattering leads to the following expression for $R_{G,app}$ of particles composed of two materials having different refractive increments

$$R^2_{G,app} = R^2_{G,A} W_A\ (dn/dc)_A/(dn/dc) + R^2_{G,B} W_B\ (dn/dc)_B/(dn/dc) + \Delta^2\ W_A W_B (dn/dc)_A (dn/dc)_B/(dn/dc)^2 \tag{10}$$

Here $R_{G,A}$ and $R_{G,B}$ are the radii of gyration of the materials A and B, W_A and W_B are their mass fractions, $(dn/dc)_A$ and $(dn/dc)_B$ are their refractive increments, (dn/dc) is the refractive increment for the composite particle and Δ

is the distance between the centers of mass of the two materials. The overall refractive increment can be expressed as

$$\mathrm{d}n/\mathrm{d}c = W_A\,(\mathrm{d}n/\mathrm{d}c)_A + W_B\,(\mathrm{d}n/\mathrm{d}c)_B \qquad (11)$$

For random copolymers, the spatial distribution of both monomeric units is on average the same and the apparent radius of gyration is equal to the true one. In block copolymers dissolved in solvents good for both blocks, the blocks may occupy different regions and all three terms in equation (10) are significant. Well-developed block copolymer micelles are spherically symmetrical (the symmetry could be violated in some types of micellar clusters), the Δ value is zero, and only the first two terms of equation (1) survive.

For micelles, in which exists a well-defined core-shell boundary, the gyration radii of the two components may be expressed as a function of the radius of the core R_c and of the whole micelle R_m. We will designate the insoluble block as A and the soluble one as B and assume that the density of polymer segments within both the core and the shell is homogeneous. This is a good assumption for the core; its validity for the shell is only fair.

$$R^2_{G,A} = (3/5)\,R^2_c \qquad (12)$$

$$R^2_{G,B} = (3/5)\,(R^5_m - R^5_c)\,/\,(R^3_m - R^3_c) \qquad (13)$$

Thus, equations (10)-(13) relate $R_{G,app}$ with R_c and R_m.

3. Dynamic Light Scattering

The method of dynamic light scattering (DLS) is complementary to SLS. The autocorrelation curve provides an information about the relaxation of concentration fluctuations of the scattering particles. In dilute solutions, this relaxation is routinely interpreted in terms of translational diffusion coefficients of the scattering particles and their hydrodynamic radii R_H. A homogeneous solute yields an autocorrelation curve that is a single exponential and its polydispersity u defined as

$$u = \Gamma_2 / \Gamma_1^{\,2} \qquad (14)$$

is close to zero. Here, Γ_1 and Γ_2 are the first and second cumulant of the autocorrelation function, respectively. A polydisperse solute yields a curve that is a composite of exponentials for each component. The contribution of individual components is weighted by their scattering power. The polydispersity analysis is most valuable for paucidisperse systems with well separated relaxation times. Sophisticated computer programs are now capable of separating the individual contributions, provided that the relaxation times differ at least by a factor of two [8].

3.1. UNIMER-MICELLE EQUILIBRIA

When studying BCM, the majority of copolymer molecules is under most experimental circumstances aggregated into micelles and only a small fraction remains in the form of unimers. With the aggregation number being in tens or hundreds, the scattering power of the micelles is so much larger than that of the unimers that the contribution of the unimers to the autocorrelation curve is virtually obscured by the much more intense scattering of the micelles. Consequently, DLS is not a very useful technique for studying unimer-micelle equilibria.

3.2. ANOMALOUS MICELLIZATION

When some micelle forming copolymers are dissolved in a solvent that is thermodynamically close to the boundary between being a selective solvent and a solvent good for both blocks, a small amount of very large particles is formed. These particles may or may not survive when the solvent is gradually transformed into a well-defined selective one. The scattering power of these large particles is enormous and they influence the autocorrelation curve even at extremely low concentration. Computer separation of the two contributions provides a valuable information in such situations [9].

3.3. POLYDISPERSE MICELLES

Polydisperse solutes yield autocorrelation curves that are sums of the curves for the individual components, each curve being governed by the appropriate diffusion coefficient. The diffusion coefficient is inversely proportional to the cube root of the hydrodynamic volume of the component and the latter is more or less proportional to the mass of the particles. Well-behaved equilibrium micellar systems usually follow the model of closed association that predicts rather narrow distribution of aggregation numbers and even narrower distribution of diffusion coefficients. In this situation, not only the relaxation times cannot be separated into individual components, but the autocorrelation curve is very close to a single exponential that provides a well defined hydrodynamic radius. The polydispersity u is very small, usually equal to zero within an experimental error.

It follows that solutions exhibiting a substantial polydispersity u (in our experience $u > 0.05$) must either contain micelles of two or more substantially different types or, more probably, must contain micellar clusters. In our experience (vide infra), even a substantial clustering may lead to rather moderate values of u.

Comparison of the experimental values of the radius of gyration $R_{G,app}$ and the hydrodynamic radius R_H is a much more sensitive technique for the detection of micellar clustering. To demonstrate this approach we will calculate their ratio $r \equiv R_{G,app}/R_H$ for several simple model situations. For homogeneous spheres $r = (3/5)^{1/2} = 0.775$.

For micellar systems, we will employ the model described by equations (10) - (13). For simplicity we will assume that $(dn/dc)_A = (dn/dc)_B = (dn/dc)$ and $W_A = W_B = 0.5$. In order to represent the fact that the shell has much smaller segment density than the core, we will assume $R_H = R_m = 2R_c$. This implies the ratio of segment densities equal to eight. Δ is obviously equal to zero. For this model we have found $r = 0.638$. When the refractive increment of the shell is smaller than that of the core, the r value decreases further.

We will now model a cluster of two micelles as two spheres just touching. We estimate that this dumbbell would have R_H equal to about $1.3R_m$. (This value is based on a model of an ellipsoid with the volume of the double micelle and the axial ratio equal to two.) Radius of gyration of bodies consisting of several parts can be calculated as a sum of two terms: 1. Radius of gyration R^{*2}_G that is calculated as if the mass (actually the scattering power in LS applications) of each part is concentrated at its center of mass (scattering power) and 2. the sum of the radii of gyration of individual parts $R^2_{G,i}$ properly weighted by their mass (scattering power) fraction w_i .

$$R^2_G = R^{*2}_G + \Sigma_i \, w_i \, R^2_{G,i} \tag{15}$$

For dumbbells consisting of homogeneous spheres $R^2_G = (8/5)R^2_m$ and the r value becomes 0.973. For dumbbells consisting of two-layered spheres described above, $r = 0.912$. For larger micellar clusters, the r value is even larger.

The scattering power of micellar clusters increases sharply with their size. Thus, even a small extent of clustering leads to an appreciable increase of the measured radius of gyration (R^2_G is a z-average) and of the r value. As a consequence, the ratio r is a very sensitive indicator of the presence of micellar clusters.

We have utilized this technique for studying BCM formed by styrene-methacrylic acid copolymers in dioxane-water mixtures. For micellar solutions that were not clustered according to independent evidence, we have measured values of r in the range 0.37-0.64. Samples that were clustered exhibited values of r between 0.74 and 1.23 [2].

4. Sedimentation velocity

The method of measurement of sedimentation velocity in the ultracentrifuge, while not used frequently in the last years, is capable of providing a wealth of information about micellar systems. The number of sedimentation boundaries and their shapes provide an unique insight into the pauci- and polydispersity of the dissolved particles. Physical characteristics of the micelles may be derived from the sedimentation coefficients and their concentration dependence. Finally, the method is eminently suited for following slow changes in the properties of the system.

4.1. PAUCIDISPERSE SYSTEMS

Under the conditions of sedimentation velocity experiments, particles that participate in unimer-micelle equilibria form two sedimenting boundaries that are easily observed by schlieren optics. Because the unimers and the micelles have the same refractive increment, the ratio of the areas of the two peaks (after a correction for the radial dilution) yields directly the mass ratio of these two components [4]. Sedimentation velocity is probably the only method capable of yielding this information.

Anomalous micellization, whenever present, produces small amounts of very massive particles with extremely large sedimentation coefficients. These particles often escape detection in routine experiments: they may sediment completely to the bottom before the full rotor speed is reached and the first photograph is taken. However, an alert observer would notice them if he watches the schlieren screen during the acceleration period.

Micellar clusters have sedimentation coefficients that are distinctly larger, yet of the same order of magnitude, compared to those of individual micelles. Thus, any shoulder or broadening of the leading edge of the sedimenting peak is the most direct and sensitive proof of micellar clustering.

When two distinctly different micellar species are present in a solution, two well-behaved sedimentation peaks are usually observed. If there is some slow process occuring in the system, repetitive sedimentation runs on the aliquots of the solution may provide a good insight into the mechanism and kinetics of this process. We will treat this situation in detail in another chapter of this monograph.

4.2. SEDIMENTATION COEFFICIENTS

Block copolymer micelles usually provide relatively narrow sedimentation peaks and their sedimentation coefficients s can be evaluated in a routine way. Their value, extrapolated toward vanishing concentration of the solute, s_0 is also obtained routinely using the standard relation

$$1/s = (1/s_0)\,(1 + k_s\,c) \tag{16}$$

where k_s is a characteristic coefficient of concentration dependence.

Sedimentation coefficient may be combined with the diffusion coefficient D (obtained from the DLS measurement) to provide an alternative access to the molecular weight M_s using the Svedberg equation

$$s_0/D = M_s(1 - v\,\rho)/RT \tag{17}$$

where v is the partial specific volume of the micelles, ρ is the density of the solvent, and RT has its usual meaning. It should be noted that, for micelles in mixed solvents, the partial specific volume should be measured at constant chemical potential (i. e., after dialysis) as the refractive increment does.

5. Intrinsic Viscosity

Viscometry has been a trusted characterization method for polymeric solutions since the beginning of polymer science. It is technically simple, does not require expensive equipment, and is very precise. It has two major drawbacks: 1. It needs usually at least 50 mg of the material for a single determination; this may be a prohibitively large quantity for some more expensive materials. (That is why biochemists use viscometry only rarely, the same applies to some more advanced block copolymers.) 2. The information obtained from measurement of viscosity is rather limited if it is not combined with any other method. However, in combination with other methods, it becomes extremely valuable.

Intrinsic viscosity $[\eta]$ is related to the ratio of the hydrodynamic volume of a particle, V_H, and its dry volume V_{dry} as

$$[\eta] = \mu\, v\, V_H \,/\, V_{dry} \tag{18}$$

where μ is a shape dependent factor that for spheres has a value 2.5; v is the specific volume of the dry particle. Thus, small values of $[\eta]$ that are observed for most micellar solutions reveal that not only the micellar cores but also the shells contain a rather large volume fraction of the shell forming polymer. It follows that the polymeric properties of the shells must be treated in terms of semidilute or possibly even concentrated solutions.

When we assume that the particles are spherical , we may substitute for the hydrodynamic volume

$$V_H = 4\pi R^3_H/3 \tag{19}$$

and for the dry volume

$$V_{dry} = Mv/N_A \tag{20}$$

where N_A is the Avogadro number. Then combination of equations (18) - (20) with μ =2.5 leads to

$$M_H = 10\pi R^3_H N_A/(3[\eta]) \tag{21}$$

Here we have used the symbol M_H to indicate the molecular weight obtained by a combination of two hydrodynamic measurements: the intrinsic viscosity and the hydrodynamic radius measurable by DLS.

5.1. THERMODYNAMIC PROPERTIES OF MICELLES

Thermodynamic properties of polymeric materials are closely related to the second virial coefficient of their solutions A_2. In solutions of linear polymers, A_2 is dominated by the polymer/solvent contact interactions; in good solvents, A_2 is typically of the order 10^{-4} mol mL/g^2. For compact particles, a model based on excluded volume theories is more appropriate. According to this model, A_2 for hard noninteracting particles is related with the molar mass of the particles M as

$$A_2 = uN_A/(2M^2) \tag{22}$$

where u is the volume excluded by the particle for centers of other particles. For impermeable spherical particles, u is given as

$$u = 8\ V_H \tag{23}$$

where V_H is the quantity defined by equation (19). Thus, for particles behaving as impermeable spheres, combination of equations (19), and (21) - (23) leads to

$$MA_2/[\eta] = 1.6 \tag{24}$$

Consequently micelles, that have intrinsic viscosities of the order of 20 mL/g and molecular weights of the order of millions, have second virial coefficients of the order 10^{-5} mol mL/g^2. For micelles with particle weight in tens of millions, the expected order of A_2 is 10^{-6} mol mL/g^2. While the necessary correction for turbidity during the measurement of A_2 by light scattering made the measured values for very large micelles somewhat ambiguous, we have found that equation (24) described our measurements quite satisfactorily [2]. Thus, micelles behave thermodynamically like hard spheres.

5.2. HYDRODYNAMIC PROPERTIES OF MICELLES

Combination of intrinsic viscosity data with sedimentation velocity measurements provides an insight into the hydrodynamic nature of micellar solutions as well. The characteristic coefficient of the concentration dependence of sedimentation coefficients k_s (cf. equation (16)) is related to the backflow of the solvent in the sedimentation cell. It is therefore related to the volume of the forward moving sedimenting particles, i.e., also to their intrinsic viscosity. Hydrodynamic coefficient κ is frequently defined as

$$k_s = \kappa\,[\eta] \tag{25}$$

Hydrodynamic theories for impermeable spheres predict that κ is about 2.8, while various theories for polymer coils predict values between zero and 1.66 [10]. In our experiments on polystyrene-*block*-poly(methacrylic acid) micelles in a mixed dioxane-water solvent we have found values of κ between 2 and 5 [10]. Such values again point toward the applicability of models of impermeable particles for our micelles.

6. Size Exclusion Chromatography

On the first look, size exclusion chromatography (SEC) is a promising method for finding the fraction of the unimers in the micellar solution, for studying the dynamics of unimer-micelle equilibrium, and for determining the size and the polydispersity of the micelles. Unfortunately, the rather involved interaction of the micellar systems with the size exclusion support does not allow for the fulfilment of this promise. Two experimental situations are of a practical interest.

1. Elution of a micellar system that, due to its thermodynamic properties, contains only a negligible concentration of free unimers. We would

expect that the micelles will be separated by the usual size exclusion mechanism and that we will be able to estimate their molecular weight and their polydispersity. However, the estimate of the molecular weight (more precisely, of the molecular volume) is based on a calibration that is routinely performed using polymers that form molecular coils. Now, the entry of coiled molecules into the pores of the exclusion matrix is certainly subject to different rules than the entry of more or less hard micellar spheres. For example, reptation of chains into and out of the pores may play a role in the former case but not in the latter. The same applies to the possibility that the entering coiled particle may be deformed and squeeze into a pore into which a solid sphere of the same hydrodynamic volume cannot enter. Thus, the lack if a suitable calibration makes the estimates of molecular weights and polydispersities impractical to say the least.

2. Elution of a micellar system in which the micelles are in equilibrium with their unimers. (The equilibrium may be considered to be established when the characteristic time for entry or escape of the unimer from the micelle is much shorter than the experimental time of the SEC experiment that is typically more than ten minutes.) In such a case, the micelles move ahead of the unimers and in this front zone they are partially dissociated to reestablish the equilibrium. The unimers left behind the micellar zone are typically below their CMC and do not micellize any more. As a result, the elution profile consists of a leading micellar peak that decreases in magnitude along the column and may eventually disappear. This peak is than followed by a plateau of the unimer. Seemingly, the analysis of such a profile should lead a wealth of information about the nature of the system. However, in real experiments another phenomenon enters into the play and spoils it. When several unimers enter the same pore in the exclusion matrix, their insoluble blocks may entangle to form a rudimentary micelle inside the pore. The size of the pore may not be sufficient for allowing a disentanglement of this micelle and it is trapped on the column. This is easily shown both by the mass balance on the column and by the fact that the lost copolymer may be recovered when a zone of a solvent good for both blocks is injected onto the column [11].

7. Conclusions

While the methods of static light scattering, dynamic light scattering, sedimentation velocity, and viscometry each provide valuable information about micellar systems, the most detailed results are obtained by combination of several methods. For example, the anomalous micellization that totally distorts the SLS measurements and pretends that the amount of the high aggregates is very high when DLS is employed, is assigned its real significance by the sedimentation velocity experiments showing a very small mass fraction of these high aggregates. The fact of the micelles behaving thermodynamically and hydrodynamically like hard spheres is obtained by the combination of viscosity measurements with other techniques. Similarly, the combination of hydrodynamic and gyration radii provides a very sensitive test of micellar clustering. Size exclusion chromatography gives us very interesting insights into the behavior of micellar systems but is not very useful for their characterization.

Acknowledgment. This research was supported by the U.S. Army Research Office (Grants DAAL03-90-G-0147, DAAH04-93-G-0405, and DAAH04-95-10127), by the Collaborative Research Grant 920166 from the Scientific and Environmental Division of the North Atlantic Treaty Organization, and by the U.S.-Czech Science and Technology Joint Fund No 95010.

8. References

1. Khougaz, K., Gao, Z., and Eisenberg, A. (1994) Determination of the Critical Micelle Concentration of Block Copolymer Micelles by Static Light Scatterring, *Macromolecules* **27**, 6341-6346.
2. Qin, A., Tian, M., Ramireddy, C., Webber, S. E., Munk, P.,and Tuzar, Z. (1994) Polystyrene-Poly(methacrylic acid) Block Copolymer Micelles, *Macromolecules* **27**, 120-126.
3. Tuzar, Z., Konák, C., Stepánek, P., Plestil, J., Kratochvíl, P., and Procházka, K. (1990) Dilute and Semidilute Solutions of ABA Block Copolymer in Solvents Selective for A or B Blocks: 2. Light Scattering and Sedimentation Study, *Polymer* **31**, 2118-2124.
4. Tuzar, Z., Petrus, V., and Kratochvíl, P. (1974) Sedimentation and Light Scattering Study of Block Copolymer Association, *Makromol. Chem.* **175**, 3181-3192.
5. Munk, P. and Tian, M. (1995) Neglected Turbidity Correction in Light Scattering Experiments - a Source of Erroneous Values of the Second Virial Coefficient, *Polymer*, **36**, 1975-1978.
6. Kratochvíl, P. (1987) *Classical Light Scattering from Polymer Solutions*, Elsevier, Amsterdam.
7. Tian, M., Arca, E., Tuzar, Z., Webber, S.E., and Munk, P. (1995) Light Scattering Study of Solubilization of Organic Molecules by Block Copolymer Micelles in Aqueous Media, *J. Polym. Sci., Polym. Phys. Ed.*, **33**, 1713-1722.
8. Provencher, E. W. (1979) Inverse Problems in Polymer Characterization: Direct Analysis of Polydispersity with Photon Correlation Spectroscopy, *Makromol. Chem.* **180**, 201-209.
9. Tuzar, Z., Kratochvíl, P., Procházka, K., and Munk, P. (1993) Block Copolymer Micelles in Aqueous Media, *Collect. Czech. Chem. Commun.* **58**, 2362-2369.
10. Tian, M., Ramireddy, C., Webber, S. E., and Munk, P. (1993) Study of Sedimentation Velocity of Block Copolymer Micelles, *Collect. Czech. Chem. Commun.* **58**, 71-76.
11. Procházka, K., Bednár, B., Tuzar, Z., and Kocirík, M. (1989) Size Exclusion of Associating Systems. II A Model Describing the Hindered Release of Solute from the Stationary Phase, *J. Liq. Chromatogr.*, **12**, 1023-1041.

DEVELOPMENT OF SYNCHROTRON SAXS/DIFFRACTION INSTRUMENTATION

Benjamin Chu
Departments of Chemistry and of Materials Science and Engineering
State University of New York at Stony Brook
Stony Brook, Long Island, New York 11794-3400 U.S.A.

Abstract

The high brilliance and wavelength tunability of synchrotron x-rays require modified designs in x-ray optics and x-ray detectors so as to optimize the experimental setups in small angle x-ray scattering (SAXS)/diffraction instrumentation. This lecture summarizes some of the instruments which have been constructed by the Chu group on the State University of New York (SUNY) X3 A1/A2 beamline of the National Synchrotron Light Source (NSLS) at Brookhaven National Laboratory (BNL). Descriptions on modified Kratky slit collimator, pinhole geometry and Bonse-Hart ultrasmall angle x-ray diffractometer (USAXD) as well as linear intensified photodiode and area (intensified) CCD x-ray position sensitive detectors are presented. Designs of the x-ray facility which tries to take into account of the synchrotron beam characteristics and of the ease in operation, as well as provisions for combining the x-ray instrumentation with other techniques for simultaneous detection of correlated physical parameters, are emphasized.

1. Introduction

The availability of synchrotron radiation together with recent advances in position-sensitive detectors (PSD) has provided a renewed interest in small angle x-ray scattering (SAXS). Synchrotron radiation has a high power density and a small beam divergence, especially in the vertical direction, and it is tunable, yielding a well-defined energy of the desired x-ray wavelength after monochromatization. By combining the desirable characteristics of the synchrotron x-rays with PSD capabilities, time-resolved SAXS (or x-ray diffraction) and anomalous SAXS (or x-ray diffraction) become readily accesible and are powerful tools to study the kinetics of material structural changes and the structures of specific elements in multicomponent materials. The high power density in the x-ray intensity with small beam divergence also permits experiments with small sample sizes, such as a single fiber filament or a very small crystal. In this respect, the main requirement of SAXS, i.e., a well defined incident x-ray beam with low parasitic scattering background, is of equal importance to x-ray diffraction. Therefore, we have designed

S.E. Webber et al. (eds.), Solvents and Self-Organization of Polymers, 383–407.

several instruments capable of performing simultaneous SAXS and x-ray diffraction experiments.

In the following sections, the Kratky block-collimation system, the pinhole collimator and the Bonse-Hart ultra SAXS instrument will be described. Furthermore, we shall summarize our developments on the intensified photodiode array x-ray linear position sensitive detector (LPSD) and the CCD x-ray area detector with and without the use of an intensifier. It should be noted that no single design in the x-ray optics can be achieved to optimize opposing requirements in accessible (small) angle and wavelength ranges, as well as resolution. Therefore, several collimation systems were designed for synchrotron SAXS. Yet, not all the instruments could be operated at the same time. Thus, we have adapted our x-ray instrumentation for use with a rotating anode x-ray generator, as shown in Fig. 1. Secondly, the x-ray instruments contain many interchangeable modular components so as to reduce cost and to increase reliability. During synchrotron operation, there is no time for repair. A sharing of interchangeable components has turned out to be extremely useful and has increased the efficiency of our x-ray operations. Example SAXS results will be described in the next lecture.

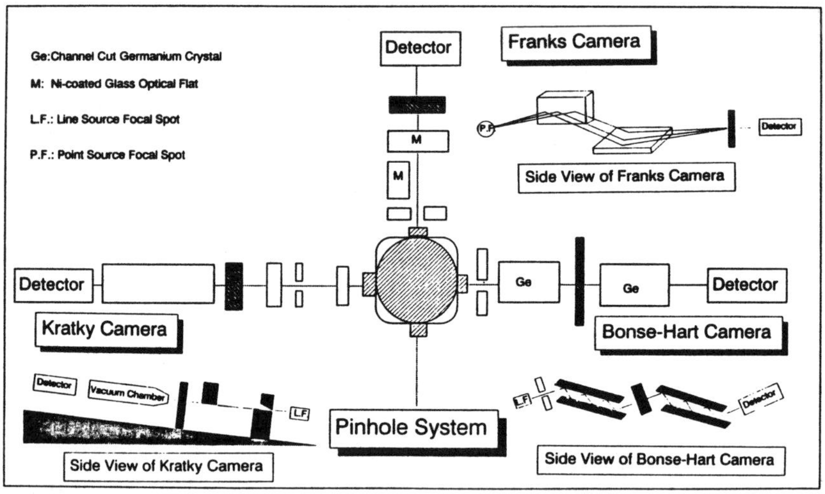

Figure 1. Schematic diagram of the synchrotron x-ray instruments which have been adapted to a rotating anode x-ray generator at Stony Brook. The Kratky and Bonse-Hart cameras, as well as the pinhole system can be used with the SUNY X3 A1/A2 beamline at NSLS.

2. Kratky Block-Collimator[1]

A block slit collimation system of the Kratky design concept has been modified and adapted for synchrotron radiation. The building block small-angle x-ray diffractometer (SAXD) not only retains the essential advantages of the Kratky camera, i.e., accessibility at small values of q[= $(4\pi/\lambda)\sin(\theta/2)$ with λ and θ being the x-ray wavelength and the scattering angle, respectively] and ease of alignment, but also provides portability and low construction cost. With our SAXD operating at the SUNY X3A2 beamline of the NSLS, we were able to reach $\theta \sim 1$ mrad corresponding to q ~ 0.04 nm^{-1} (or a Bragg spacing of 150 nm) using an incident slit width (d) of 0.5 mm and λ = 0.15 nm. At d = 0.5 mm, ~10% of the monochromatic synchrotron x-ray radiation passed through the block collimation system and a parasitic intensity to main beam intensity ratio of ~10^{-5} at θ = 1 mrad was achieved. With d ~ 0.1 mm, it is anticipated that q ~ 0.01 nm^{-1} can be accessible. By changing the vacuum tube length between the sample chamber and the beam stop from ~1400 to ~200 mm, we could reach an intermediate q range of ~0.04 < q < 10 nm^{-1} using a 5-cm-long linear position-sensitive detector with a resolution of ~100 μm/channel. A description of the SAXD and its operation is presented.

The SUNY small-angle x-ray diffractometer (SAXD)[2] without the Kratky block collimator[3] had very limited q ranges. Based on the SUNY X3A2 beam optics, as listed in Table 1, the incident x-ray beam at ~10 m from A2 mirror of Fig. 2, had very strong parasitic scattering in the vertical direction. Although the shape and the cross section of the focused primary x-ray beam depend on the bent torroidal mirror, the use of a slit (or pinhole diaphragm) to alleviate the problem of small-angle scattering generated by the lead collimator and the A2 mirror (Fig. 2) based on the four-slit camera geometry designed by Beeman et al.[4,5] proved to be difficult to operate. Generally, the beam shape changes with distance, regardless of mirror orientation. After collimation by the modified Kratky block collimator, the well-defined primary x-ray beam has a cross section of ~0.5 x 3 mm.

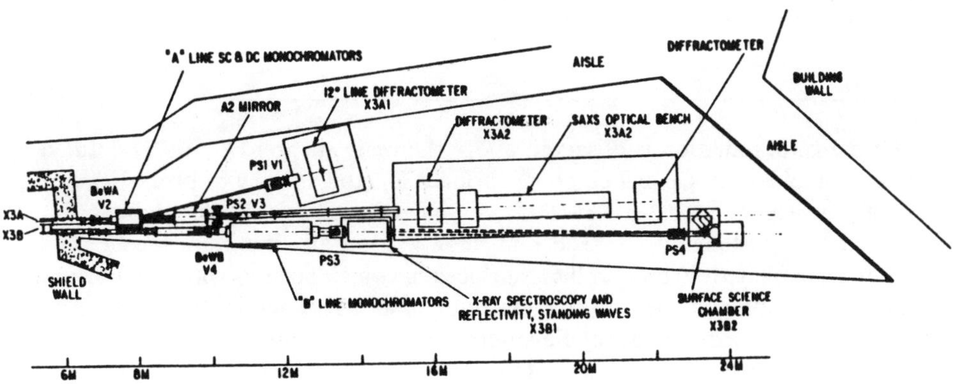

Figure 2. Schematic diagram of SUNY X3 beamline at NSLS.

Figure 3(a) shows a schematic diagram of the modified Kratky block collimator for synchrotron radiation. The basic Kratky concept remains unchanged,

TABLE 1. SUNY X3 A2 beam optics[2]

Source	X3 beam port at NSLS x-ray ring
Beamline	X3 A2 16-mrad branch line
Source size	$4\sigma_v \sim$ 0.4 mm; $4\sigma_{v'}$ (incl. single electron emission angle) ~0.3 mrad; $4\sigma_H \sim$ 1.2 mm; $4\sigma_{H'} \sim$ 0.8 mrad
Type of optics	monochromator-focusing mirror optics
Monochromator	double-crystal fixed exit rapidly tunable monochromator with Si(111), Ge(111) crystals available Bragg angle range 2° - 45°
Mirror	bent torroidal quartz mirror (60 cm long, 6 cm radius), gold coated
Source-monochromator distance	7.4 m
Monochromator-mirror distance	1.5 m
Mirror-sample distance	~10 m
X-ray beam divergence	vertical 0.2 mrad, horizontal < 7 mrad
Expected x-ray intensity	Si(111)
Intensity[a] (photon/s)	~5 x 10^{12}
Energy resolution (eV)	4.8

[a]With loss factor of ~2.5 due to mirror, monochromator, He atoms, and windows already taken into account.

---

i.e., no parasitic radiation is observed above the main section H (Fig. 3) defined by an exact coincidence of polished plane surfaces O_1 (top surface of block B_1) and O_2 (bottom surface of block B_2) in their mutual extension. Blocks B_1 and B_2 are separated by a distance b = 402.0 mm, mounted on an aluminum base ~13 mm thick, and prealigned with O_1 and O_2 surfaces in coincidence to within ~50 μm, the best which can be achieved by our machine shop. The condition for this crucial prealignment is relaxed because of an increase in b and a more directed synchrotron beam. A larger separation distance (b) is introduced so that we can accommodate the broader synchrotron beam. For example, if we take b = 400 mm and beamwidth in the vertical direction to be d, then $\delta \sim d/b$ = 2.5 mrad for d = 1 mm. A block length of about half a meter is a compromise, since the size of the vacuum box, which demands δ, ϕ, ψ angular adjustments as shown in Fig. 3(b), should not be too heavy. The vacuum box which holds the Kratky collimation system has a ψ adjustment. The sample chamber and the front end of the vacuum tube are mounted on an optical rail with ϕ and δ adjustments. The optical rail is mounted

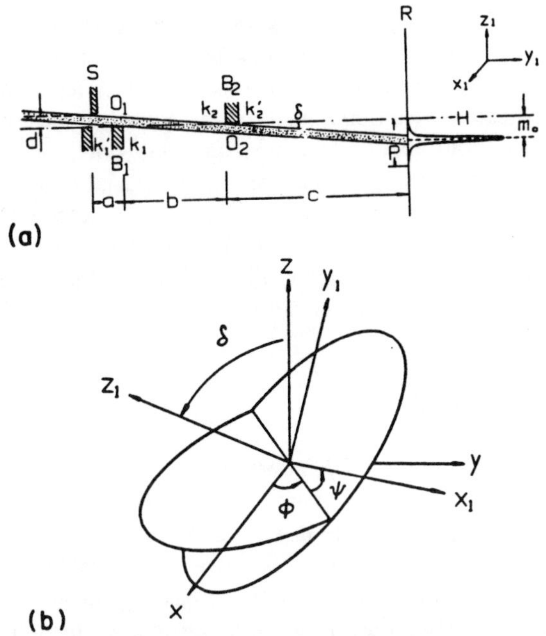

Figure 3. (a) Section along the (y) direction of the synchrotron beam perpendicular to the (x,z) plane. Vertical (z_1) scale is multifold stretched compared to the horizontal y_1 axis. Variable entrance slit S and blocks B_1 (middle block) and B_2 (bridge) are perpendicular to the (x_1,z_1) plane. In the plane of registration R, the primary beam does not have a triangular (vertical) beam profile. A Gaussian-shaped intensity beam profile is presented schematically to emphasize that the synchrotron radiation (as denoted by the dotted area) is highly collimated. m_o is the distance from the maximum intensity to the plane H. The Gaussian-shaped beam has fairly symmetrical long tails, mainly due to scattering by the edges k_1 and k_2. O_1 and O_2 represent the top surface of block B_1 and the bottom surface of block B_2, respectively. The SAXD dimensions are a = 16.7 mm, b = 402.0 mm, c, = variable from ~0.2 to 1.5 m. Notations of Fig. 1 in Ref. 3, except for a,b,c,d, and δ are preserved for a proper comparison. The subscript 1 denotes the coordinate system ($x_1y_1z_1$) for the SAXD, where (xyz) denotes the coordinate system for the synchrotron primary beam after mirror. We want $x \| x_1, z \| z_1$ and then tilt y_1 with angle δ between y and y_1 using x as the rotation axis. (b) Rotations defining Eulerian angles. ϕ is a coarse adjustment (~6 x 10^{-2} deg/division) which aligns synchrotron beam towards the y_1z_1 plane. ψ defines the horizontal (x_1y_1) plane for the primary beam and has a micrometer adjustment with a sensitivity of 6 x 10^{-3} deg/division. ψ becomes more important for line-shaped primary beams. Variable slit S is prealigned to be parallel to the block surfaces O_1 and O_2. δ is the tilt adjustment for alignment of k_2S to the (y) direction of propagation of the primary synchrotron beam. The tilt adjustment has a sensitivity of ~10^{-3} deg/division. δ changes with variable slit width d. All adjustments can be locked. (Ref. 1)

on a precision jack with fine adjustments so that the primary x-ray beam intensity can be maximized. In the present configuration, slight movements of the synchrotron beam (σ of beam cross section) have little effect on the performance of our SAXD. Furthermore, alignment can be achieved easily even under less favorable operating conditions at a synchrotron beam port. We have opted for the building block system in order to accommodate variations in sample chamber requirements and in vacuum tube lengths before the detector. The increase in dimension in our modified Kratky block collimator has also made alignment less critical in absolute magnitudes of linear dimensions. The extended Kratky block collimator is for use with synchrotron radiation only.

3. Laser-Aided Pinhole Collimator[6]

We have constructed a pinhole SAXS instrument at the SUNY X3A2 beamline, NSLS, BNL. The three pinholes were mounted in a thick-walled stainless steel pipe and pre-aligned by using a portable laser source and a CCD area detector. After the pre-alignment, incorporation of the collimator to the synchrotron source required only maximization of the incident x-ray intensity passing through the pinholes, which could be done easily by using a scintillation counter after proper attenuation. The entire synchrotron SAXS instrument setup took only a few hours even without stepping motor control for the pinhole collimator unit. By combining this collimator with a CCD-based x-ray area detector which could be assembled by using commercially available components, the SAXS instrument showed good performance for structural scales up to an order of 100 nm.

The pinhole collimator[7-24] offers more versatility because of its ability to investigate anisotropic systems. It also makes the smearing effect less serious. However, the alignment of a pinhole collimator is more delicate at very small scattering angles. Furthermore, synchrotron beamlines are usually shared by users with different instrumentation requirements. With tight beam-time schedules, it is desirable to minimize the time for instrumentation setup. One of the unique features of our Kratky collimation system is that the collimator is pre-aligned as one integral unit. A successful implementation of the same concept, i.e., to construct a pre-aligned pinhole collimation system for the synchrotron source, is described below.

Figure 4 shows a schematic diagram of a three-pinhole collimation geometry. The first and second pinholes define the incident beam. The guard pinhole blocks the parasitic scattering due to the edge scattering from the second pinhole. It is noted that due to intrinsic convergence of the synchrotron beam, the incident flux is often controlled mainly by the size of the first pinhole. In order to achieve a high angular resolution, the guard pinhole must be designed and placed in such a way that it is close, but not touching the incident beam.[25]

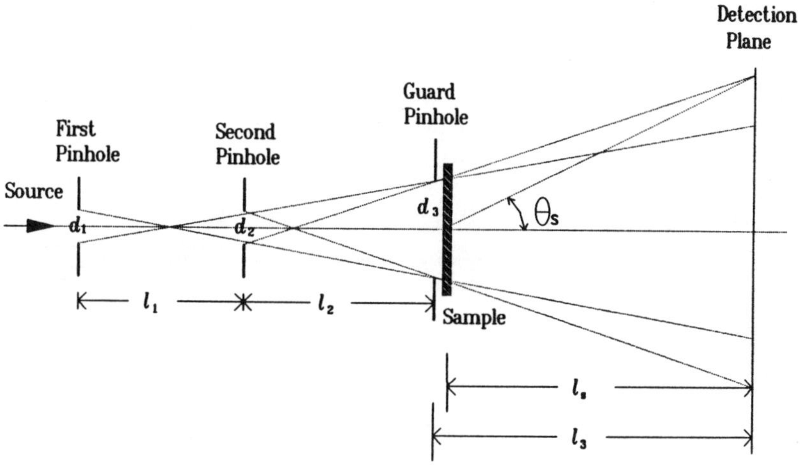

Figure 4. Schematic diagram of a pinhole collimation geometry. (Ref. 6)

The smallest angle one could reach without a serious parasitic scattering problem, θ_s, can be described by the relation:

$$\theta_s = \frac{d_2 + d_3}{2\, l_2} + \frac{d_3}{2\, l_3} + \frac{d_s}{2\, l_3} \tag{1}$$

where d_s denotes a finite size for the detector element, and d_3 ($=(d_1 + d_2)l_2/l_1 + d_2$) depends on l_1, l_2, d_1 and d_2. Based on the nature of the samples of interest and a compromise between the incident intensity and the angular resolution, we chose the following parameters: $l_1 = l_2 = 609$ mm, $d_1 = d_2 = 0.3$ mm. Therefore, d_3 should be very slightly larger than 0.9 mm. We used $d_3 = 1.0$ mm and $l_s \approx l_3 = 1030$ mm. Theoretically, the pinhole collimation geometry should yield a θ_s of 1.62 mrad for our CCD-based area detector ($d_s = 0.135$ mm) and a θ_s of 1.58 mrad for our Braun detector ($d_s = 0.046$ mm).

A laser drilling technique was used to produce slightly tapered pinholes with smooth edges. The first and third tantalum pinholes could be placed directly into the opposite ends of a ~1.2 m long stainless steel pipe (i.d. = 6.35 mm and o.d. = 19.05 mm). Threaded end caps with 3.2 mm diameter openings and rubber o-rings were used to secure the pinholes. A slot was cut into the center of the pipe to accommodate the second adjustable pinhole insert. Six set screws were used for fine tuning and securing the insert.

Figure 5 shows a schematic diagram for the pre-alignment of the pinhole collimator. The CCD area detector could provide two-dimensional data for both the beam shape and its intensity pattern.

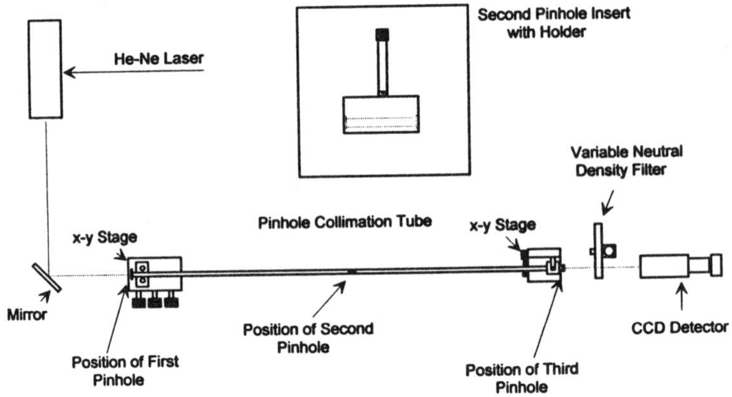

Figure 5. Schematic diagram of pre-alignment of a block pinhole collimator by using a laser source (Spectra-Physics Model 133 He-Ne laser with λ = 632.8 nm) and a CCD detector (Electrim Corp. EDC-1000). (Ref. 6)

The pre-alignment procedure is described as follows. With the end caps in place, the empty tube was mounted on two x-y positioning stages and coarsely aligned. The guard-pinhole was inserted into the tube, and its position was set by maximizing the intensity and, at the same time, creating a circular, symmetric image close to a Gaussian beam pattern. The insertion of the first pinhole may reduce the total intensity slightly. The first pinhole was positioned by again creating a symmetric beam pattern whose position is coincident with the peak position without the first pinhole in place. Finally, the second pinhole was inserted. Its position was adjusted by creating a symmetric Gaussian image with its peak position coincident with the one without the second pinhole in place. The tube was then sealed with thin-film windows for vacuum operation. It is noted that certain polymer films, such as Kapton, have their own SAXS profiles. Thus, the window material should be tested before use. This tube now becomes a self-contained, transportable, pre-aligned pinhole collimator unit, which weighs about 4 lbs. and is ready for the synchrotron setup.

Incorporation of the pre-aligned collimator to the synchrotron source could be performed by maximizing the x-ray intensity (using a scintillation counter) and by reconfirming the beam shape (using a CCD-based x-ray area detector as discussed in section V.B). The total setup time took only a few hours on the first try even without remote x-y position controls.

Figure 6 shows a schematic diagram of the pinhole SAXS instrument at the SUNY X3A2 beamline, NSLS, BNL. A Kratky rail could be used to accommodate

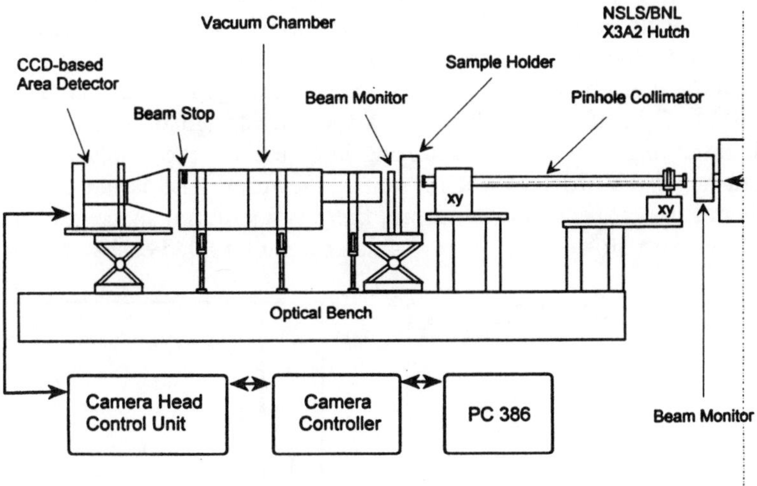

Figure 6. Schematic diagram of the pre-aligned pinhole SAXS instrument at the SUNY X3A2 Beamline, National Synchrotron Light Source, Brookhaven National Laboratory. (Ref. 6)

a series of temperature-controlled sample holders. In addition, a modified tensile stretching device could be introduced into the setup for time-resolved studies of films or fibers during stretching. A tantalum beam stop was mounted inside the vacuum chamber. A CCD-based area detector, an imaging plate or a Braun linear position-sensitive detector could be used to collect the scattering data. A wavelength of 0.154 nm was used.

The pinhole system was tested by using both isotropic and anisotropic samples with either a Braun or a CCD-based x-ray area detector (as discussed in section V.C). The one-dimensional profile for the pinhole system was compared with a profile of the same sample previously obtained by using a Kratky camera at the Stony Brook x-ray facility. The smallest scattering angle we could reach was ca. 1.89 mrad. If we took the effects of the beam stop into account, the performance of the instrument met the theoretical design specifications.

Figures 7a and 7b show the scattering patterns of a duck tendon with Fig. 7c being the 2-D image of the diffraction pattern of 7b. The spacing calculated from the sixth peak was 54.2 nm, demonstrating the fine performance of the instrument.

4. Bonse-Hart Ultrasmall Angle X-Ray Scattering (USAXS) Instruments

In a small angle x-ray scattering (SAXS) experiment, the larger the inhomogeneity size in a system, the smaller the scattering angle or the longer the x-ray wavelength is needed in order to determine the structures. A Kratky collimation system could usually reach a smallest scattering angle of $\sim$1 mrad, corresponding to a q value of

4×10^{-2} nm^{-1}, where $q = 4\pi \sin(\theta/2)/\lambda$ with θ and λ being the scattering angle and the x-ray wavelength at 0.154 nm. Utilization of longer wavelength x-rays is usually impractical due to the rapidly increased absorption coefficient of most materials of interest. On the other hand, routine laser light scattering (LLS) can cover a q range from 4×10^{-3} to 3.6×10^{-2} nm.[26] By using a specially designed apparatus, such as a prism-cell small angle laser light scattering instrument,[27] the smallest accessible q value could be of the order of 8×10^{-4} nm^{-1}. However, LLS is not suitable if the material is not transparent. Furthermore, there is a gap in q space between SAXS and LLS, from 3.6×10^{-2} to 4×10^{-2} nm^{-1}, which cannot be filled easily. In combining

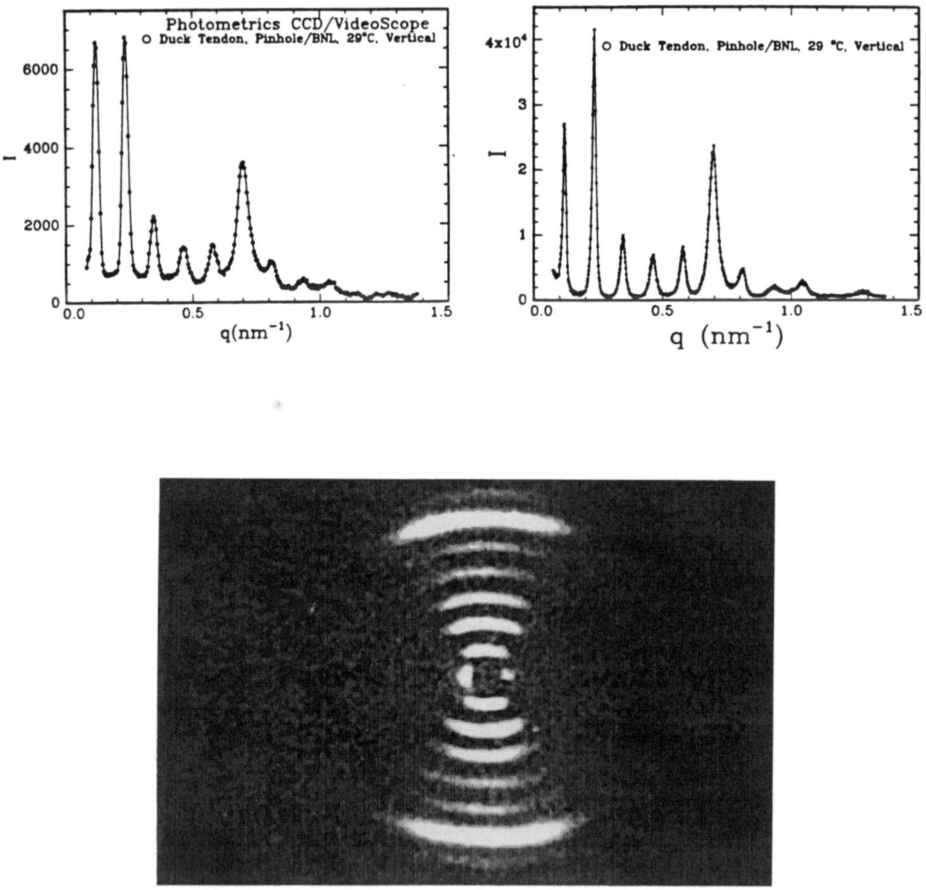

Figure 7. Scattering patterns of a piece of duck tendon by using (a) a Braun linear position-sensitive detector (measurement time 300 sec) and (b) a CCD-based area detector (measurement time 1000 sec). Note that the lower resolution of the CCD-based area detector has distorted the base line and broadened the peak. (c) 2-D image of the diffraction pattern of 7(b). (Ref. 6)

LLS and SAXS curves,[28] one has to arbitrarily shift one of the scattering curves since the magnitude of scattering is based on refractive index increment in LLS and electron density difference in SAXS. Without a good scattering curve which covers a wide q range, it is often difficult to achieve conclusive results in a comparison of the structure factor between experiment and theory. The key difficulty for SAXS to reach low enough q values is to get an x-ray beam with a small enough angular divergence without sacrificing the incident x-ray beam intensity and to determine the angular distribution of the scattered intensity at very small angles without stray x-rays. As early as the 1930's, Fankuchen and Jellinek[29] and others[30,31] proposed a SAXS setup which used one crystal to monochromatize the x-ray beam and another one as an analyzer. However, if the x-rays were reflected only once in each of the crystals, the angular resolution was not good enough. In 1966, Bonse and Hart[32] increased the number of the reflection in each of the crystals with a corresponding dramatic increase in the angular resolution.

Several Bonse-Hart ultrasmall angle x-ray scattering (USAXS) instruments,[33-35] including the use of Super Invar for mechanical elements which permit high-temperature operations and of silicon channel-cut crystals which permit higher energy incident x-ray wavelength operations were constructed. The instrument could be used to determine the structure of systems with inhomogeneity sizes on the order of ~1000 nm. The characteristics of the instrument by using synchrotron and conventional x-ray sources were compared. The use of synchrotron radiation showed much improved features not only in scattered intensity, but also in angular resolution. The scattered intensity with synchrotron x-rays was increased by a factor of about 20 when compared with a rotating anode x-ray generator. Therefore, weaker scattering systems could be investigated. By using the synchrotron radiation, the deficiency of the Bonse-Hart camera, which did not permit simultaneous multiple angle detection, could be compensated. An angular scan containing ~30 data points in the scattering angle region smaller than ~2 mrad with reasonable signal-to-noise ratio could be completed within 5-10 min for samples with reasonable scattering power. Therefore, kinetic studies could be possible if the half time of structural development in a strong scattering system is of the order of hours. For static experiments, a conventional x-ray source could be used more conveniently to obtain a scattering curve with similar quality as that by means of synchrotron radiation partially because the channel-cut crystals were not optimized for the synchrotron beam divergence and partially because alignments could be accomplished at a more leisurely pace using the conventional x-ray source.

A schematic diagram of the instrument setup is shown in Fig. 8. At Stony Brook (SB), we used an Enraf-Nonius FR-571 rotating-anode x-ray generator operating at 40 kV x 60 mA. A bellow coupling was used to connect the x-ray port to the defining slits of the instrument. This coupling also acted as a radiation shield. The slit-defined x-ray beam was monochromatized and collimated by the first crystal (monochromator). We used a channel-cut Ge crystal operating at the (220) reflection plane with six reflections. The advantage of an even number of reflections is that the x-ray beam coming out from the crystal is parallel to the incident x-ray beam. The sample was illuminated by the monochromatized and collimated incident

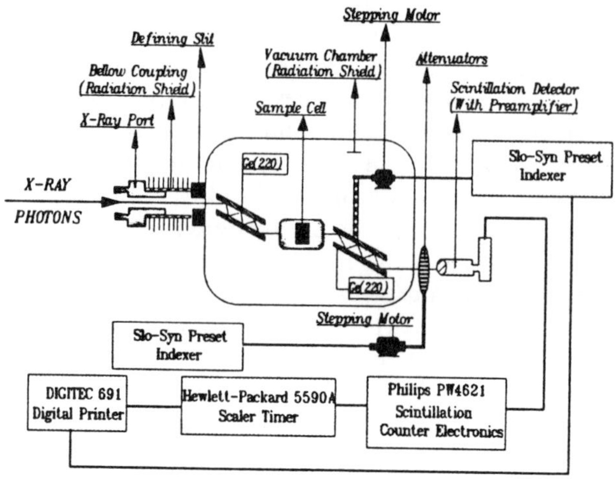

Figure 8. Schematic diagram of a Bonse-Hart camera setup. (Ref. 33)

x-ray beam. The scattered beam was analyzed by using a second channel-cut Ge crystal also operating at the (220) reflection plane with six reflections, known as the analyzer. Both crystals had independent rocking, tilting, and height adjustments. The crystals could be rocked and tilted without repositioning them in the x-ray beam because the axis of rotation of those adjustments coincided with the first bounce position in both crystals. The two crystals and the sample were inside a radiation shield which acted as a vacuum chamber. A Philips PW4621 scintillation counting system was used to collect the scattered x-rays. The scintillation counter had a relatively large collecting area which was insensitive to the changes in the scattering beam position. As we had to determine the scattered x-ray intensity from that of the main beam down to very small q values, a set of carefully calibrated multiple Nickel foil x-ray attenuators was used. The attenuation coefficients of the Nickel foils ranged from 10 to $\sim 10^5$. The attenuators were motor controlled by a Slo-Syn preset indexer. The second crystal was motor controlled by another Slo-Syn preset indexer and was capable of being rotated around the center of the sample by means of a large, precisely calibrated barrel micrometer. This indexer, as well as the output of the scintillation counting system, was connected to a Hewlett-Packard 5590A scaler/timer. A DIGITEC 691 digital printer (United System corp.) was connected to the scaler/timer to print out the intensity data. The whole system was automatic once the motor steps and the counts (or time) had been preset. We preferred preset counts because this gave us a scattering curve with similar signal-to-noise ratio at different q values. Synchrotron USAXS experiments were performed at the SUNY X3A1 and X3A2 beamlines, NSLS, BNL. The whole measurement could be controlled from outside the hutch. It is noted that many counting systems can be used to achieve the task outlined. Furthermore, the use of photodiodes having a large dynamic range offers more flexibility and is preferred, since its use may avoid the need for an x-ray attenuator.

After alignment, the instrument was stable and realignment was not necessary for a period of at least one month with a rotating anode x-ray generator. Beyond this, simple alignment checking could be performed to correct any minor distortion of the setup.

Figure 9 shows the rocking curves of the instrument measured at SB and BNL. The measurements at SB and BNL throughout this work were all performed at an x-ray wavelength of 0.154 nm. The FWHM of the rocking curve at SB is 23 s, while it is only 9 s at BNL. The reduction of FWHM at BNL is probably due to the installation of a double-crystal monochromator in front of the hutch, which already monochromatizes the synchrotron x-ray beam. The beam at BNL was also

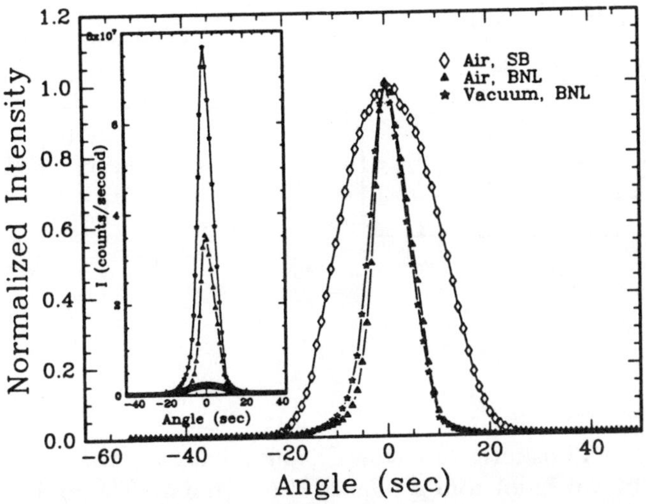

Figure 9. Rocking curves measured at SB and BNL, with and without vacuum. (Ref. 33)

focused by a gold-coated quartz mirror before entering the first crystal. Therefore, the incident beam had much lower angular divergences. Based on Fig. 9 lower q values could be reached at NSLS. The rocking curve was also determined at NSLS/BNL in vacuum. No significant changes were observed except that a higher beam intensity was obtained in vacuum.

The instrument was further calibrated by using an aqueous/alcohol latex suspension (3.2 vol%), as shown in Fig. 10. A Dow Chemical latex suspension in water/alcohol with a volume ratio of 1:2 and a claimed diameter of 176 nm was used. Although the scattered intensity is lower, it still shows three peaks. The

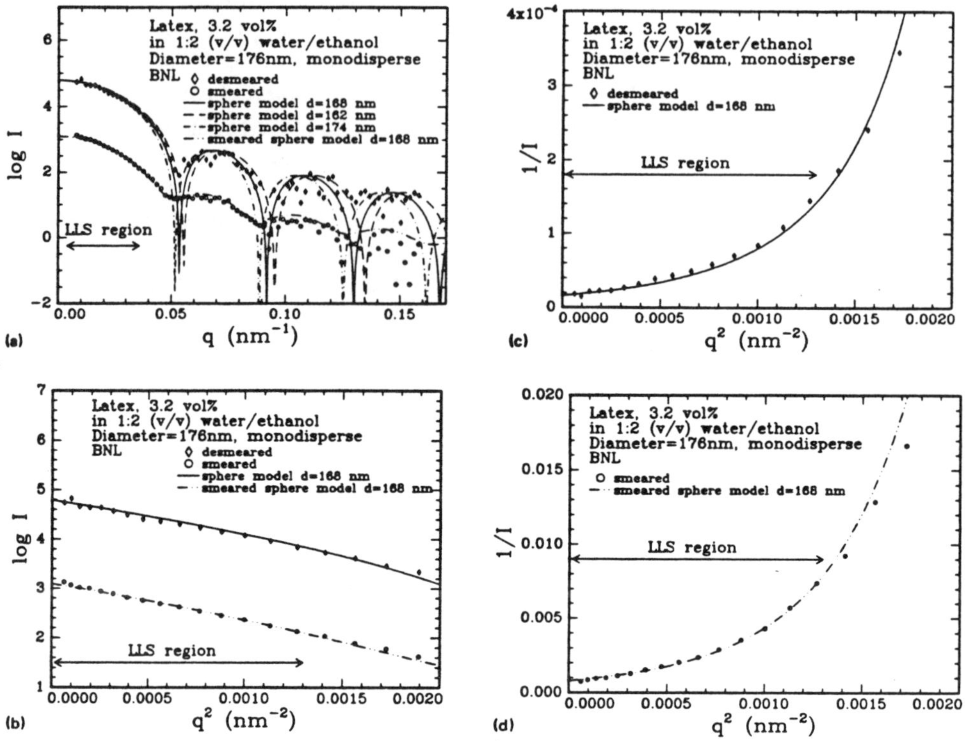

Figure 10. (a) USAXS patterns of a Dow Chemical latex suspension (3.2 vol %) in water/ethanol (1:2 vol ratio) with a claimed d of 176 nm. The experiments were performed without vacuum. Both the smeared and the desmeared curves have been corrected for the structure factor. The sphere model with d = 168 nm gives the best overall fitting for both smeared and desmeared experimental curves. A typical LLS range is marked in the figure. The LLS low q limit was calculated by using $\theta = 15°$, $\lambda_o = 633$ nm, and n = 1.5, where $\lambda = \lambda_o/n$ with n being the refractive index of the solution and λ_o being the laser light wavelength in vacuum. The high q limit is calculated by using $\theta = 140°$, $\lambda_o = 488$ nm, and n = 1.5. (b) Guinier plot of the curve presented in (a). (c) Zimm plot of the desmeared curve presented in (a). (d) Zimm plot of the smeared curves presented in (a). (Ref. 33)

scattered intensity is related to the structure factor S(q) and the form factor P(q) by the equation

$$I(q) = KCMS(Q)P(q), \quad (2)$$

where K is a constant related to the instrument setup, C and M are the concentration and the molaı mass of the particles, respectively. S(q) was calculated following Vrij et al.[36] Both the smeared and the desmeared curves shown in Fig. 6(a) have been corrected for S(q).

The room temperature Bonse-Hart was further improved by changing the mechanical elements to Invar so that high temperature operation could be achieved. Furthermore, larger gap silicon channel-cut crystals were adapted so that a range of x-ray wavelengths became accessible. The more penetrating higher energy x-rays permitted the use of thin-walled glass cylindrical cells for simultaneous laser light scattering and SAXS measurements. The reader is referred to the recent publications (references 34 and 35).

5. Position Sensitive Detectors

Small-angle x-ray scattering (SAXS) and wide-angle X-ray diffraction (WAXD) have become powerful techniques in elucidating the structure and dynamics of solids and liquids. In particular, the advantages of synchrotron radiation have made SAXS and WAXD even more powerful.[37] However, in order to fully utilize the SAXS and WAXD techniques, a good detection system becomes a critical factor besides a proper collimator for the incident x-ray beam. X-ray position-insensitive detectors, such as proportional counters and scintillation counters have been used extensively in early times. Unfortunately, such detectors restrict the possibility of time-resolved experiments due to the point taking technique. One-dimensional position-sensitive detectors could give a scattered intensity distribution in one dimension and have been widely used for the study of isotropic systems.[38] For anisotropic systems, a two-dimensional area detector is much more efficient. Even for isotropic systems, an area detector may yield much improved signal-to-noise ratio due to their larger receiving geometry. The increase in efficiency could reduce the measurement time and make time-resolved experiments possible even with conventional x-ray sources.

There have been many reports on a variety of two-dimensional detectors. Among them, the charge-coupled device (CCD) based area detector shows promising properties such as good spatial resolution, high count rate, high stability, high dynamic range and good sensitivity.[39-55] However, most reported detectors are expensive to reproduce. Therefore, their utilization has remained somewhat limited.

We have been working on developing CCD based x-ray area detectors for the SAXS facility at Stony Brook and the SUNY X3A2 beamline, NSLS, BNL. One of our goals has been the development of a state-of-the-art two-dimensional CCD based x-ray area detector at relatively low cost. A CCD x-ray area detector has been reported from this laboratory.[44] It was built by combining a fiber-optic face plate (FOFP) which was deposited with a thin layer of phosphor, a camera lens and a slow scan CCD system. The detector showed good linearity of response, good spatial linearity and minimal damage due to main beam exposure. Although it was an ideal detector for extended X-ray absorption of fine structure (EXAFS) measurements, the quantum efficiency of the detector was too low for SAXS or WAXD measurements.

Here, I shall describe three PSDs: a linear PSD and two CCD area detectors which have been designed and constructed for time-resolved SAXS and SAXS studies of anisotropic systems. In terms of dynamic and accessible angular range, the importance of imaging plates should be recognized. However, the imaging plates are usually not suitable for time-resolved measurements. Furthermore, its size and weight limitations have made the detector less maneuverable when compared with photodiode array or CCD PSDs.

5.1. LINEAR PHOTODIODE ARRAY DETECTOR[56]

A linear photodiode array position-sensitive detector for synchrotron small-angle x-ray scattering was constructed by coupling a fiber-optic face plate coated with a thin layer (~40 μm) of phosphor (Y_2O_2S:Tb) onto an electrostatically focused, intensified photodiode array detector (EG&G, PARC, Model 1422) originally designed for visible light applications. The x-ray detector was thoroughly evaluated. It showed excellent linearity of response to the incident x-ray intensity (less than 1%), stable pixel uniformity (an average of ±5%), good spatial linearity (±1 per100 pixels), and a net gain of ~17 counts/x-ray photon at 8 keV. The detector quantum efficiency was about 1 for a signal-to-noise ratio of greater than ~5%. No damage to individual pixels was found after exposure to a main beam of ~10^{10} x-ray photons/mm^2/s. A finite afterglow (> 5s) was observed from the phosphor inside the intensifier tube. The afterflow that could limit fast time-resolved experiments could be significantly improved by replacing the present intensifier with one using a fast-decay phosphor. The detector usable length could be increased from the present 25 to 50 mm by replacing the fiber-optic face plate with a 2:1 taper. Figure 11 shows a schematic diagram for the x-ray LPSD.

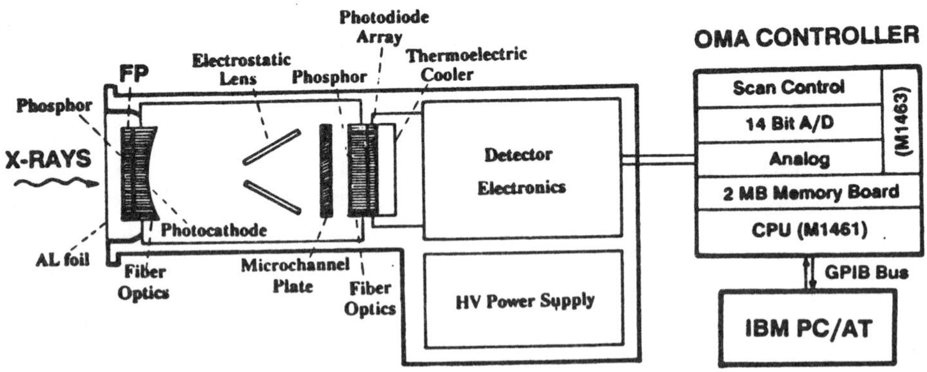

Figure 11. A schematic diagram for the x-ray photodiode array detector for small angle x-ray scattering. FP denotes the fiber-optical faceplate which has a ~40-μm Y_2O_2S:Tb phosphor deposit and is coupled to the intensifier by an index-matched grease. (Ref. 56)

5.2. CCD X-RAY AREA DETECTOR (WITHOUT INTENSIFIER) FOR FOCUSING AND ALIGNMENT PURPOSES[57]

A schematic representation of the area detector is shown in Fig. 12; a list of the components is enumerated in Table 2. The x-ray beam passes through a foil of aluminum (25 μm thick) that blocks visible light. X-ray impinges on the (Gd_2O_2S:Tb) phosphor which converts the high-energy photons to visible light. The

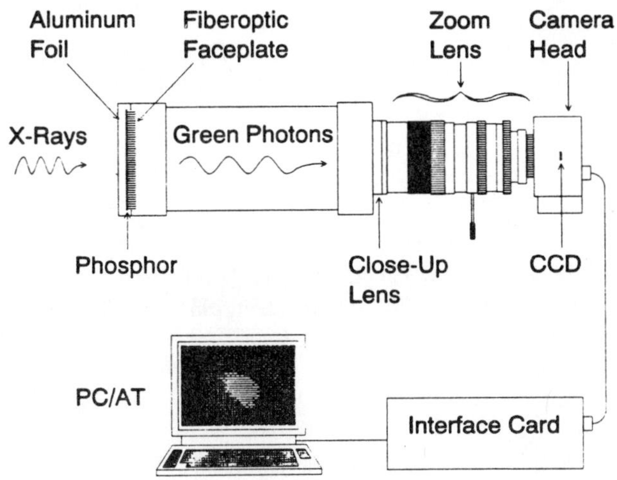

Figure 12. Schematic representation of the CCD area x-ray detection system. A hollow tube, which contains the coated fiber-optic faceplate at one end, is attached at the other end to the zoom lens system. The exposure time can be controlled via a PC computer over a range of 0.01 and ~60 s, beyond which the dark noise saturates the detector. After readout, the CCD image is displayed on the computer monitor. (Ref. 57)

picture object formed at the back of the faceplate is transferred by the zoom lens system onto the CCD. The combination of a zoom lens and a close-up lens reduces the distance required between the lens system and the optical fiber faceplate. The magnification ability of the zoom lens allows us to vary the image/object ratio. The effective CCD area could vary from 37 mm x 37 mm to 3.1 mm x 3.1 mm which corresponds, in the interlaced mode, to a pixel resolution of 0.19 mm x 0.11 mm to 16 μm x 9.4 μm, respectively. The CCD readout is performed by an interface card, plugged into a PC compatible. The card amplifies the voltage output of the CCD, converts it into digital numbers by an 8-bit ADC, and generates a VGA signal which can be diplayed on a computer monitor.

TABLE 2. Components of x-ray area detector

Components[a]	Manufacturer[b]	Specifications[c]
1. (Gd_2O_2S:Tb) Phosphor	GTE/Sylvania Thomas Electronics (100 Riverview Drive, Wayne,NJ) applied the phosphor on the fiber-optic faceplate.	P43 type 10 mg/cm^2, 6 μm grain size protective coating: potassium silicate
2. Fiber-optic faceplate	Galileo Electro-Optics (Galileo Park,Sturbridge,MA)	thickness: 6mm diamter: 60 mm
3. Lenses	no brand (made in Japan) Hoya	Zoom lens (12.5-75 m m) Closeup (+4)
4. CCD camera	Electrim Corp (P.O. Box 2074, Princeton, NJ)	EDC-1000/TE, Peltier Cooling System (35°C below ambient T).lens mount:C
5. CCD	Texas Instruments	TLTC211 size: 2.64 mm x 2.64 mm pixel array: 192 (column)x 165 (rows) pixel size: 192 x 330 (interlaced) 13.75 x 16 μm^2 13.75 x 8 μm^2 (interl.)
6. Interface card	Electrim Corp. (P.O. Box 2074, Princeton, NJ)	8-bit ADC output: VGA signal readout time: 75 ms (Compaq 386/25)

[a]Estimated cost: \$300 (1 + 2) + \$150 (3) + \$750 (4 + 5 + 6) ≃ \$1200.
[b]Not an endorsement.

A utilization of the detector to observe the main beam profile is schematically represented in Fig. 13. The SUNY X3A2 beamline at the National Synchrotron Light Source was equipped with a Kratky block collimator for SAXS.[1] The CCD detector was placed after the long vacuum chamber. The beam stop was removed so the detector could be illuminated by the main beam. The lower right image of Fig. 13 shows a two-dimensional representation of the main beam profile after

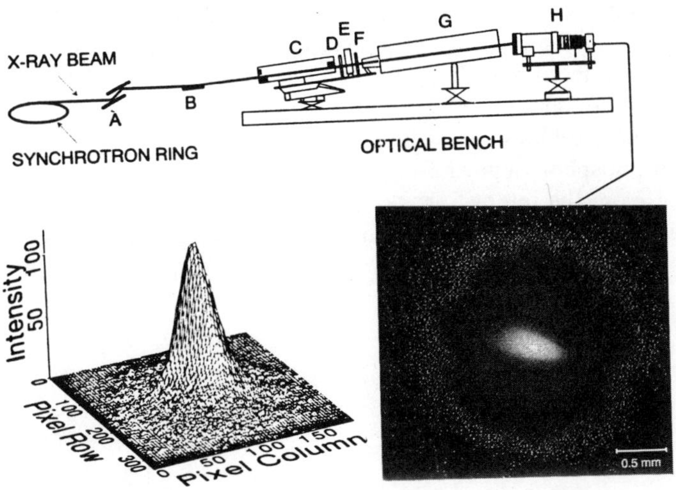

Figure 13. Schematic representation of the SUNY X3A2 beamline at Brookhaven, including the Kratky block collimation small-angle x-ray diffractometer, the x-ray area detector, and an image of the main beam taken after the long vacuum chamber. An amplified video of the image is shown in the right lower part. At the bottom left is a three-dimensional representation of the main beam image. (A) monochromator, (B) mirror, (C) Kratky block collimator, (D) first ionization chamber, (E) sample holder, (F) second ionization chamber, (G) vacuum chamber, (H) CCD area x-ray detection system. (The schematic is not drawn on scale.) The beam stop was removed in order to observe the main beam directly. The image at the bottom right was taken with an attenuation of ten Al foils (250 μm thick) and an exposure time of 20 s. The image was saved in a TIFE format and converted into PCL code as shown here. The lower right corner of the image corresponds to the readout of the pixel row 1 and column 1. The intensity unit of the three-dimensional representation of the main beam image is arbitrary. In both images, the circular shape in low intensity shoulder could be due to phosphor blooming. One can see that the beam profile was not symmetric. (Ref. 57)

attenuation by ten Al foils (250 μm thick), and the lower left image depicts the same picture in three dimensions. The circular shading observed in both images could be related to phosphor blooming. It should be noted that the beam profile was non uniform and its intensity did not follow a Gaussian profile.

Desmearing procedures that account for the finite incident beam length and width are often based on the assumption that the beam length is very large when compared with the beam width. Moreover, the width desmearing is often negligible. Knowledge of the beam profile as shown in Fig. 13 could be used to validate the usual assumptions on the incident beam profile.

5.3. INTENSIFIED CCD X-RAY AREA DETECTOR[58]

Figure 14 shows a schematic diagram of the CCD x-ray area detector. A 2:1 fiber-optic taper with a 30-μm thick layer of phosphor deposit was in optical contact (via a refractive index matching grease) with a slightly modified image intensifier. A 25 μm thick aluminum foil was mounted in front of the phosphor screen with an O-ring in order to block the visible light from entering the detector system and to protect the phosphor deposit from direct attack by moisture and dust. The image transferred from the intensifier was focussed onto the CCD chip by a camera lens which could change the observation field. The CCD output signal could be processed

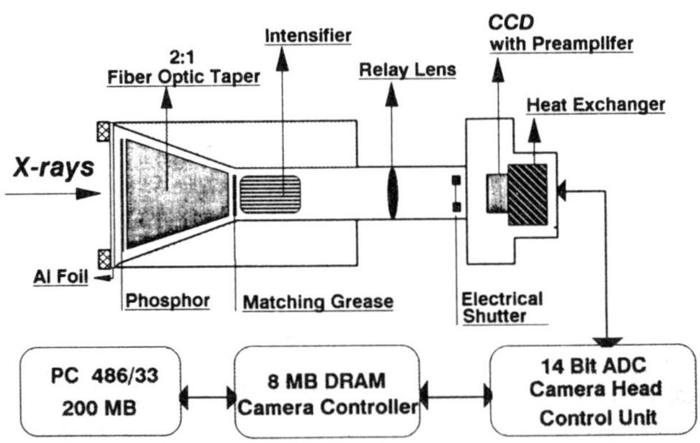

Figure 14. Schematic diagram for the x-ray area detector system based on an intensified CCD unit. (Ref. 58)

and stored (up to 20 full frames, limited by the 8 Mb RAM size) by the Photometric CE200 camera electronics unit and the CC200 camera controller. Finally the images could be transferred to a personal computer for permanent storage, display, and image manipulation. A detailed description of the x-ray area detector is listed in Table 3.

Table 3. Main components of x-ray area detector

Component	Vendor	Specification
1. Phosphor	GTE Type 2620 Sylvania Chemicals/Metals	Y_2O_2S:Tb 8.4 μm grain size Density of 5.4 g/cm^3
2. Fiber-Optic Taper	Galileo Electro-Optics Co.	6 μm diameter fibers 2.03:1 system overall magnification
3. Image Intensifier	VARO Inc.	Type II, non-gated 18 mm aperture[a] full view wafer tube with P-46 phosphor
4. Power Supply for Image Intensifier		See Note
5. Lens	AF MICRO NIKKOR	f=55 mm, 1:2.8
6. CCD Camera[b]	Photometric LTD.	series 200 LC 200 cooling unit CH 200 camera head CC 200 controller CE 200 electronics unit 14 bit ADC 8 Mb DRAM
7. CCD Chip	Thompson CSF	TH7882CDA 23 μm × 23 μm pixel size, 384 × 576 pixels CCD format

Note: Modified from an EG&G/PAR Model 1420 power supply

It is anticipated that SAXS and WAXD optics, as well as area detectors,[59] will improve with time in order to achieve larger dynamic ranges, higher spatial resolutions, and faster time frames, especially in view of the completion of third-generation synchrotrons. With the new level of high brilliance, even x-ray photon correlation could be within reach.[60]

Acknowledgement

BC gratefully acknowledges support of this work by the Department of Energy (DEFFG0286ER45237.011), the Polymers Program, National Science Foundation (DMR9301294), and the U.S. Army Research Office (DAAH0494G0053).

REFERENCES (Numbers in parentheses denote key references)

(1) Chu, B., Wu, D.-Q., and Wu, C. (1987) Kratky block-collimation small-angle x-ray diffractometer for synchrotron radiation, *Rev. Sci. Instrum.*, **58**, 1158-1163.

2. Chu, B., Phillips, J., and Wu, D.-Q. **in** Russell, T. P., and Goland, A. N., *Polym. Res. at Synchrotron Radiation Sources*, BNL 51847, UC-25 (Materials-TIC-4500), National Technical Information Service, U.S. Department of Commerce, Springfield, VA, 1985, pp. 126-132.

3. Kratky, O., and Stabinger, H. (1984) X-ray small angle camera with block-collimation system an instrument of colloid research, *Colloid. Polym. Sci.*, **262**, 345-360.

4. Ritland, H. N., Kaesberg, P., and Beeman, W. W. (1950) Double Crystal and Slit Methods in Small Angle X-Ray Scattering, *J. Appl. Phys.*, **21**, 838-841.

5. Anderegg, J. W., Beeman, W. W., Shulman, S., and Kaesberg, P. J. (1955) An Investigation of the Size, Shape and Hydration of Serum Albumin by Small-angle X-ray Scattering, *J. Am. Chem. Soc.*, **77**, 2927-2937.

(6) Chu, B., Harney, P. J., Li, Y.-J., Linliu, K., Yeh, F.-J., and Hsiao, B. (1994) A laser-aided prealigned pinhole collimator for synchrotron radiation, *Rev. Sci. Instrum.*, **65(3)**, 597-602.

7. Glatter, O., and Kratky, O. (1982) *Small Angle X-ray Scattering*, Academic Press, New York.

8. Leigh, J. B., and Rosebaum, G. (1974) An Report on The Application of Synchrotron Radiation to Low-Angle Scattering, *J. Appl. Cryst.* **7**, 117-121.

9. Hendrix, J., Koch, M. H. J., and Bordas, J. (1979) A Double Focusing X-ray Camera for Use with Synchrotron Radiation, *J. Appl. Cryst.* **12**, 467-472.

10. Hasegrove, J. C., Faruqi, A. R., Huxley, H. E., and Arndt, U. W. (1977) The design and use of a camera for low-angle x-ray diffraction experiments with synchrotron radiation, *J. Phys. E.* **10**, 1035-1044.

11. Stephenson, G. B., Ph.D. Thesis, (Stanford University, 1982).

12. Stephenson, G. B. (1988) Time-Resolved X-Ray Scattering using a High-Intensity Multilayer Monochromator, *Nucl. Instrum. Methods* **A266**, 447-451.

13. Amemiya, Y., Wakabayashi, K., Hamanaka, T., Wakabayashi, T., Matsushita, T., and Hashizume, H. (1983) Design of a Small-Angle X-ray Diffractometer using Synchrotron Radiation at the Photon Factory, *Nucl. Instrum. Methods* **208**, 471-477.

14. Tchoubar, D., Rousseaux, F., Pons, C., and Lemonnier, M. (1978) Small Angle Scattering setting at LURE: Descriptions and Results, *Nucl. Instrum. Methods* **152**, 301-305.

15. Dubuisson, J. M., Dauvergne, J. M., Depautex, C., Vachette, P., and Williams, C. E. (1986) ASAXS Spectrometer, *Nucl. Instrum. Methods* **A246**, 636-640.

16. Wakatsuki, S., Hodgson, K. O., Eliezer, D., Rice, M., Hubbard, S., Gillis, N., Doniach, S., and Spann, U. (1992) Small-angle x-ray scattering/diffraction system for studies of biological and other materials at the Stanford Synchrotron Radiation Laboratory, *Rev. Sci. Instrum.* **63**, 1736-1740.
17. Zachmann, H. G., and Wutz, C. (1992) New Insights into the Mechanism of Crystallization by Means of Time Resolved SAXS Measurements, *Polym. Prepr.* **33(1)**, 261-262.
18. Cogan, K. A., Gast, A. P., and Capel, M. (1991) Stretching and Scaling in Polymeric Micelles, *Macromolecules* **24**, 6512-6520.
19. Stuhrmann, H. B. (1985) Resonance Scattering in Macromolecular Structure Research, *Adv. Polym. Sci.* **67**, 123-163.
20. Bordas, J., Koch, M. H. J., Clout, P. H., Dorrington, E., Boulin, C., and Gabriel, A. (1980) A synchrotron radiation camera and data acquisition system for time resolved x-ray scattering studies, *J. Phys. E.* **13**, 938-944.
21. Koch, M. H. J., and Bordas, J. (1983) X-ray Diffraction and Scattering on Disordered systems using Synchrotron Radiation, *Nucl. Instrum. Methods* **208**, 461-469.
22. Bras, S., Craievich, A., Sanchez, J. M., Williams, C., and Zanotto, E. D. (1983) SmallAbgle X-ray Scattering Study of Phase Separation in Glasses using a New Position Sensitive Detector, *Nucl. Instrum. Methods* **208**, 489-494.
23. Caffrey, M., and Bilderback, D. H. (1983) Real-Time X0ray Diffraction using Synchrotron Radiation: System Characterization and Applications, *Nucl. Instrum. Methods* **208**, 495-510.
24. Bras, W., Derbyshire, G. E., Ryan, A. J., Mant, G. R., Felton, A., Lewis, R. A., Hall, C. J., and Greaves, G. N. (1993) Simultaneous time resolved SAXS and WAXS experiments using synchrotron radiation, *Nucl. Instrum. Methods* **A326**, 587-591.
25. Guinier, A., and Fournet, G. (1955) *Small-Angle Scattering of X-rays*, Wiley, New York.
(26) Chu, B. (1991) *Laser Light Scattering*, Academic, New York.
(27) Chu, B., Xu, R., Maeda, T., and Dhadwal, H. S. (1987) Prism laser light-scattering spectrometer, *Rev. Sci. Instrum.*, **59**, 716-724.
28. Li, Y., and Chu, B. (1991) Structure of aggregates of P2BCMU in dilute THF/toluene solutions, *Macromolecules*, **24**, 4115-4122.
29. Fankuchen, I., and Jellinek, M. H. (1945) Low angle x-ray scattering, *Phys. Rev.*, **67**, 201.
30. DuMond, J. W. (1947) Method of correcting low angle x-ray diffraction curves for the study of small particle sizes, *Phys. Rev.*, **72**, 83-84.
31. Beeman, W. W., and Kaesberg, P. (1947) X-ray scattering at very small angles, *Phys. Rev.*, **72**, 512.
32. Bonse, U., and Hart, M. (1966) Small Angle X-Ray Scattering by Spherical Particles of Polystyrene and Polyvinyltoluene, *Z. Phys.*, **189**, 151-162; in *Small-Angle X-Ray Scattering*, edited by H. Brumberger (Gordon & Breach, New York, 1966), p. 121.
(33) Chu, B., Li, Y.-J., and Gao, T. (1992) A Bonse-Hart ultrasmall angle x-ray scattering instrument employing synchrotron and conventioinal x-ray sources, *Rev. Sci. Instrum.*, **63**, 4128-4133.
(34) Chu, B, Li, Y.-J., Harney, P. J., and Yeh, F.-J. (1993) A high-temperature Bonse-Hart ultrasmall-angle x-ray scattering (USAXS) instrument, *Rev. Sci. Instrum.*, **64**, 1510-1514.
(35) Chu, B., Yeh, F.-J., Li, Y.-J., Harney, P. J., Rousseau, J., Darovsky, A., and Siddons, D. P. (1994) Construction of a high-energy Bonse-Hart ultrasmall-angle x-ray scattering instrument, *Rev. Sci. Instrum.*, **65**, 3233-3237.

36. Vrij, A., Jansen, J. W., Dhont, J. K. G., Pathmamanoharan, C., Kops-werkhoven, M. M., and Funaut, H. M. (1983) Light scattering of colloidal dispersions in non-polar solvents at finite concentrations, *Faraday Discuss. Chem. Soc.*, **76**, 19-35.
37. Russell, T. P. **in** Brown, G., and Moncton, D. E. (1991) *Handbook on Synchrotron Radiation*, Vol 3, Elsevier, New York.
38. Arndt, U.W. (1986) X-ray Position-Sensitive Detectors, *J. Appl. Cryst.* **19**, 145-163.
39. Widom, J., and Feng, H.-P. (1989) High peformance x-ray area detector suitable for small-angle scattering crystallographic, and kinetic studies, *Rev. Sci. Instrum.* **60**, 3231-3238.
40. Epperson, P.M., Sweeder, J.V., Bilhorn, B.B., Sims, G.R., and Deuton, M.B. (1988) Applications of Charge Transfer Devices in Spectroscopy, *Anal. Chem.* **60**, 327A-335A.
41. Epperson, P.M., and Deuton, M.B. (1989) Binning Spectral Images in a Charge-Couple Device, *Anal. Chem.* **61**, 1513-1519.
42. Collcott, T.A., Tsang, K.-L., Zhang, C.-H., Ederer, D.L., and Arakawa, E.T. (1988) Area Detector for X-ray Spectroscopy, *Nucl. Instrum. Methods. Phys. Res.* **A 266**, 578-585.
43. Eikenberry, E.F., Tate, M.W., Bemonte, A.L., Lowance, J.L., Bilderback, D., and Gruner, S.M. (1991) A Direct-Coupled Detector for Synchrotron X-Radiation Using a Large Format CCD, *IEEE Trans. Nucl. Sci.* **38**, 110-118.
(44) Fuchs, H.F., Wu, D.Q., and Chu, B. (1990) An area x-ray detector system based on a commercially available CCD-unit, *Rev. Sci. Instrum.* **61**, 712-716.
45. Rodricks, B., Clarke, R., Smither, R., and Fontaine, A. (1989) A virtual phase CCD detector for synchrotron radiation applications, *Rev. Sci. Instrum.* **60**, 2586-2591.
46. Eikenberry, E. F., Gruner, S. M., and Lowrance, J. L. (1986) A Two-Dimensional X-ray Detector with o Slow-Scan Charge-Couple Device Readout, *IEEE Trans. Nucl. Sci.* **83**, 542-545.
47. Turner, L. K., Mantus, D. S., Ling, Y.-C., Bernius, M. T. and Morrison, G. H. (1989) Development and characterization of a charge-couple device detection system for ion microscopy, *Rev. Sci. Instrum.* **60**, 886-894.
48. Clarke, R. (1990) Use of CCD Detectors for Spectroscopy and Scattering Experiments, *Nucl. Instrum. Methods Phys. Res.* **A291**, 117-122.
49. Gruner, S. M. (1989) CCD and vidicon x-ray detectors: Theory and practice(invited), *Rev. Sci. Instrum.* **60**, 1545-1551.
50. R. Clarke, W. P. Lowe, R. A. MacHarrie, C. Brizard and B. G. Rodricks, *Rev. Sci. Instrum.* **63**, 802 (1992).
51. Strauss, M. G., Naday, I., Sherman, I. S., Kraimer, M. R., Westbrook, E. M., and Zaluzec, N. J. (1988) CCD sensors in synchrotron x-ray detectors, *Nucl Instrum. Methods Phys. Res.* **A266**, 563-577.
52. Naday, I., Strauss, M. G., Sherman, I. S., Kraimer, M. R., and Westbrook, E. M. (1987) Detector with charge-coupled-device sensor for protein crystallography with synchrotron x-rays, *Opt. Eng.* **26**, 788-794.
53. Amemiya, Y., Kishimoto, S., Suzuki, M., and Ohnishi, Y. (1992) X-ray area detector based on CCD readout for diffraction study (abstract), *Rev. Sci. Instrum.* **63**, 659.
54. Allinson, N. M., Colapietro, M., Moon, K. J., Thompson, A. W., and Weisgerber, S. (1992) Charge-coupled imagers for time-resolved macromolecular crystallography, *Rev. Sci. Instrum.* **63**, 664-666.
55. Brizard, C., and Rodricks, B. (1992) Programmable CCD imaging system for synchrotron radiation studies, *Rev. Sci. Instrum.* 63, 802-805.

(56) Chu, B., Wu, D.-Q., and Howard, R. L. (1989) Evaluation of a linear photodiode array detector for synchrotron small-angle x-ray scattering measurements, *Rev. Sci. Instrum.*, **60**, 3224-3230.

(57) Chu, B., Rousseau, J., and Gao, T. (1992) Economical x-ray area detector for focussing and alignment purposes, *Rev. Sci. Instrum.*, **63**, 4000-4002.

(58) Gao, T., Li, Y.-J., Rousseau, J., Linliu, K., and Chu, B. (1993) An economical x-ray detector based on an intensified CCD unit, *Rev. Sci. Instrum.*, **64**, 390-396.

59. Cooled CCD cameras with fiber optic tapers bonding directly to the face of the CCD are able to avoid the use of image intensifiers due to its efficient optical coupling efficiency. With 3:1 magnification, it is within reach to detect individual x-ray photons with a resolution of 60 μm x 60 μm covering a spatial range of 1024 x 1024 pixels by using appropriate phosphors at sufficient high x-ray input energies.

60. Chu, B., Ying, Q. C., Yeh, F.-J., Patkowski, A., Steffen, W., and Fischer, E. W. (1995) An x-ray photon correlation experiment, *Langmuir*, Letter, **11**, 1419-1421.

STRUCTURE AND DYNAMICS OF POLYMERS BY SYNCHROTRON SAXS

BENJAMIN CHU
Departments of Chemistry and of Materials Science and Engineering
State University of New York at Stony Brook
Stony Brook, Long Island, New York 11794-3400, U.S.A.

Abstract

Laser light scattering (LLS), small angle x-ray scattering (SAXS) and small angle neutron scattering (SANS) are complementary to one another. Together, they become unmatched among the physical techniques which can investigate the structure and dynamics of polymeric materials over a large range of time and length scales. In this lecture, the application of synchrotron x-rays to some of the problems in polymer physics is presented. The work performed by Chu and collaborators on phase transition in polymer blends, structure of ionomers, conformation of polymers and supramolecular structures in solution, epoxy polymerization and crystallization processes, as well as anisotropic scattering by a single filament are discussed.

1. Introduction

One should do synchrotron x-ray experiments only when a conventional x-ray source, such as a rotating x-ray generator, does not have sufficient intensity to carry out the intended measurements. To this end, there are and shall probably continue to have abuses because the high flux of a synchrotron x-ray beam can shorten the measurement time by a factor of 10^2 - 10^4 when compared with conventional x-ray sources. In other words, one can often complete a one-month x-ray experiment in an afternoon. Thus, if a student has an access to synchrotron instrumentation, he or she seldom ever wants to use the conventional x-ray sources again. However, this observation does not mean that the conventional x-ray sources are not useful, as I have discussed briefly in the first lecture. Advances in the position sensitive detector and x-ray optics can be coupled with a conventional x-ray source to yield very powerful x-ray instrumentation suitable for many polymer physics problems. Furthermore, in a synchrotron environment, there is little time for refinements in experimentation. Tests on instrument modification and preliminary experiments can best be undertaken by using a conventional x-ray source in a laboratory. For certain experiments where the scattered intensity is very strong and the sample equilibration time between measurements is very long, such as studies on phase transitions in polymer blends, the use of a conventional x-ray source is quite adequate but permits ample time for temperature equilibration and better temperature control under a laboratory environment.

S.E. Webber et al. (eds.), Solvents and Self-Organization of Polymers, 409–431.

In this lecture, the two main features of synchrotron x-rays: high brilliance and wavelength tunability, will be used to carry out time-resolved and anomalous x-ray experiments. The high brilliance also permits measurements of samples with small dimensions, such as a single filament of fiber or a microcrystal. In the following section, and based largely on the work performed by the Chu group, I shall describe some experiments on II. phase transition in polymer blends, III. structure of ionomers, IV. conformation of polymers and supramolecular structures of block copolymers in solution, V. epoxy polymerization processes, VI. time-resolved SAXS, and VII. anisotropic scattering. Review on the physical techniques used to study the self-assembly of triblock copolymers [copoly(oxyethylene-oxypropylene-oxyethylene), also known as Pluronic·polyols] have been presented as a Langmuir lecture on structure and dynamics of block copolymer colloids [B. Chu, *Langmuir*, **11**, 414-421 (1995), and see references therein] and thus will not be included in this chapter.

Laser light scattering (LLS), synchrotron small angle x-ray scattering (SAXS) and small angle neutron scattering (SANS) are complementary in nature, although the source of scattering is different: with refractive index increment for LLS, electron density difference for SAXS, and scattering length difference for SANS. However, the key parameter which governs the interference effect is the scattering wave vector whose magnitude $K \equiv q = (4\pi/\lambda)\sin(\theta/2)$ with λ and θ being the wavelength in the scattering medium and the scattering angle, respectively. Table 1 shows the overlapping momentum transfer (K) ranges of the three complementary techniques.[1] While anomalous x-ray scattering/diffraction can be used to investigate the structures of selective elements in multicomponent materials, the hydrogen-deuterium substitution approach available in neutron scattering/diffraction remain unmatched in elucidating the structure of many polymeric systems since most polymers contain hydrogen.

2. Phase Transition in Polymer Blends[2]

In polymer blend studies, the practical aspect deals with structure/property relationship of materials which can form controlled microphase domains in a phase separation process. In this respect, crystallization (see 6.2) is a form of phase transition. Here, we shall deal with the second-order transition near the critical mixing point of a polymer blend and ask how such a phase transition phenomenon may be different from that of critical binary liquid mixtures of small molecules. It is a form of self-assembly from a broader viewpoint since the characteristic length scales can be controlled by thermodynamic and kinetic processes.

While critical binary fluids are known to exhibit Ising critical properties, blends of high molecular weight and flexible-chain polymers usually show mean field behavior, as predicted by de Gennes.[3] It is therefore intriguing to investigate the crossover from mean field to Ising behavior with decreasing polymer chain length. As the width of the Ising range is reciprocally related to the polymer chain length, one approach in the search for the crossover behavior is, indeed, by varying the molecular weight, i.e., the chain length, of one of the two polymer components in the polymer binary mixture (or blend).

TABLE I.

	LLS[a]			SAXS[b]		SANS[c]
Momentum transfer						
Representative						
wavelength (nm)		300		0.15		0.4
Scattering angle (rad)	Min $5x10^{-2}$	Typical $\pi/2$	Max π	Min $5x10^{-4}$	Typical 10^{-2}	Typical 10-2
K (nm^{-1})	$1.05x10^{-3}$	$2.96x10^{-2}$	$4.19x10^{-2}$	$2.09x10^{-2}$	0.419	0.157
K^{-1} (nm)	952	33.8	23.9	47.9	2.39	6.37
d($=2\pi/K$,nm)	$5.98x10^{3}$	212	150	301	15.0	40.0
Energy transfer						
Typical		Typical				
$\Delta\upsilon$ (wave number,cm^{-1})		$8x10^{-9}$				$8x10^{-3}$
$\hbar\Delta\omega$(eV)		10^{-12}				10^{-6}
τ (μsec)		$4x10^{3}$				$4x10^{-3}$

[a]For light scattering, we have taken $5.00x10^{-2}$ rad or 2.86° as the lowest accessible scattering angle, while ordinary light scattering spectrometers can usually reach $\theta > \sim 15°$. The maximum scattering angle is assumed to be π rad or 180°, while in practice $\theta < 180°$.

[b]For SAXS, a minimum scattering angle of $5x10^{-4}$ rad is assumed. A value of 0.5 mrad can be achieved using a Kratky block collimation system under favorable conditions or a Bonse-Hart camera. By comparing K_{max} from LLS and K_{min} from SAXS, we see that the scattering curve can be made to overlap experimentally using the two techniques. The high scattering-angle limit for SAXS overlaps with x-ray diffraction for atomic and molecular structure studies.

[c]For SANS, the neutron wavelengths, which are longer than those of x-rays, permit easier study of larger structures. Thus, SANS covers comparable ranges to SAXS, and the scattering curves from LLS and SANS can also be made to overlap experimentally.

The polymer blend, poly(vinyl methyl ether)/deuteropolystyrene (PVME/d-PS),[4] has been investigated by neutron scattering and has shown a transition from mean field to Ising behavior at ~2.4°C below the lower critical solution temperature.[5,6] Bates, et al[7] and Stepanek, et al[8] reported a crossover behavior using a relatively low molecular weight polyisoprene and partially deuterium labeled poly(ethylene-propylene) (PI/partially-d-PEP) blend over a temperature range close to 30 °C above the upper critical solution temperature. From the above two experiments, one can reasonably conclude that the crossover from mean field to Ising critical behavior indeed exists and that the Ising range can be controlled by polymer molecular weight.

Synchrotron small-angle x-ray scattering (SAXS) of poly(2-chlorostyrene)/polystyrene (P2CℓS/PS)blends in the immediate neighborhood of the critical mixing point can be achieved by first locating the critical mixing point precisely.[9] By combining a phase volume method with SAXS, the critical mixing point for P2CℓS (M_w = 2.33x10^5, M_w/M_n = 1.19)/PS (M_w = 3.7x10^4 , M_w/M_n < 1.06) in the presence of 22.6 wt % di-n-butyl phthalate was found to have a critical weight fraction of PS, $W_{PS,c}$ = 0.52, as shown in Fig. 1. A lower critical mixing temperature T_c = 134.30°C,where W_{PS} + $W_{P2CℓS}$ was set to equal one, could clearly reveal a crossover behavior in the static susceptibility and the correlation length from mean field to the Ising-like behavior, with critical exponents γ changing from 1.00 (mean field) to 1.23 (Ising) and ν changing from 0.50 (mean field) to 0.63 (Ising). The crossover occurs at 2.6 K below the critical mixing temperature, as shown in Fig. 2.

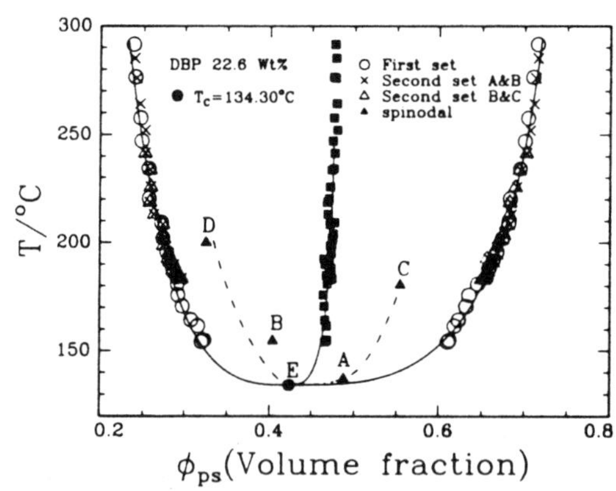

Figure 1. Coexistence curve of P2ClS/PS blends in the presence of 22.6 wt% DBP (M_w(PS) = 3.7x10^4, M_w/M_n < 1.06; M_w (P2CℓS) = 2.33x10^5 , M_w/M_n = 1.19). The coexistence curve was determined by a combination of phase volume ratio and SAXS experiments with T_c (at E) = 134.30°C and the critical exponent $\beta \simeq 0.33$ for the coexistence curve. Circles, crosses and triangles denote the sets of data determined with different initial concentrations. Filled rectangles denote the midpoints of the coexistence curve and denote the arithmetic mean of the fitting curve of the coexistence curve. (Ref. 2)

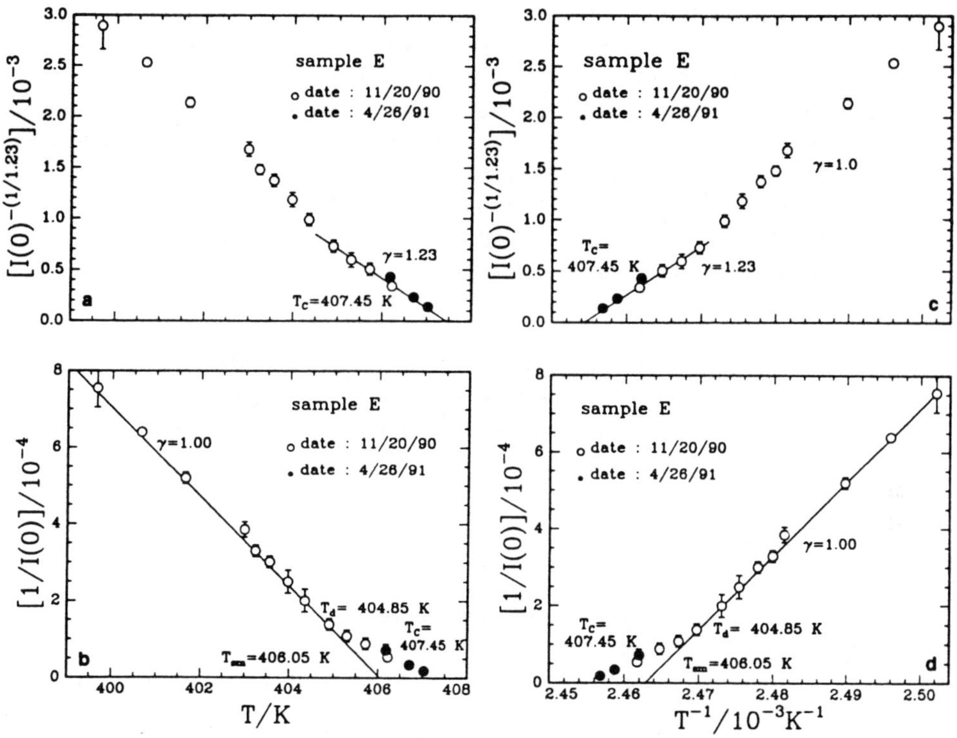

Figure 2. Plots of $I^{-1/\gamma}(0)$ versus temperature T: (a) γ = 1.23; (b) γ = 1.00. Plots of $I^{-1/\gamma}(0)$ versus 1/T: (c) γ = 1.23; (d) γ = 1.00. The plots are used to emphasize the crossover from mean field to Ising behavior in the relative static susceptibility for the polymer blend at the critical solution concentration (sample E). T_c = 407.45 K = 134.30°C.The crossover point is expressed as T_d = 404.85 K. The dates show two separate experiments with independent samples. (Ref. 2)

3. Structure of Ionomers

The morphology of ionomers has been studied extensively by many workers,[10] primarily using SAXS[11-20] and small angle neutron scattering (SANS).[20-23] In the past few years, Cooper and his co-workers made direct observation of ionic aggregates in sulfonated polystyrene ionomers[24] and observed an aggregate size of approximately 3 nm in Ni^{2+} and Zn^{2+} neutralized sulfonated polystyrenes (SPS). Register et al. used extended x-ray absorption fine structure (EXAFS) spectroscopy and found the local coordination structure about the Mn^{2+} cation to be unaltered by ionic aggregation.[25] More importantly, Ding et al.[26a] and Register and Cooper[26b] used anomalous small angle x-ray scattering (ASAXS) to study nickel-neutralized ionomers. The SAXS profiles were generally acquired at 50 and 5 eV below the

Ni^{2+} K-edge. The difference SAXS profile formed from two different x-ray energies was superimposable with the single energy SAXS pattern after appropriate scaling. Their observations proved that both the ionomer peak and the upturn near zero angle were due to an inhomogeneous distribution of ionic repeat units throughout the material and not to precipitated neutralizing agent or microvoids. Register and Cooper extended the ASAXS studies to a nickel-neutralized poly(ethylene-co-methacrylic acid) ionomer,[27] which was consistent with a three-phase morphological model, incorporating lamellar crystallites, interlamellar amorphous polymeric material, and ionic aggregates residing within the amorphous layers.

Morphological studies of sulfonated polystyrenes by using SAXS and ASAXS showed the presence of local clustering, as well as long-range inhomogeneities of ions.[28] The observed morphology could be attributed only to the metal ionic structures, as can be proven by means of ASAXS. However, with the SAXS patterns by SPS being much stronger than those of polystyrene or SPS in acid form, the excess SAXS patterns of the metal ions can be measured with greater precision by SAXS than by ASAXS. Figure 3 shows a comparison of the difference SAXS profile with the ASAXS results.

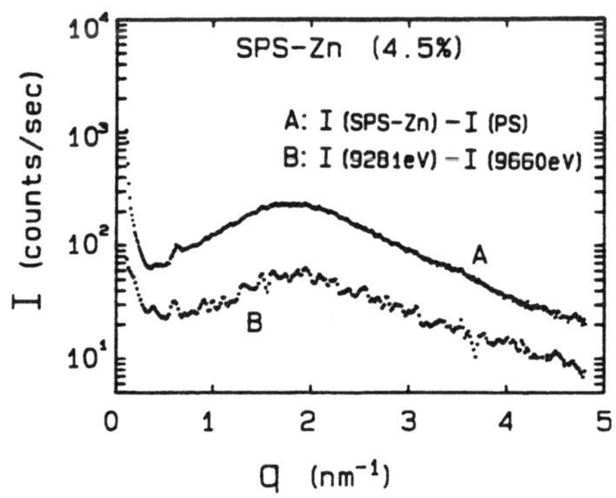

Figure 3. Comparison of difference SAXS profiles from [I(SPS-Zn) - I(PS)], curve A in Fig. 3, and [I(9281 eV) - I(9660 eV)], curve B in Fig. 3. I(PS) denotes the SAXS profile of neutral polystyrene. (Ref. 28)

The structure of ionic aggregates and the spatial arrangement of ions can best be elucidated by the correlation function analysis of precise SAXS profiles which have a wide scattering angular range and a high spatial resolution.[29] Electron density-density correlation functions of sulfonated polystyrene (SPS) sodium and zinc salts (4.5 mol% of sulfur, 100% neutralization, and M_w/M_n = 1.2) have been computed from the SAXS profiles. The results showed that the short-range strucure could be described by the liquid-like model, while the long-range structure could be fitted approximately (and empirically) by invoking a Debye-Bueche type random inhomogeneity model.

In order to study the long range inhomogeneiteis of the ionic aggregates, two Bonse-Hart ultra-small-angle x-ray scattering (USAXS) cameras were constructed and calibrated by using a polystyrene latex suspension and laser light scattering. The instruments were used to investigate the long-range inhomogeneities in a series of sulfonated polystyrene ionomers. The USAXS measurements were mainly centered in the region of $q < 0.1$ nm^{-1} (where q is the magnitude of the scattering vector), which is beyond the reach of most conventional small-angle x-ray scattering (SAXS) instruments.[30] All the scattering profiles of ionomers showed a small-angle upturn which was much higher than their parent pure polystyrene. The scattering behavior in this q region was not strongly dependent upon the counterions (except for the intensity change in amplitude), the compression-molding condition, temperature, and annealing. The small-angle upturn decreased with increasing ionic content for zinc sulfonated polystyrene ionomers, and increased with increasing ionic content for rubidium sulfonated polystyrene ionomers. By means of the Debye-Bueche model, an inhomogeneity length of around tens of nanometers was detected in the usual q range accessible by conventional SAXS (0.045 to 10 nm^{-1}). However, the USAXS curves based on the Debye-Bueche model showed a consistent downward curvature in the light-scattering q range, as shown in Fig. 4. A Guinier plot also failed to show

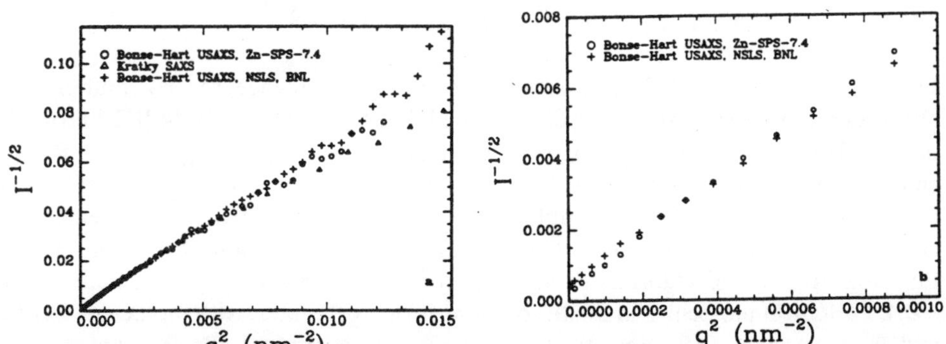

Figure 4. Debye-Bueche plot of the desmeared USAXS profile (a) and the Kratky SAXS profile (b) of the ionomer sample used in the previous Communication.[31] (Ref. 30)

a simple characteristic length. Surprisingly, the upturn could better be described by a power law $I \sim q^{-\alpha}$, with α ranging from 2.4 to 3.6 for the ionomers in this study, suggesting that, in the absence of long-range order, the structure could be very polydisperse and irregular. Figure 5 shows typical Zn-SPS samples with fast and slow cooling. The majority of the upturn could result from chemical compositional inhomogeneities of the ionomers and the processing procedure used in forming the ionomer films for various physical measurements such as SAXS.

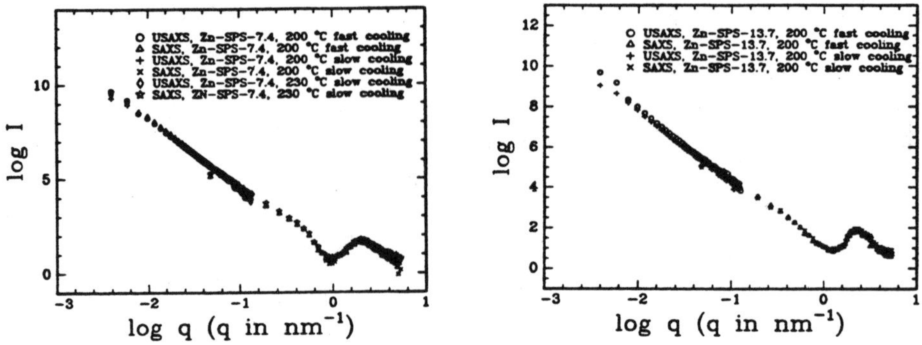

Figure 5. USAXS and Kratky SAXS profiles of ionomer samples. The temperature following the sample names in each part is the molding temperature. The time for fast cooling is ~3 min; the time for slow cooling is ~200 min. (Ref. 30)

4. Conformation of Polymers and Supramolecular Structures in Solution

4.1. POLYDIACETYLENE IN DILUTE SOLUTION[32]

Laser light scattering (LLS), transient electric birefringence (TEB), and optical absorption were used to study polydiacetylene P4BCMU (poly[1,2-bis[4-[[[2-butoxy-2-oxoethyl)amino]carbonyl]oxyl]-butyl]-1-buten-3-yne-1,4-diyl]) in dilute chloroform/toluene solutions with various solvent compositions. Two different molecular weight P4BCMU samples were investigated. P4BCMU molecules in chloroform are wormlike coils with an average persistence length of ~16 nm. When the mole fraction of chloroform in a chloroform/toluene mixture is smaller than ~0.4, P4BCMU molecules in dilute solutions form aggregates, with molecular weight and size varying according to the solvent composition. Profile analysis of autocorrelation functions yields a broad size distribution when P4BCMU is dissolved in chloroform. However, much narrower size distributions are found once the aggregates are formed. Various models are applied to fit the experimental results, yielding mostly fairly stiff rodlike structures for the aggregates.

By combining the SAXS measurements with our light-scattering results, we are able to distinguish the form factors of various models over a scattering vector range of 5×10^{-2} nm^{-1} < q < 1.5×10^{-1} nm^{-1}. Only an arrangement of 1 row x 14 columns could yield the best fitting, i.e., the particle configuration is ribbonlike as shown by the solid line in Figure 6.

Further LLS and SAXS studies on the structure of aggregates of P4BCMU in dilute THF/toluene solutions[34] showed that the structure of P4BCMU aggregates can best be described by ribbonlike unsymmetrical elliptic cylinders, in agreement with the conclusion from the chloroform/toluene solution.

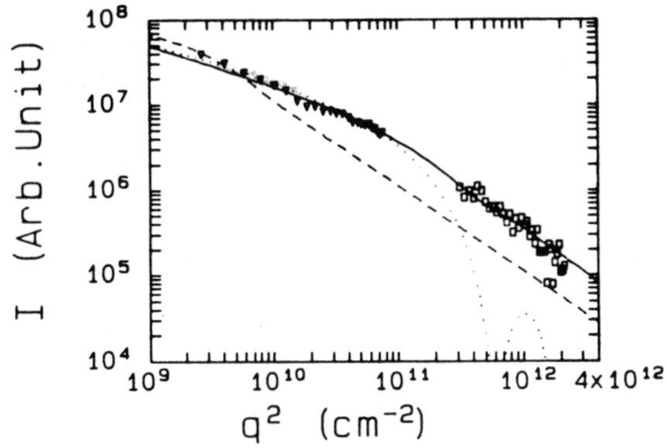

Figure 6. Excess scattered intensity from laser light scattering (hollow inverted triangles measured at λ_o = 632.8 nm, C ~ 1 x 10^{-6} g/g) and SAXS (hollow squares, measured at λ = 0.154 nm, C ~ 6 x 10^{-6} g/g) for P4BCMU in dilute toluene solution at room temperatures (~24°C)(Ref. 33). The three curves represent the theoretical simulations according to various models. Solid line: 14 parallel rods in a 1 row x 14 columns configuration with the effective diameter of each individual rod being 4 nm and the effective center-to-center distance being 7 nm. Dashed line: an infinitely thin circular disk. Dotted line: a single rod, with the diameter D being 100 nm. The lengths (diameters) of the rod (disk) aggregates were determined by using light-scattering and TEB results based on a rigid-rod model.[32]

4.2. SUPRAMOLECULAR STRUCTURES OF BLOCK COPOLYMERS IN SOLUTION[35]

Synchrotron small angle x-ray scattering has been performed on solutions of a poly(styrene-isoprene) AB block (PS/PIP) copolymer in aniline. The copolymer forms aggregates in this particular solvent. Earlier investigations[36] could not determine whether the aggregates were micelle-like or vesicle-like. In order to distinguish between these two possibilities, the form factor has to be known over a wide range of the scattering vector (q) to within a satisfactory precision. Due to the high intensity of synchrotron radiation, it was possible to obtain SAXS results with a sufficient precision, which allowed the conclusion that the aggregates were *vesicle-like* when the PS block was sufficiently short. The effect of polydispersity on the form factor has also been taken into account. Figure 7 shows plots of log(P(q)) versus q.

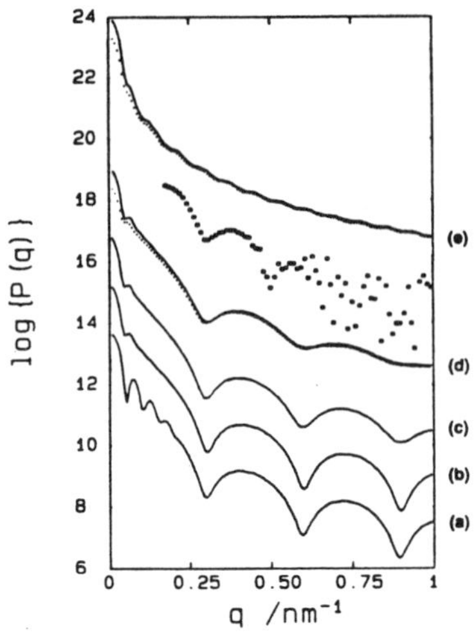

Figure 7. Logarithm of the form factor (multiplied by an arbitrary number) as a function of scattering vector. Curves (a) to (d) are form factors of hollow spheres with a mean outer radius (R_o) of 69 nm and a mean shell thickness (d) of 21 nm. Curve (a), $\sigma(R_o)$ = 5 nm; σ(d) = 0; curve (b), $\sigma(R_o)$ = 15 nm; σ(d) = 0; curve (c), $\sigma(R_o)$ = 15 nm; σ(d) = 1 nm; curve (d) = 2 nm. The hollow squares are the experimental data and curve (e) is the form factor of a solid sphere with a mean radius (R) of 77.5 nm and σ(R) = 15 nm. The dotted lines demonstrate the effect of smearing due to the finite slit height. (Ref. 35)

V. Epoxy Polymerization Process[37]

Laser light scattering and small angle x-ray scattering studies have been made of the curing of epoxy resins from 1,4-butanediol diglycidyl ether with cis-1,2-cyclohexane dicarboxylic anhydride. The epoxy resin before its gel point is soluble in methyl ethyl ketone, and the scattering techniques can be used to determine the molecular weight, the fractal dimension, and the molecular weight distribution of the branched epoxy polymer during each stage of the initial polymerization process. After gelation, SAXS remains a useful technique for studying the branched polymer structure in terms of the concept of fractal geometry for random systems.

The curing of epoxy resins from 1,4-butanediol diglycidyl ether (DGEB) with the curing agent cis-1,2-cyclohexanedicarboxylic anhydride (CH) in the presence of a small amount of catalyst benzyldimethylamine (CA) represents a cross-linking polymerization process, as shown schematically in Figure 8. The cross-linking reaction involves roughly n mol of DGEB with 2n mol of CH. The chemical reaction is known to cluster at catalytic centers, and the formation of branched structures eventually link together to form loops or polymer networks.

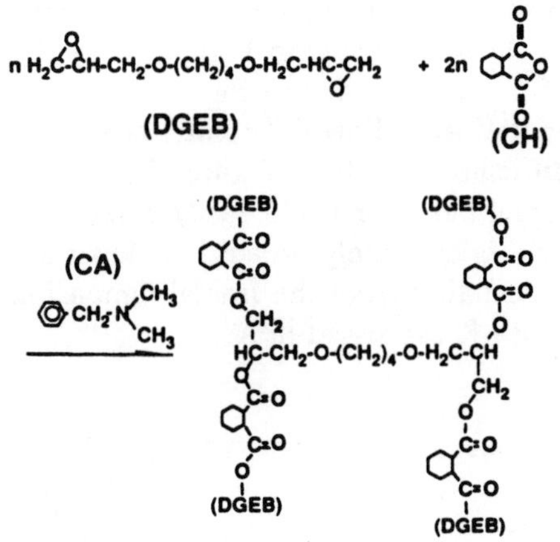

Figure 8. Cross-linking reaction of DGEB with CH in the presence of CA. (Ref. 37)

Figure 9 shows static structure factors S(K) [= I/CH'] from SAXS as a function of percent CH conversion. In plotting the scattering curves, we have scaled

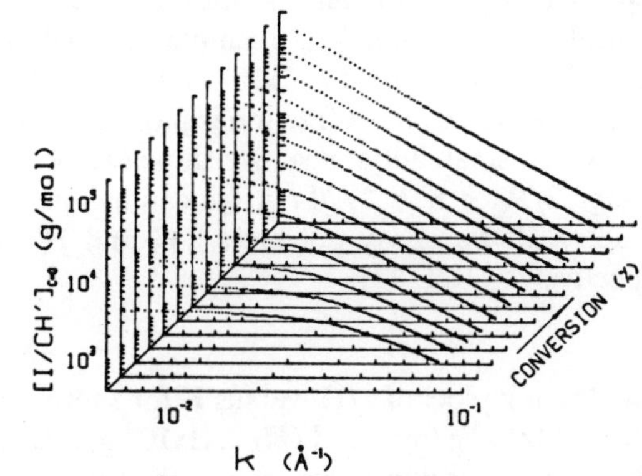

Figure 9. Structure of epoxy polymers as a function of percent CH conversion. $[I/CH']_{C=0}$ represents absolute SAXS intensity at infinite dilution in units of g/mol. Thus, $\lim_{K \to 0} [I/CH']_{C=0} = M_w$. The scattering curves are numbered with increasing percent CH conversion. Properties of the 13 samples representing epoxy polymers during different stages of the curing process (see Table 1 in reference 37).

the intensities. SAXS measures mainly the fractal geometry of the branched epoxy polymers in solution, as shown typically in Figure 10 by the hollow diamonds for sample 13 with a M_w of 5 x 10^5 g/mol. In Figure 10, we have also included LLS measurements (also denoted by hollow diamonds) at much smaller values of K(< $4x10^{-3}$ Å^{-1}). It may be difficult to see from Figure 10 that the initial slope from LLS exhibiting almost horizontal behavior could easily determine the R_g value. As the K range covered by SAXS is extremely broad, i.e., down to monomer dimensions, we would also expect a deviation from the fractal dimension beyond $2/R_g < K <$ 1/20 Å^{-1}, i.e., 20K < 1 with K expressed in Å^{-1}.

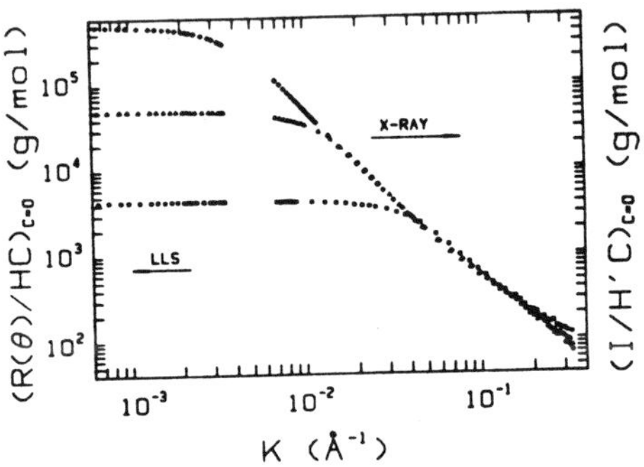

Figure 10. Log-log plots of R/HC for light scattering and I/H'C for SAXS as a function of K: (□) sample 1, (△) sample 7, (◇) sample 13. Light scattering K range: ~ $7x10^{-4}$ < K < $4x10^{-3}$ Å^{-1}. SAXS K range: $7x10^{-3}$ < K < $3.5x10^{-1}$ Å^{-1}. Each SAXS curve has 817 data points, of which only a fraction is plotted. The horizontal portion of the scattering curve with a small initial negative slope can be related to the radius of gyration. In the range $2/R_g < K < 1/20$ Å^{-1}, we can use the scattering curves to examine fractal geometry of the epoxy polymers. H and H' are light and x-ray optical constant, respectively. (Ref. 37)

Figure 11a shows log-log plots of I/H' versus K for sample 10 in MEK at C = (♦) $7.19x10^{-4}$ g/mL, (△) $1.14x10^{-3}$ g/mL, and (■) $2.41x10^{-3}$ g/mL. As we have a total of 817 data points for each scattering curve, only a fraction of the data is plotted. It should be noted that the scattering curves are slightly curved. Figure 11b shows a log-log plot of scaled scattered intensities of the three scattering curves of Figure 11a. Overlapping of the three scattering curves with zero adjustable parameters is clearly demonstrated.

By combining LLS[37-39] with SAXS,[37] we can study the branching kinetics of epoxy polymerization process.[40]

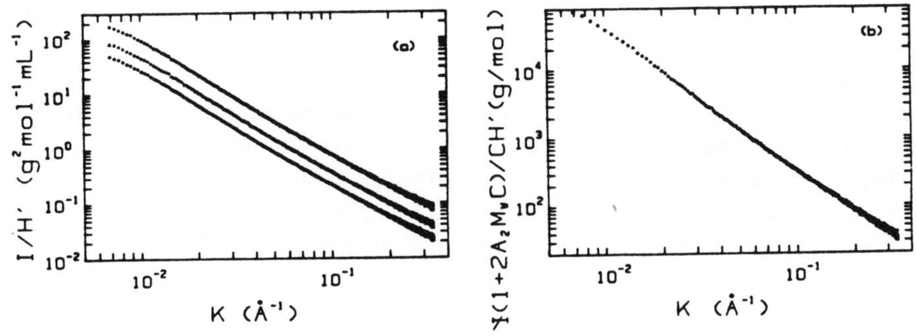

Figure 11. Concentration effect on fractal dimension of epoxy polymers (a) log-log plots of I/H' versus K for sample 10 in MEK: (♦) C = 7.19×10^{-4} g/mL; (▲) C = 1.14×10^{3} g mL; (■) C = 2.41×10^{-3} g/mL. There are 817 data points on each scattering curve. Only a fraction of data points is plotted. (b) log-log plot of a scaled intensity curve based on the scattering curves from three different concentrations as shown in Figure 11a. Same symbols as in Figure 11a with M_w = 1.42×10^{5} g/mol and A_2 = 3.63×10^{-4} mL mol g^{-2}. Again only a small fraction of data points is plotted. (Ref. 37)

6. Time-Resolved SAXS

6.1. LDPE/HDPE BLEND

Small-angle x-ray scattering has been used successfully to study the structure of polymer blends. A natural extension is to use synchrotron radiation for measurements of time-resolved structural changes during early stages of crystallization.[41-43] Time-resolved SAXS profiles of a blend of low-density polyethylene (LDPE) and high-density polyethylene (HDPE) and of the respective homopolymers have been carried out.[44,45]

The HDPE/LDPE blend, which was cooled slowly at a cooling rate of 0.3 °C/min, showed an interfibrillar scale separation within the spherulites that contained both the HDPE and the LDPE component. On the other hand, the same polymer blend, which was cooled rapidly at a colling rate of 110°C/min, appeared to show separation of HDPE and LDPE on an interlamellar scale. When the blend was quickly cooled to two successive temperatures, the scattering profiles were similar in nature to those obtained from the same sample that was directly quenched. The two-step rapid cooling represented quenching (1) from ~125 °C, which was above the HDPE crystallization temperature but higher than the LDPE crystalization temperature, and (2) from 110 to 100 °C, which was below the LDPE crystallization temperature, while the direct rapid cooling was from 135 to 65 °C.

Figure 12 shows a schematic diagram of thermal block for temperature-jump measurements.[46]

Figure 13a shows selected Lorentz-corrected SAXS intensities measured during the first step of the crystallization at 110 °C. The curves show only a single peak, which grows in the usual manner. After the temperature was lowered to 100 °C, the LDPE began to crystallize. The resulting scattering profiles are shown typically in Fig. 13b. The shift in peak position from small to larger scattering angles suggests that LDPE is being crystallized between the previously formed HDPE lamellae. A

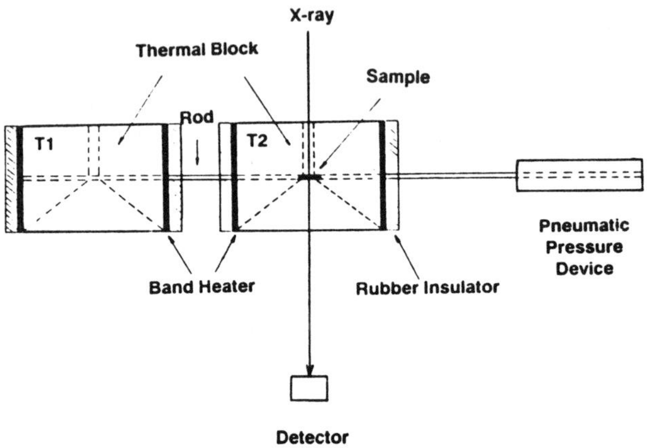

Figure 12. Schematic diagram of thermal block for temperature-jump measurements. Ref. 46)

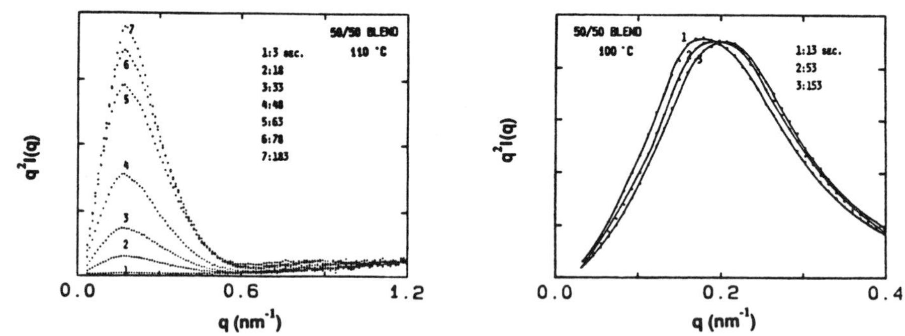

Figure 13. (a) Lorentz-corrected intensity of the 50/50 HDPE/LDPE blend measured during the isothermal crystallization at 110°C. Each SAXS curve represents a 5-s collection time. (b) Lorentz-corrected intensity of 50/50 HDPE/LDPE blend measured during the isothermal crystallization at 100 °C, after the blend has been crystallized at 110 °C for 45 min. Each SAXS curve represents a 5-s collection time. (Ref. 45)

similar shift was observed for the same sample that was rapidly cooled from ~135 to 65 °C. Thus, the same conclusion can be drawn for a rapid two-step crystallization as for the rapid one-step crystallization; i.e., during rapid quenching lamellae of HDPE and LDPE are mixed together, while there is evidence for the formation of bundles of HDPE lamellae and LDPE lamellae in the slow cooling process (0.3 °C/min).

6.2. CRYSTALLIZATION KINETICS OF PEEK[46]

Time-resolved synchrotron small angle x-ray scattering was employed to investigate crystallization behaviors at elevated temperatures and fast isothermal crystallization kinetics around ~230°Cfor poly(aryl ether ether ketones)(PEEKs). The crystalline structure parameters were evaluated by means of one-dimensional correlation function analysis. The long period, the lamellar thickness and the one-dimensional crystallinity as a function of temperature were compared between two samples which had different repeat units in the polymer backbone. For the two samples, the long period and the lamellar thickness were found to increase with increasing crystallization temperature simultaneously. For the crystallization kinetics study, the initial non-isothermal condition due to the temperature retardation in samples was corrected by using a multi-step T-jump process. A single exponential behavior of the invariant change during isothermal crystallization was observed and interpreted by the diffusion-controlled mechanism. The activation energies of the segmental diffusion were estimated from the data under the non-isothermal condition. The crystallization rates were compared between the two samples.

The PEEKs used in our study have the following structural formulas:

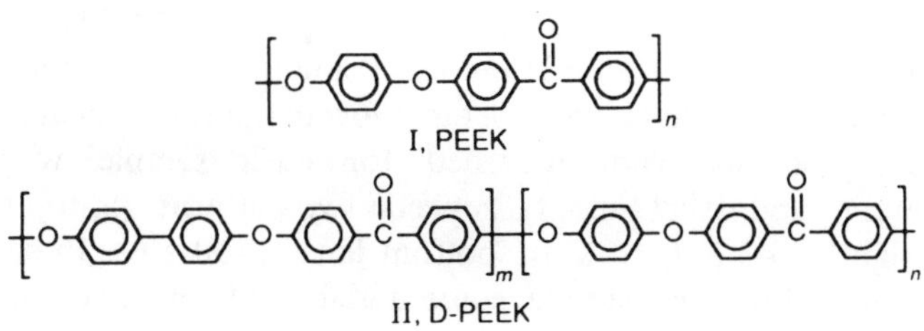

I, PEEK

II, D-PEEK

The samples of structure I and of structure II are denoted as, respectively, PEEK and D-PEEK. Figure 14 illustrates a three-dimensional SAXS pattern following the crystallization kinetics of PEEK and D-PEEK. All the SAXS patterns presented here were smoothed by using a points-add-up method and the cubic spline smoothing technique. In the initial stage in each set of patterns, one can see only the exponential-like decay curves without a notable peak associated with the periodic electron density fluctuations. With a further increase in time, a maximum at q~0.4 nm^{-1} appears and becomes more pronounced and finally pertains no remarkable change.

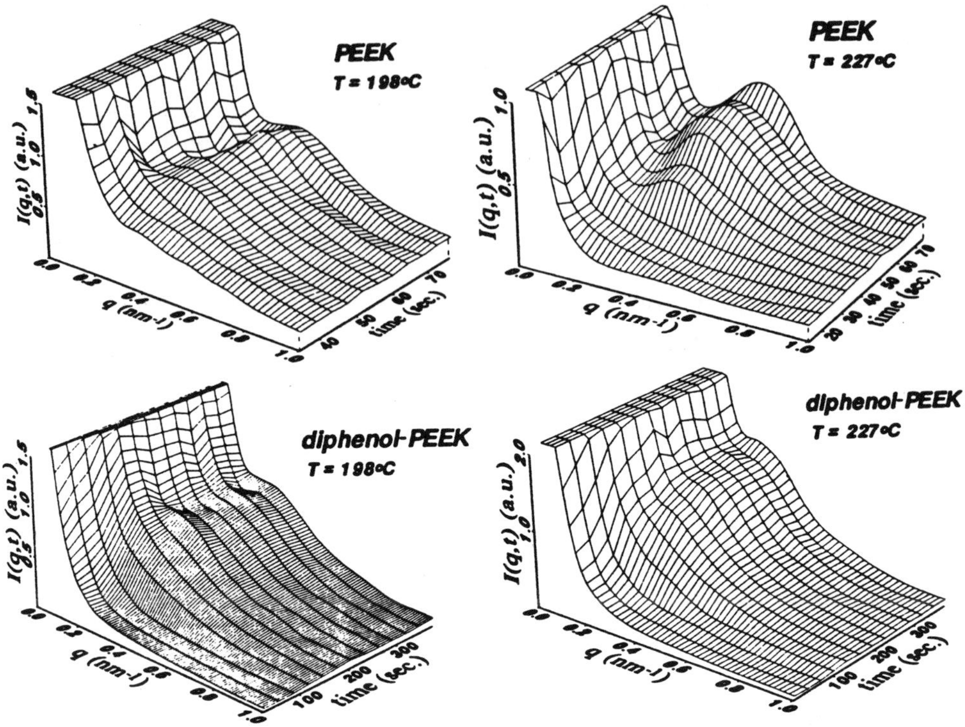

Figure 14. Time evolution of SAXS patterns after a temperature-jump from 50 °C to higher temperatures, which are indicated on the plot, for PEEK and D-PEEK from the amorphous state. The elapsed time was counted right after the sample was jumped into a high temperature chamber from a low temperature chamber. Fifty or more consecutive patterns were recorded for each sample with different accumulation time, t_a, and delay time, t_d, between two patterns: (a, top left) $t_a=3$ s, $t_d=0$ s; (b, top right) $t_a=3$ s, $t_d=0$ s; (c, bottom left) $t_a=10$ s $t_d=15$ s; (d, bottom right) $t_a=10$ s, $t_d=10$ s. For each sample, some initial and later stage patterns which display no significant change are not presented here for a better view. (Ref. 46)

6.3. PHASE-SEPARATION KINETICS IN A SEGMENTED POLYURETHANE[47,48]

Extensive SAXS studies have been carried out on segmented polyurethanes, including effects of temperature and annealing,[49] and of hard-segment flexibility[50], as well as the relation of multiphase structure with spherulite formation.[51]

SAXS was used to investigate the kinetics of phase separation of a segmented polyurethane which was quenched from the homogeneous melt state to lower

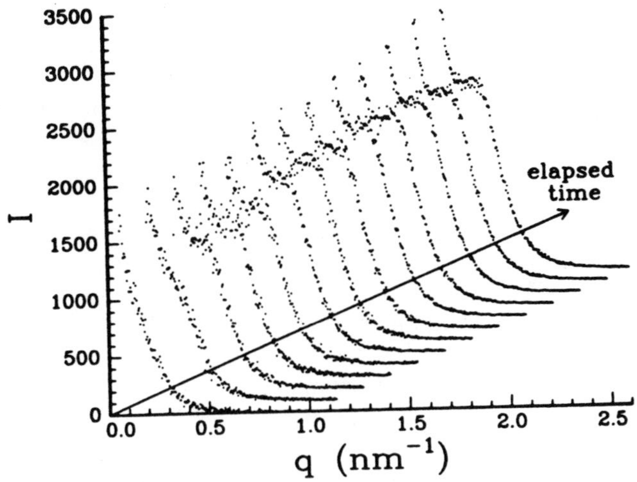

Figure 15. Time dependence of SAXS profiles by quenching the homogeneous melt of PTMO-2000-PU-39 from 220 to 131°C. The scattered intensity showed a monotomic increase with increasing elapsed time. The elapsed time was counted immediately after the sample was jumped from the melting cell (220°C) to the measuring cell (131°C). From left, the curves represent, respectively, the SAXS profiles at the elapsed time of 10, 35, 56, 79, 103, 145, 209, 370, 516, 816, 1151, 2310 and 3522 s. After several preliminary tests, we found that phase separation did not proceed at the same rate during the whole process. Therefore, we chose to use varying exposure times in order to get the best signal-to-noise ratio. At the initial stage of the phase separation, the exposure time was 20 s to accommodate the rapid phase separation process. The exposure time was increased (up to 300 s) as the phase separation slowed down considerably at later stages. The elapsed time indicated above represents the time period between jumping the sample to the lower annealing temperature and the midpoint time of the data collection period in the SAXS profile. (Ref. 48)

temperatures. The phase-separation process in the segmented polyurethane was very slow at most of the annealing temperatures, owing to the high viscosity of the system and low mobility of the hard segment. The phase-separation rate depended strongly on the annealing temperature, with the maximum rate being at an annealing temperature of around 80-107°C. Further increase in the annealing temperature reduced the rate. The rate decrease could be attributed to an increase in the hard- and soft-segment compatibility at higher temperatures. The interdomain spacing did not change throughout the whole annealing process, a characteristic of segmented copolymers. However, the interdomain spacing increased with increasing annealing temperature. Both the phase-separation process and the peak intensity changes could be analyzed according to an equation of relaxation, which indicated that the pahse separation in the segmented polyurethane system behaved like a relaxation process.

SAXS was used to investigate the microphase structure and microphase separation kinetics of two segmented polyurethanes with 4,4'-diphenylmethyl diisocyanate (MDI) and 1,4'-butanediol (BD) as the hard segment and poly(tetramethylene oxide)(PTMO) and poly(propylene oxide) end-capped with poly(ethylene oxide)(PPO-PEO)(M_n ~ 2000) as the soft segments.

7. Anisotropic Scattering.[52,53]

Fibers of poly(p-phenylene terephthalamide) (PPTA) show high tensile strength, high Young's modulus, high thermal stability, and low creep, due to the fully extended macromolecular chains. Unlike the usual crystalline polymers, the chains are highly extended with poor lateral order, in the so-called paracrystalline structure even in the noncrystalline zone. Wide-angle x-ray diffraction (WAXD) and small-angle x-ray scattering (SAXS) have been used to study the lattice structure and the relation between the lattice structure, the morphology, and the macroscopic properties, with or without stress.[54-57] However, these studies employed fiber bundles exclusively. Only a very small angular change in the x-ray diffraction pattern could be observed because PPTA fibers attain very limited deformation before they are broken. Therefore, the stress distribution in each filament and the slight misalignment of the filaments could easily smear the results of WAXD.

A schematic diagram of the geometrical arrangement for the x-ray setup is shown in Figure 16, which also depicts an actual 2-D x-ray diffraction pattern of a bundle of ~25 filaments using the Siemens/Nicolet area detector. The shades show different intensity zones, with the darkest shade having the highest intensity.

Figure 17 shows equatorial x-ray diffraction patterns at fixed meridional angles computed from the Siemens/Nicolet 2-D area x-ray detector. To insure accurate measurement of intensity and positonal information, the area detector was calibrated on site using both ^{55}Fe flood-field source and the 0.154 nm x-rays from the beamline. The flood-field correction uses the uniform intensity distribution of the ^{55}Fe source to create a set of intensity and positional mapping arrays. These arrays, when loaded into the detector's computer memory, correct the incoming data for any intensity or positional anisotropy that may occur due to wire irregularity or preamplifier gains. The detector electronics were also optimized for the 0.154 nm synchrotron beam energy by adjusting the detector anode bias so as to insure maximum electron detection from the gas mixture. For the 2-D x-ray diffraction pattern, we used an incident x-ray beam cross section of 0.5 x 0.5 mm^2. Each curve was integrated over 1.5° in the meridional direction. The measurement for the entire 2-D x-ray pattern took 300 s. The diffraction data as denoted by the dots were noisier. The solid curves were constructed by using 3-point smoothing of data 10 times. From Fig. 18, we see the 110 and 200 reflections at 2θ = 20.50° and 22.57°, respectively, as well as the 211 double "spots" at 29.29°, with θ being the Bragg angle. If a bundle of 25 filaments can be measured over a 300 s time period with 1/10 the optimal x-ray power density, a single PPTA filament of 28 μm diameter should take ~1000 s, which is easily within reach in our experimental setup. The spread of the 110 and

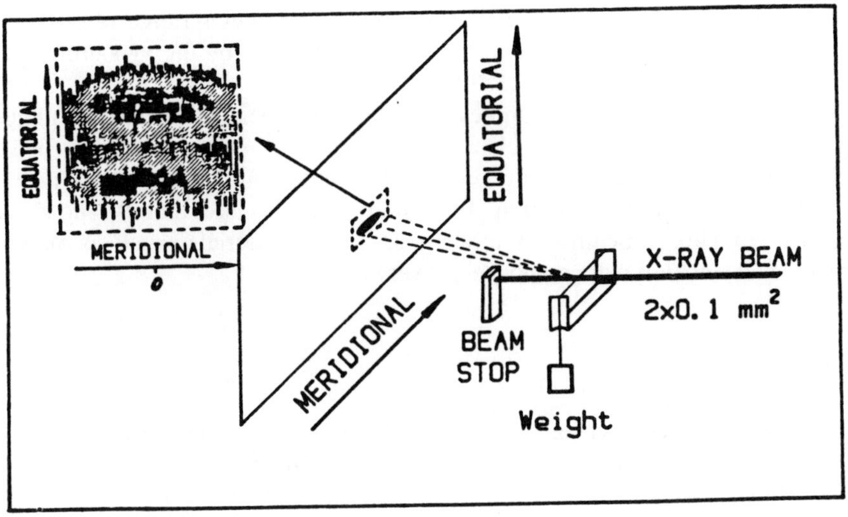

Figure 16. Schematic diagram of the geometrical arrangement for the x-ray setup. The diffraction pattern inside the dashed lines on the upper left corner was measured on a bundle of ~25 filaments by using a Siemens/Nicolet 2-D position-sensitive area detector over a 300 s time period. (Ref. 52)

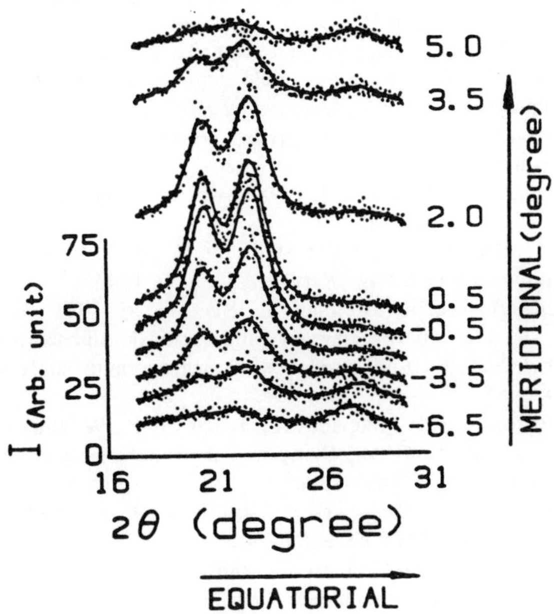

Figure 17. Equatorial x-ray diffraction patterns of a bundle of ~25 PPTA filaments at fixed meridional angles computed from the Siemens/Nicolet 2-D area x-ray detector. Each curve was integrated over 1.5° in the meridional direction., Dots denote real experimental data, and the solid lines are constructed by smoothing the actual data 10 times.

200 reflections in the meridional direction covers an angular range of ~10°. The PPTA fibers have not shown noticeable structure damage by the synchrotron x-ray beam after short exposures of tens of minutes.

I have shown you some examples of polymer physics which can be studied by using synchrotron x-rays. It is clear that synchrotron x-rays is a very powerful technique for structural and kinetic studies. By combining synchrotron x-rays with LLS and SANS (including neutron diffraction), one can see that the potential to investigate complex structures in modern advanced materials, such as polymers, polymer blends and polymer composites, is very promising.

Acknowledgement

BC gratefully acknowledges support of this work by the Department of Energy (DEFG0286ER45237), the Polymers Program, National Science Foundation (DMR9301294), and the U.S. Army Research Office (DAAH0494G0053).

References (Numbers in parentheses denote key references on the topics discussed.)

1. Chu, B. (1991) *Laser Light Scattering, Basic Principles and Practice*, 2nd Edition, Academic Press, New York; p.10.
(2.) Chu, B., Ying, Q., Linliu, K., Xie, P., Gao, T., Nose, T., and Okada, M., (1992) Synchrotron SAXS study of mean field and ising critical behavior of poly(2-chlorostyrene)/polystyrene blends, *Macromolecules* **25**, 7382-7388.
3. deGennes, P.-G. (1979) *Scaling Concepts in Polymer Physics*; Cornell University Press, Ithaca,NY; Chapter 4.
4. Herkt-Maetzky, C., and Schelten, J. (1983) Critical fluctuations in a binary polymer mixture, *Phys. Rev. Lett.* **51**, 896-899.
5. Schwahn, D., Mortensen, K., and Yee-Maderira, H. (1987) Mean-field and ising critical behavior of a polymer blend, *Phys. Rev. Lett.* **58**, 1544-1546.
6. Schwahn, D., Mortensen, K., Springer, T., Yee-Maderira, H., and Thomas, R. (1987) Investigation of the phase diagram and critical fluctuations of the system polyvenylmethylether and *d*-polystyrene with neutron small angle scattering, *J. Chem. Phys.* **87**, 6078-6087.
7. Bates, F. S., Rosedale, J. H., Stepanek, P., Lodge, T. P., Wiltzius, P., Frederickson, G. H., and Hjelm Jr., R. P. (1990) Static and dynamic crossover in a critical polymer mixture, *Phys. Rev. Lett* **65**, 1893-1896.
8. Stepanek, P., Lodge, T. P., Kedrowski, C., and Bates, F. S. (1991) Critical dynamics of polymer blends, *J. Chem. Phys.* **94**, 8289-8301.
(9.) Chu, B., Linliu, L., Ying, Q.-C., Nose, T., and Okada, M. (1992) Coexistence curve of polystyrene/poly(2-chlorostyrene) blends, *Phys. Rev. Lett.* **68**, 3184-3187.
10. See, for examples, Eisenberg, A., King, M. (1977) *Ion-Containing Polymers*, Academic Press, New York. MacKnight, W. J., Earnest Jr., R. R. (1981) Structure and properties of ionomers, *J. Polym. Sci., Macromol. Rev.* **16**, 41-122.
11. MacKnight, W. J., Taggart, W. P., and Stein, R. S. (1974) A model for the structure of ionomers, *J. Polym. Sci., Polym. Symp.* **45**, 113-128.

12. Roche, E. J., Stein, R. S., Russell, T. P., MacKnight, W. J. (1980) Small-angle x-ray scattering study of ionomer deformation, *J. Polym. Sci., Polym. Phys. Edn.* **18**, 1497-1512.
13. Gierke, T. D., Munn, G. E., and Wilson, F. C. (1981) The morphology in Nafion perfluorinated membrane products as determined by wide- and small-angle x-ray studies, *J. Polym. Sci., Polym. Phys. Edn.* **19**, 1687-1704.
14. Yarusso, D. J., and Cooper, S. L.: (1983) Microstructure of ionomers: interpretation of small-angle x-ray scattering data, *Macromolecules* **16**, 1871-1880; (1985) Analysis of SAXS data from ionomer systems, *Polymer*, **26**, 371-378.
15. Weiss, R. A., and Lefelar, J. (1986) The influence of thermal history on the small-angle x-ray scattering of sulphonated polystyrene ionomers, *Polymer* **27**, 3-10.
16. Fitzgerald, J. J., Kim, D., and Weiss, R. A. (1986) The effect of diluents on the ionic interactions in sulphonated polystyrene ionomers, *J. Polym. Sci., Polym. Lett. Edn.* **24**, 263-268.
17. Gebel, G., Aldeberg, P., and Pineri, M., (1987) Structure and related properties of solution-cast perfluorosulfonated ionomer films, *Macromolecules* **20**, 1425-1428.
18. Fujimura, M., Hashimoto, T., and Kawai, H., (1981) Small-angle x-ray scattering study of perfluorinated ionomer membranes. 1. Origin of two scattering maxima, *Macromolecules* **14**, 1309-1315; Fujimura, M., Hashimoto, T., and Kawai, H., (1982) Small-angle x-ray scattering study of perfluorinated ionomer membranes. 2. Models for ionic scattering maximum, *Macromolecules*, **15**,136-144.
19. Galambos, A. F., Stockton, W. B., Koberstein, J. T., Sen, A., Weiss, R. A., and Russell T. P., (1987) Observation of cluster formation in an ionomer, *Macromolecules* **20**, 3091-3094.
20. Roche, E. J., Stein, R. S., and MacKnight, W. J. (1980) Small-angle x-ray and neutron scattering studies of the morphology of ionomers, *J. Polym. Sci., Polym. Phys. Edn.* **18**, 1035-1045.
21. Roche, E. J., Pineri, M., Duplessix, R., and Levelut, A. M. (1981) Small-angle scattering studies of Nafion membranes, *J. Polym. Sci., Polym. Phys. Edn.* **19**, 1-11.
22. Earnest, R. T., Higgins, J. S., and MacKnight, W. J., (1982) Small-angle neutron scattering from polypentenamer sulfonate ionomers, *Macromolecules* **15**, 1390-1395.
23. Clough, S. B., Cortelek, D., Nagabhushanam, T., Salamone, J. C., and Watterson, A. C. (1984) Small angle scattering from ampholytic styrene ionomers, *Polym. Eng. Sci.* **24**, 385-390.
24. Li, C., Register, R. A., and Cooper, S. L. (1989) Direct observation of ionic aggregates in sulphonated polystyrene ionomers, *Polymer* **30**, 1227-1233.
25. Register, R. A., Sen, A., Weiss, R. A., and Cooper, S. L., (1989) Effect of thermal treatment on cation local structure in manganese-neutralized sulfonated polystyrene ionomers, *Macromolecules* **22**, 2224-2229.
26. (a) Ding, Y. S., Hubbard, S. R., Hodgson, K. O., Register, R. A., and Cooper, S. L., (1988) Anomalous small-angle x-ray scattering from a sulfonated polystyrene ionomer, *Macromolecules*, **21**, 1698-1703.
(b) Register, R. A., and Cooper, S. L., (1990) Anomalous small-angle x-ray scattering from nickel-neutralized ionomers. 1. Amorphous polymer matrices, *Macromolecules* **23**, 310-317.
27. Register, R. A., and Cooper, S. L., (1990) Anomalous small-angle x-ray scattering from nickel-neutralized ionomers. 2. Semicrystalline polymer matrices, *Macromolecules* **23**, 318-323.

(28) Wu, D.-Q., Chu, B., Lundberg, R. D., and MacKnight, W. J., (1993) Small-angle x-ray scattering (SAXS) studies of sulfonated polystyrene ionomers. 1. Anomalous SAXS, *Macromolecules*, **26**, 994-999.

(29) Wu, D.-Q., Chu, B., Lundberg, R. D., and MacKnight, W. J., (1993) Small-angle x-ray scattering (SAXS) studies of sulfonated polystyrene ionomers. 2. Correlation function analysis, *Macromolecules*, **26**, 1000-1007.

(30) Li, Y.-J., Peiffer, D. G., and Chu, B. (1993) Long-range inhomogeneities in sulfonated polystyrene ionomers, *Macromolecules*, **26**, 4006-4012; see also Wang, J., Li, Y.-J., Peiffer, D. G., and Chu, B. (1993) Small-angle x-ray scattering investigation of temperature influence on microstructures of an ionomer, *Macromolecules*, **26**, 2633-2635.

(31) Chu, B., Wang, J., Li, Y.-J., and Peiffer, D. G., (1992) Ultra-small-angle x-ray scattering of a zinc sulfonated polystyrene, *Macromolecules*, **25**, 4229-4231.

(32) Xu, R.-L., and Chu, B., (1989) A polydiacetylene in dilute solutions, *Macromolecules*, **22**, 3153-3161.

(33) Chu, B., Xu, R.-L., Li, Y.-J., and Wu, D.-Q., (1989) High flux x-ray scattering of polydiacetylene P4BCMU in dilute solutions, *Macromolecules*, **22**, 3819-3821.

(34) Li, Y.-J., and Chu, B., (1991) Structure of aggregates of P4BCMU in dilute THF/toluene solutions, *Macromolecules*, **24**, 4115-4122.

(35) Hilfiker, R., Wu, D.-Q., and Chu, B. (1990) Synchrotron SAXS measurements on solutions of poly(styrene-isoprene) AB block copolymer in aniline, *J. Colloid & Interface Sci.*, **135**, 573-579.

(36) Hilfiker, R., Chu, B., and Xu, Z. (1989) Micelle forming properties of polystyrene/polyisoprene AB block copolymers, *J. Colloid & Interface Sci.*, **133**, 176-184.

(37) Chu, B., and Wu, C., (1988) Structure and dynamics of epoxy polymers, *Macromolecules*, **21**, 1729-1735.

(38) Wu, C., Zuo, J., and Chu, B., (1989) Molecular weight distribution of a branched epoxy polymer: 1,4-butanediol diglycidyl ether with *cis*-1,2-cyclohexanedicarboxylic anhydride, *Macromolecules*, **22**, 633-639.

(39) Wu, C., Zuo, J., and Chu, B., (1989) Laser light scattering studies of epoxy polymerization of 1,4-butanediol diglycidyl ether with *cis*-1,2-cyclohexanedicarboxylic anhydride, *Macromolecules*, **22**, 838-842.

(40) Wu, C., Chu, B., and Stell, G. (1991) Branching kinetics of epoxy polymerization of 1,4-butanediol diglycidyl ether with *cis*-1,2-cyclohexanedicarboxylic anhydride, *Makromol. Chem. Symp.*, **45**, 75-86.

41\. Russell, T. P., and Koberstein, J. T. (1985) Simultaneous differential scanning calorimetry and small-angle x-ray scattering, *J. Polym. Sci., Polym. Phys. Ed.*, **23**, 1109-1115.

42\. Eisner, G., Riekel, C., and Zachmann, H. G. (1985) Synchrotron radiation in polymer science, *Adv. Polym. Sci.*, **67**, 1-57.

43\. Ungar, G., and Keller, A. (1986) Time-resolved synchrotron X-ray study of chain-folded crystallization of long paraffins, *Polymer*, **27**, 1835-1844.

(44) Song, H. H., Stein, R. S., Wu, D.-Q., Ree, M., Phillips, J. C., LeGrand, A., and Chu, B., (1988) Time-resolved SAXS on crystallization of a low-density polyethylene/high-density polyethylene polymer blend, *Macromolecules*, **21**, 1180-1182.

(45) Song, H. H., Wu, D.-Q., Chu, B., Satkowski, M., Ree, M., Stein, R. S., and Phillips, J. C., (1990) Time-resolved small-angle x-ray scattering of a high density polyethylene/low density polyethylene blend, *Macromolecules*, **23**, 2380-2384.

(46) Wang, J., Alvarez, M., Zhang, W.-J., Wu, Z.-W., Li, Y.-J., and Chu, B., (1992) Synchrotron small angle x-ray scattering of crystalline structures and isothermal kinetics of poly(aryl ether ether ketones), *Macromolecules*, **25**, 6943-6951.

(47) Li, Y.-J., Gao, T., and Chu, B., (1992) Synchrotron SAXS studies of the phase-separation kinetics in a segmented polyurethane, *Macromolecules* **25**, 1737-1742.

(48) Chu, B., Gao, T., Li, Y.-J., Wang, J., Desper, C. R., and Byrne, C. A., (1992) Microphase separation kinetics in segmented polyurethanes: Effects of soft segment length and structure, *Macromolecules* **25**, 5724-5729.

(49) Li, Y.-J., Gao, T., Liu, J., Linliu, K., Desper, C. R., and Chu, B., (1992) Multiphase structure of segmented polyurethanes: Effects of temperature and annealing, *Macromolecules* **25**, 7365-7372.

(50) Li, Y.-J., Reu, Z., Zhao, M., Yang, H., and Chu, B., (1993) Multiphase structure of segmented polyurethanes: Effects of hard-segment flexibility, *Macromolecules* **26**, 612-622.

(51) Li, Y.-J., Liu, J., Yang, H., Ma, D., and Chu, B. (1993) Multiphase structure of segmented polyurethanes: Its relation with sperulite structure, *J. Polym. Sci., Part B, Polym. Phys.*, **31**, 853-867.

(52) Chu, B., Wu, C., Li, Y.-J., Harbison, G., Roche, E. J., Allen, S. R., McNulty, T. F., and Phillips, J. C. (1990) Synchrotron x-ray diffraction of a single filament and a bundle of poly-(p-phenylene terephthalamide) filaments, *J. Polym. Sci. C, Polym. Lett.* **28**, 227-232.

(53) Li, Y.-J., Wu, C., and Chu, B. (1991) Synchrotron wide-angle x-ray diffraction of single-filament PPTA fibers under stress, *J. Polym Sci. B, Polym. Phys.*, **29**, 1309-1311.

54. Tashiro, K., Kobayashi, M., and Tadokoro, H., (1977) Elasti moduli and molecular structures of several crystalline polymers, including aromatic polyamides, *Macromolecules* **10**, 413-420.

55. Ii, T., Tashiro, K., Kobayashi, M., and Tadokoro, H., (1987) X-ray study of lattice tensile properties of fully extended aromatic polyamide fibers over a wide temperature range, *Macromolecules* **20**, 347-351.

56. Gaymans, R. G., Tijssen, J., Harhama, S., and Bantjes, A. (1976) Elastic modulus in the crystalline region of poly-(p-phenylene terephthalamide), *Polymer*, **17**, 517-518.

57. Northolt, M. G., and van Aartsen, J. J. (1973) On the crystal and molecular structure of poly-(p-phenylene terephthalamide), *J. Polym. Sci., Polym Lett. Ed.*, **11**, 333-337.

FROM THE RULER TO THE SEXTANT. DIPOLE-DIPOLE ELECTRONIC ENERGY TRANSFER AS A TOOL FOR PROBING POLYMER CONFORMATION AND MORPHOLOGY

AHMAD YEKTA AND MITCHELL A. WINNIK*
Department of Chemistry and Erindale College,
University of Toronto, Toronto, Ontario, Canada M5S 1A1

1. Introduction

More than four decades have passed since the seminal work of Förster [1,2] on direct nonradiative electronic energy transfer [DET] between chromophores. The technique of DET has now advanced into an important tool for characterizing structure on a nanometer-scale. Systems investigated include self-organized structures such as micelles and vesicles, as well as Langmuir-Blodgett layers, and various self-assembled microstructures formed from block copolymer melts. In some instances, DET is employed in characterizing interfaces between dissimilar materials, as in polymer blends or the surface of a vesicle, and other instances to try to infer the dimensions of domain of small size. About three decades ago Stryer coined the term "Spectroscopic Ruler" to the DET technique for measurements of nanometer distances in chromophore-labeled molecules of biological interest [3,4]. Since then progress has made it possible to measure the distribution of distances in a way that allows one to map out the size and shape of domains in two and three dimensions. As a result of the recent advances one can now think of the technique in these applications as a nano-scale "Spectroscopic Sextant" capable of characterizing the morphology and interface structure of complex systems [5].

By a morphological characterization one means a determination of the size, shape, and density profile of each polymer component across the self-assembled domain. One can draw analogy to the activity of a land surveying engineer in the production of kilometer-scaled topographical maps. He or she positions staff poles at a number of points of reference and measures their relative distances to a selected group of other points of interest in the mapping area. The resulting data are then summarized into a map showing the relative distances, geometric shapes and character of the sections of interest. To mimic this activity on a nano-scale, one may first prepare polymers which self-assemble containing well-chosen dyes attached to specific sites. The dyes would become localized during self-assembly, effectively labeling the microdomain of interest as microscopic staff poles of reference. Next, DET measurements of the distances between the dye chromophores should, in principle, provide information about the size and shape of the labeled domains. In addition, how the dyes are distributed in the space of the microdomain should furnish information on the local density profiles. Such experiments would be of interest to the polymer physics community because they could provide new information about polymer-polymer and polymer-solvent interfaces. The

S.E. Webber et al. (eds.), Solvents and Self-Organization of Polymers, 433–455.

studies would also be important to spectroscopists interested in reaction dynamics in restricted geometries because they would provide access to new examples of restricted spaces in which to study energy transfer kinetics.

2. Synopsis

In order to provide an introduction to the basics of DET, and in order to classify the voluminous literature covering more than four decades of theoretical and experimental work on the subject, this chapter is organized into 7 sections. We focus attention on the most important measurable parameter, the time-dependent donor fluorescence decay and its analysis. First we introduce the basic principles of DET, deferring some of the details to an appendix. Next we categorize various DET studies, distinguishing intramolecular from intermolecular experiments. The intramolecular class is subdivided into Cases Ia and Ib, where the chromophores are positioned either at a fixed distance of separation, or where there is a distribution of interdistances. The intermolecular processes span a much broader range of experiments which can in turn be subdivided into two general classes. In the simpler situation (Case IIa), i.e. the case treated by Klafter and Blumen, DET occurs between donors (all of whom are identical because they are related by symmetry) and uniformly distributed acceptors. The more complex category (case IIb) involves DET between donors and acceptors which have a non-uniform distribution in some region of space. It is this situation which is the most difficult to treat theoretically, and which has the greatest potential for application to the description of the interface morphology in self-assembled systems, as well as to polymer diffusion across interfaces.

3. Basics of DET

Consider a collection of donor molecules (D) excited to D* by a short-lived pulse of light ($h\nu_{ex}$). In the absence of self-interaction, the D^* decay back to D with a natural decay rate $1/\tau_D^\circ$, or a mean lifetime τ_D°. Experimentally, one observes that the fluorescence emission ($h\nu_{em}$) from the sample, $I_D(t)$, decays exponentially, i.e.

$$I_D(t) = I_D(0) \exp\{-t/\tau_D^\circ\} \tag{1}$$

Now consider the following scheme.

$$\begin{array}{ccccc} D + A & \xrightarrow{h\nu_{ex}} & D^* + A & \xrightarrow{w(r)} & D + A^* \\ & & \downarrow [1/\tau_D^\circ] & & \\ & & D + h\nu_{em} \text{ (observed)} & & \end{array}$$

Here each D^* is separated by a distance r from another molecule which we call A for acceptor. The excited D^* can emit fluorescence as before, but is also subject to the possibility of transferring its electronic excitation energy to the neighboring molecule.

The type of energy transfer we speak of does not require formation of a D-A contact pair and the consequent electron exchange through the overlap of electron clouds. In the classical picture of the DET phenomena, we picture the oscillating electron cloud of D^* as a transmitting antenna which sets up an electric field in its proximity that oscillates at some frequency ν_o. If the electron cloud of the neighboring A molecule is also able to oscillate at the same frequency (resonance), then it will act as a receiving antenna, drawing the excitation energy of D^*, to form A^*. In the quantum mechanical view, one states that there is a finite probability, given by eq. (2), that the excitation energy will end up in A, forming A^*.

$$\frac{w(r)}{w(r) + [1/\tau_D^{\circ}]} \tag{2}$$

Here w(r) stands for the rate constant of the DET process which depends on the separation distance r. In other words, DET offers an extra channel of decay to D^*, thus shortening its lifetime from τ_D° to some value τ_D, given by

$$\frac{1}{\tau_D} = \frac{1}{\tau_D^{\circ}} + w(r) \tag{3}$$

Note that this transfer of energy does not involve the emission of a photon by D^* followed by its reabsorption by A, a competing process (deemed "trivial energy transfer") which would not affect the lifetime of D^*.

The physics of DET was first described by Förster, who showed that the most prevalent type of donor-acceptor interaction is dipole-dipole in nature, and the rate of energy transfer can be expressed as

$$w(r) = \frac{1}{\tau_D^{\circ}}(R_o/r)^6 \tag{4}$$

In this relation, R_o (also called the Förster Radius) is a constant that determines the length scale of the DET experiment. The value of R_o, which depends on the specific choice of the D/A pair, ranges typically from 1 nm to 8 nm, and is calculated from the overlap of the emission spectrum of the donor with the absorption spectrum of the acceptor (see Appendix A). In fact eq. (4) represents a simplified description of the DET rate law because w(r) also depends on the relative orientation of the transition moments of the D^*/A pair. This point is discussed in Appendix A. Here we neglect orientation effects to focus on other factors affecting fluorescence decay analysis. We note that in most instances, the magnitude of the orientation factor is a number close to unity, and in specific cases is equal to unity.

Take note that the r^{-6} dependence of the rate law limits effective measurement range of the DET ruler to within $2R_o < r < 0.5R_o$. Outside this range the rate is either so slow as to make little detectable influence, or so fast as to leave little donor fluorescence intensity for any practical measurement. Nevertheless the user is not limited in the choice of donors and acceptors as a large number of chromophores have been studied spectroscopically and the relevant values of R_0 for each pair tabulated [6].

The essence of eq. (4) is that it relates the measurement of time (i.e. 1/rate, scaled by τ_D°) to the measurement of distance (scaled by R_0). In real systems, however, DET takes place among D/A pairs separated by various distances. The underlying matrix morphology affects the distribution of distances and this in turn leads to a distribution of rates of decay. If the dyes are confined to regions of space of finite volume, integration of the rate over the pair distribution can in principle provide information about the size and shape of these domains. Our challenge then becomes one of extracting information about morphology through analysis of the fluorescence decay rates of the total collection of donors in the system.

3.1 SIMPLIFYING ASSUMPTIONS

To simplify our presentation we consider only systems that comply with certain requirements that are relatively easy to achieve experimentally. First, we require that the matrix be sufficiently viscous that within the lifetime τ_D° no translational diffusion of D^* or A takes place, i.e. r is independent of time. We categorize the variety of systems according to the way they lead to discrete distributions of D/A pairs. We focus our discussion mostly on the donor fluorescence decay function, $I_D(t)$, and consider to a smaller extent, the quantum yield of donor fluorescence, ϕ_D. Also, for reasons related to experimental difficulties, we do not discuss the acceptor fluorescence decay function, $I_A(t)$.

Next, we require that the local donor concentration C_D is sufficiently dilute that donor-donor DET cannot compete with donor-acceptor DET. Fayer and coworkers [7,8] have shown that this condition can be met if $C_D/C_A \ll (R_0/R_0^{D\text{-}D})^3$. That is we assume that no D-D energy migration can take place. The theoretical approach to the analysis of data in the presence of D-D energy transport was taken up by Gochanour, Andersen, and Fayer and later applied to polymeric systems by the same workers, in addition to Frank, Fredrickson and others. Their approach is different from those considered in the present writing and the interested reader is referred to excellent reviews that have appeared on the topic [9-12].

4. Case Ia. Intramolecular DET Between Isolated D^*-A Pairs Separated by a Uniformly Fixed Distance R

Latt [13], Stryer [3,4] and others considered the case in which the donor/acceptor moieties are attached to the backbone of a rigid molecule, separated by a fixed distance R. Experimentally, such samples should be sufficiently dilute that no intermolecular DET among the "isolated" pairs can take place. Because all excited D^* are equivalent, with a common D-A separation R, all rates of decay are equal. As a result, $I_D(t)$ is characterized by a single exponential of lifetime t_D given by

$$I_D(t) = I_D(0) \exp\{-t/t_D\} \tag{5a}$$

$$\frac{1}{\tau_D} = \frac{1}{\tau_D^\circ} + \frac{1}{\tau_D^\circ}(R_o/R)^6 \tag{5b}$$

It is then clear that a measurement of the decay time τ_D leads to the determination of the distance R, hence coinage of the term "Spectroscopic Ruler."

The general relation for the donor fluorescence yield (ϕ_D°) in the absence and presence of the acceptors (ϕ_D) is given by

$$\frac{\phi_D}{\phi_D^\circ} = \frac{1}{\tau_D^\circ} \int_0^\infty I_D(t)\, dt \tag{6}$$

Eq. (6) is a fundamental relation that is valid for all fluorescence decay profiles associated with DET. When applied to (5a), we find

$$\frac{\phi_D^\circ}{\phi_D} = 1 + (R_o/R)^6 \tag{7}$$

Only for this simplest of all Case Ia situations can eq. (7) also be derived from a consideration of the probability of DET for each isolated pair (cf. eq. (2)). In all other cases one must, as in eq. (6), integrate over the decay profile to calculate the DET efficiency.

While the steady-state determination of ϕ_D may appear easy, measurement of decay profiles are to be preferred because, even for the simple Case Ia, steady-state measurements of relative donor fluorescence yield are beset by the difficulty that the acceptor might absorb parts of the donor fluorescence and/or intensities of the excitation beam.

5. Case Ib. Intramolecular DET Between Isolated D^*-A Pairs Separated by a Distribution of Distances P(r)

In this case D and A are covalently bound to specific positions of a flexible molecular backbone, rendered immobile in a high viscosity medium, and the system is sufficiently dilute such that intermolecular DET does not take place. Although the distance between each D/A pair is fixed, the allowed conformations of the backbone lead to a distribution of interdistances. In Figure 1a we show a particular conformation of a D/A end-labeled backbone. Alternatively, one might have a random distribution of labeling sites on the backbone, schematically represented in Fig. 1b.

To find $I_D(t)$, we need to sum terms that are similar to eq. (5a), each weighted for the population of the various conformational interpair distances. Each separation distance leads to its own discrete rate of decay, and summing over the distribution leads to a nonexponential decay profile expressed by eq. (8).

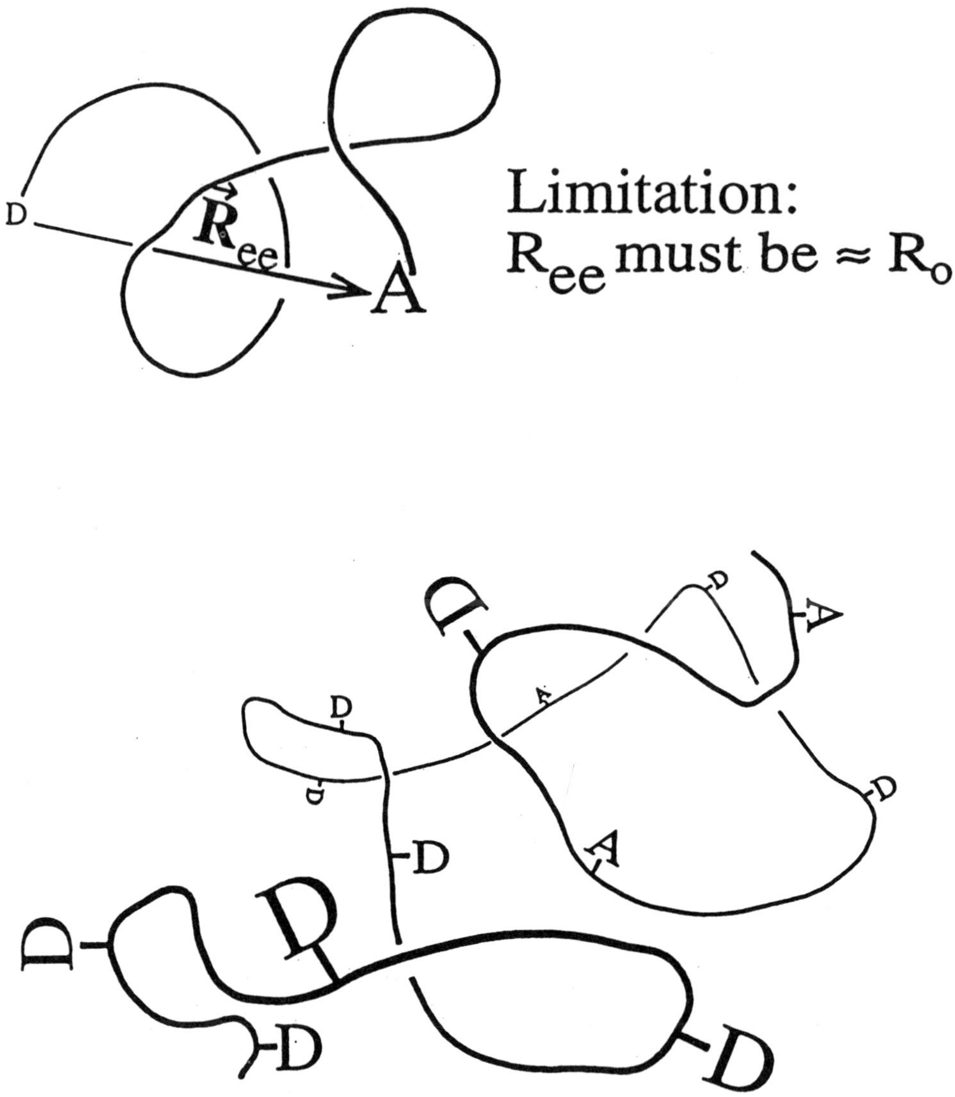

Figure 1. Schematic representation of polymer backbones labeled with donor (D) and acceptor (A) chromophores: (a) top, specific D/A labeling at the chain ends; (b) bottom, random labeling along the chain.

$$I_D(t) = I_D(0) \exp\{-t/\tau_D^\circ\} \int_0^\infty \exp[-tw(r)]\, P(r)\, dr \tag{8}$$

Here w(r) is given by eq. (1) and P(r) dr is the probability of finding a given molecular conformation with the D-A distance lying between r and r + dr.

In eq. (8), and other similar equations that follow, we integrate over separation distances from 0 to infinity, which implicitly allows for the possibility of D-A pairs with $r = 0$. Some workers correct for this by taking the lower limit of the integration as $r = \sigma$, where σ is the sum of the D/A van der Waals' radii, typically 0.3-1.0 nm. By taking $\sigma = 0$ we gain analytical simplicity; and the error introduced for analyzing experimental data is negligible because the very high rate of DET ensures that no significant emission is detectable from such closely spaced pairs.

Note also that if the backbone is rigid and all D/A pairs are separated by a fixed distance $r = R$, then $P(r) = \delta(r - R)$, where the delta function symbolizes the sharpness of the distribution function P(r). In this case from eq. (8) we recover eq. (5).

One obvious attraction of eq. (8) is that it allows one to attempt to determine the end-to-end distribution P(r) of a polymer of a given length by synthesizing a labeled polymer with a donor at one end and an acceptor at the other end (c.f. Fig. 1a). The reader is referred to excellent reviews by Haas [14] and Liu [15] of their work and those of Steinberg, Katchalski, Lakowicz and others in experiments involving polypeptides, which were dissolved in a viscous medium to suppress motion of the polymer, and examined at sufficient dilution that all DET was intramolecular. The usual procedure is that one fits the data to eq. (8) making an apriori assumption about the nature of P(r). For example, for a gaussian chain one may assume $P(r) \propto r^2 \exp\{-3 r^2/2\langle R_{ee}^2\rangle\}$, where R_{ee} is the root-mean-square end-to-end distance of the tagging sites. Recent work from Liu suggests that if the analysis of data is performed carefully, one may recover P(r) without having to make any prior assumptions about the functional form which might be involved [15].

One of the limitations of this approach (apart from the difficulties in synthesizing the doubly-labeled polymer) is that the length of polymer that can be investigated is rather limited (cf. Fig. 1a). The critical Förster Radius R_o is too small (2-8 nm) compared with the end-to-end distance of most polymers of interest. Nevertheless, there are times when this experiment might be very interesting. For example, a block copolymer which forms spherical micelles is predicted to have distorted chain conformations for both the block confined to the spherical insoluble domain and for the solvent-swollen soluble block. Fig. 2 shows a representation of a block copolymer micelle, containing one molecule synthetically labeled with D at the junction and A at one end, and introduced as a tracer into a sample of similar but unlabeled block copolymer. An experiment like this has recently been proposed by Procházka, where one could monitor the D-A pair separation in the micellar system [16]. The problem to overcome here is fluorescence detection from low bulk concentrations of the donor required to ensure no double occupation of the micelles by the labeled polymers.

We emphasize again that the methodologies considered in this writing are restricted to the condition of low donor concentrations. However, Fayer, Frank, Fredrickson et. al. pioneered a different approach to the study of chain conformation involving transport of energy among a high local concentration of donors [9-11]. One prepares a polymer chain containing donors at either end only, or donors and acceptors ("traps") distributed along the backbone, as exemplified in Fig. 1b. This distribution can be either regular or random. Here significant DET occurs if the mean D-D and D-A separation is on the order of the respective R_o. Because of the occurrence of D-D energy migration, one now samples much larger distances, i.e. that of the coil dimensions. A key parameter of measurement is the polarization decay of the donors following polarized excitation of the sample. Here one assumes that a single step transfer from D* to D leads to a loss of

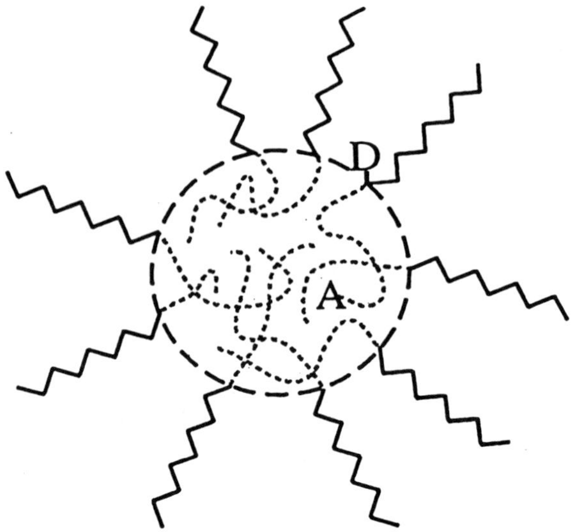

Figure 2. Schematic representation of the chains of a block copolymer self-assembled into a micellar structure. The insoluble block (dotted curve) forms the micellar core. The soluble block forms the micellar corona. For simplicity, only eight of the ca 100 aggregated chains are shown. A single chain is specifically junction-labeled with the donor and end-labeled with the acceptor.

polarization. For the case of the isolated polymer coil in a rigid medium one can recover the root-mean-square end-to-end distance when there is only one donor label at each end of the chain (cf. Fig. 1a), or determines the characteristic ratio (chain stiffness, C_∞), and radius of gyration (R_g) when there is random labeling of the chain (cf. Fig. 1b).

6. Intermolecular DET Between D^* and Randomly Distributed Acceptors. Cases IIa and IIb

Most DET experiments involve intermolecular energy transfer. DET can occur between small molecules dissolved in a continuous matrix, solubilized in phase-separated domains as in micelles, adsorbed to surfaces, or attached to polymers in such a way that the dyes become localized when the system separates into discrete phases. The size, shape and distribution of these phases constitute the "morphology" of the system. In this section we examine DET with the idea that in some instances, where D and A are confined to relatively small spaces, with one or more dimensions on the order of R_0, the kinetics of DET become sensitive to the size and shape of the confining space. One speaks of this situation as involving DET in a restricted geometry.

Many of the experimental difficulties associated with evaluating the decay profile for intramolecular DET (Case I) are alleviated in intermolecular DET (Case II). For example, instead of having to synthesize position-specific doubly-labeled polymers needed for intermolecular DET studies, one prepares polymers containing a single dye attached to a specific site. In Case II systems, one no longer has to carry out experiments at high dilution. In fact, the acceptor concentration C_A becomes an important experimental variable that can be used to test various theoretical models by data fitting procedures. From our earlier discussions we know that significant information from DET is obtained only when the mean D-A separation distance is on the order of R_o. Since the mean separation goes as $(C_A)^{-1/3}$, one can deduce that meaningful experiments involve values of C_A in the range of 5 mM when R_o = 8 nm, to 50 mM when R_o = 2 nm.

We will encounter two distinct categories of restricted geometries. First is the type examined in detail by Klafter, Blumen, Drake, and coworkers [17-19], in which all the donors are related by translational symmetry and are thus equivalent, and which we discuss under Case IIa. Next are the situations in which there is a distribution of donor environments which we discuss under Case IIb. In the first case, one can evaluate the donor decay profile by averaging over the D/A pair distribution. In the second case, one must also consider the distribution of donors, and for each set of donors located in a given environment, average over the distribution of D/A pair distances. One key feature of many of the latter systems is that the distribution of D/A pairs near the periphery of the confining space is different that than away from the edges.

The comparison of Cases IIa and IIb is depicted in Figures 3, 4 and 5. In Fig. 3 we have three situations where the D's and A's are distributed uniformly (but randomly) on the *surfaces* of a sphere, an infinitely extended cylinder, or a finite rectangular plane. In Figs. 3a and 3b all donors have equivalent loci in the sense that any given D is equally likely to have the same distribution of A's about it (translational symmetry). In contrast, in Fig. 3c we note the "edge effect" where on the average, the D's on the edges, when compared to those in the center, see a different distribution of A's about them. On the other hand, in Fig. 4 we depict three situations where the D's and A's are distributed nonuniformly (but randomly) in systems with spherical or planar symmetry. The degree of shading represents the concentrations (or probability of finding) of the donors and acceptors. Fig. 4a may model the diffuse surface morphology of a micelle forming junction-labeled block copolymer micelle. Fig. 4b is similar to 4a except that the block copolymer is donor-labeled at the junction, and acceptor-labeled at the end of the insoluble block. Fig. 4c models the diffuse morphology of a labeled bilayer-forming system. Fig. 5 depicts more detailed features of block copolymer chains for situations pertaining to Figs. 3a and 4a.

7. Case IIa. Intermolecular DET. All Donors are Equivalent

If all donors are equivalent, then any given D is equally likely to have the same distribution of A's about it. In general this requires a translational symmetry in the geometry of the space, which may come about from the underlying geometry of the container and a uniform concentration profile for the acceptors. For example, this condition is always satisfied if D/A are uniformly distributed in infinite media in one, two, or three dimensions, such as on a thin rod, on a large flat plane or in a continuous matrix. In this context, "infinite" implies a space in which its characteristic dimensions are all much larger than R_o, e.g. > 100 R_o. This condition is also satisfied if D and A are uniformly distributed over the surface of a sphere or over the surface of a long thin cylinder (cf. Figs 3a and 3b). In contrast, the condition of equivalent donor loci is not satisfied if the A's are non-uniformly distributed over the spherical surface (cf. Figs 4a and 4b), or if both are distributed uniformly, but over the surface of, say, an ellipsoid.

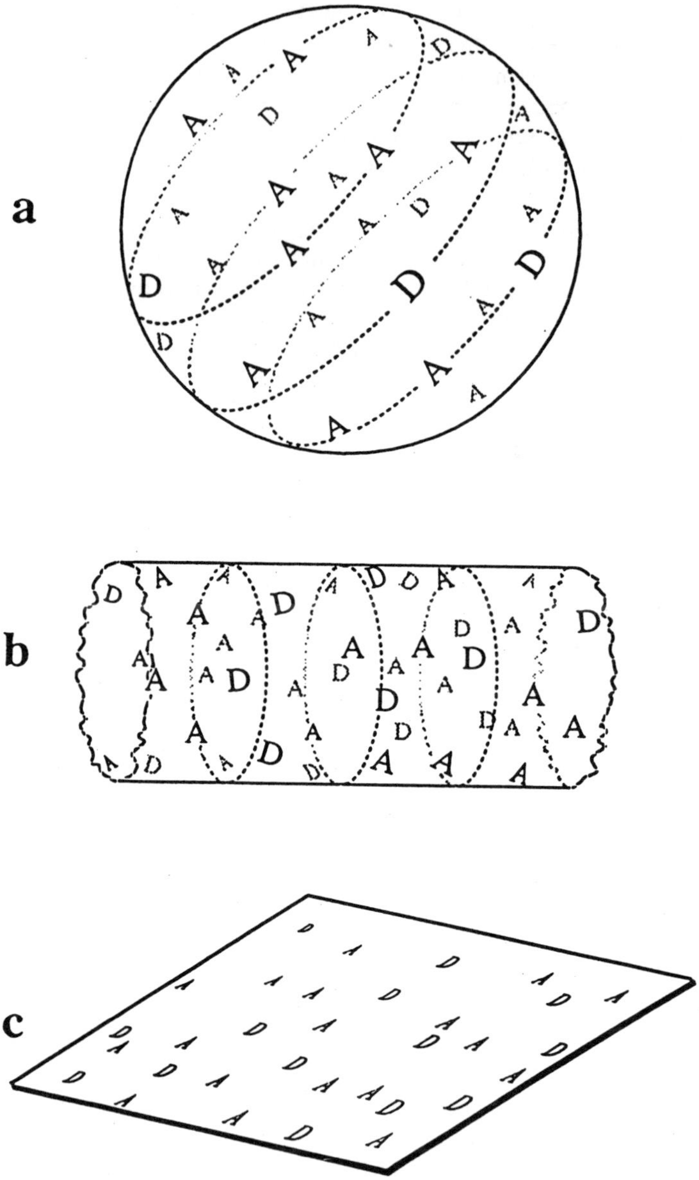

Figure 3. Schematic representations of random and uniformly distributed donors and acceptors on the surface of (a) a sphere; (b) an infinitely long cylinder; (c) a finite rectangular sheet. Situations (a) and (b) belong to Case IIa systems because the donors occupy environments that are translationally invariant. Situation (c) belongs to Case IIb systems because the donors near the edges of the plane occupy environments different from those near the center.

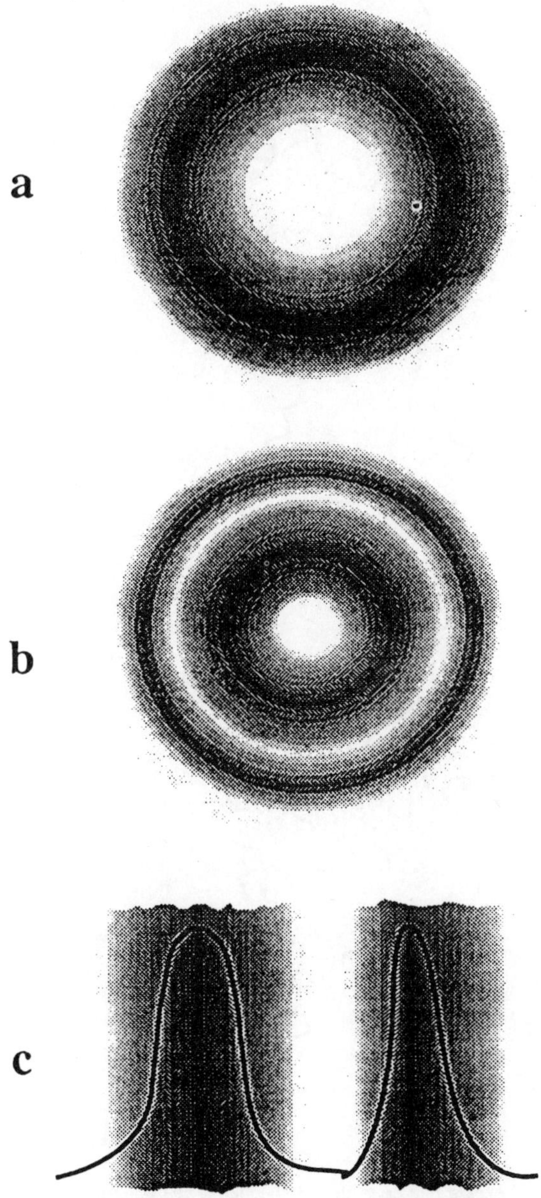

Figure 4. Examples of Case IIb situations showing the donor and acceptor concentration profiles (graduated shading) for random and nonuniform distributions in the interior of spheres and a planar bilayer. Situation (a) can model the diffuse interface of a block copolymer micelle where each chain is labeled at the junction with either a D or an A (cf. Fig. 5b). Situation (b) can model the micelle in (a) except here each chain is labeled at the insoluble end with an A and at the junction with a D (cf. Fig. 2, with every chain labeled).

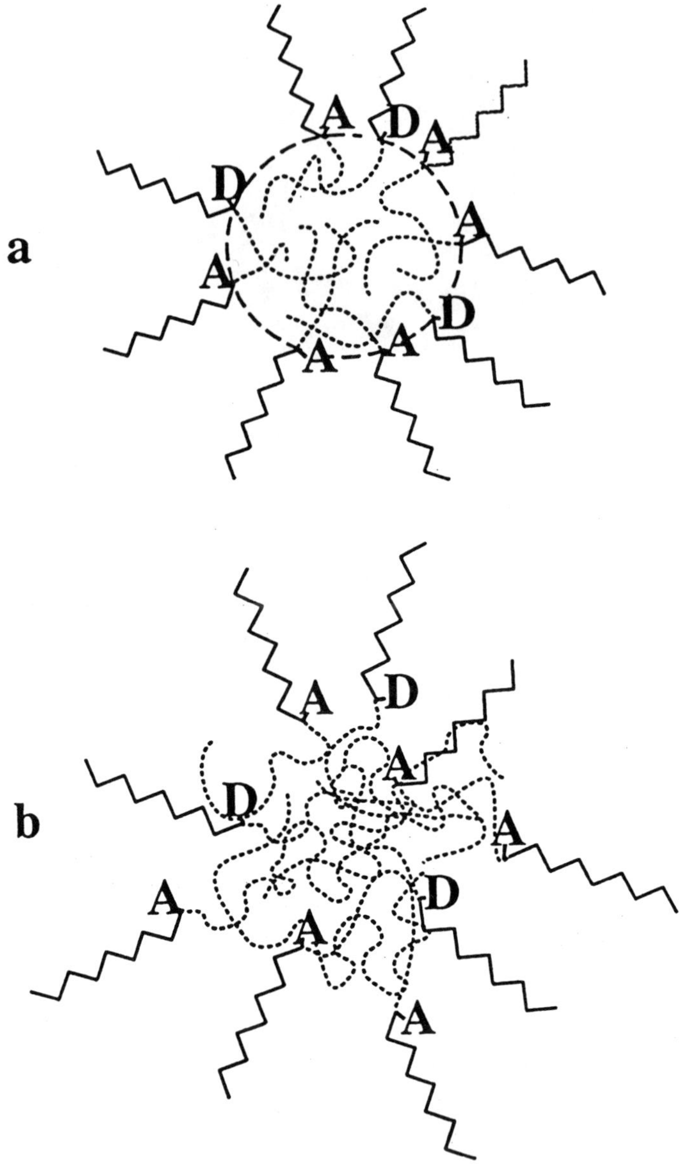

Figure 5. As in Fig. 2 except that each chain is labeled at the junction with either an A or a D. In situation (a) the micellar core is a compact sphere with a smooth surface similar to that modeled in Fig. 3a. In situation (b) the micellar core forms a diffuse interface with the soluble blocks, similar to that represented in Fig. 4a.

7.1 THE KB METHODOLOGY

Klafter and Blumen (KB) and coworkers developed a generalized method for evaluating decay profiles related to Case IIa in which all D are related by symmetry. They derived formal expressions for evaluating $I_D(t)$ which we write in the form

$$I_D(t) = I_D(0) \exp\{-t/\tau_D^{\circ}\}\ \varphi(t, \vec{\mathbf{r}}_0) \tag{9a}$$

$$\varphi(t, \vec{\mathbf{r}}_0) = \exp\{-g(t)\} \tag{9b}$$

$$g(t) = p \int_V [1 - \exp[-tw(r)]]\ \rho(r)\, dV \tag{9c}$$

We separate eq. (9) into three parts to facilitate discussion. Eq. (9a) is quite general because $I_D(t)$ is always the product of two terms, each expressing the probability of independent events. The first term, $\exp\{-t/\tau_D^{\circ}\}$, which is directly observed when $C_A = 0$, describes the survival probability of D^* due to all natural modes of decay except DET. The second term, involving DET, leads to the nonexponential decay profile of the donor. In this term $\varphi(t, \vec{\mathbf{r}}_0)$ expresses the probability that D^* will survive the passage of a time t without transferring energy to any neighboring acceptor molecule, in the absence of other modes of decay, i.e. the DET-related survival probability. The position vector $\vec{\mathbf{r}}_0$ is written only to emphasize that the rate of the DET process depends on the location of D^*. Because the KB methodology assumes equivalent loci for all donors, $\vec{\mathbf{r}}_0$ does not enter into consideration here.

Eq. (9b) shows the usual form for the appearance of $\varphi(t)$, i.e. a decaying exponential of some function of time, that we call the *exponent function* and write as g(t). There are two underlying reasons for the appearance of this particular mathematical form of $\varphi(t)$. First is the assumption, clearly valid, that the probabilities of DET from any D^* to various neighboring A's are independent. Second is the assumption that donors and acceptors behave as randomly distributed points, occupying space independent of one another. This can be justified as follows: In bulk, a typical acceptor has a molar volume of ca 0.1 to 1 L/mol. The highest C_A that one can meaningfully study in a DET experiment is ca 50 mM. That is, at most, about 0.5% to 5 % of the sample volume is occupied by the acceptors. Therefore, the probability of two or more A's occupying the same space is negligibly small. The type of randomness described above satisfies a Poisson distribution for the probability of a certain number of acceptors occupying some volume about a donor. From these assumptions, plus the rate-law eq. (1), and a count of the number of acceptors surrounding a given D^* at a distance between r and r + dr, one can derive the KB expression (9).

In eq. (9c), the integration is carried over the geometry-dependent volume of the container system. In it p is defined (rather vaguely) as the probability that an acceptor fills an occupation site. We note that p is a constant proportional to C_A and dependent on the geometry of the system. The function $\rho(r)$ is a geometry-dependent site-density distribution function. When normalized, $\rho(r)\, dr$ expresses the probability of finding a given D/A pair, separated by a distance between r and r + dr. The function is evaluated from geometric considerations pertaining to each problem. Later in this writing we will encounter examples of $\rho(r)$ in Eqs. (15a), (16a), and (17). One gets a better appreciation of the meaning of each term by noting that for situations where D/A are contained within a restricted volume (e.g. micelles),

$$\langle n_A \rangle = p \int_V \rho(r)\, dV \tag{10}$$

where $\langle n_A \rangle$ is the mean number of acceptors within each container.

7.2. EXAMPLES OF $I_D(t)$ IN INFINITELY EXTENDED SYSTEMS

By infinite systems, we mean that the sample dimensions are much larger than R_o. It is not straight forward to decide how much larger than R_o is necessary for the system to be considered "infinite." In reality, all samples are of finite size, and D/A pairs at the edges have a different distribution than those in bulk. What is essential is that the fraction of donors near the edges be sufficiently small as to make a negligible contribution to the $I_D(t)$ signal in an experiment.

We begin our analysis by reminding the reader that according to Eqs, (9) the expression for $I_D(t)$ is given as a product of terms of the form

$$I_D(t) = I_D(0) \exp\{-t/\tau_D^\circ\} \exp\{-g(t)\} \tag{10}$$

in which, once the exponent function g(t) is known, the problem is solved. For this reason, in the following examples we only write the expression for the exponent function g(t).

7.2.1. *The 3-Dimensional Förster Decay*

The simplest example, the Förster Decay, was originally derived by Förster, and was later verified in various experiments. It describes DET in an infinitely-extended 3-dimensional matrix. Here $\rho(r)dr \propto 4\pi r^2 dr$, i.e. the volume of a thin spherical shell, of radius r and thickness dr, surrounding a given D^*. After substitution in (9c) one obtains

$$g(t) = 2\, g_3\, (t/\tau_D^\circ)^{1/2} \quad ; \quad g_3 = \frac{2}{3}\pi^{3/2}\, C_A R_o^3 \quad ; \quad \frac{2}{3}\pi^{3/2} = 3.71222 \tag{11}$$

Note that C_A has units of number density. The factor 2 preceding the constant g_3 was introduced by Förster, and as a matter of consistency with previous literature, we follow suit.

7.2.2. *The 2-Dimensional Flatland*

The next example, originally derived by Hauser et al describes DET for D and A embedded on an infinitely-extended 2-dimensional flatland [20] (also for a one-dimensional wire). This form of decay has been applied to analysis of fluorescence decay data from donor/acceptor-doped bilayers, and to DET within a single layer of a Langmuir-Blodgett film [21]. In this case $I_D(t)$ is found from the exponent expression

$$g(t) = 2\, g_2\, (t/\tau_D^{\circ})^{1/3} \quad ; \quad g_2 = \frac{\pi}{2}\Gamma(2/3)\, C_A R_o^2 \quad ; \quad \frac{\pi}{2}\Gamma(2/3) = 2.12705 \qquad (12)$$

7.2.3. *The Fractal Decay*

KB showed that for infinitely extended media Eqs. (11) and (12) are two special cases of the general 'stretched' exponential equation

$$g(t) = 2\, g_\Delta\, (t/\tau_D^{\circ})^{\Delta/6} \qquad (13)$$

where the constant g_Δ is proportional to the acceptor concentration, and Δ represents one of the three Euclidean dimensions 1, 2, or 3. Klafter and Blumen showed that if D and A are embedded in a fractal lattice, then Δ is the Hausdorff (fractal) dimension of the lattice [18]. The term "fractal" implies that the structure satisfies the symmetry requirements of self-similarity. While scattering techniques span a range of length scales sufficient to assess whether real systems are self-similar, DET experiments do not. It is a rare case where a fractional exponent (i.e. Δ value) obtained in fitting experimental data to eq. (13) may be properly interpreted as a fractal dimension. More commonly, Δ represents an apparent dimensionality (i.e. Δ_{app}) with no obvious physical meaning. Such values arise either because of significant edge effects, crossover effects, or more likely, nonuniform distributions of D and A confined to a restricted geometry.

Eq. (13) is commonly employed as a phenomenological expression in treating data involving DET in restricted geometries. For example, in the absence of any prior knowledge about the underlying morphology of a system, one usually fits the donor fluorescence data to eq. (13), and often finds that one recovers acceptable fits (goodness-of-fit $\chi^2 \approx$ 1-2). Furthermore, when the experiment is repeated for various values of C_A, one finds that the plot of Δ versus C_A is flat-linear, as expected. In these experiments, one often finds fractional values of Δ, ranging from 1.5 to 2.8. This is one signature of the existence of domains within the morphology where D/A are confined.

El-Sayed et al. were the first to emphasize that a fractional value of Δ does not imply a fractal morphology [22]. Edge effects can lead to a crossover, as for example for donors on the surface and acceptors in the interior of a thin cylinder. At short times, D* transfers energy to nearby A, which have a 3-dimensional distribution [23,24]. At long times, surviving D* transfer energy to A which are further away than the thickness of the cylinder, and have a 1-dimensional-like distribution. Understanding edge effects in restricted geometries is a difficult problem. The most effective way to appreciate these effects is through simulations of the DET experiment.

7.3. EXAMPLES OF $I_D(t)$ IN CONFINED SYSTEMS

As stated above, for many restricted geometry experiments, when analyzed by the fractal decay expression (13), the systems are characterized by a crossover of dimensionality defined with the help of eq. (14) as

$$\Delta(t) \equiv 6t \frac{d}{dt} \mathrm{Ln}[g(t)] = 6 \frac{d\,\mathrm{Ln}[g(t)]}{d\,\mathrm{Ln}(t)} \tag{14}$$

That is, a plot of Ln[g(t)] vs. Ln(t) would show transitions from one value to another with Δ dominated by one distribution at early times (i.e. closely spaced pairs), and a different distribution at later times. The fit of fluorescence decay data to (13), of course, only gives some intermediate value or apparent dimensionality Δ_{app}.

7.3.1. *Donors and Acceptors on the Surface of a Sphere (Radius R_S)*

Here we deal with surface-to-surface DET as depicted in Fig. 3a. Of course, for any D^*-A pair, DET can take place through the inside of the sphere. However, we shall soon see why for most systems of interest, one still obtains an effective dimensionality of 2.0. The expression describing the normalized pair distribution function of distances of points over the surface of a sphere is given by

$$\rho(r)\,dr = \frac{1}{2}(r/R_S)\,dr/R_S \tag{15a}$$

The decay expression was found by Levitz et. al. [25] by substitution of (15a) in (9c). They showed that within the time window of $0 \le t \le 2\,(R_S/R_O)^6\,\tau_D^{\circ}$, the resulting exponent-expression g(t) is given by

$$g(t) = 0.339 <n_A> (R_O/R_S)^2\,(t/\tau_D^{\circ})^{1/3}\,[1 + 0.0231\,(R_O/R_S)^4\,(t/\tau_D^{\circ})^{2/3} + \ldots]$$
$$\frac{\Gamma(2/3)}{4} = 0.339 \tag{15b}$$

where the mean number of acceptors ($<n_A>$) per sphere (defined in eq. (10)) equals $4\pi R_O^2\,C_A$. Note that for DET on a sphere with $R_S > R_O$, the above time window is satisfied because the detection sensitivity of a normal decay-measuring apparatus is limited to a measurement time window of $0 \le t \le 6\,\tau_D^{\circ}$. Furthermore we take note of the fact that for $R_S > R_O$ the "cross-over terms" in the brackets are negligible and eq. (15b) shows a dimensionality of 2.0, identical to eq. (12) for DET on an infinitely-extended flatland (i.e. $R_S \to \infty$).

It is instructive to see how eq. (15b) was utilized to analyze DET data from *junction*-labeled block copolymer micelles of PS-PMMA in a solvent comprised of 30 wt% methanol in dioxane [26]. The labels are such that R_O = 2.3 nm. The micellar hydrodynamic radius (R_H) obtained from dynamic light scattering was 19 ± 1 nm, which, combined with intrinsic viscosity results, gave a micellar aggregation number (N_{agg}) of 140. If the micelle core were pure PS, the calculated core radius would be R_S = 8.4 nm. Fits of the DET data to a model of Figs. (3a) and (5a), that is eq. (15b), led to R_S = 14 nm. The discrepancy could be explained by presuming that the micellar core was partially swollen by solvent. However this presumption would imply that D/A are distributed over a diffuse interphase, similar to that modeled by Figs. (4a) and (5b), in which case Eqs. (9) and (15) would be inapplicable. Further indication of a diffuse interphase was obtained by fitting the data to the fractal decay equation (13). In this case, Duhamel et. al. recovered an apparent dimensionality $\Delta_{app} = 2.3 \pm 0.1$, suggesting a 3-dimensional DET perturbed by edge effects associated with the diffuse periphery of the interphase. A more complete analysis of the data would require a deeper knowledge of the detailed geometry of the interphase thickness and its radial profile (i.e. Case IIb).

7.3.2. *Donors on the Surface and Acceptors Inside of a Sphere (Radius R_S)*

Blumen et. al. derived expressions describing DET between donors on the surface of a sphere and acceptors distributed randomly inside [27]. Again, here all the donors are related by translational symmetry. The decay expression was found by substitution of the normalized pair distribution function of distances

$$\rho(r)\,dr = \frac{3}{2}(r/R_S)^2\,[1 - \frac{1}{2}(r/R_S)]\,dr/R_S \tag{16a}$$

in (9c) to yield

$$g(t) = g_3\,(t/\tau_D^\circ)^{1/2}\,[1 - 0.567\,(R_o/R_S)\,(t/\tau_D^\circ)^{1/6} + \ldots.] \tag{16b}$$

where g_3 was defined in eq. (11). For situations where the sphere size is much bigger than R_o, i.e. $R_o/R_S << 1$, the extra terms inside the brackets become negligible. Note that the term outside of the bracket is one-half that of the Förster expression (11). This is because any D^* on the surface of a large sphere sees acceptors distributed towards only one half of available space. Also note that if the sphere is large, a fit of such data to the fractal decay expression (13) would yield an apparent dimensionality $\Delta = 3.0$. For smaller sized spheres, however, the extra terms in the bracket of eq. (16b) become significant (the cross-over effect) and a fit of such data to the fractal expression would result in a value of Δ_{app} different from 3.0.

7.3.3. *Other Geometries*

KB et. al have considered a few other geometries [24,27], all where the donors occupy equivalent loci, and derived the related $I_D(t)$. These include situations for surface-to-inside DET and surface-to-surface DET in a long cylinder (cf. model of Fig. 3b). However even for an important and simple case such as a sphere randomly loaded with a few donors and acceptors (inside-to-inside DET) a solution has not been available. This is despite the fact that one assumes uniform values for C_A and C_D, and the fact that an expression has been derived [28,29] for the pair distribution function $\rho(r)$ of distances inside of a sphere of radius R_S.

$$\rho(r)\,dr = 3\,(r/R_S)^2\,[1 - \frac{1}{2}(r/R_S)]^2\,[1 + \frac{1}{4}(r/R_S)]\,dr/R_S \tag{17}$$

In this case not all donor positions inside the sphere are equivalent. Some donors lie closer to the surface of the sphere and have a distribution of A's around them different from donors closer to the center. The assumptions of Case IIa (eq. (9)) are no longer applicable and one deals with a Case IIb problem.

8. Case IIb. Intermolecular DET. Nonuniform Distribution of D and A

This is the most general case some examples of which are shown in Figs. 3c and 4 and 5b. Here each D senses a different environment (lack of translational symmetry) due to the way the acceptors are distributed about it, either because C_A is nonuniform in space, and/or because the acceptors are distributed uniformly but in a geometry lacking sufficient symmetry. Many natural systems of interest are expected to possess some element of symmetry. For example, mono- or multi-layered systems formed from self assembly of block copolymers and other amphiphilic compounds possess a planar symmetry with the axis perpendicular to the plane. Micellar and many other colloidal systems are known to possess spherical symmetry, while certain micellar and block copolymer systems possess string-like morphologies with cylindrical symmetry.

Recent theoretical work in our laboratories has shown that if the space of the system has some axis of symmetry, the inherent difficulties involved in formulating $I_D(t)$ can be overcome [5]. That is, it is possible to develop a rigorous and relatively simple methodology. Without going into the details, in this section we use heuristic arguments to show the avenues of thought that go into development of the general formulation.

If the donors and acceptors label specific positions of the self-assembling system, then the concentrations of D and A will vary in space according to the underlying morphology. If one has a system of planar symmetry, the concentrations are given by $C_D(z)$ and $C_A(z)$, where z is chosen as the axis perpendicular to the plane. On the other hand, for systems of spherical or cylindrical symmetry, the concentrations would be given by $C_D(R)$ and $C_A(R)$, with R measuring the radial distance from the center of the sphere or central axis of the cylinder, respectively.

Let q stand for the generalized coordinate (z or R). The partial symmetry of the system makes some subset of D^* equivalent to each other. The main idea is to evaluate the element of donor fluorescence emission $[\delta I_D(q, t)]$ from this subclass of donors, and then sum all such intensities emanating from various positions q in the sample, to find the total observable $I_D(t)$ One argues that in a planarly symmetric system, all donors lying in a slice between two sheets at z and z + dz, possess equivalent loci in the sense discussed earlier. Hence, the subset of D^* follow the KB formalism (eq. (9)) for evaluation of their local DET-related survival probability, $\varphi(z, t)$. The same can be said about the other two symmetries, where now all D lying between two spherical (or cylindrical) sheets positioned at R and R + dR would posses equivalent loci, and the KB formalism would describe the local survival probability due to DET-related events, $\varphi(R, t)$. The emission intensity from these thin cuts of the sample is proportional to $\varphi(q, t)$ and also the number of D^* initially formed upon excitation, which is in turn proportional to the number of light-absorbing ground-state D's in the cut. That is, one writes

$$I_D(t) = \exp\{-t/\tau_D^\circ\} \int C_D(q, t)\, \varphi(q, t)\, dV \tag{18a}$$

where q stands for the generalized coordinate of symmetry, z or R, respectively.

$$dV = dz, \text{ planar symmetry} \tag{18b}$$
$$dV = 4\pi R^2\, dR, \text{ spherical symmetry} \tag{18c}$$
$$dV = 2\pi R\, dR, \text{ cylindrical symmetry} \tag{18d}$$

To evaluate $\varphi(q, t)$ one uses the KB approach of Eqs. (9), keeping in mind the fact that the distribution of A's about each D^* is nonuniform, but that all A's within a distance range of r to r + dr accept energy with the same rate (cf. eq. (1)). One then needs to count the total (mean) number of A's lying between two spherical shells distanced at r and r + dr from the central D^*.

At this point it is instructive to see how the KB Eqs. (9) lead to the description of $I_D(t)$, or more simply the exponent function g(t) in eq. (11) for the simplest of all cases, which is a homogeneous 3-dimensional solution of donors and acceptors described by the Förster Decay. This is obtained simply via eq. (9c) with

$$g(t) = \int_0^\infty [1 - \exp[-tw(r)]]\; C_A\, (4\pi r^2\, dr) \tag{19}$$

where w(r) is given by eq. (1). Note that because C_A is a constant, it can be taken out of the integral sign, so that integration [30] by parts leads to eq. (11) for g(t). Now the way we have written eq. (19) above is intended to be foretelling. $4\pi r^2$ dr is the volume of a thin spherical shell of thickness dr and radius r, about a given D^*. $C_A\, (4\pi r^2\, dr)$ is the total number of acceptors within this volume. Similarly, one can argue that for the general Case IIb situations one would have a relation similar to (19), i.e.

$$\varphi(q, t) = \exp\{-g(q, t)\} \tag{20}$$

$$g(q, t) = 4\pi \int_0^\infty [1 - \exp[-tw(r)]]\; \langle C_A(q, r)\rangle\, r^2\, dr \tag{21}$$

where $(4\pi r^2\, dr)\, \langle C_A(q, r)\rangle$ now stands for the total number of acceptors within a spherical shell of radius r and thickness dr, surrounding a D^* located at q (i.e., z or R). The averaging sign $\langle\rangle$ indicates that $\langle C_A(q, r)\rangle$ is an averaged concentration in the homogenizing sense. That is $C_A(q)$ is nonuniform in space. If one counts the total number of acceptors within a given shell and divides this number by the volume of the shell $(4\pi r^2\, dr)$, one obtains a *locally* homogenized (mean) concentration, all members of which have the same rate of energy transfer from the central D^* under consideration.

Thus from geometric considerations alone, one can evaluate $<C_A(q, r)>$ for the various symmetries of interest. One finds [5],

planar symmetry, $$<C_A(z, r)> = \frac{1}{2r} \int_{(z-r)}^{(z+r)} C_A(z')\, dz' \qquad (22a)$$

spherical symmetry, $$<C_A(R, r)> = \frac{1}{2r\,R} \int_{|R-r|}^{(R+r)} C_A(R')\, R'\, dR' \qquad (22b)$$

where the concentration profiles C_A, and C_D would be specified by the physics of the system under consideration.

Case IIb formulation is quite general and can apply to Case IIa situations in restricted spaces or continuously extended concentration profiles. Furthermore and more conveniently, in this formulation there is no longer a need for an apriori knowledge of the pair distribution function $\rho(r)$, as was required with Case IIa situations in eq. (9).

In summary, one calculates the "homogenized" acceptor concentration $<C_A(q, r)>$ from eq. (22), substitutes the result in Eqs. (21) and (20) to evaluate the respective local exponent function $g(q, t)$, and local survival probability $\varphi(q, t)$. The latter is in turn substituted in eq. (18a) with a proper consideration of the symmetry to finally evaluate the desired $I_D(t)$.

9. Appendix A

A more complete representation of eq. (1) can be written as

$$w(r) = \frac{3}{2}\frac{\kappa^2}{\tau_D^\circ}(R_o/r)^6 \qquad (A1)$$

where κ^2 is a dimensionless quantity that depends on the relative orientation of the transition moment dipoles of the D^*/A pair. As originally described by Förster, it is given by

$$\kappa = \cos\theta_{DA} - 3\cos\theta_D \cos\theta_A \qquad (A2)$$

Here θ_{DA} is the angle between the D^* and A dipoles; θ_D, and θ_A are the angles between each of the respective dipoles and the vector $\vec{r}$ joining their centers. Later work by Steinberg [31] simplified the relation by showing that only two angles are required to completely specify Eq. (A2).

The Critical Förster Radius is expressed through the spectroscopic measurables as

$$R_o^6 = \frac{3000\,\mathrm{Ln}(10)\,\phi_D^\circ}{64\,\pi^5\,n^4\,N}\;\frac{\int_0^\infty F_D(\nu)\,\varepsilon_A(\nu)\,\nu^{-4}\,d\nu}{\int_0^\infty F_D(\nu)d\nu} \qquad \text{(A3)}$$

where $F_D(\nu)$ is the donor emission spectrum; ν is frequency in wave numbers (cm^{-1}); $\varepsilon_A(\nu)$ is the acceptor absorption spectrum in $M^{-1}\,cm^{-1}$; ϕ_D° is the donor emission quantum yield in the absence of any acceptors; n is the refractive index of the medium; and N is Avogadro's number. The overlap integral of eq. (A3) implies that R_o can be significant only if the peak absorption coefficient of $\varepsilon_A(\nu)$ is large (10^4-10^6 $M^{-1}\,cm^{-1}$), and has significant spectral overlap with the donor emission spectrum $F_D(\nu)$.

Some workers absorb the orientation factor into the definition of R_o by multiplying expression (A3) by the factor $\frac{3}{2}\kappa^2$. With this definition, eq. (A1) reduces to eq. (1) of the text. However, in this case R_o is no longer a true molecular constant and a value must be assumed for κ^2. In the original derivation, Förster showed that if the dipoles rapidly randomize by reorienting within the duration of τ_D°, then κ^2 can be replaced by its mean value, 2/3, and again eq. (A1) transforms into eq. (1). The assumption of rapid reorientation may not be so appealing when we consider that the systems we deal with assume a rigid medium as far as translational diffusion is concerned. However, Blumen, Baumann/Fayer, and Berberan-Santos/Prieto have considered the orientation factor explicitly (cf. eq. (A2)) for a variety of systems. In the cases considered thus far, the "effective" value of (3/2) κ^2, which depends on the dimensionality and excitation polarization, is still a number close to unity [32-34].

10. Acknowledgment

The authors thank NSERC Canada and the Ontario Center for Materials Research for their support of this work.

11. References

1. Förster, Th. (1948) Intermolecular energy transference and fluorescence, *Ann. Phys. (Leipzig)* **2**, 55-75.
2. Förster, Th. (1949) Experimental and theoretical investigation of intermolecular transfer of electron activation energy, *Z. Naturforsch.* **4a**, 321-327.
3. Stryer, L. and Haughland, R. P. (1967) Energy transfer: a spectroscopic ruler, *Proc. Natl. Acad. Sci. U.S.A.* **58**, 719-726.
4. Stryer, L. (1987) Fluorescence energy transfer as a spectroscopic ruler, *Annu. Rev. Biochem.* **47**, 819-846.

5. Yekta, A., Duhamel, J., and Winnik, M.A. (1995) Dipole-dipole energy transfer. Fluorescence decay functions for arbitrary distributions of donors and acceptors: systems with planar geometry, *Chem. Phys. Lett.* **235**, 119-125.
6. Berlman, I.B. (1973) *Energy Transfer Parameters of Aromatic Compounds*, Academic Press, New York.
7. Loring, R. F., Andersen, H. C., and Fayer, M. D. (1982) Electronic excited state transport and trapping in solution, *J. Chem. Phys.* **76**, 2015-2027.
8. Miller, R. J. D., Pierre M., and Fayer, M. D. (1983) Electronic excited state transport and trapping in disordered systems, *J. Chem. Phys.* **78**, 5138-5146.
9. Frank, C.W., Fredrickson, G.H., and Andersen, H.C. (1985) Electronic excitation transport as a tool for the study of polymer chain statistics, in M.A. Winnik (ed.), *Photophysical and Photochemical Tools in Polymer Science*, D. Reidel Publishing Co., Dordrecht, pp. 495-522.
10. Peterson, K.A., Stein A.D., and Fayer M.D. (1989) Electronic excitation transport in Restricted geometries, in J. Klafter and J.M. Drake (eds.), *Molecular Dynamics in Restricted Geometries*, John Wiley and Sons, New York, pp. 39-75.
11. Yeng, Y.H. and Frank C.W. (1989) Electronic excitation transport on polymer chains, in J. Klafter and J.M. Drake (eds.), *Molecular Dynamics in Restricted Geometries*, John Wiley and Sons, New York, pp. 77-98.
12. Byers, J.D., Friedrichs, M.S. Friesner R.A., and Webber S.E. (1989) Polymer structure and down-chain electronic energy transfer, in J. Klafter and J.M. Drake (eds.), *Molecular Dynamics in Restricted Geometries*, John Wiley and Sons, New York, pp. 39-75.
13. Latt, S.A., Cheung, H.T., and Blout, E.R. (1965) Energy transfer. A system with relatively fixed donor-acceptor separation, *J. Am. Chem. Soc.* **87**, 995-1003.
14. Haas, E. (1985) Folding and dynamics of proteins studied by non-radiative energy transfer measurements, in M.A. Winnik (ed.), *Photophysical and Photochemical Tools in Polymer Science*, D. Reidel Publishing Co., Dordrecht, pp. 325-350.
15. Liu, G. (1993) Determination of intermolecular distance distribution functions using the "spectroscopic ruler". 1. Theoretical feasibility, *Macromolecules* **26**, 1144-1151.
16. Procházka, K. (1995) A Monte Carlo study of tethered chains in spherical volumes: chain conformations and nonradiative energy transfer in systems with end-tagged chains, *J. Phys. Chem.* **?**, ?-?.
17. Klafter, J. and Blumen, A. (1984) Fractal behavior in trapping and reaction, *J. Chem. Phys.* **80**, 875-877.
18. Klafter, J., Blumen, A., and Drake, J. M. (1989) Relaxation and diffusion in Restricted geometries, in J. Klafter and J.M. Drake (eds.), *Molecular Dynamics in Restricted Geometries*, John Wiley and Sons, New York, pp. 1-22.
19 Drake, J.M., Klafter, J., and Levitz, P. (1991) Chemical and biological microstructures as probed by dynamic processes, *Science* **251**, 1574-1579.
20. Hauser, M., Klein, U.K.A., and Gösele, U. (1976) Extension of Förster's theory of long-range energy transfer to donor-acceptor pairs in systems of molecular dimensions, *Z. Phys. Chem.* **101**, 255-266.
21. Yamazaki, I., Tamai, N., and Yamazaki, T. (1990) Electronic excitation transfer in organized molecular assemblies, *J. Phys. Chem.* **94**, 516-525.
22. Yang, C.L., Evesque, P., and El-Sayed, M.A. (1985) "Fractal-like", but nonfractal behavior of one-step dipolar energy transfer on regular lattices with excluded volume, *J. Phys. Chem.* **89**, 3442-3444.
23 Drake, J.M. and Klafter, J. (1990) Dynamics of confined molecular systems, *Phys. Today* **43**, 46-55.

24 Levitz, P., Drake, J.M., and Klafter, J. (1988) Direct energy transfer in pores: geometrical crossovers and apparent dimensionality, *Chem Phys. Lett.* **148**, 557-561.

25. Levitz, P., Drake, J.M., and Klafter, J. (1988) Critical evaluation of the application of direct energy transfer in probing the morphology of porous solids, *J. Chem. Phys.* **89**, 5224-5236.

26. Duhamel, J. , Yekta, A., Ni, S., Khaykin, Y., and Winnik, M. A. (1993) Characterization of the core of polystyrene block poly(methyl methacrylate) polymer micelles by energy transfer, *Macromolecules* **26**, 6255-6260.

27. Blumen, A., Klafter, J., and Zumofen, G. (1986) Influence of restricted geometries on the direct energy transfer, *J. Chem. Phys.* **84**, 1397-1401.

28. Berberan-Santos, M. (1986) On the distribution of the nearest neighbor, *Am. J. Phys.* **54**, 1139-1141.

29 Barzykin, A.V. (1991) Spatial distribution of probes in a micelle: effects on the energy transfer time-resolved measurables, *Chem. Phys.* **155**, 221-231.

30. In dealing with DET problems one frequently encounters integrals of the type

$$\int_0^\infty [1 - \exp[-tw(r)]]\, r^m\, dr = \frac{\Gamma[(5 - m)/6]}{(m + 1)}\, R_o^{(m + 1)}\, (t/\tau_D^\circ)^{(m + 1)/6}$$

where m is a whole number $-1 < m < 5$, and the Euler Γ numbers can be found tabulated as $\Gamma[1/6] = 5.5663$; $\Gamma[2/6] = 2.6789$; $\Gamma[3/6] = \pi^{1/2} = 1.7724$; $\Gamma[4/6] = 1.3541$; $\Gamma[5/6] = 1.1288$.

31. Steinberg, Z. (1968) Nonradiative energy transfer in systems in which rotatory Brownian motion is frozen, *J. Chem. Phys.* **48**, 2411-13.

32. Blumen, A. (1981) On the anisotropic energy transfer to random acceptors, *J. Chem. Phys.* **74**, 6926-6933.

33. Baumann, J. and Fayer, M.D. (1986) Excitation transfer in disordered two-dimensional and anisotropic three-dimensional systems: effects of spatial geometry on time-resolved observables, *J. Chem. Phys.* **85**, 4087-4107.

34. Berberan-Santos, M. and Prieto, M.J.E. (1988) Monte Carlo simulation of orientational effects on direct energy transfer, *J. Chem. Phys.* **88**, 6341-6349.

USE OF FLUORESCENCE METHODS TO CHARACTERIZE THE INTERIOR OF POLYMER MICELLES

S. E. WEBBER
Department of Chemistry and Center for Polymer Research
The University of Texas at Austin, Austin, Texas, USA 78712

1. Introduction

In the examples of polymer micelles known to date the driving force in their formation has been the use of selective solvents which forces one component into a collapsed and self-associating condition. Many of the properties of these micelles are deduced from classical methods, i.e. light, X-ray, or neutron scattering, viscosity, sedimentation, etc. These methods yield the properties of the whole micelle, such as hydrodynamic diameter or radius of gyration, core radius, and aggregation number. However these methods provide little insight as to the modification of the conformation of chains located within the core or the core-corona interfacial region. None of the classical methods yield information about chain dynamics or small molecule diffusion within the micelle core. This latter subject will be taken up in a different chapter by Arca.

The use of polymers with covalently attached fluorophores can help elucidate some of these properties because the local environment of the fluorophore is examined. Because fluorescence is a local phenomenon, such experiments should not be used to deduce large-scale micellar properties. It always has to be kept in mind that the fluorescent tag may perturb the local environment, such that the fluorescence might not report on typical polymers. Therefore fluorescence studies should always be combined with as many other techniques as possible to avoid misleading conclusions.

While fluorescence methods can be applied to the corona region as well as the core, we confine our discussion in this chapter to the more elusive core and core-corona interfacial region. The specific issues we will address are: 1) following the formation of micelles by fluorescence; 2) the mobility in a core

S.E. Webber et al. (eds.), Solvents and Self-Organization of Polymers, 457–478.

swollen by the presence of a good solvent for the core polymer, either during the formation of the micelle or by adding such a solvent to a micelle solution; 3) insight as to the smoothness of the core-corona interface using fluorescence quenching or energy transfer.

The organization of this chapter will be as follows:

(1) A brief review of the methods for tagging diblock or graft copolymers with fluorescent moieties.

(2) A brief review of some of the specific features of fluorescence spectroscopy with emphasis on their application to self-organized polymers.

(3) Examples of studies that address the issues described above taken from the literature or on-going work in our laboratory.

2. Chemical Methods

This discussion is by no means comprehensive but gives some examples that illustrate different strategies. In the following B refers to the core-forming polymer.

2.1 CHEMICAL METHODS FOR TAGGING A-B BLOCK COPOLYMERS

a) One Fluorophore at the A-B Junction

The method described in reaction sequence (i) has been applied by Quirk and Mattice [1], Riess [2] and Winnik [3] primarily for polystyrene-polyethylene oxide polymers but has also been applied to poly(styrene)-*block*-poly(methyl methacrylate) or poly(styrene)-*block*-poly(*t*-butyl methacrylate) [4]. This method takes advantage of the fact that 1,1 diarylethylene derivatives will not polymerize but can mediate an anionic polymerization between styrene and methacrylates. In step (1) R^-M^+ is typically sec-butyl lithium or cumyl potassium. The fluorophore (= fluor) in step (2) has been 9-phenanthryl or 2-anthryl in work reported so far although it seems likely that a wide range of species can be used so long as they do not have a sufficiently reactive proton that can terminate the anion reaction. R' in step (3) can be an easily hydrolyzible group such as t-butyl or trimethylsilyl to produce a hydrophilic block after the polymerization.

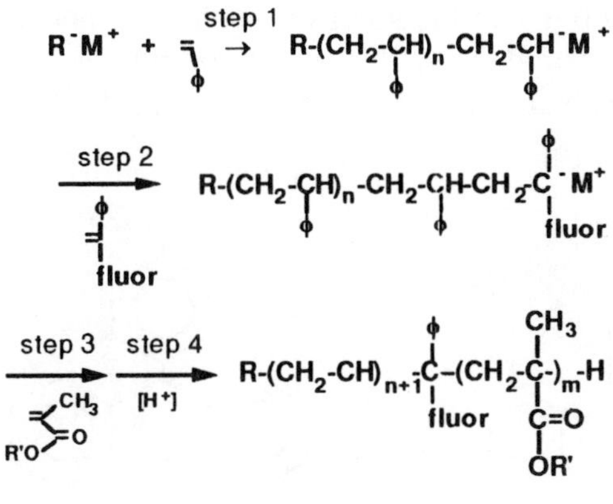

(i)

b) One Fluorophore at the Beginning of the B Block

This can be accomplished by incorporating the fluorophore as part of the initiator in step (1). The former has been accomplished in our lab [4] using a species analogous to cumyl potassium (**1**).

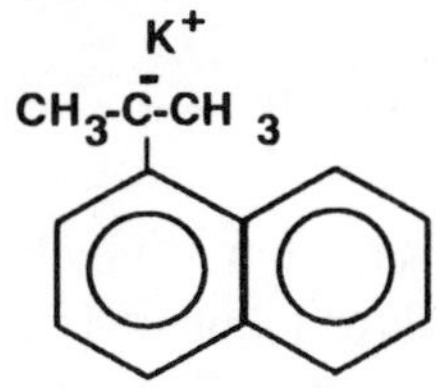

1

It would be anticipated that the reactivity of other species like **1** would decrease dramatically for larger aromatic chromophores.

c) Short Blocks of Fluorophores at the Beginning or End of a Block

Step (1) can be modified to include a sequential addition of fluorophore:

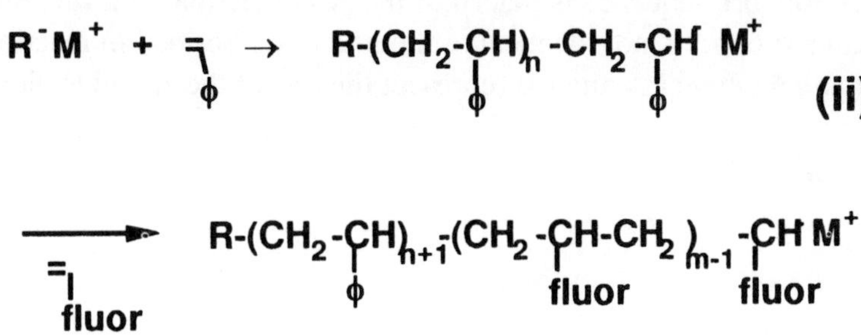

(ii)

Assuming that the reactivity of the vinyl fluorophore is not too low the order of addition can be reversed producing a block of fluorophores at the beginning of the sequence. This has been accomplished with 2-vinylnaphthalene (**2**) [5] and 9,10 diphenyl anthracene (**3**) [6].

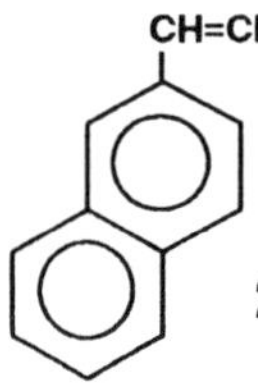

2

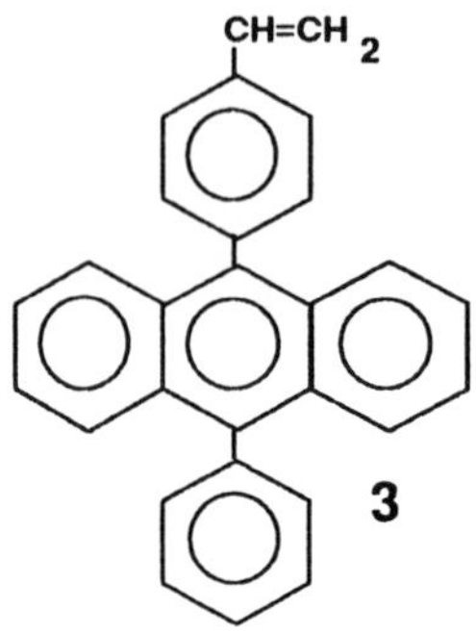

3

Unlike the previous methods this approach produces a distribution of sequence lengths. For the ideal case a Poisson distribution is expected,

$$P(m) = \langle m\rangle^{m} \exp(-\langle m\rangle)/m! \qquad (1)$$

where $\langle m\rangle$ is the average sequence length, which can be controlled by the mole ratio of vinyl fluorophore to initiator ($\langle m\rangle$ = [fluor]/[initiator]) and P(m) is the probability of a sequence of length m. Consequently if one wishes to have no more than one fluorophore per chain then $\langle m\rangle \ll 1$ is required, which means that a significant fraction of chains will not have any label. In many cases dilution of the tagged chains by untagged chains may be an advantage to avoid energy transfer (see later). The reaction sequence described in (ii) for $\langle m\rangle > 1$ has the advantage that excimer formation can be used as an alternative method of monitoring the local mobility of the chain. This method is relatively easy to apply but lacks the precision of the methods described in section (a) and (b). Also note that if a fluorophore sequence is placed at the A-B junction with this method that the chemical structure is not identical to the process described in reaction (i), i.e., compare **4** and **5** where PA and PB represent the rest of the A and B block.

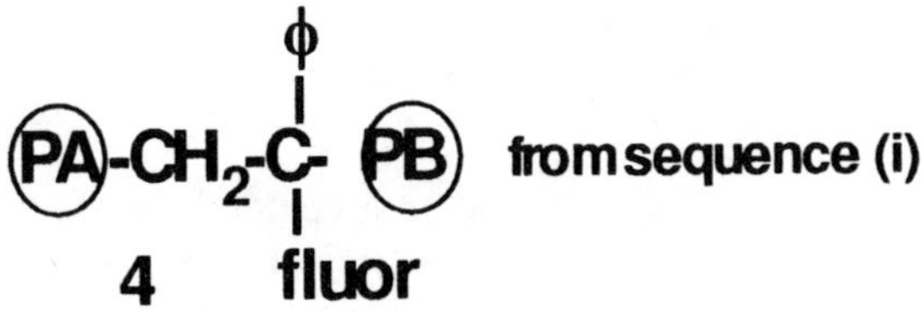

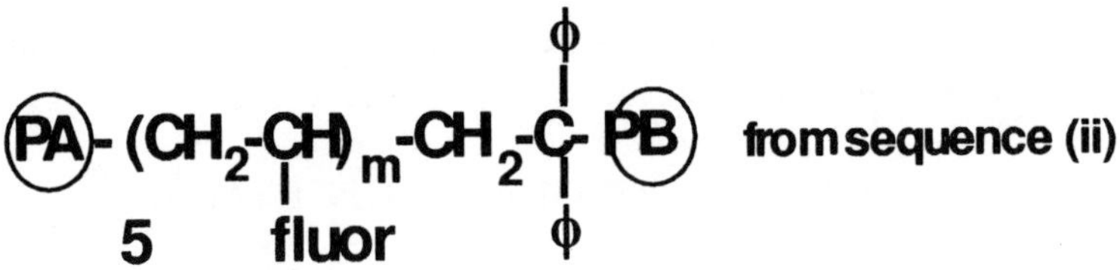

d) Doubly-tagged Polymers

By combining the tactics described in the preceding section one can place two fluorescence probes on the polymer chain. By using the initiator **1** and a sequence like (i) we have prepared [4]

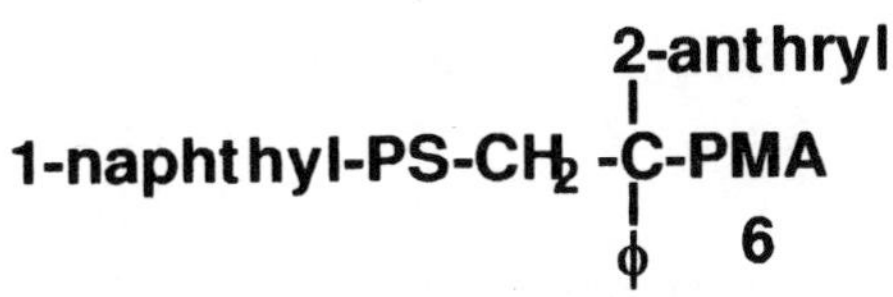

In **6** PS refers to a poly(styrene) block and PMMA and PMA refer to poly(methyl methacrylate) and poly(methacrylic acid) respectively. The primary motivation for preparing these doubly-tagged polymers has been to take advantage of Förster energy transfer (FET). The rate of donor energy transfer to the acceptor (naphthalene and anthracene respectively in **6**) is given by

$$k_{FET} = \frac{1}{t_0{}^D}\left(\frac{R_0{}^{DA}}{R}\right)^6 \qquad (2)$$

where $t_0{}^D$ is the normal donor excited state lifetime, $R_0{}^{DA}$ is a parameter that depends on the donor acceptor pair and R is the spatial separation of D and A. For the naphthalene-anthracene donor-acceptor pair $R_0{}^{DA} \approx$ 20-25 Å. The values of $R_0{}^{DA}$ are usually tabulated for the case that the fluorophores freely rotate during their excited state lifetimes [7]. If the fluorophores are randomly oriented with respect to each other but held rigidly in place during the donor excited state lifetime, k_{FET} has to be slightly modified [7]. Other cases and geometries can be treated by Förster theory so long as the orientation of the donor-acceptor pair is known. Theory has also been developed for the cases that the donor-acceptor separation is changing with time [8]. Thus by monitoring the rate of donor fluorescence sensitization of the acceptor one can obtain a measure of the distribution of D-A separations. Such experiments have been reported for doubly-tagged polymers in the solution phase [9,10.] Experiments in the solid and micellar phases are in progress in our laboratory.

2.2 CHEMICAL METHODS FOR TAGGING GRAFT COPOLYMERS

Our fluorescence studies of micelles formed from graft copolymers are the only ones that we have encountered in the literature, and in our work the ultimate bulk solvent is water (after complete dialysis from a mixed solvent). Therefore in the following amphiphilic polymers will be emphasized. In one case an alternating polymer is prepared from copolymerization of the fluorophore (2-vinylnapthalene) with maleic anhydride, followed by imidization with a $-NH_2$ terminated polystyrene in which the molecular weight of the polystyrene is ca. 3000 (prepared by anionic methods [11]). Finally the polymer is hydrolyzed to produce structure 7 [12].

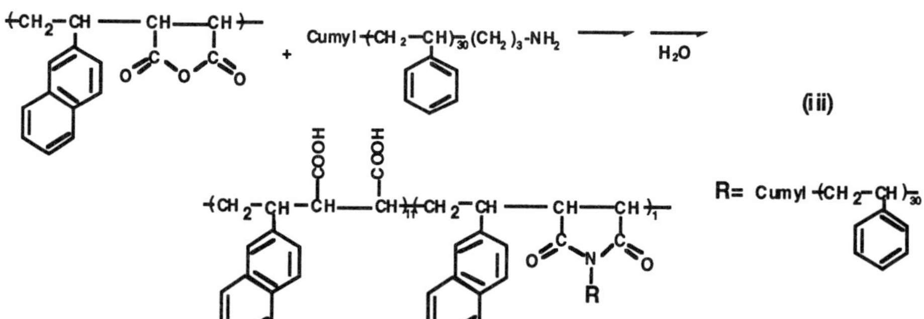

(iii)

7

Structure 7 represents a polymer with hydrophobic grafts. A related polymer (**8**) was produced by the following sequence [13]:

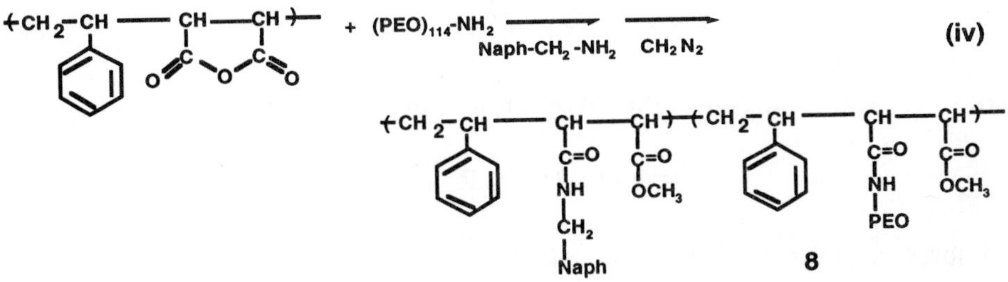

(iv)

8

In this case the hydrophilic grafting density was varied to elucidate the micelle properties as a function of this parameter.

An interesting observation for both of the above graft polymers is that the micelles they form are well-behaved, even though the backbone polymer was prepared by free-radical polymerization with $M_w/M_n \geq 1.5$ although the grafts were relatively monodisperse.

3. Fluorescence methods

3.1 ENERGY TRANSFER AND CHROMOPHORE EXCHANGE BETWEEN MICELLES

Förster energy transfer (FET) is a non-radiative process that deactivates an excited state [14],

$$D^* + A \xrightarrow[k_{FET}]{} D + A^* \qquad (3)$$

where k_{FET} is the Förster energy transfer rate given by eq. (2). The distance over which this process can occur is significantly larger than collision diameters (typically 15-25 Å [7]) and the net result is that the donor fluorescence from D* is replaced by that of the acceptor, A*. FET among donor moecules also occurs at sufficiently high donor concentratons and leads to fluorescence depolarization (which will be discussed in section 3.4).

For homogeneous solution Förster derived the probability of energy transfer to be given by

$$P(x) = \pi^{1/2}\, x\, e^{x^2}\, [1 - \mathrm{erf}(x)] \tag{4}$$

where $x = c/c_0$ (c is the concentration of the acceptor) and c_0 is given by

$$c_0 = 3000/(2\pi^{3/2} N_0 R_0^3) \tag{5}$$

The fluorescence decay of the donor is given by

$$I_D(t) = \exp[-t/t_0 - \frac{3}{4}\,\pi^{3/2}\, c\, R_0^3 \sqrt{t/t_0}\,] \tag{6}$$

These equations are modifed in restricted geometries [15] and will be discussed elsewhere with applications to elucidating polymer interfaces and polymer associations (see the Chapter by Yekta and Winnik). FET can also be exploited with doubly-tagged chains to assess polymer conformations [9], [10] (see section 2.1(d)). We have applied this idea to a polymer that forms micelles, as well be discussed later in this Chapter.

The dynamics of small molecule or polymer exchange between micelles can be characterized by mixing micelles composed of polymers that are similar or identical except that they are tagged with different fluorophores, one of which can act as an energy donor to the other. Thus when the two micelles are initially mixed and the donor is excited, primarily donor fluorescence is observed. As chromophores are exchanged between micelles the donor will sensitize the fluorescence of the acceptor. An example of this kind of study is polystyrene-*block*-hydrogenated polyisoprene tagged with carbazole (donor) or anthracene (acceptor). In this study a relatively high mole fraction of the polystyrene groups are tagged (> 0.4) [16]. The exchange of polymers was observed upon allowing a micelle solution to cool from 90° C to ambient. Mattice et al. [17] have studied polystyrene-chromophore-polyethylene oxide where chromophore is either the donor (naphthalene) or the acceptor (pyrene). These chromophores are placed at the mid-point of the diblock polymer using a reaction sequence like (i). The molecular weights of the polystyrene blocks were relatively low (4500) but even with such short segments no exchange was observed in room temperature aqueous solutions. At 60° C equilibrium seems to require approximately two days with an apparent first order rate constant for exchange of $1.7 \times 10^{-5}\ \mathrm{sec}^{-1}$. Winnik and coworkers [18] have also used mid-tagged polymers prepared with methods like reaction (i) to prepare polystyrene-probe-poly(methyl methacrylate) polymers with the probe either a phenanthrene (energy donor) or anthracene (energy acceptor).

These polymers form micelles in 30:70 dioxane:methanol mixed solvents. The kinetics of energy transfer were analyzed in terms of a fractal geometry for the interface and it was concluded that the interfacial region is diffuse.

Some work in our laboratory has also used fluorescence to monitor the exchange of small molecules between micelles [19] or the absorption [20] and release [21] of small molecules. In the latter case a methodology has been developed to permit the estimate of the diffusion constant of the small molecule fluorophore inside the micelle. The estimated diffusion constants are extraordinarily low. For phenanthrene diffusion out of the glassy polystyrene core we obtain $D \approx 10^{-18}$ cm^2/sec. This will be discussed in a later Chapter by Arca.

3.2 TAKEUP OF SMALL MOLECULES TO DETERMINE CMC

Winnik and coworkers [22,23] used the uptake of pyrene by micelles to characterize the cmc of polystyrene-*block*-polyethylene oxide. In this case the fluorescence properties of pyrene in a polystyrene core are quite different than in the aqueous phase. By means of an analysis of the excitation and emission spectra and the fluorescence decay as a function of polymer concentration the cmc was found to be in the range of 1-3 mg/L, with the exact value depending on the molecular weight of the blocks which was different for diblock and triblock polymers.

3.3 FLUORESCENCE QUENCHING

Fluorescence quenching provides another method for characterization of the exposure of a fluorophore to the aqueous phase. Polystyrene-*block*-poly(methacrylic acid) polymers with naphthalene probes at the beginning of the polystyrene block or near the mid-point (prepared by reaction sequence (ii)) form micelles readily in dioxane:water mixtures. An ionic quencher like Tl^+ is confined to the aqueous phase and it was found that naphthalene groups located at the beginning of the polystyrene block could not be quenched by Tl^+ [24]. This demonstrates that the polystyrene core is not microporous with respect to water. On the other hand, the naphthalene at the polystyrene-poly(methacrylic acid) junction could be quenched, but the normal Stern-Volmer equation, eq 7, was not obeyed. Instead at higher quencher concentrations a negative deviation from equation 7 was observed and the quenching could be fit to eq 8 [25].

$$I_0/I(Q) = 1 + K_{SV}[Q] \tag{7}$$

$$\frac{I_0}{I_0 - I(Q)} = \frac{1}{f_a} + \frac{1}{f_a K_{SV}[Q]} \tag{8}$$

In this latter equation f_a corresponds to the fraction of fluorophores that are accessible to the quencher and K_{SV} corresponds to the quenching constant of the exposed fluorophores.

3.4 TIME-DEPENDENT FLUORESCENCE DEPOLARIZATION

Monitoring the time dependence of fluorescence is common practice (see O'Connor et al. for an excellent discussion of all aspects of this technique [26]). The use of time-dependent fluorescence to study donor-acceptor separations will be discussed elsewhere (see the Chapter by Yekta and Winnik)).

For time-dependent depolarization one excites a sample with pulsed polarized light and detects the fluorescence emission at polarizations parallel and perpendicular to the excitation polarization. The decays from the two polarizations are usually convolved individually using up to 35 fixed exponential functions [26], [27] and the fluorescence anisotropy is computed from the resultant decays channel by channel using equation 9:

$$r(t) = [I_{\|}(t) - G\,I_{\perp}(t)] \,/\, [I_{\|}(t) + 2\,G\,I_{\perp}(t)] \tag{9}$$

where $I_{\|}(t)$ and $I_{\perp}(t)$ are the emission collected through the parallel and perpendicular polarizers respectively. The G factor (the instrumental anistropy) in eq 6 is required to account for the different sensitivities of both the monochromator grating and to a lesser extent the detector to horizontally and vertically polarized light. Fayer at al. circumvent the need to correct for the monochromator bias by rotating the polarization of the excitation light using a Pockel cell rather than rotating a polarizer between the sample and monochromator [28].

Steady-state fluorescence depolarization is similar to that of the time-dependent experiment. The sample is excited with vertically polarized light and the fluorescence emission is detected through a polarizer at parallel and perpendicular orientations to the excitation light. The G factor again must be considered and computed to insure that the true ratio of fluorescence is measured. Interpretation of the steady-state fluorescence anisotropy data will depend on an assumed decay law [29]. This places obvious limits on the interpretation of the data and relies heavily on assumptions since r(t) is not directly observed during the steady-state experiment. Unlike time-dependent fluorescence depolarization, this technique does not provide detailed information on probe motions in three dimensions.

Upon irradiating the sample with ultrashort plane-polarized monochromatic light of the appropriate wavelength excitation of a fraction of the molecules occurs. Preferential excitation occurs for molecules with the transition dipole moment for absorption, $\boldsymbol{\mu}_{ab}$ parallel to the excitation polarization [30]. The non-equilibrium distribution of excited molecules starts to re-equilibrate immediately. Several simultaneous and competing processes contribute to the overall relaxation of the system: (a) Radiative electronic relaxation of excited molecules. This process is controlled by optical selection rules. (b) Non-radiative relaxation, which includes a number of competitive processes (both monomolecular and bimolecular), such as vibrational internal conversion, intersystem crossing, energy transfer, etc. These processes depend on various factors such as the structure of the fluorophore and on its mobility, internal flexibility, and interaction with microenvironment. (c) Rotational diffusion of the fluorophore contributes to the orientational relaxation of the system. (d) Other processes, such as solvent relaxation, which proceed on the picosecond and sub-picosecond time-scale and are of less importance for the dynamics of polymer chains. Fluorescence from an optically allowed $S_1 \rightarrow S_0$ transition is usually strongly polarized at early times after the polarized excitation since the absorption and emission transition moments are nearly parallel to each other for many fluorophores. For fluorophores with a low symmetry, those dipole moments may form a fixed angle and the fluorescence polarization is lower at early time. In dilute fluorophore systems r(t) monitors the rotational diffusion of fluorophores and is related to the time-dependent correlation function of the orientation of the emission transition moment, $\boldsymbol{\mu}_{em}(t)$, at the time t and the absorption transition moment, $\boldsymbol{\mu}_{ab}(0)$, at the instant of excitation

$$r(t) = (2/5) \langle P_2(\boldsymbol{\mu}_{em}(t) \bullet \boldsymbol{\mu}_{ab}(0)) \rangle \qquad (10)$$

where P_2 is the Legendre polynomial of the second order and the brackets denote ensemble averaging [31,32].

4. Experimental Examples

In the following we consider some published and recent unpublished results that illustrate the application of fluorescence methods to characterize polymer micelle cores.

4.1 FORMATION OF POLYMER MICELLES

In some of the prior work on polymer micelles by Winnik et al.[22],[23] the takeup of pyrene that accompanied micellar formation was used to establish the CMC. In our own work with polymers like those prepared in reaction sequence (ii) excimer fluorescence of the naphthalene diminished dramatically when micelles were formed [27],[5(c)] (a similar phenomenon occurs when the core is swollen by a selective solvent, i.e. excimer fluorescence increases, see section 4.2).

In recent work micelle formation using polymers like **8** with different grafting densities was followed as a function of solvent composition via the monomer and excimer fluorescence (see Fig. 1 and 2) which correlated very well with the change in the intensity of scattered light (Fig. 3).[13] The naphthalene chromophore is very sensitive to environment and has a strong tendency to form excimers, especially in water [33,34]. The photophysics reflect the change in polymer configuration that occurs by varying the composition of the dioxane/water mixed solvent. The near-isobestic point at approximately 360 nm in Figure 1 indicates the presence of only two fluorescent species, monomer and excimer.

In Figure 2 we plot the monomer to excimer ratio as a function of solvent composition. It is seen there that the monomer to excimer ratio decreases dramatically until the water content reaches 30% and decreases more gradually thereafter. This trend is similar for all graft copolymer micelles and we infer that the environment of the backbone in mixed solvents does not vary with the number of grafts. Combining these complementary results suggests that the backbone first collapses with the addition of ca. 30% water to produce excimer fluorescence and then these collapsed polymers form micelles as the percent of water increases.

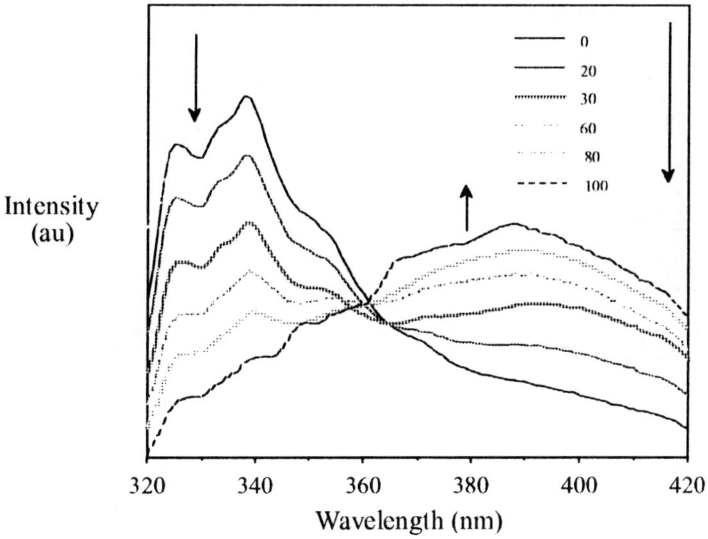

Figure 1 Steady state fluorescence spectra of graft copolymer micelle as a function of percentage water in dioxane solutions.

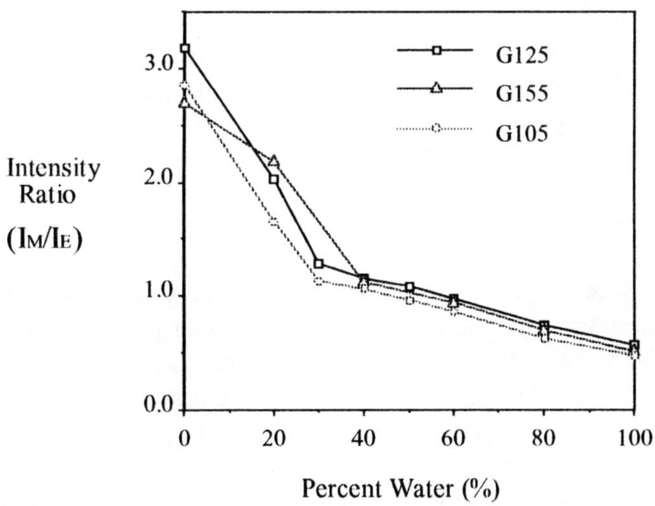

Figure 2 Monomer to excimer emission intensity ratio as a function of the percentage water in dioxane solutions for three graft copolymer micelles.

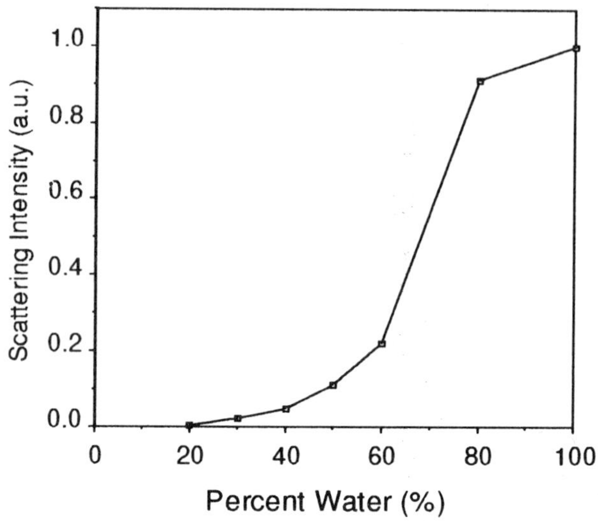

Figure 3 Scattering intensity as measured by QELS for a typical micelle as a function of percentage water in dioxane solutuion.

One also obtains insight to the behavior of polymer **8** as a function of solvent from the time resolved fluorescence decays for both monomer and excimer. It is interesting that there is a significant rise time for excimer formation at all solvent compositions. In Fig. 4 we plot the average decay of the monomer and the rise time of the excimer as a function of solvent composition.

The rise time can be attributed to two phenomena that contribute to the excimer emission: (1) segmental diffusion or rotation during the naphthalene singlet excited state lifetime to a position of co-facial overlap with a neighboring naphthalene, which favors excimer formation; (2) energy migration (presumably through the Förster mechanism) to some pre-formed excimer site [35]. Since the lifetime of the singlet state is longer at low concentrations of water, the former mechanism is likely to dominate. The relatively long rise times at these low concentrations of water are reasonably assigned to segmental motion. At higher concentrations of water it seems reasonable to expect that the naphthalenes are collapsed into close contact, maximizing the probability for Förster energy transfer [36]. Finally at 100% water the naphthalene moieties are locked into the glassy micelle core where segmental motion is limited, and energy migration to an excimer forming site is likely to dominate, in addition to direct excitation of excimer sites. Energy transfer/migration can occur on a much faster time scale than segmental motion, consistent with the shorter rise times observed [37].

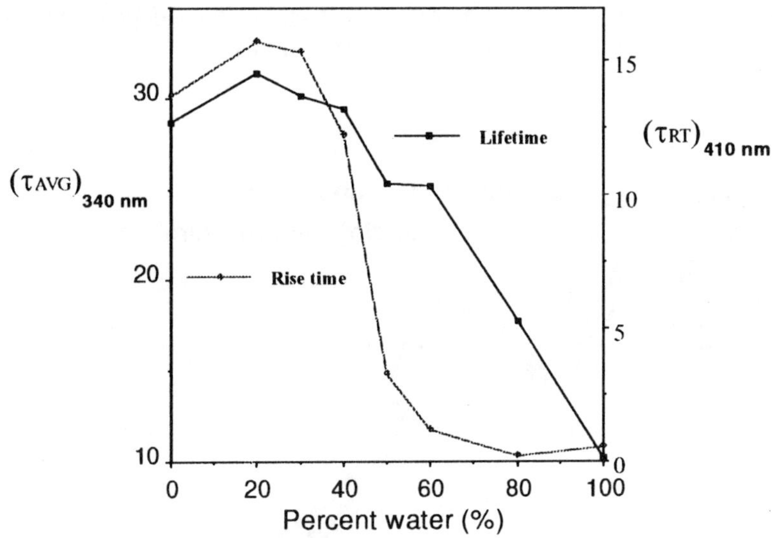

Figure 4 Average lifetime for the monomer (at 340 nm) and rise time for the excimer (at 410 nm) for a typical graft copolymer micelle as a function of percentage of water in dioxane/water solutions

The doubly tagged polymer **6** (with and without the polymethacrylic acid block) was prepared in order to examine FET from the naphthalene to the anthracene for the unimer and micelle with different solvent compositions. This data was expected to provide elucidation of the end-to-midpoint distribution of separations at different stages of polymer self-organization. However it was found that the naphthalene moiety was so sensitive to its environment that an interpretation of the shortening of the naphthalene fluorescence lifetime was precluded. An unexpected observation was made that the naphthalene spectrum changes dramatically as a function of solvent composition during micelle preparation (Fig. 5). This effect was observed for very low mole fractions of tagged polymers co-micellizing with untagged polymer, such that it cannot be ascribed to excimer fluorescence (also the wavelength is different, see the following). Attempts to reproduce this spectral modification in environments such as dry or dioxane-swollen polystyrene films were unsuccessful. Other model systems such as an analogous, "all-methyl" system (1-methyl naphthalene in 50% acetic acid/toluene) also failed to show similar red shifted emission. However, when any of our polymers is dissolved in a concentrated (ca. 70 wt. %) dioxane solution of the unlabeled copolymer, the same broadened, red-shifted naphthalene emission is observed (Figure 6). This is not a result of excimer formation because the emission maximum is ca. 350 nm (not 410 nm) and there is minimal and constant concentration of N-pS-A. The red-shift does exhibit a dependence on the concentration of the unlabled pS-pMA polymer (Figure 6).

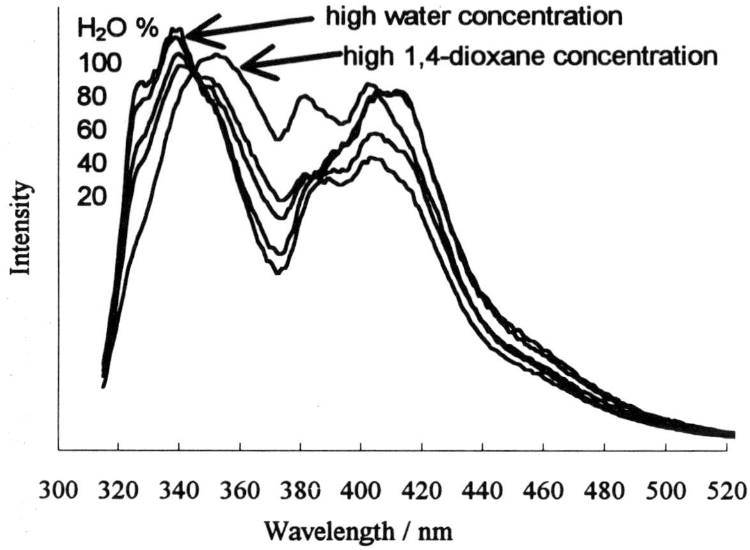

Figure 5. Steady-state emission (λ_{ex} = 293 nm) for N-pS-A-pMA/pS-pMA comicelles recorded at 20% increments in dioxane/water concentration from 80% to 0%

Thus we conclude that the red-shifted naphthalene fluorescence is a result of the interaction with pMA segments in a dioxane-rich environment (note that this experiment could not be carried out with pure pMA because of its poor solubility in dioxane). Since the analogous effect was not observed for a model compound in acetic acid/toluene, we speculate that this must be the result of the simultaneous interaction of many acid groups with the fluorophore. However the more important implication is that at dioxane contents down to 30 volume % the local environment of the micelle must be similar to a concentrated PS-pMA solution. In other words, the strong segregation into a core and corona structure has not occurred. Thus we envision a structure much more like a "droplet" containing the same number of chains as the final micelle but without the strong phase separation of the naphthalene units from the polymethacrylic acid part of the polymer.

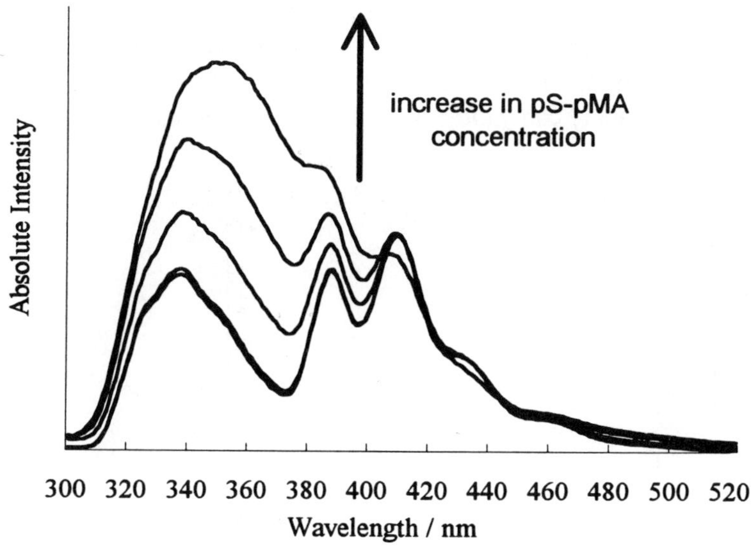

Figure 6. Steady-state emission (λ_{ex} = 293 nm) spectra as a function of pS-pMA concentration in dioxane. For a constant N-pS-A concentration. The pS-pMA concentrations are 0.00, 0.44, 8.97, 23.45 and 72.40 mg/ml.

4.2 MOBILITY IN SWOLLEN CORES

Poly(2-vinylnaphthalene)-*block*-polystyrene-*block*-poly(methacrylic acid) was prepared with a sequence like (ii) (denoted N_n-S-MA in the following). Micelles formed from N_n-S-MA offered the opportunity to study the swelling of the polystyrene core by the addition of trace amounts of benzene to the aqueous solution. The increase in excimer fluorescence for N_4-S-MA tracked beautifully the decrease in r_∞ for mixed micelles containing N_1-S-MA (see Figure 7). (r(t) as a function of benzene to styrene ratios is shown in Figure 8). The fast dynamic part should be susceptible to molecular dynamics modeling, assuming that all energy transfer effects have been removed by dilution (swollen cores show less energy transfer than collapsed cores). However this has not been attempted because of its complexity. The qualitative information from these experiments is clear. At a benzene to styrene ratio of ca. 3 the core becomes rubbery with rapid motion of pendent naphthalenes. Despite this mobility the micelles remain stable and non-aggregated up to a benzene to styrene ratio of 5-6 and are stable over shorter times for a weight ratio up to ca. 9. Note that the stability of the micelle:benzene system depends dramatically on the rate of benzene addition. Adding benzene too quickly can yield aggregates and precipitation.

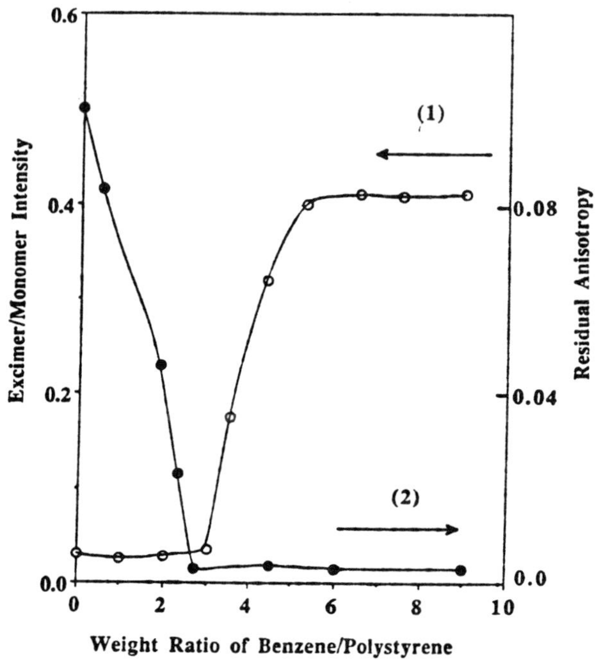

Figure 7 Curve 1: Steady state excimer to monomer emission; Curve 2; r_∞ for N1-S-MA as a function of added benzene

4.3 SMOOTHNESS OF THE CORE-CORONA INTERFACE

Micelles can be formed from the polymer polystyrene-*block*-poly(vinyl 9,10-diphenylanthracene)-*block*-poly(methacrylic acid) (denoted S-vDPA-M) produced via sequence (ii). Micelles can also be produced by diluting this probe polymer with untagged polymer so long as the degree of polymerization for the hydrophobic blocks is similar. The quenching of the diphenylanthracene fluorescence by Tl^+ is not very effective and is only modestly dependent on pH (see Figure 9). The quenching curve has negative curvature indicative of hindered access by the Tl^+ to the fluorophore (see equation 8) and when the data is fit to this equation the fraction of accessible chromophore is in the range 0.41-0.47 for all pH values examined. For a linear polyacid with a small fraction of covalently bound chromophore the quenching by a monovalent ion is a very strong function of pH [38,39]. These quenching results for the micelle imply an indistinct interfacial region and relatively little deprotonation of the poly(methacrylic acid) in the region near the core-corona interface.

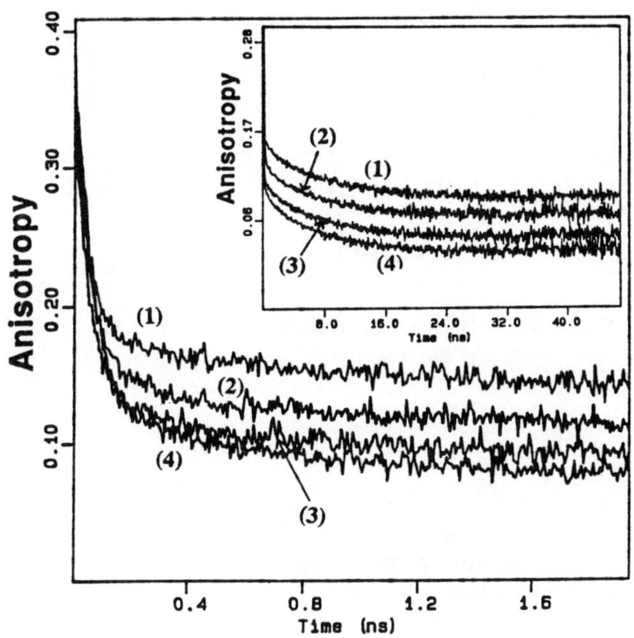

Figure 8 r(t) of N1-S-MA in water with the following weight ratios of benzene to styrene: Curve 1, 0; Curve 2, 0.5; Curve 3, 1.2; Curve 4, 1.8. Insert: same curves on a longer time scale

Very similar observations were made for micelles formed from S-N_1-PMA using Tl^+ as a quencher [24], or using I^- to quench 8 [13]. Thus we conclude that a wide range of environments exists for these mid-polymer fluorophores and a significant fraction are not in contact with the aqueous phase. Because this quenching process requires close contact of the fluorophore and Tl^+, the root-mean-square roughness of the interface cannot be established from these results. In future work we will attempt to estimate this property from energy transfer experiments.

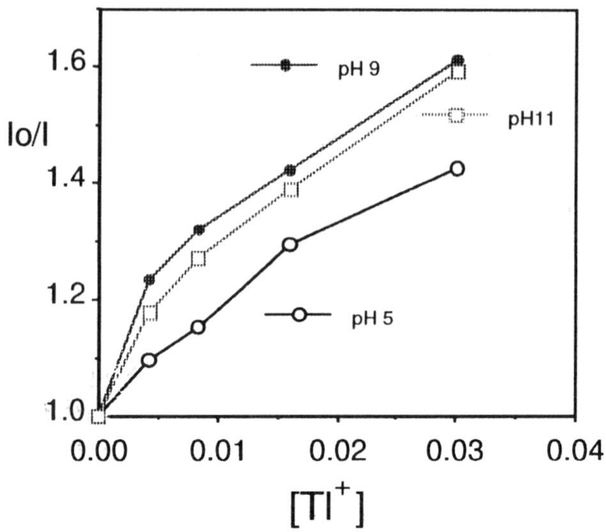

Figure 9 Quenching curves of S-vDPA-M by Tl^+ at different pHs.

Acknowledgments

Our research on polymer micelles has been supported over the years by the Army Research Office (Grant DAAH04-95-1-0127), the National Science Foundation Polymers Program (Grant DMR-9308307), and the Robert A. Welch Foundation (Grant F-356), and is gratefully acknowledged.

References

1. (a) Quirk, R.P.; Perry, S.; Medicuti, F.; Mattice, W.L. (1988) *Macromolecules* **21**, 2294; (b) Quirk, R.P.; Zhu, L. (1989) *Makromol. Chem.* **190**, 487; (c) Quirk, R.P. Schock, L.E. (1991) *Macromolecules* **24**, 1237
2. Caldérava, F.; Mruska, Z.; Hurtrez, G.; Lerch, J.-P.; Nugay, T.; Riess, G. (1994) *Macromolecules* **27**, 1210
3. (a) Ni, S.; Juhué, D.; Moselhy, J.; Wang, Y.; Winnik, M.A. (1992) *Macromolecules* **25**, 496; (b) Chan, L.; Winnik, M.A.; Al-Takrity, E.T.A.; Jenkins, A.D.; Welton, D.R.M. (1987)

Makromol. Chem. **108**, 2621; (c) Hruska, Z.; Vuillemin, B.; Riess, G.; Katz, A.; Winnik, M.A. (1992) *Makromol. Chem.* **193**, 1987

4. Martin, T. J.; Webber, S. E.(1995) *Macromolecules* **28**, 8845
5. (a) Tuzar, Z.; Webber, S.E.; Ramireddy, C.; Munk, P. (1991) *Polymer Prepr., Am. Chem. Soc., Div. Polym.Chem.* **32(1)**, 525; (b) Cao, T.; Munk, P.; Ramireddy, C.; Tuzar, Z.; Webber, S.E. (1992) *Macromolecules* **24**, 6300; (c) Procházka K.; Kiserow D.; Ramireddy C.; Tuzar Z., Munk P.; Webber S.E. (1992) *Macromolecules* **25**, 454
6. Sowash G.; Webber, S. E. (1988) *Macromolecules* **21**, 1608
7. I.B. Berlman, (1973) *Energy Transfer Parameters of Aromatic Compounds*, Academic Press, New York,
8. (a) Yokota, M.; Tanimoto, O. J. (1967) *Phys. Soc. Jpn.* **22**, 779; (b) Liu, G.; Guillet, J.E. (1990) *Macromolecules* **23**, 2973 and references therein.
9. (a) Ni, S.; Juhué, D.; Moselhy, J.; Wang, Y.; Winnik, M.A. (1992) *Macromolecules* **25**, 496; (b) Chan, L.; Winnik, M.A.; Al-Takrity, E.T.A.; Jenkins, A.D.; Welton, D.R.M. (1987) *Makromol. Chem.* **108**, 2621; (c) Hruska, Z.; Vuillemin, B.; Riess, G.; Katz, A.; Winnik, M.A. (1992) *Makromol. Chem.* **193**, 1987
10. (a) Liu, G.; Guillet, J. E. (1990) *Macromolecules* **23**, 1388; (b) *ibid.* 1393.
11. This material was prepared by standard anionic polymerization methods using cumyl potassium as an initiator. (Ramireddy, C.; Tuzar. Z.; Procházka, K.; Webber, S.E.; Munk, P. (1992) *Macromolecules* **25**, 2541). The amino termination was accomplished using a protected a-halo-w-aminoalkane (Ueda, K.; Hirao, A.; Nakahama, S. (1990) *Macromolecules* **23**, 939).
12. Cao, T.; Yin, W.; Webber, S. E. (1994) *Macromolecules* **27**, 7459-
13. Eckert, A; Webber, S. E. (1996) *Macromolecules* **29**, 560.
14. (a) Förster, T. (1948) *Ann. Phys.* **2**, 55; (b) Förster, T Z. (1949) *Naturforsch. A* **4**, 321; (c) Förster, T. (1959) *Discuss. Faraday Soc.* **27**, 7
15. *Molecullar Dynamics in Restricted Geometries*, Eds. J. Klafter and J. M. Drake, (1989) John Wiley & Sons: New York
16. Procházka, K.; Vajda, S.; Fidler, V.; Bednár, B.; Mukhtar, E.; Almgren, M.; Holmes, S. J. (1990) *Mole. Struct.* **219**, 377
17. Wang, Y.; Balaji, R.; Quirk, R.; Mattice, W.L. (1992) *Polymer Bulletin* **28**, 333
18. Duhamel, J.; Yekta, A.; Ni, S.; Khaykin, Y.; Winnik, M.A. (1993) *Macomolecules* **26**, 6255
19. Cao, T.; Munk, P.; Ramireddy, C.; Tuzar, Z.; Webber, S.E. (1991) *Macromolecules* **24**, 6300
20. "Fluorescence Studies of Pyrene Capture by Naphthalene Labeled Diblock Copolymer Micelles in Aqueous Media", S.L. Fox, J. Chan, D. Kiserow, C. Ramireddy, P. Munk, and S.E. Webber (1996) *Hydrophilic Polymers, Advances in Chemistry, No. 248*: J.E. Glass, Ed.; American Chemical Society, Washington, D.C. , Chap. 9, p. 141-162.
21. Arca, E.; Tian, M.; Webber, S.E.; Munk, P. (1996) *Polymer Analysis and Characterization* **2**, 31
22. Zhao, C-L; Winnik, M.A .; Riess, G.; Croucher, M.D. (1990) *Langmuir* **6**, 514
23. Wilhelm, M.; Zhao, C.-L.; Wang, Y.; Xu, R.; Winnik, M.A.; Mura, J.-L.; Riess, G.; Croucher, M.D. (1991) *Macromolecules* **24**, 1033.
24. Cao, T.; Yin, W.; Armstrong, J.L.; Webber, S.E. (1994) *Langmuir* **10**, 1841
25. Lakowicz, J.R. (1986) *Principles of Fluorescence Spectroscopy*; Plenum Press: New York.
26. O'Connor, D. V; Phillips, D. (1984) *Time-Correlated Single Photon Counting*;, Academic Press: Florida,

27. Kiserow, D. J.; Prochazka, K.; Ramireddy, C.; Tuzar, Z.; Munk, P.; Webber, S. E. (1992) *Macromolecules* **25**, 461.
28. Quitevis, E. L.; Marcus, A. H.; Fayer, M. D. (1993) *J. Phys. Chem.* **97**, 5762
29. Anufrieva, E. V.; Gotlib, Y. Y. (1981) *Advances in Polymer Science (Luminescence)*, **40**, , Springer-Verlag Berlin Heidelberg, New York
30. Fleming G. R. (1986) *Chemical Applications of Ultrafast Spectroscopy*. Oxford University Press, New York
31. Michl J., Thulstrup E.W. (1986) *Spectroscopy with Polarized Light.*, VCH Publishers, New York
32. Procházka K., Medhage B., Almgren M., Mukhtar E., Svoboda P., Trnaná J., Bednár B. (1993) *Polymer* **34**, 103
33. Kiserow, D.; Chan, J.; Ramireddy, C.; Munk, P.; Webber, S. E. (1992) *Macromolecules* 25, 5338.
34. (a) Morishima, Y.; Kobayashi, T.; Nozakura, S-I.; Webber, S. E. (1987) *Macromolecules* **20**, 807; (b) Morishima, Y.; Lim, H. S.; Nozakura, S-I.; Sturtevant, J. L. (1989) *Macromolecules* **22**, 1148
35. Morishima, Y.; Tominaga, Y.; Nomura, S.; Kamachi, M. (1992) *Macromolecules* **25**, 861.
36. From ref. [7] for naphthalene-naphthalene transfer R_0 = 11 Å. In ref.[35] R_0 is estimated to be 8.3 Å for substituted naphthalenes.
37. (a) Byers, J. D.; Friedrichs, M.; Friesner, R. A.; S.E. Webber (1988) *Macromolecules* **21**, 3402; (b) Byers, J. D.; Friedrichs, M.; Friesner, R. A.; S.E. Webber (1989) in ref. [15]. p. 99; (c) Byers, J. D.; Parsons, W. S.; Friesner, R. A.; Webber, S.E. (1990) *Macromolecules* **23**, 4835; (d) Byers, J.D.; W.S. Parsons; Webber, S.E. (1992) *Macromolecules* **25**, 5935
38. Chu, D-Y.; Thomas, J.K. (1984) *Macromolecules* **17**, 2142-
39. Delaire, J. A.; Rodgers, M.A.J.; Webber, S. E. (1984) *J. Phys. Chem.* **88**, 6219.

FLUORESCENCE DEPOLARIZATION METHOD TO STUDY POLYMER CHAIN DYNAMICS

MASAHIDE YAMAMOTO, KEIKO ONO, AND SHINZABURO ITO
Division of Polymer Chemistry, Graduate School of Engineering
Kyoto University
Sakyo-ku Kyoto 606-01, Japan

1. Introduction

The dynamic property of a polymer chain is important in understanding the macroscopic physical properties. The molecular motion of polymer chains in dilute solution consists of the Rouse mode and segmental mode. The relaxation time of the former is in the sub-microsecond to millisecond order and that of the latter is in the nano- and subnano-second order. The nature of segmental motion has been extensively studied [1-3], but still the details remain unknown.

The local motion in a dilute solution has been studied by neutron scattering [4], NMR [5], ESR [3,5], dielectric relaxation [6,7], dynamic light scattering [8], and fluorescence depolarization [9-12]. In our laboratory, the local motion of various polymers has been studied by the fluorescence depolarization method, where a fluorescent probe is usually introduced in a specific part of a polymer chain to measure the relaxation time of the local motion.

Recently, we have reported the dynamic behavior of poly(alkyl methacrylate)s [13], polystyrenes [14], and poly(cis-isoprene) [15]. The effects of molecular structure, stereoregularity, molecular weight, solvent quality on the local motions were studied. The relaxation time of local motion of polymer chains depends so much on these factors. Herein we examined the relaxation times of polystyrene (PS) by the fluorescence depolarization method. First, we reviewed the

S.E. Webber et al. (eds.), Solvents and Self-Organization of Polymers, 479–498.

fluorescence depolarization method as a technique to measure the relaxation times of local chain motions. Second, we examined the relationship between the reduced relaxation time and the local segment density for PS in various solvents [14]. Here, we used the solvents with a viscosity less than 1 cP at 20 °C, and examined the effects of chain expansion on the relaxation time and the activation energy. Finally, we studied the local motion of PS in high viscosity solvents [16]. High viscosity has an anomalous effect on the activation energies of the reduced relaxation times of polymer local motions. The apparently abnormal activation energy values have been appropriately treated and the nature of segmental motion of polymer chains is discussed.

2. Fluorescence Depolarization Method

2-1. PRINCIPLE OF FLUORESCENCE DEPOLARIZATION [9-11]

When both light absorption and fluorescence emission belong to the same electronic transition, namely S_0 -S_1, the absorption oscillator and the emission oscillator can be assumed to coincide and they can be represented by a transition moment, *M*,

If the plane-polarized light, *P*, is incident on a solution of fluorescent molecules with *M*, the

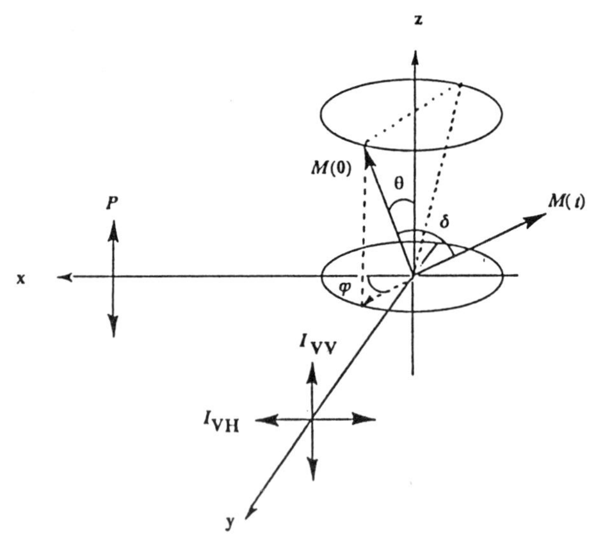

Figure 1. Optical geometry of fluorescence depolarization method.

probability of excitation, $p(\theta)$, is given as

$$p(\theta) \propto \cos^2\theta \quad (1)$$

where θ is the angle between P and M (Figure 1). Then, the emission from an excited molecule is represented by the transition moment M. The emission vector is randomized with time by the rotational Brownian motion of the molecule and fluorescence depolarization occurs.

In the steady state measurements, the degree of polarization, p, is defined as

$$p = (I_{VV} - I_{VH}) / (I_{VV} + I_{VH}) \quad (2)$$

or the averaged anisotropy ratio, r, is defined as

$$r = (I_{VV} - I_{VH}) / (I_{VV} + 2I_{VH}) \quad (3)$$

where I_{VV} and I_{VH} are the emission intensities of parallel and perpendicular components to P, respectively.

In the time-resolving measurements, the time-dependent anisotropy ratio $r(t)$ is defined by

$$r(t) = (I_{VV}(t) - I_{VH}(t)) / (I_{VV}(t) + 2I_{VH}(t)) \quad (4)$$

This anisotropy ratio is directly related to the orientational autocorrelation function of the transition moment of the fluorescent probe, $\Phi(t)$ as

$$r(t) = r_0\, \Phi(t) \quad (5)$$

where r_0 is the initial anisotropy ratio (ideally $r_0 = 0.4$). Here, the autocorrelation function $\Phi(t)$ is defined as

$$\Phi(t) = \langle P_2[\boldsymbol{M}(0)\cdot\boldsymbol{M}(t)]\rangle = (3\langle\cos^2\delta\rangle - 1) / 2 \quad (6)$$

where P_2 is a Legendre polynomial of the second order, δ is the angle between $M(0)$ and $M(t)$, and the bracket represents ensemble average.

Eq (5) shows that the orientational autocorrelation function can be obtained by the fluorescence depolarization method without any assumption of models. In addition, when the lifetime of a fluorescent molecule is longer than the relaxation time of investigating molecules, the relaxation time estimated by $r(t)$ is accurately obtained.

The averaged anisotropy ratio r is related to $\Phi(t)$ as

$$r = r_0 \int_0^\infty \Phi(t)\, I(t)\, dt \,/\, \int_0^\infty I(t)\, dt \tag{7}$$

where $I(t)$ is the normalized total fluorescence intensity which is evaluated as $I_{VV}(t) + 2I_{VH}(t)$.

If $I(t)$ decays in a single exponential term, i.e., $I(t) = \exp(-t/\tau)$ with the lifetime τ of a fluorescent probe,

$$r = \tau^{-1} \int_0^\infty r(t) \exp(-t/\tau)\, dt \tag{8}$$

Furthermore, if $\Phi(t)$ decays in a single exponential term, i.e., $\Phi(t) = \exp(-3t/\rho)$ with the rotational relaxation time ρ, eq (8) becomes,

$$r = r_0 / (1 + 3\tau/\rho) \tag{9}$$

or

$$(p^{-1} - 1/3) = (p_o^{-1} - 1/3)(1 + 3\tau/\rho) \tag{10}$$

where p_0 is the limiting value of p for an infinitely long relaxation time. This is the case when the fluorescence molecule can be regarded as a sphere and the relation $\rho = 3\eta V/kT$ is valid (η; medium viscosity, V; hydrodynamic volume). Most real fluorescent molecules are far from a spherical body, and in general, time-dependent measurement of $r(t)$ becomes desirable.

2.2. EXPERIMENTAL TECHNIQUE

The time-resolved anisotropy ratio, $r(t)$, has been measured by using a single photon counting (SPC) system [17]. Figure 2 shows the instrumental scheme, which consists of a light source, 100MHz discriminator (Ortec model 436), delay box (Ortec DB463), microchannel plate-photomultiplier tube (MCP-PMT, Hamamatsu Photonics), a constant fraction discriminator (Ortec model 583), time-to-amplitude converter (TAC, Ortec model 457), and multichannel analyzer (MCA, Norland Ino-Tech 5300). We used a diode laser (Hamamatsu Photonics Picosec Light Pulser, PLP-01) as a light source. Its wavelength is fixed at 411 nm. The diode laser has many merits, i.e., high and stable power, high repetition rate up to 10 MHz, time-independent pulse profile, short pulse width less than 50 ps, simple operation, and a compact form convenient for optical arrangements. Although this

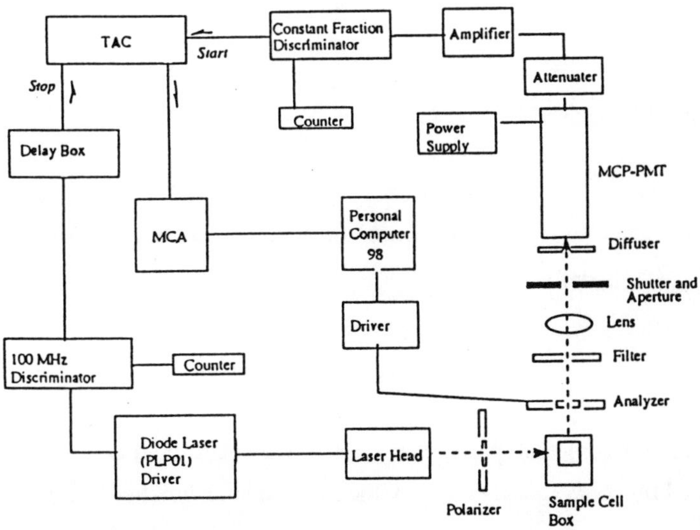

Figure 2. Block diagram of a single-photon-counting system.

pulse is vertically polarized, a polarizer (Glan-Thompson prism) was placed between the laser head and cell box. In front of the detector, an analyzer (Polaroid HNP'B) and a suitable set of cutoff filters were inserted. It is necessary to obtain the profile of the instrumental response function, by which the decay curve of fluorescence is deconvoluted. The laser pulse profile was measured by using Ludox as a scatterer for estimation of instrumental function. The pulse drift was less than 5 channels

(0.136 ns / channel) in several hours, and the anisotropy ratio of one sample is measured within ca. 30 min. Therefore, the influence of drift for the data is negligible in this time range. HOYA Y-44 and V-42 were used as filters to measure the fluorescence from the samples, and SC-41 was used to measure the scattering of Ludox solution. The analyzer was controlled and rotated 90° every 100 s by a computer to measure I_{VV} and I_{VH} alternately and to avoid the time dependence of the two intensities, I_{VV} and I_{VH}. MCP - PMT was used at 2.9 kV. FWHM of the instrumental function by this MCP - PMT was ca. 350 ps. The SPC system was used at the inverse mode; the photon detected by MCP - PMT as the start signal and the electric trigger signal of the diode laser as the stop one. The temperature of the sample was controlled by circulation of water from a thermoregulated bath with the temperature controller TTM-104 (TOHO) and Pt100 resistance, and its deviation was ±0.1 ℃.

2.3. ANALYSIS [17, 18]

The emission intensities in the vertical direction ($I_{VV}(t)$) and horizontal direction ($I_{VH}(t)$) were collected as the function of time, t. The decay of total fluorescence intensity, $I_{obsd}(t)$, is defined as

$$I_{obsd}(t) = I_{VV}(t) + 2G\, I_{VH}(t) \tag{11}$$

where G is the instrumental parameter. Here, we determined G = 1.0094. $I_{calcd}(t)$ was evaluated by convolution of $I(t)$ by the instrumental function, $P(t)$.

$$I_{calcd}(t_i) = \int_0^{t_i} P(T)\, I(t_i - T)\, dT \tag{12}$$

where eq 13 was employed as the function for $I(t)$.

$$I(t) = a_1 \exp(-t/t_1) + a_2 \exp(-t/t_2) \tag{13}$$

The parameters of $I(t)$ could be estimated by fitting $I_{calcd}(t)$ to $I_{obsd}(t)$ by the method of nonlinear-least-squares with weighting factors by propagating the Poisson error as

$$w_i = [\, I_{VV}(t_i) + 4G^2 I_{VH}(t_i) \,]^{-1} \tag{14}$$

The reduced sum of the squares of the residuals, χ^2, was evaluated as a criterion for goodness of fitting, and it is defined by eq 15,

$$\chi^2 = \{1 / (n - m)\} \; \Sigma w_i \, [\, I_{obsd}(t_i) - I_{calcd}(t_i) \,]^2 \tag{15}$$

where n is the number of data points and m the number of variable parameters.

The observed time-resolved anisotropy ratio is given by

$$r_{obsd}(t) = (I_{VV}(t) - G\, I_{VH}(t)) \,/\, (I_{VV}(t) + 2G\, I_{VH}(t)) \tag{16}$$

As for Φ(t), we used an empirical equation, eq 17, .

$$\Phi(t) = \Sigma \; g_i \exp(-t/T_i) \tag{17}$$

Practically we use eq (18) with the sum of two exponentials.

$$\Phi(t) = g_1 \exp(-t/T_1) + g_2 \exp(-t/T_2) \tag{18}$$

Eq 18 shows a good agreement with the anisotropy decay data in a wide viscosity range.

$r_{calcd}(t_i)$ was evaluated by convolution as

$$r_{calcd}(t_i) = r_0 \int_0^{t_i} P(T)\, \Phi(t_i - T)\, I(t_i - T)\, dT \,/\, \int_0^{t_i} P(T)\, I(t_i - T)\, dT \tag{19}$$

$r_{calcd}(t_i)$ was also fitted to $r_{obsd}(t_i)$ by the method of nonlinear-least-squares with weighting factors by propagating the Poisson error as

$$w_i = 3I_{obsd}(t_i) \,/\, \{1 + G + 3Gr_{obsd}(t_i) - 3[r_{obsd}(t_i)]^2 - 2(2G - 1)[r_{obsd}(t_i)]^3\}$$

(20)

$$\chi^2 = \{1/(n-m)\} \; \Sigma w_i \, [r_{obsd}(t_i) - r_{calcd}(t_i)]^2 \tag{21}$$

As a criterion for goodness of fitting, χ^2 defined by eq 21 was used. The parameters g_1, g_2, T_1, T_2, and r_0 were determined as the best-fit ones.

The mean relaxation time is defined as

$$T_m = r_0^{-1} \int_0^\infty r(t)\, dt = \int_0^\infty \Phi(t)\, dt = \Sigma g_i T_i \tag{22}$$

Then, when i =2, we get eq 23.

$$T_m = g_1 T_1 + g_2 T_2 \tag{23}$$

In our work, the sum of two exponentials is sufficient to fit the simulated curve to the observed one. Figure 3 shows a fitting example of r(t) with eq 18. The parameters, T_1 and T_2, may reflect two kinds of representative motional modes, e.g., an isolated transition and the cooperative counter-rotation in the Hall and Helfand model [19]. However, at this stage the physical meaning of these parameters is not known. In the following discussion, we use the mean relaxation time, T_m.

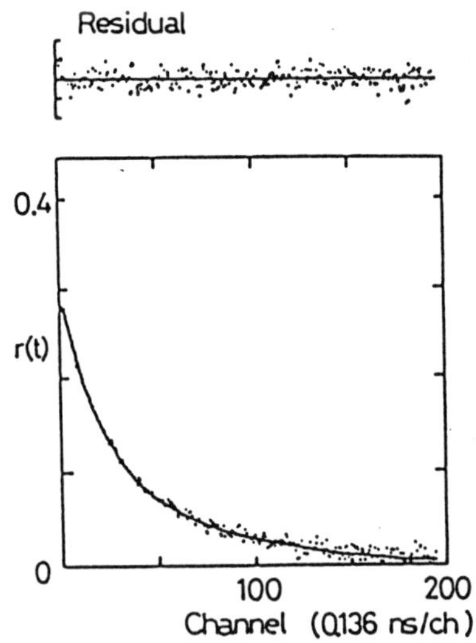

Figure 3. Decay curve of anisotropy ratio measured for PS in cyclohexane at 31.5 °C (···) and its fitting curve by the method of nonlinear-least-squares (—).

2.4. EVALUATION OF E^* BY THE THEORY OF KRAMERS' DIFFUSION LIMIT

We first evaluated the apparent activation energy of local motion E^* by the theory of Kramers' diffusion limit [19]. When a particle with a frictional coefficient ζ passes over an energy barrier with the height E^*, the velocity coefficient k is presented by eq 24,

$$k \propto \zeta^{-1} \exp(-E^*/RT) \tag{24}$$

where R is the gas constant and T is absolute temperature. T_m is proportional to the reciprocal of k and by using the Stokes law $\zeta \propto \eta$, the following equation is derived.

$$T_m = A\,\eta \exp(E^*/RT) \tag{25}$$

Therefore the apparent activation energy of the local motion of polymer, E^*, can be obtained from the slope of the plot of $\ln(T_m/\eta)$ vs. $1/T$ as seen from eq 26.

$$T_m/\eta = A \exp(E^*/RT) \tag{26}$$

Solvent viscosity is represented by eq 27,

$$\eta = \eta_0 \exp(E_\eta/RT) \tag{27}$$

where E_η is the activation energy of the solvent viscosity. Then, eq 26 can be modified to eq 28.

$$T_m = A\,\eta_0 \exp\{(E^*+E_\eta)/RT\} \tag{28}$$

The slope of the plot of $\ln(T_m)$ against $1/T$ gives (E^*+E_η). Then the value of E^* can be obtained by subtracting E_η from (E^*+E_η).

3. Local Chain Dynamics of Polystyrene

3.1. SAMPLE PREPARATION

Anthracene-labeled polystyrene (PS) chains were prepared by anionic polymerization initiated by n-butyl lithium and the living end was deactivated by a bifunctional terminator 9,10-bis(bromomethyl)anthracene (Figure 4). The prepared polymer was fractionated with GPC into two parts: the PS labeled at the middle of main chain, and that terminated by anthracene at the chain end. The former is employed for fluorescence depolarization measurements. The obtained PS has a Mw of 9.7×10^4, *Mw*/*Mn* of 1.05, and racemo fraction of dyad of 0.54.

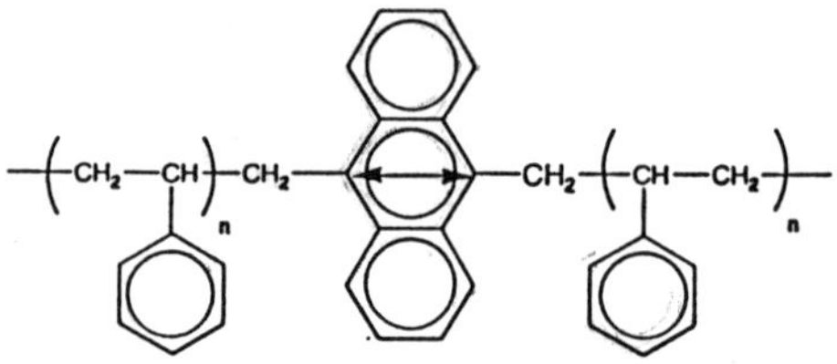

Figure 4. **Anthracene-labeled PS sample.**

TABLE 1. Solvent viscosity at 20 °C, its activation energy, and chain expansion factor at 34.5 °C for PS solution

solvent	η (cP)	E_η (kcal mol^{-1})	$\overline{\alpha_\eta^3}$
CH	0.979	2.9	1.00
MEK	0.402	1.7	1.11
EA	0.447	1.7	1.12
BA	0.740	2.4	1.26
BZ	0.652	2.5	1.66
TO	0.587	2.1	1.64
EB	0.678	2.2	1.64

The chromophore concentration of all polymer solutions was kept at less than 10^{-5} mol/l, i.e., the polymer concentration of all samples was kept at less than 3×10^{-3} g / cm^3 and the samples were degassed. The polymer sample was dissolved in seven kinds of solvents, i.e., cyclohexane (CH), methyl ethyl ketone (MEK), ethyl acetate (EA), butyl acetate (BA), benzene (BZ), toluene (TO), and ethylbenzene (EB). Table 1 shows their solvent viscosity and its activation energy. All the solvents are in low viscosity below 1 cP. By capillary viscometry, we evaluated the chain expansion parameter $\overline{\alpha_\eta^3}$, which is the ratio of the intrinsic viscosity of a sample solution to the intrinsic viscosity at the θ condition. Cyclohexane was used as the θ solvent for PS, and the θ temperature was 34.5 °C.

3.2. SOLVENT EFFECT ON THE RELAXATION TIME OF LOCAL CHAIN MOTIONS

Figure 5 shows the relationship between the reduced relaxation time, T_m/η, and the chain expansion factor, $\overline{\alpha_\eta^3}$, for PS at 34.5 °C, which is the θ temperature in cyclohexane for PS. The reduced relaxation time in the θ solvent was the longest of all PS solutions, and T_m/η became shorter as the solvent quality became better. The polymer chain moves faster in good solvents than in poor solvents.

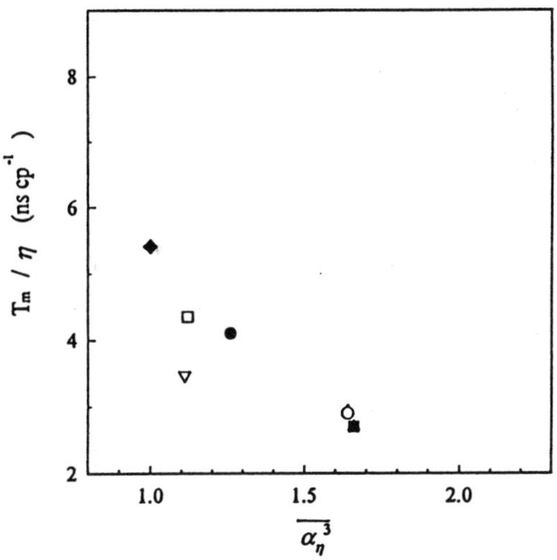

Figure 5. Plots of reduced relaxation time T_m/η vs. $\overline{\alpha_\eta^3}$ for PS solution. cyclohexane(◆); MEK (▽); ethyl acetate (□); butyl acetate (●); benzene (■); toluene (○); ethylbenzene(△).

The local conformational transition is considered to depend on the local segment density of the polymer chain through the excluded volume effect. As the solvent quality changes from poor to good, the polymer chain in the solution becomes more expanded, and the local segment density of the polymer chain becomes smaller. Ediger et al. [20] calculated the local segment density around the center segment of a polyisoprene chain in a good solvent and in a θ solvent. They emphasized that the center segment investigated is different from the center of mass of the chain. They reported that the largest change in local chain dynamics with solvent quality occurs for a high molecular weight polymer. According to their calculation, for a polymer with Mw = $1x10^5$, the local segment density in a θ solvent is 3 times that in a good solvent.

In a good solvent, the polymer chain interacts with solvent molecules more strongly than with the segments of the polymer chain itself. The local motion of the conformational transition of a polymer chain has a smaller resistance, and the relaxation time becomes shorter in good solvents than in poor solvents. The conformational transition of a polymer chain strongly depends on the segment density .

3.3. SOLVENT EFFECT ON THE ACTIVATION ENERGY OF LOCAL CHAIN MOTION

Figure 6 shows the plots of $\ln(T_m/\eta)$ vs $1/T$ for each PS solution. In a cyclohexane solution a line has an inflection point at ca. 20 °C, which is ca. 10 °C below the θ temperature. This seems to be due to the microscopic aggregation of PS. For PS solutions, the value of E^* above the θ temperature was largest in the θ solvent and became smaller with improvement of the solvent quality. These results can be explained by the fact that the segment density becomes lower as the solvent quality becomes better. For example, the value of E^* in CH solvent was 2.6 kcal mol^{-1}

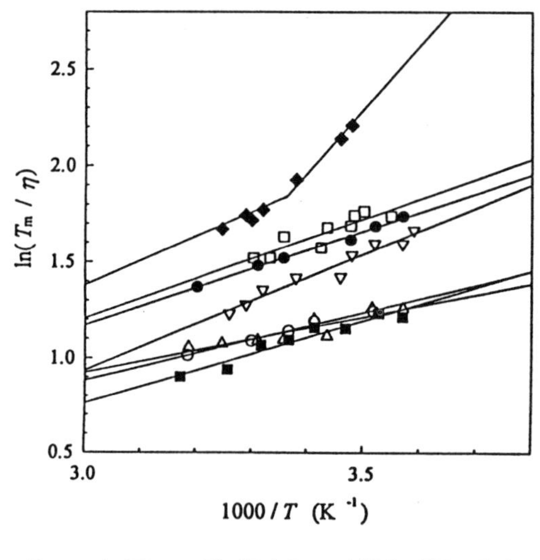

Figure 6. Plots of $\ln(T_m/\eta)$ vs. $1/T$ for PS solution. Symbols are the same as in Figure 5.

and it decreased to ca. 1.3 kcal mol^{-1} in good solvents. Figure 7 shows the relationship between E^* and $\overline{\alpha_\eta^3}$ for PS at 34.5 °C. Gronski et al. [21] reported the activation energy measured by ^{13}C-NMR for PS in toluene. They used the three bond jump model for diamond lattice chains [22]. They derived the activation energy from the plot of the correlation time characterizing the conformational jumps against $1/T$, and the value was 2.5 kcal mol^{-1}. This value is also higher than the value we obtained, 1.4 kcal mol^{-1}. The values of E^* are considered to depend on the time range used for each technique [23]; i.e., the difference among the values of E^* depends on the motional modes observed by different methods.

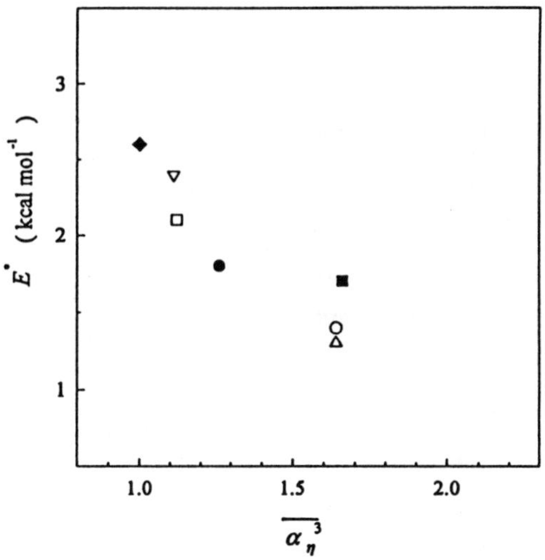

Figure 7. Plots of the activation energy $\bar{E}^*$ vs. $\overline{\alpha_\eta^3}$ for PS solution. Symbols are the same as in Figure 5.

For PS samples, the relaxation time of local chain motions decreased with the expansion of the chain coil in the good solvents. The activation energy also decreased with the chain expansion. The conformational transition of the main chain strongly depends on the segment density.

4. Local Chain Dynamics of PS in High Viscosity Solvents

4.1. APPARENT ACTIVATION ENERGY OF LOCAL CHAIN MOTION

The local motion of a polymer chain is generally evaluated by the theory of Kramers' diffusion limit, where the friction between the solvent molecule and the polymer chain is proportional to the average solvent viscosity, η. However, for a highly viscous solvent of more than 2 cP, such as diethylphthalate (DEP), cyclohexanone (CHO), tripropionin (Figure 8), the friction is not proportional to η, and the theory of Kramers' diffusion limit cannot be applied to the local motion in the highly viscous solvents [16]. In this section, we discuss the anomaly of activation energy of the local chain motion in high viscosity solvents and describe the feature of the local chain motions. The PS sample is the same as that used in section 3.

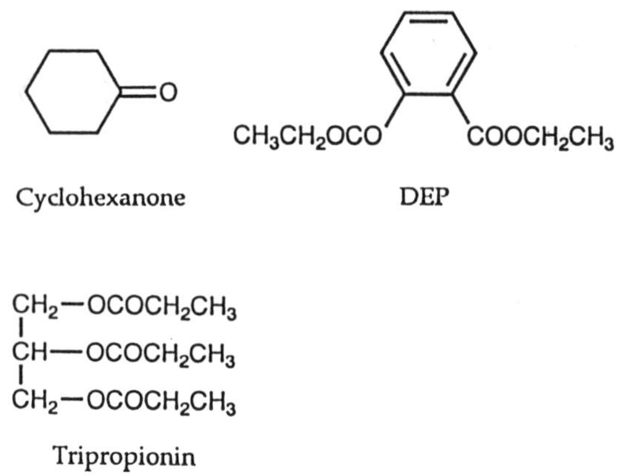

Figure 8. Chemical structure of high viscosity solvents.

The viscosity of solvents was changed by changing the mixing ratio of DEP-CHO. The mixing ratio of DEP-CHO was varied from (a) 0:5, (b) 3:2, (c) 4:1, and (d) 5:0. Table 2 shows the solvent viscosity (η) of these solvents, its activation energy (E_η), and the chain expansion factor of PS chain ($\overline{\alpha_\eta}$) measured by viscometry. The chain expansion factor shows that DEP, CHO and their mixed solvents are good solvents for PS, while tripropionin is a poor solvent with a high viscosity.

TABLE 2. Solvent viscosity at 20 °C, its activation energy, and chain expansion factor, $\overline{\alpha_\eta}$, for PS solutions at 34.5 °C

solvent	η / cP	E_η / kcal mol^{-1}	$\overline{\alpha_\eta}$
tripropionin	7.63	6.5	0.929
DEP-CHO 5:0	13.4	6.3	1.10
DEP-CHO 4:1	7.75	5.3	1.12
DEP-CHO 3:2	4.98	4.6	1.13
DEP-CHO 0:5	2.20	3.3	1.17

Figure 9 shows the plots of $\ln(T_m/\eta)$ vs. $1/T$. For CHO which is a good solvent with a relatively low viscosity, the positive E^* value of 1.0 kcal mol^{-1} was obtained. With the addition of more DEP which is also a good solvent with a high viscosity, the slope changed from a positive value to a negative value. The second row of Table 3 shows the activation energy E^*, which was determined by the slope in Figure 9. The apparent activation energy E^* was negative for high viscosity solvents: the result is anomalous and unrealistic. It is also noted that the result does not depend on the solvent quality, i.e., whether the solvent quality is good or poor, but on the solvent viscosity.

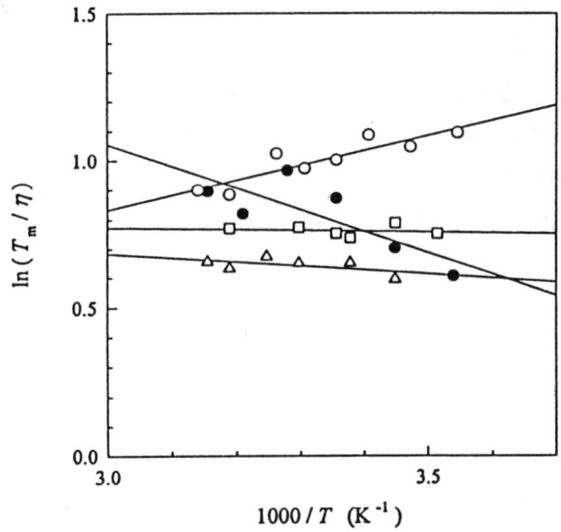

Figure 9. $\ln(T_m/\eta)$ vs. $1/T$ plots for PS solutions. CHO (○); DEP:CHO =3:2 (□); DEP:CHO = 4:1 (△); DEP (●).

TABLE 3. Activation energy evaluated by Kramers' equation, Grote-Hynes' equation, and Robinson's equation for PS solutions

solvent	E^* /kcal mol^{-1}	E^*_{GH} /kcal mol^{-1}	E^*_R /kcal mol^{-1}
tripropionin	-1.9		
DEP-CHO 5:0	-1.4	-0.1	1.9
DEP-CHO 4:1	-0.5	0.6	2.0
DEP-CHO 3:2	0.0	1.0	2.2
DEP-CHO 0:5	1.0	1.7	2.3

4.1.1. *Evaluation of E_{GH}^* Based on Grote-Hynes Theory*

To interpret the above anomalous result, we evaluated the activation energy E_{GH}^* of the relaxation time by the Grote-Hynes Theory. Grote and Hynes extended Kramers' theory by considering the frequency dependence of the friction [24]. In high viscosity solvents, the shear viscosity of the solvent does not adequately represent the friction experienced by the solute, and Grote-Hynes theory takes into account this effect. We evaluated the activation energy of a polymeric local motion by eq 29 based on Grote-Hynes theory.

$$T_m = A'\eta^{\alpha} \exp(E_{GH}^*/RT) \quad (0 < \alpha < 1) \tag{29}$$

where E_{GH}^* represents an activation energy.

The assumption for the determination of α is that α is a constant value independent of both solvent and temperature. The value of α can be obtained from the slope in the log-log plots of T_m vs. η based on this assumption. In this study, the α value of 0.79 at 25 °C was obtained. With $\alpha = 0.79$ we determined the apparent activation energy of local motion, E_{GH}^*. The third row of Table 3 shows E_{GH}^* determined by this procedure. E^* is still negative for the highest viscosity solvent, that is, E^* depends on the kind of solvent. This means that the treatment of solvent viscosity in eq 29 is not enough for the evaluation of the local motions of polymer chains and that the assumption of α = constant is not valid.

4.1.2. *Evaluation of E_R^* Based on the Equation of Robinson*

We also considered a different approach to evaluate the correct activation energy value. To interpret the rate of photoisomerization of stilbene, Robinson et al. [25] proposed a relationship between friction ζ and viscosity η, which is given by eq 30.

$$F(\zeta) = A_1[1-1/(A_2/\eta+A_3)]^{1/2} \tag{30}$$

By extending the Kramers theory, we get eq 31.

$$T_m = A_1'[1 - 1/(A_2/\eta + A_3)]^{-1/2} \exp(E_R^*/RT) \tag{31}$$

In eq 31, A_1', A_2 and A_3 are adjustable parameters, and the plot of T_m vs. η

was made in each solution at 13 °C to obtain these values. These parameters are assumed to be independent of both solvent and temperature. The parameters, A_2 and A_3, were evaluated by fitting eq 31 to these experimental data by the method of nonlinear-least-squares, where the initial setting values of A_2 and A_3 were 0.1 and 1 ($A_3 > 1$, by definition). Then, we could determine the parameter values, A_2: 0.0286 cP, A_3: 1.000 by fitting a curve to the data in the whole range from the low viscosity solvents to the high viscosity ones.

Robinson et al. [25] reported that this expression also includes a frequency dependence, and this form of friction fits both experimental data and molecular dynamics simulation quite well. As an example, they obtained the parameter values A_2 = 0.0182 cP and A_3 = 1.001 for the photoisomerization of 3, 3'-diethyloxadicarbocyanine iodide in the alcohols in the viscosity range 0.2−14 cP. In our system, similar values were obtained as shown above.

The fourth row of Table 3 shows the activation energies calculated by using eq 31. The activation energy E_R^* was 1.9−2.3 kcal mol^{-1} for all four samples. This indicates that in high viscosity solvents such as DEP an appropriate positive value of E_R^* is obtained by consideration of a solvent friction based on Robinson's equation.

Now let us compare the activation energy E_R^* with that reported by others. Ediger et al. [26] studied the local motion of PS in various dilute solutions by the same fluorescence depolarization method. They used the Generalized Diffusion and Loss (GDL) model for the analysis different from ours. They did not observe the complete decay of the anisotropy in high viscosity solvents above 4 cP, and the GDL model did not provide an ideal fit to their data. They made a reasonable guess about the shape of the correlation function in order to estimate the average relaxation times. They used the average of the τ_2/τ_1 values obtained from the data which decayed to < 0.05 to constrain the shape of the correlation function for the solvents above 4 cP. The average activation energy obtained in good solvents was 2.6 ± 0.7 kcal mol^{-1}.

Here, the E_R^* values obtained in this study are comparable to those obtained for PS in good solvents with low viscosity below 1 cP, in which the average activation energy is ca. 2 kcal mol^{-1} (see Figure 7). We consider that the activation energy of ca. 2 kcal mol^{-1} obtained in this study is a reasonable value for the height of the internal barrier of C-C bonds.

4.2. LOCAL CHAIN MOTION IN HIGH VISCOSITY SOLVENTS

We discussed three treatments of E* in high viscosity solvents. The former two treatments gave unrealistic results and the third gave positive activation energy values. Therefore, for evaluation of the dynamics of the middle segments of the polymer chain in high viscosity solvents, the equation of Kramers' diffusion limit is not appropriate. Probably the friction of a polymer chain with solvent molecules differs in high viscosity solvents, depending on the kind of polymers and motional modes. In the middle of the PS chain, the difference between macro-viscosity and micro-viscosity becomes pronounced above a solvent viscosity of 2 cP, and the friction of a polymer chain with solvent molecules is not proportional to macro-viscosity. A similar anomaly of polymer chain dynamics in high viscosity solvents has also been reported by Ediger et al. [28]. They reported that the theory of Kramers' diffusion limit gave a negative value for an activation energy of poly(isoprene) in dioctyl phthalate.

The specific intermolecular interaction of the solvents is considered to be the cause of the anomalous effect. For example, Khudayarov et al. [27] demonstrated by dielectric relaxation that glycerol esters having high viscosity have a wide distribution of relaxation time, which is derived from the ordering of the liquid structures by the specific molecular interactions. In the present study, the strong dipole-dipole interaction of ester groups forms a network structure and raises the viscosity of the solvent: then under such a circumstance, the middle part of a polymer chain moves in such ordered solvents and simple application of Kramers' diffusion limit gives an erroneous activation energy. That is, under such a condition, bulk viscosity and local friction can operate over different length scales and an effective viscosity for a local segment must be taken into account even for low molecular weight solvents.

In conclusion, solvents having high viscosity seem to form a network structure with their intermolecular interaction. In such a solvent network, a polymer chain behaves like a plasticizer in polymer solids: activation energy of local chain motion is smaller than that of the solvent viscosity.

5. References

1. Ferry, J.D. (1980) *Viscoelastic Properties of Polymers* 3rd.ed., John Wiley & Sons, New York.

2. Doi, M. and Edwards, S.F. (1986) *The Theory of Polymer Dynamics,* Oxford Univ. Press, New York.

3. Bailey, R.T., North, A.M., and Pethrick, R.A. (1981) *Molecular Motion in High Polymers* Clarendon Press, Oxford.

4. Higgins, J.S. and Maconnachie, A. (1987) in *Methods of Experimental Physics,* Sköld, K., Price, D. L., Eds., Academic Press, New York, Vol. 23, Part C, p 287.

5. Glowinkowski, S., Gisser, D.J., and Ediger, M. D. (1990) Carbon-13 nuclear magnetic resonance measurements of local segmental dynamics of polyisoprene in dilute solution: Nonlinear viscosity dependence, *Macromolecules* **23**, 3520-3530.

6. Mashimo, S., Winsor, P., IV, Cole, R.H., Matsuo, K., and Stockmayer, W.H. (1986) Conformation and dynamics of poly(olefin sulfones) in solution. 1. High-frequency dielectric relaxation and carbon-13 NMR relaxation, *Macromolecules* **19**, 682-688.

7. Adachi, K. (1990) A molecular model for cooperative local motions in amorphous polymers, *Macromolecules* **23**, 1816-1821.

8. Chu, B. (1991) *Laser Light Scattering,* 2nd ed., Academic Press, New York.

9. Nishijima, Y. (1973) Fluorescence method in polymer research in S.Onogi, K.Uno (eds.) *Progress in Polymer Science*, Japan, Kodansha, Tokyo , Vol. 6 pp. 199-251.

10. Anufrieva, E.V. and Gotlib, Y.Y. (1981) Investigation of polymers in solution by polarized luminescence, *Adv. Polym. Sci.* **40**, 1-68.

11. Winnik, M. Ed. (1986) *Photophysical and Photochemical Tools in Polymer Science,* D. Reidel Publishing Company, Dordrecht, .

12. Ediger, M.D. (1991) Time-resolved optical studies of local polymer dynamics, *Annu. Rev. Phys. Chem.* **42**, 225-250.

13. Sasaki, T., Arisawa, H., and Yamamoto, M. (1991) Fluorescence depolarization study on local motions of anthracene-labeled ply(alkyl methacrylate)s in dilute solutions and evaluation of their chain stiffness, *Polym. J.* **23**, 103-115.

14. Ono, K., Okada, Y., Yokotsuka, S., Sasaki, S., and Yamamoto, M. (1994) Chain dynamics of styrene polymers studied by the fluorescence depolarization method, *Macromolecules* **27**, 6482-6486.

15. Ono, K., Ueda, K., and Yamamoto, M. (1994) Local chain dynamics of poly(cis-1,4-isoprene) in dilute solutions studied by the fluorescence depolarization method, *Polym. J.* **26**, 1345-1351.

16. Ono, K., Okada, Y., Yokotsuka, S., Ito, S., and Yamamoto, M. (1994) Chain dynamics of polystyrene in high viscosity solvents studied by the fluorescence depolarization method, *Polym. J* . **26**, 199-205.

17. O'Conner, D.V. and Phillips, D. (1984) *Time-Correlated Single-Photon Counting,* Academic Press, London.

18. Sasaki, T., Yamamoto, M., and Nishijima,Y. (1988) Chain dynamics of PMMA in dilute solutions studied by the fluorescence depolarization method, *Macromolecules* **21**, 610-616.

19. Helfand, E. (1971) Theory of the kinetics of conformational transitions in polymers, *J. Chem. Phys* **54**, 4651-4661.

20. Waldow, D.A., Johnson, B.S., Hyde, P.D., Ediger, M.D., Kitano, T., and Ito, K. (1989) Local segmental dynamics of polyisoprene in dilute solution: Solvent and molecular weight effects, *Macromolecules* **22**, 1345-1351.

21. Gronski, W., Schafer, T., and Peter, R. (1979) ^{13}C Relaxation studies of ^{13}C enriched polystyrene in solution, *Polym.Bull.* **1**, 319-324.

22. Valeur,B., Jarry, J.P., Geny, F., and Monnerie, L. (1975) Dynamics of macromolecular chains. I. Theory of motions on a tetrahedral lattice, *J.Polym. Sci., Polym. Phys. Ed.* **13**, 667-682.

23. Friedrich, C., Lauprêtre, F., Noë l, C., and Monnerie, L. (1981) Polystyrene dynamics in dilute solution: A further investigation by electron spin resonance. Comparison with other techniques, *Macromolecules* **14,** 1119-1125.

24. Grote, R.F. and Hynes, J.T. (1980) The stable states picture of chemical reactions.II. Rate constants for condensed and gas phase reaction models, *J. Chem. Phys* **73**, 2715-2732.

25. Lee, J., Zhu, S.B., and Robinson, G.W. (1987) An extended Kramers equation for photoisomerization, *J. Phys. Chem.* **91**, 4273-4277.

26. Waldow, D.A., Ediger, M.D., Yamaguchi, Y., Matsushita, Y., and Noda, I. (1991) Viscosity dependence of the local segmental dynamics of anthracene-labeled polystyrene in dilute solution, *Macromolecules* **24**, 3147-3153.

27. Khudayarov, U.V. and Khudaiberdyev, V.N. (1990) Dipole relaxation in glycerol esters, *Russian J. Phys. Chem.* **64** , 1015-1018.

28. Adolf, D.B., Ediger, M.D., Kitano, T., and Ito, K. (1992) Viscosity dependence of the local segmental dynamics of anthracene-labeled polyisoprene in dilute solution, *Macromolecules* **25**, 867-872.

SUBJECT INDEX